# Environmental Nexus for Resource Management

This book gives detailed information about how soil, water and wastes can be managed to overcome the various global issues via possible nexus thinking. The emphasis is on the environmental resource perspective of global climate change-related issues. It provides stepwise information on climate change and adaption strategies, urbanization and its impact and management strategies, environmental nexus approaches to cope with global challenges and recourses conservation and ecological approaches to restore the damaged ecosystem.

Features:

- Compiles the possible nexus approaches that contribute to managing the atmospheric environmental variables in sustainable ways.
- Focuses on environmental resources perspective of the global change.
- Covers how soil, water and waste may be managed in a nexus.
- Explains modern strategies to manage the present environmental situation that are feasible and safe to the environment.
- Discusses environmental nexus for judicious resource management.

This book is aimed at researchers and graduate students in environmental sciences and engineering and sustainable development.

# ENVIRONMENTAL NEXUS IN WASTE MANAGEMENT

**Series Editors: Vineet Kumar,** *Department of Microbiology, School of Life Sciences at the Central University of Rajasthan, Rajasthan, India* **and Sunil Kumar,** *researcher in Environmental Engineering and Science*

The book series on "Environmental Nexus for Waste Management" addresses novel approaches and techniques linked to sustainable development paths with an interdisciplinary focus on waste resources monitoring, assessment, management, water/wastewater reclamation, and recycling as well as discussing solutions for more sustainable use of natural resources. It is a broad compendium of waste management related to research and development and discussions on the most up-to-date tackling strategies. The series of books constitutes an information source and facilitator for the transfer of knowledge, focusing on practical solutions, and better understanding towards achieving sustainable development.

**Solid Waste Treatment Technologies**
Challenges and Perspectives
*Edited by Pratibha Gautam, Vineet Kumar and Sunil Kumar*

**Microbial Nexus for Sustainable Wastewater Treatment**
Resources, Efficiency, and Reuse
*Edited by Vineet Kumar, Sunil Kumar, Pradeep Verma and Sartaj Ahmad Bhat*

**Environmental Nexus for Resource Management**
*Edited by Hanuman Singh Jatav, Tatiana Minkina, Satish Kumar Singh, Bijay Singh and Vishnu D. Rajput*

For more information about this series, please visit: www.routledge.com/Environmental-Nexus-in-Waste-Management/book-series/CRCENWM

# Environmental Nexus for Resource Management

Edited by Hanuman Singh Jatav,
Tatiana Minkina, Satish Kumar Singh,
Bijay Singh and Vishnu D. Rajput

CRC Press
Taylor & Francis Group
Boca Raton  London  New York

CRC Press is an imprint of the
Taylor & Francis Group, an **informa** business

Designed cover image: ©Shutterstock Images

First edition published 2025
by CRC Press
2385 NW Executive Center Drive, Suite 320, Boca Raton FL 33431

and by CRC Press
4 Park Square, Milton Park, Abingdon, Oxon, OX14 4RN

*CRC Press is an imprint of Taylor & Francis Group, LLC*



ISBN: 978-1-032-41453-9 (hbk)
ISBN: 978-1-032-41455-3 (pbk)
ISBN: 978-1-003-35816-9 (ebk)

DOI: 10.1201/9781003358169

Typeset in Times
by Apex CoVantage, LLC

*Dedicated to my parents*
*Late. Shri Babulal Jatav Smt. Supedi Devi*
*To my beloved parents,*
*whose unwavering support and endless love have been my guiding light.*
*This work is a testament to your sacrifices, encouragement, and belief in me.*
*Thank you for always being my pillars of strength.*

# Contents



**Chapter 1**  Management of Soil, Waste and Water in the Context
of Global Climate Change ................................................................. 1

*Kankan Datta, Swarnavo Chakraborty,
and Aryadeep Roychoudhury*



**Chapter 2**  Potential Eco-Friendly Techniques for Water, Soil,
and Waste Management ................................................................... 27

*Muhammad Farhan Qadir, Khuram Shehzad Khan,
Muhammad Naveed, Taqi Raza, and Neal S. Eash*

**Chapter 3** Potential Role of Healthy Soil for Global Environment ................... 45


*Muhammad Farhan Qadir, Khuram Shehzad Khan,
Muhammad Naveed, Noman Younas, Muhammad Yousuf,
Muhammad Hamza Khalil, Azmat Qadir Maria, Taqi Raza,
and Neal S. Eash*

**Chapter 6**    Making the Water-Soil-Waste Nexus Wor: Framing the
Boundries of Resource Flows......................................... 114

*Laila Shahzad, Faiza Sharif, Arifa Tahir, Umair Raiz,
Maryam Ali, Masooma Batool, and Ahmad Mahmood*



**Chapter 7**    Nexus Approach: Resource Management for Soil Productivity....... 135

*Rajneesh Thakur*

*Laila Shahzad, Umair Raiz, Sadia Yousaf, Nimra Riaz, Rabiya Nasir, and Qaiser Fareed Khan*

**Chapter 9**     Soils' Role in Meeting Global Demand for Food, Water,
and Bioenergy.............................................................................. 169

*Chinmaya Kumar Swain and Kiran Kumar Mohapat*

**Chapter 10**  The Nexus between Environmental Damage, Poverty, and
Climate Change in Hard-to-Reach Areas: A Somber
Tale of the 21st Century ................................................. 188

*Amitava Panja, Punam Bhattacharjee, Subhradip Bhattacharjee,*
*Dinesh Kumar, Govind Makarana, Manish Kushwaha,*
*M.R. Yadav, Rakesh Kumar, and Vishnu D. Rajput*

**Chapter 11** Urban Soil Management for Improved Resource-Efficiency of Developing Cities .................................................................................. 223

*Mahima Dixit, Immanuel Chongboi Haokip,
Bhabani Prasad Mondal, Muskan Porwal, Debabrata Ghoshal,
Abhishek Mandal, and B.P. Dhyani*

**Chapter 12**  Potential Nexus Approach for Sustainable Soil Productivity Resource Management ................................................. 243


*Prithwiraj Dey, B.S. Mahapatra, Biplab Mitra, Arya Kumar Sarvadamana, Arnab Roy Chowdhury, Abhishek Rana, and Diyan Mandal*

**Chapter 13** The Ecosystem Approach and Environmental Justice Nexus in Natural Resource Management ..................................... 274

*Vikas and Rajiv Ranjan*



**Chapter 14** Soil: A Potential Source for Mitigating Food, Water and Bioenergy Crisis ............................................................ 287

*Kumar Chiranjeeb, Sachin Kumar, and Ranbir Singh Rana*

**Chapter 15**  Role of Soils for Satisfying Global Demands for Food, Water,
and Bioenergy.......................................................................... 302

*Bhabani Prasad Mondal, G. Bhupal Raj, Mahima Dixit,*
*Tithli Sadhu, Vaibhav Pandit, Laxmi Prasanna,*
*Arghya Chattopadhyay, Suman Dutta, and Suchith Kumar*



**Chapter 16**  Agriculture, Rural Poverty, and Natural Resource Management
in Less Favored Environments: Revisiting Challenges and
Conceptual Issues .................................................................... 317

*L. Devarishi Sharma, Rahul Sadhukhan, Hiren Das,*
*L. Banarjee Singh, and Animesh Sarkar*

**Chapter 19**  Engineered Nanoparticles for Plant-Originated Environmental
Consequence.......................................................................... 363

*Sushmita Thokchom, Guntamukkala Sekhar,
Prem Kumar Bharteey, and Sumit Rai*

# Preface

The human population is increasing at a tremendous rate, and at the same time, people from rural areas are migrating to cities for their livelihood. It added an additional burden to food production. As per the UN's latest report, in third world countries, a further approx. 2.1 billion populaces will be deriding in cities by 2030. The fact itself says that the population is burgeoning at an alarming rate coupled with the problems of climate change and food security and global environmental issues. This compels us for the adoption of an intensive farming system to supply the food requirements and address the challenges which could be faced by 2050. The living body, i.e., soil is a non-renewable resource because its formation takes thousands of years. It may be considered that the native inherent capacity of this versatile and worthy natural resource may get exhausted in order to fulfill our daily food demands. In the coming era, it is necessary to adopt new technological interventions or cropping systems to meet increasing demands of food. Thus, the intensive cultivation is one of the promising approaches. *Environmental Nexus for Resource Management* gives detailed information about how soil, water and wastes can be managed to overcome the various global issues. The recourses should be managed in such a way so our environment could not deteriorate. This new book has the brief information on possible nexus thinking to tackle these global issues. The main emphasis of the book has been given on environmental resources perspective of the global change such as urbanization, climate change and environmental management–related issues. The book also gives information on the possible nexus approaches that could help to contribute to manage environment in sustainable ways. The whole book has stepwise information briefly on climate change and adaption strategies, urbanization and its impact and management strategies, environmental nexus approaches to cope with global challenge and recourses conservation and ecological approach to restore the damaged ecosystem. The whole book briefly compile the different aspect in different chapter form as Chapter 1: Management of Soil, Waste and Water in the Context of Global Climate Change; Chapter 2: Potential Eco-Friendly Techniques for Water, Soil, and Waste Management; Chapter 3: Potential Role of Healthy Soil for Global Environment; Chapter 4: Soil-Based Implication Approach for Environmental Nexus; Chapter 5: Soil Security to Address Potential Global Issues; Chapter 6: Making the Water-Soil-Waste Nexus Wor: Framing the Boundaries of Resource Flows; Chapter 7: Nexus Approach: Resource Management for Soil Productivity; Chapter 8: Climate Change, Profligacy, Poverty and Destruction: All Things Are Interconnected; Chapter 9: Soils' Role in Meeting Global Demand for Food, Water, and Bioenergy; Chapter 10: The Nexus between Environmental Damage, Poverty, and Climate Change in Hard-to-Reach Areas: A Somber Tale of the 21st Century; Chapter 11: Urban Soil Management for Improved Resource-Efficiency of Developing Cities; Chapter 12: Potential Nexus Approach for Sustainable Soil Productivity Resource Management; Chapter 13: The Ecosystem Approach and Environmental Justice Nexus in Natural Resource Management; Chapter 14: Soil: A Potential Source for Mitigating Food, Water and Bioenergy Crisis; Chapter 15: Role of Soils for Satisfying Global Demands for Food, Water, and Bioenergy; Chapter 16: Agriculture, Rural Poverty, and Natural Resource Management in Less Favored Environments: Revisiting Challenges and Conceptual Issues; Chapter 17: Environmental Resource Management with Reference to Global Climate Change; Chapter 18: Management of Urban Polluted Soil to Enhance Resource Use Efficiency in Developing Cities; and Chapter 19: Engineered Nanoparticles for Plant-Originated Environmental Consequence. The present *Environmental Nexus for Resource Management* book will be helpful to the researchers, scientific community, academician, business farmers and policymakers to overcome various environmental concerns.

# Acknowledgements

Every accomplishment task deserved a token of thanks for those who helped directly or indirectly to complete that work. It gives me pleasure to remember and express my feelings to those who helped me in the form of motivation to compile this outcome. I, Dr. Hanuman Singh Jatav, as an editor of this book, *Environmental Nexus for Resource Management,* acknowledge the token of thanks to Prof. Balraj Singh, Hon'ble Vice-Chancellor, Sri Karan Narendra Agriculture University, Jobner Jaipur, for his motivation to compile such work. I am deeply indebted to Dr. Amitava Rakshit, Associate Professor, Banaras Hindu University, Varanasi, for his lifelong motivation to compile the work. I also extend my thanks to Prof. K.K. Sharma (HOD-SSAC) and Dr. S.K. Dadhich for their direct and indirect support.

It's very difficult to acknowledge the token of words to Prof. S.R. Dhaka (Dean-COA Fatehpur-Sikar) who not only supported but also guided me to compile this upshot for the students. I would like to thank, from the bottom of my heart, Dr. Sanjay K. Attar, Prof. Athar Uddin, Dr. Muddser A. Khan, Dr. K.C. Verma, Dr. Kailash Chandra, Dr. Mujahid Khan, Dr. Champa L. Khatik, Dr. Shubita Kumawat, Dr. Ramu Meena Dr. Ajeet Singh and Dr. Vishnu D. Rajput for the direct and indirect support to compile this outcome.

At the outset, being a student of the great institution Banaras Hindu University, Varanasi, I am thankful to the Prof. A.K. Ghosh, Prof. Nirmal De, Prof. Surendra Singh (FISSS), Prof. Priyankar Raha, Prof. P.K. Sharma, Dr. Y.V. Singh and Dr. Ashish Latare for their support and guidance during my studies.

I am grateful to my schoolteachers: Shri Surendra Sarthak Ji, Shri Ghanshyam Ji, Shri Harkesh Ji and Shri Hanuman Ji. I am also delighted to offer thanks to my friends: Dr. Sunil Kumar, Dr. Dinesh Jinger, Dr. Sonu Kumar, Dr. Surendra Singh Jatav, Dharamveer Verma and Rajat.

I am also thankful to Prof. Bijay Singh for their regular efforts in compiling and critically evaluating this book. I am also thankful to Prof. S.K. Singh and Dr. Vishnu D. Rajput for their continued checking of the content to lessen scientific or grammatical errors. I also extend my sincere gratitude to my family member who always supports me in any difficult situation.

The graces of the Lord Hanuman Ji, Goddess Maa Sharde, along with my parents, Smt. Supedi Devi and Shri Babulal Jatav, have always blessed to me and given me patience and power to overcome difficulties that came my way in the accomplishment of this endeavour. I cannot dare to say thanks but only pray to bless me always.

**Dr. Hanuman Singh Jatav,**
**Assistant Professor, Soil Science and Agril.**
**Chemistry, COA—Fatehpur Shekhawati**
**Sikar (SKNAU—Jobner, Rajasthan)**

# About the Editors

**Hanuman Singh Jatav** is Assistant Professor in the Department of Soil Science and Agricultural Chemistry, Sri Karan Narendra Agriculture University, Jobner Jaipur (Rajasthan), India. He has more than six years of teaching and ten years of research experience in the field of sewage sludge, biochar, micro nutrients, heavy metals, and soil fertility management, including a M.Sc. (Ag.) and a Ph.D. He has completed B.Sc. (Hons.) agriculture from Maharana Pratap University of Agriculture and Technology, Udaipur, Rajasthan, India. Dr. Jatav has completed his M.Sc. (Ag.) and Ph.D. in soil science and agricultural chemistry from the Institute of Agricultural Sciences, Banaras Hindu University, Varanasi—U.P. India. At present, Dr. Jatav is an emerging person in the field of research, transforming himself to an excellent scientific level by enhancing research capacity by promoting environmental sustainability. He has published more than 120 paper(s), including research papers, review papers, books, book chapters, and technical papers from various national and international repute journals and publishers. He is serving as Academic Editor and member of subject expertise in Taylor & Francis, Frontiers, CARJ, JEB, ICAR Journals, and Science Domain for the last six years. He is serving as an active member of more than 15 various national and international scientific societies. Dr. Jatav has been awarded a certificate of gratitude as an ambassador by the IUCN: International Union for Conservation of Nature, Gland, Switzerland (www.iucn.org), in the World Conservation Congress during the Global Youth Summit. He has been also awarded the best paper and poster presentation for his work. He has also attended more than 40 conferences. Besides this, he has also completed more than 35 training sessions pertaining to his research. Considering Dr. Jatav works at an excellent level, he has been awarded as Outstanding Assistant Professor of the Sri Karan Narendra Agriculture University, Jobner Jaipur (Rajasthan), India.

**Tatiana Minkina** is Head of Soil Science and Land Evaluation Department of Southern Federal University and Head of International Master's Degree Educational Program "Management and estimation of land resources" (2015–22, accreditation by ACQUIN). Her area of scientific interest is soil science, biogeochemistry of trace elements, and environmental soil chemistry, including soil monitoring, assessment, modeling, and remediation using physicochemical treatment methods. She was awarded in 2015 with the Diploma of the Ministry of Education and Science of the Russian Federation for many years of long-term work for the development and improvement of the educational process, a significant contribution to the training of highly qualified specialists. Currently, she is handling projects funded by the Russian Scientific Foundation, Ministry of Education and Science of the Russian Federation, and Russian Foundation of Basic Researches. She is a member of Expert Group of Russian Academy of Science, the International Committee on Contamination Land, Eurasian Soil Science Societies, the International Committee on Protection of the Environment, and the International Scientific Committee of the International Conferences on Biogeochemistry of Trace Elements. Total scientific publications: 757 (English).

**Satish Kumar Singh** received his B.Sc. (Ag) degree in 1985 and M.Sc. (Ag) in 1988 and Ph.D. in soil science and agricultural chemistry in 1992 from the Banaras Hindu University, Varanasi. He started his professional career in 1990 as Lecturer of Agricultural Chemistry in Amar Singh (PG) College, Lakhaoti, Bulandshahr. Subsequently, Dr. Singh joined as Associate Professor in the year 1999 at G.B. Pant University of Agriculture and Technology, Pantnagar (Uttarakhand), and Professor in 2006 at Banaras Hindu University, Varanasi. Prof. Singh is Fellow of Indian Society of Soil Science, New Delhi, and Indian Society of Agricultural Chemists, Allahabad. He has guided 31 postgraduate and ten Ph.D. students. He has more than 250 publications to his credit that includes research papers, books, book chapters, laboratory manuals, technical bulletins, and technical folders.

**Bijay Singh**, FNA, FNAAS, FISSS, Department of Soil Science, Punjab Agricultural University. A soil chemist by training and one of the ten national professors of the Indian Council of Agricultural Research from 2006 to 2012, Dr. Bijay Singh is presently associated with Punjab Agricultural University, Ludhiana (Punjab, India), as INSA Honorary Scientist. He served as Senior Soil Chemist from 1994 to 2006 and as INSA Senior Scientist from 2012 to 2017.

During more than four decades, research efforts by Dr. Bijay Singh have resulted in a better understanding of the nitrogen balance in the soil-plant system, thereby leading to the formulation of guidelines for (i) enhancing fertilizer nitrogen use efficiency in rice, wheat, and maize; (ii) reducing fertilizer nitrogen related environmental pollution; (iii) promoting integrated management of mineral fertilizers with organic manures; and (iv) site-specific nitrogen management. Dr. Singh's research contributions have also helped in improving our understanding of the basic and applied aspects of management of fertilizers, organic manures, and crop residues and providing scientifically sound directions for achieving not only sustainable high yield levels in cereal cropping systems but also for maintaining soil health for the future generations.

In the late 1970s, when fertilizers were being introduced during the green revolution in northwestern India, Dr. Singh convincingly proved that until and unless fertilizers will be managed as recommended, groundwater in the region may get polluted with nitrates. Several fertilizer management strategies to reduce nitrate pollution of groundwater were revealed, which included balanced use of fertilizers, growing deep-rooted crops in rotations, and coordination between water management and fertilizer nitrogen management. For several years, he worked on finding means and ways to improve fertilizer nitrogen use efficiency. It was demonstrated that the nitrogen use efficiency of urea can be achieved by imparting slow-release characteristics through coating with appropriate materials. Using nitrification and urease inhibitors, it was revealed that urea applied to irrigated wheat grown in coarse-textured soils was lost preferentially via ammonia volatilization rather than leaching and nitrification-denitrification. As urea-nitrogen transported to the subsurface soil layers with irrigation water was not lost via ammonia volatilization, it was recommended that urea should be top-dressed just before an irrigation event.

In the late 1990s, Dr. Singh spearheaded the research in South Asia on managing fertilizer nitrogen in cereal crops on a field-specific basis and as per the need of the crop. Site-specific nitrogen management optimally accounted for the supply of soil nitrogen over time and space to match the requirements of crops by applying the right amount of fertilizer nitrogen at the right time during the crop growing season. He introduced the use of canopy reflectance sensors, chlorophyll meters, and its inexpensive alternative, the leaf color chart, to quickly and reliably monitor leaf nitrogen status for formulating real-time, site-specific nitrogen management strategies for rice, wheat, and maize. Dr. Singh is the pioneer in South Asia for introducing GreenSeeker crop reflectance sensor to guide site-specific management of fertilizer nitrogen in rice and wheat. Recently, an app for smartphones has been developed by his team to help farmers to apply urea to different crops on a site-specific basis using leaf color chart, chlorophyll meter, and GreenSeeker crop reflectance sensor.

During the last two decades, research efforts of Dr. Singh are becoming more focused on soil health. He has convincingly concluded that soil health is adversely influenced only when nitrogen fertilizers are applied at rates more than the optimum.

Dr. Bijay Singh has contributed over 370 publications, including 180+ research papers in journals of national and international repute, 50 reviews of which eight are in *Advances in Agronomy*, and 50+ book chapters. Dr. Singh's research on different aspects of fertilizer management, crop nutrition, and environmental issues related to fertilizers has been published in high-impact factor journals, and publications authored by him have 8,900+ citation index, h-index of 49, and i10-index of 99.

Dr. Bijay Singh is a fellow of Indian National Science Academy, National Academy of Agricultural Sciences, and Indian Society of Soil Science. He has been decorated with several awards, including the Rafi Ahmad Kidwai memorial prize of the Indian Council of Agricultural Research—the most prestigious award in agricultural sciences in India. The Fertilizer Association of India has honoured him with its Silver Jubilee award two times in 1984 and 2001. National Academy of Agricultural Science honoured him with its Recognition Award in 2000 and Dr. N.S. Randhawa memorial award in 2008. He is a recipient of Professor KS Bilgrami award (2009) and Professor B.D. Tilak lecture award (2014) of the Indian National Science Academy. The Indian Society of Soil Science honoured him with its Platinum Jubilee Commemoration award for lifetime contributions in enhancing fertilizer nitrogen use efficiency in cereal crops in 2019. In 2020, the Fertilizer Association of India awarded him the FAI Golden Jubilee Award for Excellence for the best work done in the field of plant nutrition.

Dr. Bijay Singh delivered a J.N. Mukherjee-Indian Society of Soil Science Foundation lecture in 2006, Dr. G.S. Sekhon memorial lecture in 2012, and Dr. Radha Raman Agarwal memorial lecture in 2015. He was also elected as the vice chairman of the Commission on Soil Fertility and Plant Nutrition of the International Union of Soil Sciences during 2010–14.

**Vishnu D. Rajput** is working as Leading Researcher (Associate Professor) and Head of Soil Health Lab at the Southern Federal University, Rostov-on-Don, Russia. His ongoing research is based on the toxic effects of bulk- and nano-forms of metals and investigating the bioaccumulation, bio/geo-transformations, uptake, translocation, and toxic effects of bulk- and nano-forms of metals on plant physiology, morphology, anatomy, the ultrastructure of cellular and subcellular organelles, cytomorphometric modifications, and DNA damage. He is also discovering the possible remediation approaches such as using biochar/nano-biochar–based sorbents and nanotechnological approaches. With long experience and experimental work, Dr. Rajput comprehensively detailed the state of research in environmental science in regard to how nanoparticles/heavy metals interact with plants, soil, microbial community, and the larger environment as well as possible remediation technology using nanoparticles/nano-biochar. He has published (total of 351 scientific publications), 204 peer-reviewed, full-length articles, 18 books, 70 chapters (Scopus indexed), 44 conference articles, and achieve h-index (22 by Scopus with 2,068 citations) till November 2022. He is an editorial board member of various high-impacted journals. He is a leader of several grants, including Mega-Grant, BRICS, and performer/co-PI of several national and international projects. Dr. Rajput has received the Certificate for Appreciation 2019 and 2021, Certificate of Honor 2020, and Diploma Award 2021 by the Southern Federal University, Russia, for outstanding contribution in academic, creative research, and publication activities. He has also received the "Highly Qualified Specialist" status by the Russian government and is included in the list of top 2% of Russian scientists.

# Contributors

**Ghulam Hassan Abbasi**
Institute of Agro-Industry and
    Environment, Faculty of Agriculture
    and Environment
The Islamia University of Bahawalpur
Punjab, Pakistan

**Maryam Ali**
Sustainable Development Study Center
Government College University
Katchery Rd, Lahore 54000, Pakistan

**Muhammad Ali**
Institute of Agro-Industry and
    Environment, Faculty of Agriculture
    and Environment
The Islamia University of Bahawalpur
Punjab, Pakistan

**Muhammad Ameen**
Department of Soil Science, Faculty of
    Agriculture and Environment
The Islamia University of Bahawalpur
Bahawalpur, Pakistan

**Muhammad Naveed Arshad**
Institute of Biology, Environment and
    Rural Studies
Aberystwyth University
United Kingdom

**Iqra Aslam**
Institute of Soil and Environmental
    Sciences
University of Agriculture Faisalabad
Faisalabad, Pakistan

**Tabinda Athar**
Institute of Soil and Environmental
    Sciences
University of Agriculture Faisalabad
Faisalabad, Pakistan

**Muhammad Ashar Ayub**
Institute of Agro-Industry and
    Environment, Faculty of Agriculture
    and Environment
The Islamia University of Bahawalpur
Punjab, Pakistan

**Masooma Batool**
Sustainable Development Study
    Center
Government College University
Katchery Rd, Lahore 54000,
    Pakistan

**Biswaranjan Behera**
ICAR-Indian Institute of Water
    Management (IIWM)
Bhubaneswar, Odisha 751023,
    India

**Prem Kumar Bharteey**
Department of Agricultural
    Chemistry & Soil Science
C.C.R. (P.G.) College
    Muzaffarnagar-251001
Uttar Pradesh, India

**Punam Bhattacharjee**
GS Research, Growing Seed
    Organization
Dharmanagar, Tripura 799250, India

**Subhradip Bhattacharjee**
Agronomy Section
ICAR-National Dairy Research
    Institute
Karnal 132001, India

**Devilal Birla**
Department of Agronomy
Anand Agricultural University
Anand, Gujarat, India

**Swarnavo Chakraborty**
Department of Biotechnology
St. Xavier's College (Autonomous)
30 Mother Teresa Sarani, Kolkata
    700016, West Bengal, India

**Arghya Chattopadhyay**
Institute of Agricultural Sciences
Banaras Hindu University
Varanasi, Uttar Pradesh, India

**Rukoo Chawla**
Department of GPB, Rajasthan College
    of Agriculture
MPUAT
Udaipur-313001, Rajasthan,
    India

**Kumar Chiranjeeb**
Department of Soil Science, Centre
    for Geo-Informatics Research and
    Training
CSK HPKV
Palampur, 176062, India

**Arnab Roy Chowdhury**
Department of Agronomy
Bihar Agricultural University
Sabour, Bihar, India

**Hiren Das**
YP-II, Division of DSRE
ICAR RC for NEH Region
Meghalaya, India

**Supratim Das**
ICAR-National Rice Research
    Institute
Cuttack, Odisha, India

**Amit K. Dash**
ICAR—National Academy of
    Agricultural Research
    Management
Rajendranagar, Hyderabad 500030,
    India

**Kankan Datta**
Department of Biotechnology
St. Xavier's College (Autonomous)
30 Mother Teresa Sarani, Kolkata
    700016, West Bengal, India

**Prithwiraj Dey**
Agricultural & Food Engineering
    Department
Indian Institute of Technology
Kharagpur, West Bengal, India

**B.P. Dhyani**
Sardar Vallabhbhai Patel University of
    Agriculture and Technology
Meerut, 250110, Uttar Pradesh, India

**Mahima Dixit**
Sardar Vallabhbhai Patel University of
    Agriculture and Technology
Meerut, 250110, Uttar Pradesh, India

**Suman Dutta**
Division of Genetics and Plant Breeding
Ramkrishna Mission Vivekananda
    Educational and Research Institute
Kolkata, West Bengal, India

**Neal S. Eash**
Department of Biosystems
    Engineering & Soil Science
University of Tennessee
Knoxville

**Hina Fatima**
Soil Science Laboratory, School of
    Applied Biosciences
Kyungpook National University
80 Daehakro, Bukgu, Daegu, 41566,
    South Korea

**Kumar Gaurav**
Department of Agronomy, College of
    Agriculture
GBPUA&T
Pantnagar, Uttarakhand 263145, India

**Debabrata Ghoshal**
ICAR-Indian Agriculture Research
    Institute
Delhi, 110012, India

**Immanuel Chongboi Haokip**
ICAR-Indian Institute of
    Soil Science
Bhopal, 462038, Madhya Pradesh,
    India

**Muhammad Hamza Khalil**
Institute of Soil and Environmental
    Sciences
University of Agriculture Faisalabad,
    3800 Pakistan

**Khuram Shehzad Khan**
College of Resources and
    Environmental Sciences
National Academy of Agriculture Green
    Development
China Agricultural University, Beijing
    100193, China

**Qaiser Fareed Khan**
Sustainable Development Study
    Center
GCU Katchery Rd, Lahore 54000,
    Pakistan

**Neha Khardia**
Department of SSAC, Rajasthan
    College of Agriculture
MPUAT
Udaipur 313001, Rajasthan, India

**Dileep Kumar**
Micronutrient Research Project
Anand Agricultural University
Anand, Gujarat, India

**Dinesh Kumar**
ICAR-Central Coastal Agricultural
    Research Institute
Old Goa 403402, India

**Rajeev Kumar**
Department of Agronomy, College of
    Agriculture
GBPUA&T
Pantnagar, Uttarakhand 263145,
    India

**Rakesh Kumar**
Agronomy Section
ICAR-National Dairy Research
    Institute
Karnal 132001, India

**Sachin Kumar**
Centre for Geo-Informatics Research
    and Training
CSK HPKV
Palampur, 176062, India

**Suchith Kumar**
School of Agriculture
SR University
Warangal, Telangana, India

**Manish Kushwaha**
University Seed Farm
Punjab Agricultural University
Nabha, Patiala (Punjab) 147201,
    India

**Lakshman**
Department of Agronomy
Anand Agricultural University
Anand, Gujarat, India

**B.S. Mahapatra**
Bidhan Chandra Krishi
    Viswavidyalaya
Mohanpur, West Bengal

**Ahmad Mahmood**
Department of Soil and Environmental
    Sciences
Muhammad Nawaz Shareef University
    of Agriculture
Multan, Pakistan

**Govind Makarana**
Division of Crop Research
ICAR-Research Complex for Eastern
  Region
Patna 800014, India

**Abhishek Mandal**
ICAR-Indian Agriculture Research
  Institute
Delhi, 110012

**Diyan Mandal**
School of Crop Health & Biology
  Research
ICAR-National Institute of Biotic Stress
  Management
Raipur, Chhattisgarh 493225, India

**Azmat Qadir Maria**
Institute of Soil and Environmental
  Sciences
University of Agriculture
  Faisalabad
3800 Pakistan

**Biplab Mitra**
Department of Agronomy
Uttar Banga Krishi Viswavidyalaya
Pundibari, West Bengal, India

**Kiran Kumar Mohapatra**
ICAR-National Rice Research
  Institute
Cuttack, Odisha, India

**Bhabani Prasad Mondal**
School of Agriculture
SR University
Warangal, 506371, India

**Bhabani Prasad Mondal**
Department of Agriculture, School
  of Biotechnology, Biosciences &
  Agrisciences
Brainware University
Kolkata, West Bengal, India

**Rabiya Nasir**
Department of Environmental Science
University of Lahore
54000, Pakistan

**Muhammad Naveed**
Institute of Soil and Environmental
  Sciences
University of Agriculture Faisalabad
Punjab 38000, Pakistan

**Pujyasmita Nayak**
ICAR-National Rice Research Institute
Cuttack, Odisha, India

**Neha**
Department of Dairy Science and Food
  Technology, Institute of Agricultural
  Sciences
Banaras Hindu University
Varanasi 221005, India

**Vaibhav Pandit**
School of Agriculture Science
Malla Reddy University
Maisammaguda, Dulapally, Hyderabad,
  Telangana, India

**Amitava Panja**
Dairy Extension Division
ICAR-National Dairy Research Institute
Karnal 132001, India

**Man Park**
Soil Science Laboratory, School of
  Applied Biosciences
Kyungpook National University
80 Daehakro, Bukgu, Daegu, 41566,
  South Korea

**V.K. Paswan**
Department of Dairy Science and Food
  Technology, Institute of Agricultural
  Sciences
Banaras Hindu University
Varanasi 221005, India

**K.C. Patel**
Micronutrient Research Project
Anand Agricultural University
Anand, Gujarat, India

**Muskan Porwal**
Jawaharlal Nehru Krishi
    Vishwavidyalaya
Jabalpur 482004, Madhya Pradesh,
    India

**Laxmi Prasanna**
School of Agriculture
SR University
Warangal, Telangana, India

**Muhammad Farhan Qadir**
Institute of Soil and Environmental
    Sciences
University of Agriculture Faisalabad
Punjab 38000, Pakistan

**Sumit Rai**
Center for Environment Assessment and
    Climate Change
GB Pant National Institute of
    Himalayan Environment
Kosi-Katarmal, Almora 263 643,
    Uttarakhand, India

**Umair Raiz**
Department of Soil and Environmental
    Sciences
Muhammad Nawaz Shareef University
    of Agriculture
Multan, Pakistan

**Umair Raiz**
Soil and Water Testing Laboratory for
    Research
Bahawalpure 63100, Pakistan

**G. Bhupal Raj**
School of Agriculture
SR University
Warangal, Telangana, India

**Vishnu D. Rajput**
Academy of Biology and Biotechnology
Southern Federal University
Rostov-on-Don 344090, Russia

**Shri Ram**
Department of Soil Science, College of
    Agriculture
GBPUA&T
Pantnagar, Uttarakhand 263145, India

**Abhishek Rana**
Krishi Vigyan Kendra Gopalganj
    (Dr. RPCAU, Pusa)
Gopalganj, Bihar, India

**Ranbir Singh Rana**
Centre for Geo-Informatics Research
    and Training
CSK HPKV
Palampur, 176062, India

**Rajiv Ranjan**
Department of Botany, Faculty
    of Science
Dayalbagh Educational Institute
    (Deemed to be University)
Dayalbagh, Agra 282005, India

**Md Basit Raza**
ICAR-National Academy of
    Agricultural Research
    Management
Rajendranagar, Hyderabad 500030,
    India

**Taqi Raza**
Department of Biosystems
    Engineering & Soil Science
University of Tennessee
Knoxville

**Nimra Riaz**
Sustainable Development Study Center
GC University
Katchery Rd, Lahore 54000, Pakistan

**S. Rohith**
Department of Economics
IARI, Pusa Campus
New Delhi 110012, India

**Aryadeep Roychoudhury**
Discipline of Life Sciences, School of
    Sciences
Indira Gandhi National Open
    University
Maidan Garhi, New Delhi 110068,
    India

**Yamini S.**
Department of Dairy Science and Food
    Technology, Institute of Agricultural
    Sciences
Banaras Hindu University
Varanasi 221005, India

**Tithli Sadhu**
School of Agriculture
SR University
Warangal, Telangana, India

**Rahul Sadhukhan**
MTTC & VTC (Mizoram)
    (CAU-Imphal)
India

**Anuj Saraswat**
Department of Soil Science, College of
    Agriculture
GBPUA&T
Pantnagar, Uttarakhand 263145,
    India

**Animesh Sarkar**
Technical Assistant in DBT-Biofortified
    Maize Project
ICAR-Research Complex for NEH
    Region
Tripura Centre, Lembucherra,
    India

**Arya Kumar Sarvadamana**
Grassland & Fodder Research Unit
ICAR-National Research Centre on
    Camel
Bikaner, Rajasthan, India

**Guntamukkala Sekhar**
Department of Agronomy
Centurion University of Technology and
    Management
R. Sitapur, Odisha 76121, India

**Syed Shabbar Hussain Shah**
Graduate School of Fisheries and
    Environmental Sciences
Nagasaki University
1–14 Bunkyo-machi, Nagasaki
    852–8521, Japan

**Laila Shahzad**
Sustainable Development Study Center
Institute of Risk and Disaster
    Reduction, University College
    London
GCU Katchery Rd, Lahore 54000,
    Pakistan

**Faiza Sharif**
Sustainable Development Study
    Center
Government College University
Katchery Rd, Lahore 54000,
    Pakistan

**L. Devarishi Sharma**
MTTC & VTC (Mizoram)
    (CAU-Imphal)
India

**Sonal Sharma**
Department of SSAC, Rajasthan
    College of Agriculture
MPUAT
Udaipur 313001, Rajasthan, India

**L. Banarjee Singh**
College of Agriculture
Iroisemba (CAU-Imphal)
India

**Praveen Singh**
Department of Soil Science and
    Agricultural Chemistry
Anand Agricultural University
Anand, Gujarat, India

**Chinmaya Kumar Swain**
ICAR-National Rice Research Institute
Cuttack, Odisha, India

**Arifa Tahir**
Department of Environmental Science
Lahore College for Women University
Lahore, Pakistan

**Abdel Rahman Al Tawaha**
Department of Biological Sciences
Al Hussein Bin Talal University
P.O. Box 20, Ma'an, Jordan

**Rajneesh Thakur**
Department of Plant Pathology, College
    of Horticulture
Dr. Yashwant Singh Parmar University
    of Horticulture and Forestry
Nauni, Solan HP. 173230, India

**Sushmita Thokchom**
Department of Entomology
Assam Agricultural University
Jorhat 785013, Assam, India

**Vikas**
Department of Botany, Faculty
    of Science
Dayalbagh Educational Institute
    (Deemed to be University)
Dayalbagh, Agra 282005, India

**Aisha Abdul Waris**
Department of Isotope
    Biogeochemistry
Helmholtz Centre for Environmental
    Research
UFZ Permoserstraße 15, 04318 Leipzig,
    Germany

**M.R. Yadav**
Division of Agronomy
Rajasthan Agricultural Research
    Institute
SKNAU, Jobner 302018, India

**Manish Yadav**
Department of Soil Science
Punjab Agricultural University
Ludhiana, Punjab, India

**Suwa Lal Yadav**
Department of Soil Science and
    Agricultural Chemistry
Anand Agricultural University
Anand, Gujarat, India

**Noman Younas**
Institute of Soil and Environmental
    Sciences
University of Agriculture
    Faisalabad
3800 Pakistan

**Sadia Yousaf**
Sustainable Development Study
    Center
GCU Katchery Rd, Lahore 54000,
    Pakistan

**Muhammad Yousuf**
Institute of Soil and Environmental
    Sciences
University of Agriculture Faisalabad
3800, Pakistan

# 1 Management of Soil, Waste and Water in the Context of Global Climate Change

*Kankan Datta, Swarnavo Chakraborty, and Aryadeep Roychoudhury*

## 1.1 INTRODUCTION

Rapid climatic change and subsequent alterations in the surrounding environment has become the point of concern of the current day scenario. Climate change is caused by various anthropogenic activities as well as natural phenomenon. Since preindustrial period, the global temperature has been raised by almost 1.2 degrees which can be further raised in the upcoming days (Masson-Delmotte et al., 2018). Paris agreement on climate change was successful in pen and paper. However, considerable impact has not yet been noticed (Schleussner et al., 2018). Apart from temperature increase encompassing the entire atmosphere, climate change can impose other prominent impacts, like change of precipitation pattern, arctic snow melting, rise in basal sea level and thawing of permafrost (Sokka et al., 2020). Arctic region is highly sensitive to this temperature increase and climate change, which has triggered polar ice cap melting and rise in sea levels. Global warming and climate change terms go side by side since global warming is the key player in climate change in recent days. The term global warming induces associations with temperature fluctuations, weather changes, concerning changes, man-made errors and other negative effects; whereas climate change highlights general weather pattern changes and higher degree of natural fluctuations (Figure 1.1), which ultimately culminates in changed precipitation pattern (Schuldt et al., 2011; Whitmarsh, 2009).

Soil is an important land component and can be classified as a four-dimensional body on the basis of length, width, depth and time at the atmospheric and lithospheric interface, exhibiting utmost importance to all terrestrial life. Soil degradation means loss in its capability to provide ecosystem services (Ess) of interests to all life forms possible. Principal processes of soil degradation are salinization, nutrient and carbon (C) depletion and erosion, decline in soil structure, drought and tilth. The green revolution and other technologies since the mid-20th century were developed under a stable climate assumption. However, the current condition is not the same. Perpetual mismanagement can cause replacement of the climax community in a specific biome

DOI: 10.1201/9781003358169-1

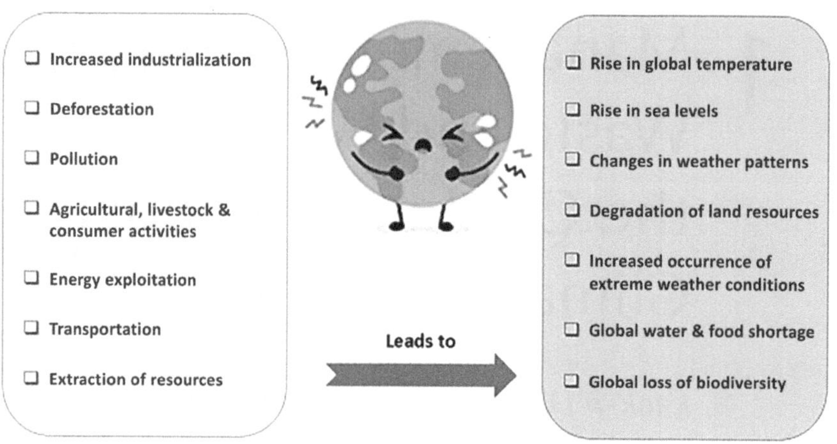


**FIGURE 1.1** Environmental and climatic alterations occurring globally due to various anthropogenic activities.

because of soil degradation. It has been reported that changes in fire regimes, land uses and climatic fluctuations are replacing the tropical humid forest by a savanna (grass) vegetation in the Amazon Basin (Veldman and Putz, 2011). Soil and water degradation goes side by side and are interlinked. Overexploitation of the fresh water from the water bodies and continuous deterioration to its quality parameters can lead to future water shortage. Wastewater treatment and government-undertaken protocols of restoring rivers and water bodies must be carefully implemented so that our future generation does not face any difficulties.

To overcome these situations, various remediation strategies have been undertaken with careful planning. However, some of them themselves can be held responsible in creating environmental disbalance and inducing further changes in its surroundings. Overreliance on fossil fuel related non-renewable energy sources is really concerning. It can provide the amount of energy necessary, but on the contrary, obnoxious greenhouse gases released subsequently can be a serious cause of the global warming. On the other hand, renewable energy sources are inefficient in providing the current energy demand. A mixture of the two can be implemented strategically in controlling the overall deteriorating effect on the environment. In addition, over usage of chemical fertilizers can cause both groundwater contamination and soil property alterations, including acidification of the soil (Kanter, 2018). Irrigated soil can prove to be effective in water holding and better crop development. Chemical fertilizer usage demands greater amount of water. Biofertilizers can sometimes also be used side by side with the inorganic ones for a better yield, but changed soil properties can render them inactive. So coupling these two can also work wonders in maintaining the overall soil property (Stewart et al., 2005), but a strict check should be imposed on the usage of chemical fertilizers.

Ever-increasing population size and increasing demands of essential commodities has led to large-scale industrialization and urbanization. This process is directly proportional to the overall waste generation. The factories and agricultural sectors are generating tonnes of waste and by-products which get emptied into the environment, causing major harm to it (Ravindran et al., 2018). Each year, around 7–9 billion tonnes of waste gets emptied into the environment, out of which around 60% gets processed properly to restore the natural balance. However, the remaining 35%–40% waste is of serious concern, as most of it is non-recyclable and starts to get accumulated in the environment over time, causing harm to it (Chen et al., 2020). Increased population size and density will make things worse in the future unless a proper management strategy is adapted. The reuse, reduce and recycle, or the 3Rs, approach might not be always useful. So creating awareness in the mankind is mandatory. The 1989 signed Basel Convention of hazardous waste generation and management strategy is yet to be implemented in its full scale (Korcheva, 2021). Reducing the reliance on fossil fuels; maintained processing of foods and cosmetics; and cutting off each day effluent from the industries are some of the methods, which will prove to be useful in future to control the ever-increasing environmental harm. Proper use of recycling approach and generation of eco-friendly by-products like biosurfactants and biofuels can simultaneously solve two major problems of environmental damage and economic instability (Sagar et al., 2018; Guan et al., 2016). In this chapter, we will focus mainly on the various sources responsible for the environmental degradation and pollution, alongside the necessary remediating measures that should be undertaken to reduce environmental burden. The soil, water pollution and waste management strategies from a biological point of view have been discussed in the upcoming sections.

## 1.2 MANAGEMENT OF SOIL IN THE CONTEXT OF GLOBAL CLIMATE CHANGE

Soil happens to be the thinnest upper layer of earth's crust, sustaining the whole biodiversity. This soil layer is currently facing all kinds of activities causing deterioration and alteration of its properties. "Soil pollution" refers to the presence of chemicals or certain substances in higher-than-normal concentration, thereby imparting adverse effects on any non-targeted organism. The Status of the World's Soil Resources report (SWSR) has listed soil pollution as one of the main soil threats affecting its properties all over the world and the ecosystems services provided by them. Soil pollution also leads to reduced food safety, as the crops grown in contaminated soil are unsafe for consumption, due to the presence of various harmful chemical substances and heavy metals (Rodríguez-Eugenio et al., 2018). Heavy metal contamination in soil can also cause soil pH alteration and decrease of overall crop yield. The crops sustaining in these contaminated soils are often a major player in bioaccumulation. Many harmful chemicals associated with contaminated soil can reach groundwater level via leaching, causing groundwater pollution. Moreover, such contamination can spread into freshwater bodies, via surface water runoffs and is one of the main causes of eutrophication. The contaminants can also harm soil microorganisms directly, leading to the loss of soil biodiversity. As a result, soil fertility gets a major hit. Soil pollution can be broadly categorized as point-source pollution and diffused pollution (Rodríguez-Eugenio et al., 2018). When soil pollution is restricted to a particular area where an event or series of events cause alterations of soil properties and sources of the pollution are easily identified, the pollution is referred to as point-source pollution. This type of pollution includes any kind of industrial effluent-based pollution, mining, fertilizer spraying over an agricultural land, etc. On the other hand, diffused pollution occurs over a larger area, and the sources are not easily identified. This type of pollution involves pollutant transportation via air-water-soil system. Different activities, ultimately culminating in soil pollution, are described in the upcoming sections.

### 1.2.1 Sources of Soil Pollution

#### 1.2.1.1 Natural Source

Soil is generally formed by weathering of rock and upper surfaces of the earth. Some of these sources themselves can be responsible for the contamination of soil. Various heavy metals can increase the baseline value of these necessary microelements, leading to widespread toxicity. Various rocks can release certain amount of radio nuclide and toxic chemicals to alter the pH and properties of the soil. For instance, arsenic (As) contamination is one of the major concerns of biogeologists. Weathering of various arsenophytic rocks, mineral ores of arsenic and volcanic eruption can increase this heavy metal concentration in the soil (Albanese et al., 2007; Díez et al., 2009). Radon (Rn) is a radioactive gas and shows a controlled penetration pattern from deep to upper soil layers. However, carbonate containing soil shows a much higher level of Rn emission, with pronounced alteration in the soil properties (Gregorič et al., 2013). Besides rock and mineral weathering, natural calamities like volcanic eruptions and forest fires are major contributors of soil pollution. Several toxic chemicals including polycyclic aromatic hydrocarbons (PAHs) and dioxin-related compounds can pose serious threats to the soil-grown flora (Deardorff et al., 2008). Volcanic eruptions can also add several heavy metals like mercury (Hg), copper (Cu), zinc (Zn), cadmium (Cd) and chromium (Cr) into the soil (Dœlsch et al., 2006).

#### 1.2.1.2 Anthropogenic Activities

Various man-made activities have promoted the soil quality to deteriorate to a large extent over the decades. Some of these activities include industrial effluents, high usage of chemical pesticides and fertilizers, mining, municipal wastes, petroleum derived products, etc. Both intentional and unintentional activities can ultimately culminate in global soil property alterations. Some of the anthropogenic sources of soil pollution are listed in the following section.

#### 1.2.1.3 Agricultural Activities

Application of chemical fertilizers, over usage of pesticides and animal manures are some of the major sources of soil pollution. These agrochemical products can increase the pH of the soil to a huge extent and cause heavy metal contamination of soil (Rodríguez-Eugenio et al., 2018). Usage of water, heavily contaminated with sewage and industrial effluents in agricultural fields, seems to be responsible for widespread soil pollution. These can pose a serious threat to food safety and human health. Agricultural fields also face point-source pollutions from accidental spills of various hydrocarbons and fuel compounds which are used during crop cultivation or harvesting process. Animal manures can cause major damage to the soil if not properly disposed. Animal excreta may contain certain microorganisms and lipolytic medicinal chemicals which can cause soil property change. These lipolytic chemicals can persist in the soil for a long time to pose risk from the food safety point of view (Zhang et al., 2015).

### 1.2.1.4 Industrial Activities

Industrial effluents are a major concern, as this is responsible for all the three pollutions: soil, water and air. Gaseous pollutants and aerosol-like pollutants directly enter the atmosphere and can precipitate down as acid rain to increase the soil pH, causing havoc damage. Industrial wastes are occasionally directly disposed of in the rivers and other water bodies. Usage of this water for agricultural purposes can lead to soil pollution. Textile factories, rubber processing industry, tannery, pharmaceutical industry and soap and detergent industries can cause salinization, a major reason behind global soil property change. Therefore, factories and industrial sectors are generating loads of detrimental effluents, which are responsible for causing major harm to the surrounding environment (Ravindran et al., 2018).

### 1.2.1.5 Waste Management and Sewage Disposal

Ever increasing global population calls for a proper management and disposal strategy to combat increased waste production. Such implementations are yet to be witnessed. The two major strategies for waste disposal are landfilling and incineration. Both are heavily responsible for soil pollution and contamination. Various PAHs, pharmaceutical chemicals and derived products produced as a result of such activities can accumulate in soil to alter its properties (Swati et al., 2014).

### 1.2.1.6 E-Waste

Technological advancement has brought the whole world in human grasp. The ever-increasing production of telecommunication-based and other electronic devices pose a major threat to the global soil. Once an electronic device becomes nonfunctional, it contributes to the E waste. These wastes cannot be treated by conventional methods. As a result, hazardous chemicals leach over years from these non-recyclable E-wastes imposing damage to the soil. Some sectors use the recyclable approach of the used metals like copper and gold. However, the simplest recyclable approach undertaken is burning of the outer covering to recycle the metallic portions. This approach causes release of various toxic gases and PAHs which ultimately pose greater threats over non-recycled ones (Perkins et al., 2014; Itai et al., 2014).

### 1.2.1.7 Urbanization and Transportation Activities

Infrastructural development and quick urbanization has a huge impact on environmental degradation. Quick landfilling and sealing of lands can lead to soil degradation and alteration. On the other hand, increased number of cars can also contribute to the soil pollution. Petrochemical spills and fossil fuel combustion can release multiple PAHs and related derived toxic substances, which can accumulate in the soil to cause its deterioration (Mirsal, 2008). Besides, lead (Pb) gasoline combustion can release this heavy metal into the atmosphere, as well as contaminate the environment, which can reach the soil to cause lead contamination. It has been documented that almost 10 million tonnes of lead has been transferred to global soil, of which United States of America bears the sole responsibility of adding around 6 tonnes (Mielke and Reagan, 1998).

## 1.2.2 MITIGATION STRATEGIES

To counteract all these kinds of soil derogatory activities, proper planning and fruitful mitigation strategies seems to be the need of the hour. From the food safety and human health point of view, soil pollution can also cause major impacts on the community. Phytoremediation and exploitation of hyperaccumulator plants can cure the heavy metal–related toxic effects on the soil. Besides afforestation, sustainable crop cultivation and strategic use of fossil fuels can also prove to be effective in controlling soil pollution.

### 1.2.2.1 Afforestation

Planting more trees in a particular area is called afforestation. Afforestation can be effective in combating point-source pollution of soil. Trees help to retain the soil fertility by preventing soil erosion and washing of minerals from the soil. Moreover, this is an eco-friendly and sustainable approach of cultivation, which can potentially cure the soil pollution. Another such strategy is crop rotation, where different crops are cultivated over the year to reduce the same type of nutrient reliance and increase the carbon content in the soil. Besides, it can also reduce the usage of chemical pesticides and fertilizers (Bowles et al., 2020). Local Indian tribes of Arunachal Pradesh and Nagaland implement another kind of eco-friendly approach known as "Jhum cultivation". In this kind of approach, after one round of cultivation is over, the land is cleared, and the remains are burnt. This can increase the overall potash and other necessary element levels. As a result, nutrient reliance on the soil can be sustained. However, burning of crop remains adds on to the atmospheric load of detrimental greenhouse gases. Hence, this method is rarely practiced.

### 1.2.2.2 Phytoremediation

Phytoremediation is an eco-friendly approach of reviving heavy metal contaminated soil. In this relatively new approach, green plants are used to remediate these contaminated soils. There are various categories of phytoremediation, termed as rhizofiltration, phytoextraction, phytovolatilization, phytostabilization, etc. In phytoextraction, the metal ions are accumulated in the aerial parts of the plants, which can later be incinerated in a controlled environment to ensure a proper disposal of these heavy metals (Napoli et al., 2019). In phytostabilization, plant roots absorb the heavy metals and collect them in the rhizosphere region (Radziemska et al., 2017). In phytovolatilization, certain toxic heavy metals like selenium and mercury can be volatilized through the foliage parts (Limmer and Burken, 2016).

Many plants have been identified, displaying phytoremediating properties (Brooks, 1998; Baker et al., 2000). These plants are termed as hyperaccumulator plants, as they accumulate heavy metals in their body parts. These hyperaccumulators can be of higher dry weight or biomass or sometimes with lower biomass with profuse rooting pattern (Landberg and Greger, 1996). Planting of *Thlaspi* sp. and *Arabidopsis* sp., being some of the extensively studied hyperaccumulator plants, have been readily implemented currently as soil pollution remediation strategies.

The hyperaccumulator plants can sequester heavy metals in various locations. For instance, in *Thlaspi caerulescens*, zinc (Zn) is sequestered as a soluble form in the epidermal vacuolar compartments (Frey et al., 2000). Puschenreiter et al. (2003) conducted an experiment where they used two metal hyperaccumulator plants, *T. arvense* and *T. goesingense,* grown in non-contaminated as well as contaminated soil. Accumulation and reduction of free Zn in the rhizosphere was observed in *T. goesingense* plants grown in polluted soil. In case of the plants from non-contaminated source, Zn accumulation occurred, but rhizosphere-free Zn level was not altered much. Nickel (Ni) is another such heavy metal, whose presence has also been found to be altered in soil since this heavy metal tends to be sequestered by *T. goesingense* only when grown in controlled microenvironment. In contrast, labile Ni level showed an increase in both the soil, suggesting the usual tendency of heavy metal (Ni) mobilization by *T. goesingense* (Puschenreiter et al., 2003). These properties can be genetic in nature and can vary from plant to plant significantly, even among different cultivars. The mechanism of heavy metal accumulation, uptake, removal, mobilization to different plant parts and compartmentalization vary with each plant species and is of utmost importance in determining its role in phytoremediation.

However, there can be certain drawbacks too. The accumulated metals can stay in the parts of the plants as it is, and after the death of the plant, the nutrient cycling can recontaminate the soil with these heavy metals. In many cases, these plants later are collected and their hyperaccumulatory parts are incinerated. The ash and aerosol formed from the incineration process can release some of these heavy metals in the environment which can later fall on the surface of earth during raining, with water droplets. So a controlled and proper disposal strategy is also mandatory (Koptsik, 2014).

### 1.2.2.3 Bioremediation

Bioremediation is a simple method of restoring natural balance by means of treating organic pollutants and biomacromolecules with microbial organisms. Rise in detrimental industrial and anthropogenic activities have led to the release of toxic effluents into the environment. These xenobiotics can exist in the soil for a prolonged time to cause major harm to the soil flora and fauna by virtue of the compromised soil properties. The bioremediation technique can be classified as in-situ bioremediation and ex-situ bioremediation. In the in-situ technique, the soil is generally not removed from its original place to achieve remediation. It is a comparatively safer, as well as cheaper, method, as the organisms used are harmless and no major processing is required (Sasikumar and Papinazath, 2003). However, the microbes must have a proper chemotactic ability since the cells will get attracted to the contaminated site to mediate the remediation procedure. On the other hand, in case of the ex-situ process, the contaminated soil portion is excavated by means of heavy machinery for remediation. Various microbes can be used in the bioremediation process to cure the polluted soil contaminated with organic pollutants, radioactive contaminants, etc.

*Geobacter metallireducens* is one such microflora which can reduce the amount of radioactive uranium from polluted soil (Gregory and Lovely, 2005). Some microorganisms of the genera *Pseudomonas* and *Corynebacterium* can also remediate the oil spills and petrochemical contamination in soil (Sasikumar and Papinazath, 2003). These microbes can also potentially reduce the level of heavy metal contaminants like selenium, etc.

### 1.2.2.4  Organic Farming

Another eco-friendly, sustainable approach to mitigate soil pollution is organic farming. This farming technique mainly focusses on controlled use of pesticides and fertilizers, an effective crop rotation strategy with legumes to increase nitrogen content in soil and controlled use of organic livestock derived manures. This can help to reduce overreliance on the residual soil nutrients after each crop cycle. Recent studies have found that these kinds of organic farming and crop rotation strategy can be effective in overall increase of crop yield and maintaining soil fertility (Smith et al., 2019).

### 1.2.2.5  Biological Control of Pathogens

Biological level of control can be exerted on certain plant pathogenic bacteria to limit their derogatory effects on crop cultivation. The controlling microbes or bacteria used are generally harmless to the plants. Species of *Bacillus* and *Pseudomonas* can be used to limit the growth of various pathogenic microorganisms. Some important biomolecules associated with the genus *Bacillus* are lipopeptides and non-ribosomal proteins, polyketide compounds, siderophores and bacteriocins. In general, the lipopeptides have a broad spectrum of antagonizing activity against certain plant pathogens like bacteria, fungi and viruses. Circular lipopeptides from iturin, surfactin and fengycin families affect the membranes of the targeted pathogenic microbes. Polymerase chain reaction and gene expression analysis showed the presence of various biosynthetic operons, including operons for iturin, fengycin, bacillomycin and surfactin in different strains of *Bacillus* (Fira et al., 2018). Similarly, some strains of fluorescent *Pseudomonads* produce secondary metabolites that are secreted in the media to inhibit the growth of various fungal pathogens. Siderophores are iron chelators that are generally involved in this type of control but do not seem to be the key player in the controlling activity (Haas and Défago, 2005). In this case also, the bio-controlling agents must be used in a proper controlled manner so that it does not interfere with the crop yield and productivity and help to limit the growth of various pathogenic microbes in the rhizosphere as well as foliar parts.

These remediating measures can be broadly categorized as in-situ, which refers to on-site remediation, and ex-situ, or off-site, remediation (Sasikumar and Papinazath, 2003). Both types of treatments can be categorized further under the physical, chemical and biological treatments. All the mentioned techniques have both advantages and disadvantages. Over-irrigation can cause major harm to soil properties and parameters, and the use of fossil fuels for the remediating measures must be carefully controlled to limit the toxic emissions. Organic farming needs greater amount of water and is expensive, as the yield is still limited and the process is still far from being implemented in larger scale. So an overall eco-friendly, economic and easy-to-implement procedure must be formulated and carefully adopted.

## 1.3  MANAGEMENT OF WASTE IN THE CONTEXT OF GLOBAL CLIMATE CHANGE

Increasing urbanization and industrialization are the main reasons behind rapid waste generation. Various industries are simultaneously indulging themselves in this reckless act and causing the environmental pollution to skyrocket. Rising population density, modernization of lifestyle and increased daily needs call for improved industrial sectors, including agro, automobile, paper, textile and so on. These industries are the major contributors in waste generation, which are primarily organic in nature. These hazardous substances are causing environmental deterioration daily (Panesar et al., 2015). Hence, an efficient channelization of these waste products and improvisation of suitable management strategies should become the prime goal currently. Some of the organic substances can be effectively converted into less toxic forms or biodegradable forms via various biological and chemical methods. Various sources of waste generation are discussed in the following section (Table 1.1).

**TABLE 1.1**
**Classification of Wastes**

| Types of waste | Components | Classification | References |
| --- | --- | --- | --- |
| Non-hazardous waste | Mainly constitute any form of rubbish, which upon recycling imposes no harm to humans and environment as a whole. The common examples are paper, glass, plastic, metal and other forms of organic materials, certain health care and packaging wastes, etc. | Agricultural waste (manure, harvest waste, etc.)<br>Industrial waste (paper, plastic, metal, glass, etc.)<br>Municipal waste (packaging products, food scraps, paper, cloth, etc.)<br>Household waste (toilet wastes, vegetable peels, etc.) | Sharrock et al., 2009; Szabo et al., 2018.<br>Sajna and Gottumukkala 2019.<br>Federici et al., 2009.<br><br>Dhar et al., 2017. |
| Hazardous waste | Contains wastes having properties, which pose potential harm to human health and environment. These wastes require more strict legal actions, storage and disposal strategies. The common examples are corrosive chemicals, pesticides, radioactive wastes, batteries, electronic and E-wastes, paints, oils, syringes, unused medicines and medical products. | Household waste (batteries, electronic products, etc.)<br>Industrial waste (radioactive wastes, paints, pesticides, chemicals, etc.)<br>Hospital waste (unused medical products, syringes, etc.) | Mor et al., 2021.<br><br>Smidt and de Vos 2004; Galt, 2008; Ravindran et al., 2018.<br>Squire, 2013. |

### 1.3.1 SOURCES OF WASTE GENERATION

- **Agricultural waste:** Agricultural industry is an ever-growing industry, which will continue to grow in the upcoming years, progressively with increase in the population density. Ever-increasing daily needs and food requirement, aided by the excessive usage of chemical and factory-based methods, can cause major derogation to the food crops. Each year, approximately 1.5 billion tonnes of agricultural waste is generated, which makes around one-third of the total agricultural productivity (Ravindran et al., 2018). Agricultural wastes constitute the remaining portions obtained after processing of raw yielded products like grains, cereals, vegetable crops, fruits, etc. This ever-increasing demand has led to huge increment in this kind of waste generation (Panesar et al., 2015). These wastes are mostly generated during farming, processing or storage operations and livestock maintenance (Mo et al., 2018).

Food processing industry and agricultural industry are linked by terms of raw materials used and products generated. However, the waste generated can be drastically different. The food processing industry produces processed food items, using the agricultural items. The agricultural industry, on the other hand, is built up using the manpower and equipment to increase the yield. The waste products include livestock waste, petroleum or fuel-based organic waste, processed agricultural wastes like hay and coconut fibers (Szabo et al., 2018; Sharrock et al., 2009).

The cereal processing produces large amounts of hay and bran as wastes. The dairy industry leads in the production of large amounts of whey, organic macromolecules, inorganic detergent traces and complex biomolecular aggregates. The poultry and meat industry produces feathers and animal waste materials. Similarly, the aqua industry produces fish and crab scales, endoskeleton remnants, etc. (Sharma et al., 2020). Most of these agrochemical wastes are biodegradable. However, owing to their biomolecular make up, these substances are well capable of sustaining and supporting microbial growth, including various pathogens. Moreover, the lignocellulosic portion of the agro waste is non-soluble and takes longer time to decompose (Girelli et al., 2019).

- **Food industry:** Most of the raw materials obtained from the agricultural industry are inedible or cannot be consumed directly without processing. Food processing industry is one of the largest and important industrial sectors currently. The food being a perishable entity needs to be processed or preserved to increase their shelf life and store them for longer durations in cold room. This industry alone is responsible for more than half of the total waste generation worldwide (Sharma et al., 2020).

The food processing industry produces large amounts of leftovers post processivity. The apple juice forming industry generates around 10%–11% solid fruit remnants post juice extraction (Dhillon et al., 2013). Similarly, fruit cutting, peeling and juice extraction can generate large amounts of wastes. Cereal and pulse processing industry generates heaping amounts of hay, kernels, raw fibrous remnants, etc. Milk processing and dairy industry generates wastewater and other micro and macro elements and detergent-like substances (Sharma et al., 2020; Dhar et al., 2017).

- **Paper and pulp industry:** Documentation and printing is done on a piece of paper. The economy of the country is also printed on a piece of paper. So the product of pulp and paper producing industry is of utmost importance in our daily life. Paper is directly produced from tree woods. Huge requirements of chemical energy, particularly water, is causing a major impact on the environment. It has been estimated that roughly 100–250 $m^3$ of fresh water is used to make only 1 tonne of paper, out of which a heaping amount of water is generated as wastewater. Roughly 75–200 $m^3$ of wastewater is released for 1 tonne of paper production (Haq and Raj, 2020). Various synthetic and artificial chemicals or substances are added to increase the productivity and durability of paper. These resin-like synthetic substances can deposit on soil to contaminate it.

The waste from the paper and pulp industry can be listed as (a) rejected materials (including glass, metals, fibrous lumps); (b) primary paper sludge (various filtered and floating waste particles); (c) secondary paper sludge (from clarifiers containing biological materials and short fibers); and (d) de-inked sludge (as colors, additives, short fibers and ink) (Sajna and Gottumukkala, 2019). The wastewater effluent from such industry contains around 200–300 different types of organic matter and chemical substances, phenolics, lignin-like fibrous materials, ascorbate organic halogens (AOX), fatty acid derivates, microplastics, etc. Some of them are non-biodegradable and are largely responsible for eutrophication, increased biological oxygen demand (BOD) and biomagnification (Mohammed et al., 2021).

- **Petrochemical and pharmaceutical industry:** Waste products generated from petrochemical and pharma industries are highly toxic and non-biodegradable in nature, owing to the presence of high concentration of organic and chemical pollutants. The waste is sometimes directly thrown in water bodies, which can create multiple problems, including reduction in marine biodiversity. The wastes from these industries contain small suspended heavy metal particles, solid sludge, plastic wastes, etc. The oil spills and toxic gases can also be released into the environment to induce pollution (Panizza et al., 2018).
- **Biomedical waste:** Health care infrastructure improvement can both prove to be a boon as well as a curse, increased biomedical waste generation being on the negative side. Accumulation of biomedical waste can overburden the environment with toxic, hazardous and noxious chemicals, which get deposited over time. The biomedical wastes can cause several problems, being neurotoxic and nephrotoxic in nature, and can even lead to cell growth cessation and death. Biomedical wastes like syringe, plastic holders, saline bottles, cotton swabs, etc., are non-biodegradable and can stay in the environment for a prolonged time period. On the other hand, the microbial

biomass coming along with it can be highly pathogenic and contaminate both surface water and soil, as well as groundwater (Squire, 2013).

- **E-waste:** Electronic equipment and gadgets have become inseparable part of our lives, and this ever-increasing dependence is also causing electronic residuals, or E-waste, to be the fastest-growing waste contributor, with a 20% annual increase each year. E-wastes comprise computer accessories, household E-scrap, telecommunication remnants, electrical wires, unused accessories, etc. These waste products are filling out the landfill fast, and small micro pollutants, like microplastics, generated can cause severe harm to the population, as well as the environment (Mor et al., 2021).

## 1.3.2　Mitigation Strategies

The waste products formed from these industries can cause major damage to the environment and can ultimately culminate in reduction of marine and terrestrial biodiversity. Soil, water and air can all be affected from industrial effluents directly or indirectly. So a proper processing and strategic disposal method should be generated as early as possible. The major mitigation strategies and related benefits are discussed in the upcoming sections.

- **Recyclable wastes:** The recent Environment Action Program of 2020 has emphasized on the recycle and reuse of waste materials (Correddu et al., 2020). Most of the agro wastes formed from the excess grain and fruit portions can be used in other sectors. For instance, the fibrous husk portions from crop grains, coconuts, corn, fruit and vegetable peels, soybean meal and sunflower meal can be used as animal feeds and reused (Federici et al., 2009; Correddu et al., 2020). The organic waste material and excreta generated are biodegradable and can be efficiently converted into biofertilizers.

Rice straw is one of the major wastes generated in huge amounts worldwide from the agro industries. Most of these straws are set on fire or used as fuel in village areas to cook food. Burning of these fibrous redundant portions can emit huge amounts of toxic gases into the environment and cause global warming. These rice straws are nowadays air-dried and, along with rock phosphate and some other chemicals, can be used to generate compost piles, which help in better growth of cultivated plants and work as a natural fertilizer. There are four phases of composting: initial brief mesophilic phase, thermophilic phase, cooling phase and, lastly, maturation phase (Correddu et al., 2020). Depending on the temperature difference in each phase, the microbial flora density of the population gets changed and dictates the physiochemical change of state of the compost. This increase can indirectly affect the carbon-nitrogen ration to make it less toxic comparatively. The compost is built slowly over the span of 2–3 weeks at 55 °C. This process ensures that no further contamination persists, and the compost is used in the organic farming (Rashad et al., 2010). Azo dyes can be used to process the effluent or wastewater portion to help make it clean. Solid waste management must be focused on a specific policy called "the 3Rs", standing for reduce, reuse and recycle (Das et al., 2019). In the recent environmental summit, another R has been added to make it the 4R policy. The last R stands for recovery. Complete recovery of a product is never possible, and a major amount is lost during processing and other important works.

- **Biofuel formation:** Agro and food wastes can be fermented to generate ethanol and other biofuel substances, which can have loads of applications. The fermentation of agro waste products as raw material and their further saccharification can generate these fuel substances (Guan et al., 2016). Intasit et al. (2020) formulated a method of forming a potassium-free, solid biofuel from palm and date pre-treatments. The palm waste pre-treatment was done with *Aspergillus tubingensis* with the help of solid-state fermentation protocol. This helped to reduce the potassium content by almost 90% while further concentrating cellulosic biomass, ultimately yielding a substrate with biodiesel-like properties.

Other agro or food-based waste products and animal waste materials can be used to generate many types of biofuel substances, including methane, ethanol, methanol, etc. (Kibbler et al., 2018). Dhar et al. (2017) showed that a combinatorial fermentation of 60% food waste and 40% manure can generate large amounts of ethanol, which burns with prominent blue flames and can generate a temperature higher than the conventional fossil fuels. So this can be used as an alternative of fossil fuels. Similarly, pyrolytic processing and gasification of food wastes can create syngas and char which have a potent capability to replace conventional fossil fuels. Hydrochar can be formed by hydrothermal carbonization (HTC), a process in which high moisture containing food wastes are fermented under high temperatures to generate this kind of biofuel (Kibler et al., 2018).

Biohydrogen is another potent biofuel having promising properties and can be formed by thermo-mechanical wastewater pulping, followed by dark fermentation under the assistance of microbes (Dessì et al., 2018). *Thermonaero* Sp. is a potent thermophilic microbe perfectly viable under 70 °C and have shown the highest yield of biohydrogen. Paper-industry-generated wastes can be processed by means of enzymatic hydrolysis method with the help of NS50013 and NS50010 to generate liquid biofuels. *Trichoderma reesei* produces the enzyme cellulose, which facilitates this process. Further processing of the effluent made by this bacterium, with the help of *Enterobacter aerogenes*, produces the biohydrogen biofuel (Lakshmidevi and Muthukumar, 2010). In addition, the paper-and-pulp-industry-generated waste products can be used to generate biogas (Ahmed and Gupta, 2009), electricity (Zalas and Cynarzewska, 2018) and biochar (Mohammadi et al., 2019).

- **Regeneration of materials:** Food and agro-industry-generated wastes are a huge reservoir of polyphenols, vitamins, carotenoids, dietary fibers, oils, etc., which can be easily extracted using different processing techniques (Sagar et al., 2018). These bioactive compounds can be used by pharma industries, food processing industries, etc.

Ketchup forming industries and other agro-industry-discarded tomato peels and remnants can be a proven source of lycopene, betacarotene, glutamate, aspartate, etc. (Szabo et al., 2018). It has been noted that food wastes like grape, orange, lemon peels and seeds of mango, avocado and jackfruit can contain up to 15% of phenolics, as compared to the fruit pulps (Sagar et al., 2018).

The lignocellulosic waste portions generated from the agricultural sector is a storehouse of substances like cellulose, pectin, lignin and hemicellulose (Ravindran et al., 2018). Agro-industry-generated wastes (like corn steep liquor, cassava bagasse, beet molasses, grape skin and pulp remnants, jackfruit seeds, sweet potato hydrolysate, soybean pomace, sugarcane bagasse, potato starch water, rice hull, jatropha seedcake, palm kernel, etc.) can be processed using either solid-state or submerged fermentation procedure in the presence of *Aureobasidium pullulans* to generate the microbial exopolysaccharide pullulan.

Pomegranate production and juice extraction and processing are huge industries and can generate high amounts of waste products. The juice yield is typically less than half of the total weight of a fruit. So the seeds and the peel portions can pile up as waste products. Solid-state fermentation of pomegranate peel waste, using microbes, can be implemented to generate citric acid in large amounts (Ullah et al., 2012; Faour-Klingbeil and Todd, 2018). Sunflower meals and seeds, wheat bran, olive oil cake, maize bran and pericarp portions can be processed in the presence of *Bacillus* Sp. to generate important enzymes like α-amylase, pectinase, protease, cellulose, etc. (Salim et al., 2017). Paper industry effluents are largely saturated with lignin, which comes as a black liquid. This substance can be used to generate various adhesives, adsorbents, dispersants, etc. (Carrott and Carrott, 2007; Gupta and Shukla, 2020). Such food, paper and agro waste management and value-added product formation is working in a dual way to reduce the overall waste amount and to generate revenue.

- **Petrochemical spills and use of superbug:** Crude oil and petrochemical spills are highly damaging towards the environmental stability, as it is a complex mixture of non-biodegradable organic pollutants and can cause major harm to the biodiversity. The petrochemical spillage and pollution are not a new concern, as several significant events have already caught attention of environmentalists and scientists. Oil shipping disasters like *Prestige* (2003), *Erika* (1999) and *Exxon Valdez* (1989) had already drawn the attention of many people towards this blooming problem (Brooijmans et al., 2009). Oil eating or hydrocarbon degrading superbugs can be implemented to clean up such contaminated sites. Obligate hydrocarbon degrading bacterium *Alcanivorax borkumensis* is worth mentioning in this regard, as this bacterium constitutes almost 80% of the overall bacterial population in the oil spillage sites (Schneiker et al., 2006). In addition, the genome study of this bacterium has revealed that it can efficiently degrade alkanes.

This bacterium produces large amounts of exopolysaccharide substances and pili structures to remain attached at the oil-water surface interface. It also produces large amounts of biosurfactants to emulsify the undissolved lipid portion in the oil spillage. The high GC-containing genome contains a cassette of *alkK* and *alkL* genes, which encodes several alkane-degrading enzymes like alkane hydrolase, alcohol dehydrogenase and cytochrome P450 (Brooijmans et al., 2009).

Polycyclic aromatic hydrocarbons (PAHs) are highly stable structures, as they contain multiple aromatic rings and are one of the major non-biodegradable substances found in petrochemical spillage. These hazardous carcinogenic chemicals can persist long in the environment, post degradation of linear chain alkanes. Various genetically modified PAH-degrading superbugs can be used to reduce the content of environmental toxic chemicals. Some of worth mentioning microbes in this regard belong to the genus of *Arthrobacter, Pseudomonas, Burkholderia, Mycobacterium, Rhodococcus, Sphingomonus.* etc. (Seo et al., 2007; Vandermeer and Daugulis, 2007; Haritash and Kaushik, 2009; Martinkova et al., 2009).

- **Biomedical and E-waste management:** The PAHs, polychlorinated dibenzofurans (PCDFs), polychlorinated biphenyls (PCBs) and various heavy metals are some of the most concerning toxic chemicals found in the biomedical wastes. We can follow a simple segregation approach to separate dangerous waste products from the other not-so-much-harmful ones. The more toxic ones must be properly sterilized to kill off all the microorganisms prior to disposal. Some of the equipment and waste products like bandages, blood-soaked cottons, etc., are burnt off in the open-site dampening process in a controlled environment to reduce the overall environmental impact of these waste materials (Squire, 2013).

Similarly, the E-wastes are also one of the major contributors of PAHs in the environment. The E-waste parts can be recycled to use in the brand-new items instead of disposing it off completely. The metal parts can be melted down to use in transistors and other electronic devices to reduce the load on the environment (Mor et al., 2021). However, a proper management strategy is yet to be implemented at its full scale.

## 1.4 MANAGEMENT OF WATER IN THE CONTEXT OF GLOBAL CLIMATE CHANGE

Water is one of the most essential daily requirements of humankind. Out of the two-thirds water portion all over the world, only 3% is consumable and usable in daily commodities. The rest 97% comprises of the oceanic salt water, which cannot be consumed (Preisner, 2020). However, this 3% is also facing major dreaded anthropogenic activities, contamination and deterioration. Increased urbanization has led to increased demand of daily necessities, and industrialization has already started to make the scenario worse. To cope with the future necessities and for the maintenance of a constant supply of pure drinking water, an efficient management strategy must be undertaken. The freshwater sources like lakes, rivers and groundwater are facing constant overexploitation. The industrial effluents are directly thrown in water bodies, thereby increasing the microbial pool in fresh water and increasing the toxic carcinogen levels in water. Both organic as well as inorganic pollutants are overburdening the environmental stability, as they lead to freshwater contamination. The constant changing environment also imposes a negative effect on the treatment strategies as well. Ever-increasing sea level and increased rainfall are increasing the wastewater pool which is damaging the pipeline system of wastewater treatment plan (WWTP). Besides, it can create corrosion in the underground drainage pipes to cause contamination in the groundwater pool. In this portion, we will discuss various pollutants and their effects on water bodies.

### 1.4.1 Sources of Water Contamination

The basic freshwater contamination can be categorized into surface water contamination and groundwater contamination (Schwarzenbach et al., 2010), both having overlapping sources and effects. Some of the major sources and outcomes are discussed here.

- **Surface water contamination by factory and agricultural waste:** Agricultural and factory-based wastes are directly thrown in the water bodies which can lead to surface water contamination. These wastes are rich in organic as well as inorganic and micro pollutants. The agricultural wastes include pesticides, insecticides, chemical fertilizers, animal manures and raw agricultural wastes (Galt, 2008). These substances are highly rich in organic matter and can cause eutrophication of the water bodies. Phosphate is the key player in eutrophication, as this microelement is of major importance in algal and cyanobacterial growth. The resulting algal bloom deprives the marine microbiota and organisms of dissolved oxygen (Werner, 2002). As a result, the biodiversity and the marine flora can get seriously harmed. Dichlorodiphenyltrichloroethane (DDT) and other forms of detergent molecules as micro pollutants that can be deposited in the marine biota. This can flow over different trophic levels to cause biomagnification. Nitrogen-rich fertilizers like urea and ammonium sulfate can reach the water bodies, via surface runoffs, to contaminate it and decrease the dissolved oxygen levels in water (Galt, 2008). Similarly, the industrial effluents from different industries, like petrochemicals, pharmaceuticals, food processing and paper and pulp industry, are rich in inorganic and ionic pollutants, microbial contamination and micelle-like lipid soluble substances which are of serious concern. These contaminants can stay in the water bodies for prolonged period to alter the water quality.
- **Pathogens and microbial contamination:** Bacteria and microbial flora can be both beneficial as well as harmful. The harmful bacteria are referred to as pathogens that can cause harm to the humans. Water bodies can be contaminated by these pathogenic microflora and become non-potable, causing dreaded diseases in humans like cholera, giardiasis and dysentery. Coliforms are one of the major indicators of the water quality index. A higher coliform number indicates that the water is contaminated with pollutants. The faecal coliforms are majorly concerning, as these can form waterborne diseases like cholera. Other such disease-forming pathogen species include bacteria of the genus *Salmonella* and *Burkholderia* and parasitic amoebae like *Entamoeba* and *Giardia*; they can cause major harm to human health if they reach the human body (Homas, 2000).
- **Groundwater contamination from hazardous wastes and spills:** Municipal solid waste landfills, accidental spills, hazardous waste sites and

unused production facilities are some of the major causes of the groundwater contamination. These wastes are largely inorganic in nature and cause major harm to the environment. Chlorinated ethenes and fuel hydrocarbons including PAHs, polychlorinated dibenzo-p-dioxins (PCDDs) from the pesticide-manufacturing industry, methylmercury, etc., are the current points of concern, as these are highly carcinogenic in nature and can sustain in the environment for prolonged period (Farhadian et al., 2008; Smidt et al., 2004; Selin, 2009). Reckless overexploitation of groundwater as a drinking water resource needs to be checked. Scrutinizing for the presence of well-known contaminating agents, proper assessment of the health risks upon exposure to chemical contaminants and implementation of well-planned, appropriate and cost-effective remediating measures are essential. These are mostly non-point-source pollution, which means that the polluting agents are hard to characterize or even quantify. As a result, a broad precautionary measure should be maintained.

- **Atmospheric pollutants and acid rain:** Small inorganic pollutants and gaseous substances remain in the atmosphere, which can reach the water bodies during rain. Gases like carbon dioxide, carbon monoxide, sulphur dioxide, nitrogen dioxide, etc., produced during fossil fuel burning, insecticide spraying, emission from coolers and refrigerators can stay in the environment as gaseous contaminants. During lightning, these substances can be turned into toxic free radicals, and their further condensation and chemical reactions with water particles can lead to the formation of various acids like nitric acid ($HNO_3$), sulphuric acid ($H_2SO_4$) and carbonic acid ($H_2CO_3$). These then come down as acid rain to contaminate various water bodies. Acid rain can alter the pH of the surface as well as groundwater, which can change various water quality parameters (Letchinger, 2000; Brian, 2008). Apart from the aforementioned parameters, several other polluting agents are also prevalent. During mining, the heavy metal–containing leachate can travel down the soil layers and get emptied into groundwater, causing groundwater contamination. Similarly, agricultural urban area runoffs contain various polluting substances that can fall into the water bodies to change the water quality and contaminate it. So a proper remediating procedure or management strategy is needed.

## 1.4.2 Mitigation Strategies

The wastewater from various sources is treated through various chemical and biological treatment strategies to purify them. The wastewater treatment is a common municipal approach to restore the water quality parameters and reduce the environmental damage.

1. **Biological wastewater treatment:** Biological wastewater treatment is a long-adapted technique, which plays a pivotal role in municipal wastewater management. The biological treatment is advantageous over the other

chemical and physical procedures, as the biological system is not much stringent from the point of view of pH, temperature and concentration of polluting substance (Metcalf et al., 1991). The microbes that are generally used in this biological wastewater treatment procedure are particularly archaebacteria and some eubacteria as well. These microbes can easily grow in most kinds of environments over a larger temperature distribution. Algae and protozoans are generally used in wastewater stabilization pond and activated sludge digestion procedures, respectively. Viruses are generally not used in the purification protocol, but these entities can be efficiently removed from the wastewater by biological treatment. During the procedure, the bacterial mass or suspension is generally mixed with other solid mass or effluents, which has marked importance as well (Metcalf et al., 1991). It provides anchorage to the bacterial culture so that they can efficiently grow, and the sludge provides necessary nutrients for the bacterial growth. The biological treatment can be categorized under three main categories:

A. **Suspended growth systems:** As evident from the name only, in this procedure, the microbes are grown in suspension in the liquid media. The procedure is carried out in either aerobic or anaerobic conditions under mixing. Sequence batch reactor (SBR), plug flow, continuous stirred tank bioreactors (CSTR), etc., use this suspended growth system of purification (Ishak et al., 2012).

B. **Attached growth systems:** In attached growth systems, a solid system like rock, slag or plastic is used, which helps in generating a continuous biofilm of the bacterial culture. The microbes produce large amounts of polysaccharide substances to form a biofilm. Bioreactors with attached biomass- or biofilm-generation abilities show a greater amount of processivity, as the large population of bacteria can efficiently metabolize the sludge or biomass (Hsien and Lin, 2005).

C. **Hybrid systems:** This system uses a combination of the aforementioned procedures. So both activated sludge and submerged media content are maintained. Fixed bed bioreactors are a typical example of such hybrid systems. Tyagi et al. (1992) studied the use of RBC-polyurethane foam (PUF) and found that for hybrid systems, it can provide better anchorage for the microbial culture to develop and efficiently digest the sludge.

Physical and chemical treatment of the wastewater also hold major importance, as it helps to remove the bulk portion of the solid waste from the untreated water. The solid bulk portion like sand, gravel, stones, etc., are removed from the water by passing the wastewater through grit chamber. The grass, grease and floatable lipid portions are removed in the next chamber—called dissolved air flotation (DAF) units. Similarly, post biological treatment, the clarifier removes or settles down the bulk portion of the sludge to allow the treated water to pass to the further treatment processes (Adin and Asano, 1998).

The physical methods separate the bulk portion of waste from the water, whereas chemical treatment enhances the overall water quality parameters. Neutralization process helps to balance the overall pH of the water. The flocculation and coagulation steps help to reduce the remaining smaller particles by coagulating them into a bigger mass. Lastly, chlorination and ozonation help in the disinfecting process and aid in the generation of potable water (Adin and Asano, 1998).

- **Ganga action plan (GAP):** Ganges is the longest river flowing through the heart of India. The increasing pollution load and continuous exploitation of this river made scientists really worried about the future of this river. In 1985, the Government of India undertook the Ganga Action Plan (GAP), which mainly focused on the restoration of the water quality and the reduction of the waste load on this river. In 2009, the river was given the status of a national river and reconstituted measures were implemented. The biological oxygen demand (BOD) is a measure of water quality index, and a higher value indicates that the water is highly contaminated with organic pollutants. During the stretch of 1985–96, the average BOD of Ganga was more than 10–12 mg $L^{-1}$, which pointed towards a very poor water quality. Post GAP by 2009, this value was reduced in the range of 3–4 mg $L^{-1}$ (Hasan, 2015). However, the water quality is yet not perfect, and proper information and data release to the local population is necessary.

Ganges and its tributaries hold major importance, as the water flowing through these streams helps in agriculture and industrialization and improves the economy of India. However, the wastes generated from various sectors again get emptied into these streams, rendering the water non-usable. Yamuna is one of the major tributaries of Ganges and one of the most significantly polluted one as well. To change this status, a bilateral management strategy was signed between India and Japan, called Yamuna action plan (YAP). YAP is divided into three stages, out of which the third stage is being implemented currently (Srivastava and Prathna, 2022). Till the 2020 COVID-19 lockdown period, the data has been calculated to notice a major reduction in BOD. The BOD and chemical oxygen demand have already been reduced by 42% and 40%, respectively, of its previous value (Patel et al., 2020).

- **Valorization and biosurfactant formation:** Industrialization have increased in a rapid manner to keep pace with ever-increasing population demands. Higher demands for food and agricultural products have started to overburden the environment with all the industrial waste products. Most of these products, being rich in nutrients, are well suited for microbial growth, some of which can be pathogenic and pose serious threats to the biodiversity (Ravindran et al., 2018). Valorization of food industry waste is an environmentally important process with huge economic significance, which helps in proper disposal and remediation of the hazardous waste products. Apart from biofuel formation with the agro-based waste products and animal excreta, there can be other applications like biofertilizer and biosurfactant formation (Rashad et al., 2010).

The wastewater effluents from the various industries can be processed to generate economically significant items, which would otherwise go on polluting the environment. These wastewater effluents are rich in free fatty acids, aromatic hydro-carbons, degumming acid (citric acid or phosphoric acid), pesticides and other toxic substances. Gudiña et al. (2016) formulated a method of making biosurfactant, using a combination of three types of waste products: corn-steep liquor, sugarcane molasses and oil mill wastewater. They kept the concentrations of corn-steep liquor and sugarcane molasses constant at around 10%. However, they observed that increasing the concentration of oil mill wastewater from 5% to 15% can drastically increase the concentration of rhamnolipid biosurfactant. Wastewater from paper industry and the activated sludge can serve as a nutritive media for the growth of marine microalgae *Nannochloropsis oculate*, which helps in synthesizing eicosapentaenoic acid by anaerobic digestion (Polishchuk et al., 2015). Oil can be separated from the oil mill wastewater by decantation procedure. Elkacmi et al. (2016) used this procedure to extract highly purified oleic acid. The less pure poor oil portion can be saponified into soaps, glycerol, etc. Similarly, acid wastewater can be valorized to make candles, soaps and other items (Welz, 2019).

## 1.5  CONCLUSION AND FUTURE PERSPECTIVES

Healthy soil, clean water and waste-free habitable environment are essential for life on earth. Healthy plant life and optimal food production are the prime require-ments of the world's fast-growing population. As a result of human intervention, the inter-linkages between the two key elements of the biosphere (viz., soil and water), via geochemical and hydrological events, have been broadly altered. As discussed in this chapter, the increasing environmental awareness and health management proto-cols emphasize more on the minimization of global waste production and call for an efficient management procedure. We must focus more on the 3R principles (reduce, reuse and recycle) to achieve a global waste minimum. Most of the organic wastes are biodegradable and thus can be efficiently channelized towards fuel produc-tion. The fibrous and non-organic waste portions can be used to make bio-friendly by-products like surfactants and economically important substances. The establish-ment of zero-waste generating industry is still far from reach, and thus, the recycle and reuse methods should be the prime management strategy to reach our goal of minimizing waste production. So an efficient approach while considering the 3R systems and the recovery of the waste material can lead towards safeguarding of the environment as well as contribute towards the global economy. In addition, for safeguarding the water supply and prevention of water quality deterioration, efficient water transfer schemes can be implemented. Moreover, climate change–induced rise in sea level often threaten acres of lands from being eroded away. Contamination of soil as a result of unsustainable land usage, improper methods of waste disposal and use of excess levels of biocides and fertilizers often lead to the release of toxic metals and organic pollutants into soil which ultimately also deteriorate the adjoining water quality drastically. More stress has to be laid on the usage of sustainable food items, generation of homemade composts, proper disposal of drugs, batteries and other xenobiotics, setting up of eco-friendly industrial sectors, stock breeding and farming techniques and curtailing the use of contaminating biocides. Hence, holistic improvi-sation of foolproof soil, waste and wastewater management strategies are mandatory.

## ACKNOWLEDGEMENTS

Financial assistance from the Science and Engineering Research Board, Government of India, through the grant EMR/2016/004799 and Department of Higher Education, Science and Technology and Biotechnology, Government of West Bengal, through the grant 264(Sanc.)/ST/P/S&T/1G-80/2017 to Dr. Aryadeep Roychoudhury are gratefully acknowledged.

## REFERENCES

Adin A, Asano T (1998) The role of physical-chemical treatment in wastewater reclamation and reuse. *Water Science and Technology* 37: 79–90.

Ahmed I, Gupta AK (2009) Syngas yield during pyrolysis and steam gasification of paper. *Applied Energy* 86: 1813–1821.

Albanese S, De Vivo B, Lima A, Cicchella D (2007) Geochemical background and baseline values of toxic elements in stream sediments of Campania region (Italy). *Journal of Geochemical Exploration* 93: 21–34.

Baker AJM, McGrath SP, Reeves RD, Smith JAC (2000) Metal hyperaccumulator plants: A review of the ecology and physiology of a biological resource for phytoremediation of metal-polluted soils. In: Terry N, Banuelos G, editors. *Phytoremediation of Contaminated Soil and Water*. Boca Raton: Lewis Publishers; pp: 85–108.

Bowles TM, Mooshammer M, Socolar Y, Calderón F, Cavigelli MA, Culman SW, Deen W, Drury CF, y Garcia AG, Gaudin AC, Harkcom WS (2020) Long-term evidence shows that crop-rotation diversification increases agricultural resilience to adverse growing conditions in North America. *One Earth* 2: 284–293.

Brian M (2008) Water pollution by agriculture. *Philosophical Transactions of the Royal Society B* 363: 659–666.

Brooijmans RJ, Pastink MI, Siezen RJ (2009) Hydrocarbon-degrading bacteria: The oil-spill clean-up crew. *Microbial Biotechnology* 2: 587.

Brooks RR (1998) *Plants That Hyperaccumulate Heavy Metals*. Wallington: Climate Action Network International; p: 379.

Carrott PJM, Carrott MR (2007) Lignin–from natural adsorbent to activated carbon: A review. *Bioresource Technology* 98: 2301–2312.

Chen DMC, Bodirsky BL, Krueger T, Mishra A, Popp A (2020) The world's growing municipal solid waste: Trends and impacts. *Environmental Research Letters* 15: 74021.

Correddu F, Lunesu MF, Buffa G, Atzori AS, Nudda A, Battacone G, Pulina G (2020) Can agro-industrial by-products rich in polyphenols be advantageously used in the feeding and nutrition of dairy small ruminants? *Animals* 10: 131.

Das S, Lee SH, Kumar P, Kim KH, Lee SS, Bhattacharya SS (2019) Solid waste management: Scope and the challenge of sustainability. *Journal of Cleaner Production* 228: 658–678.

Deardorff T, Karch N, Holm S (2008) Dioxin levels in ash and soil generated in Southern California fires. *Organohalogen Compounds* 70: 2284–2288.

Dessì P, Porca E, Lakaniemi AM, Collins G, Lens PN (2018) Temperature control as key factor for optimal biohydrogen production from thermomechanical pulping wastewater. *Biochemical Engineering Journal* 137: 214–221.

Dhar H, Kumar S, Kumar R (2017) A review on organic waste to energy systems in India. *Bioresource Technology* 245: 1229–1237.

Dhillon GS, Kaur S, Brar SK (2013) Perspective of apple processing wastes as low-cost substrates for bioproduction of high value products: A review. *Renewable and Sustainable Energy Reviews* 27: 789–805.

Díez M, Simón M, Martín F, Dorronsoro C, García I, Van Gestel CAM (2009) Ambient trace element background concentrations in soils and their use in risk assessment. *Science of the Total Environment* 407: 4622–4632.

Dœlsch E, Van de Kerchove V, Saint Macary H (2006) Heavy metal content in soils of Réunion (Indian Ocean). *Geoderma* 134: 119–134.

Elkacmi R, Kamil N, Bennajah M, Kitane S (2016) Extraction of oleic acid from Moroccan olive mill wastewater. *BioMed Research International* 2016: 1397852.

Faour-Klingbeil D, Todd EC (2018) The inhibitory effect of traditional pomegranate molasses on S. typhimurium growth on parsley leaves and in mixed salad vegetables. *Journal of Food Safety* 38: 12469.

Farhadian M, Vachelard C, Duchez D, Larroche C (2008) In situ bioremediation of monoaromatic pollutants in groundwater: A review. *Bioresource Technology* 99: 5296–308.

Federici F, Fava F, Kalogerakis N, Mantzavinos D (2009) Valorisation of agro-industrial by-products, effluents and waste: Concept, opportunities and the case of olive mill wastewaters. *Journal of Chemical Technology & Biotechnology: International Research in Process, Environmental & Clean Technology* 84: 895–900.

Fira D, Dimkić I, Berić T, Lozo J, Stanković S (2018) Biological control of plant pathogens by Bacillus species. *Journal of Biotechnology* 285: 44–55.

Frey B, Keller C, Zierold K (2000) Distribution of Zn in functionally different leaf epidermal cells of the hyperaccumulator *Thlaspi caerulescens*. *Plant, Cell & Environment* 23:675–687.

Galt RE (2008) Beyond the circle of poison: significant shifts in the global pesticide complex, 1976–2008. *Global Environmental Change* 18: 786–799.

Girelli L, Tocci M, Gelfi M, Pola A (2019) Study of heat treatment parameters for additively manufactured AlSi 10 Mg in comparison with corresponding cast alloy. *Materials Science and Engineering: A* 739: 317–328.

Gregorič A, Vaupotič J, Kardos R, Horváth M, Bujtor T, Kovács T (2013) Radon emanation of soils from different lithological units. *Carpathian Journal of Earth and Environmental Sciences* 8: 185–190.

Gregory KB, Lovely DR (2005) Remediation and recovery of uranium from contaminated subsurface environments with electrodes. *Environmental Science & Technology* 39: 8943–8947.

Guan W, Shi S, Tu M, Lee YY (2016) Acetone-butanol-ethanol production from Kraft paper mill sludge by simultaneous saccharification and fermentation. *Bioresource Technology* 200: 713–721.

Gudiña EJ, Rodrigues AI, de Freitas V, Azevedo Z, Teixeira JA, Rodrigues LR (2016) Valorization of agro-industrial wastes towards the production of rhamnolipids. *Bioresource Technology* 212: 144–150.

Gupta GK, Shukla P (2020) Insights into the resources generation from pulp and paper industry wastes: Challenges, perspectives and innovations. *Bioresource Technology* 297: 122496.

Haas D, Défago G (2005) Biological control of soil-borne pathogens by fluorescent pseudomonads. *Nature Reviews Microbiology* 3: 307–319.

Haq I, Raj A (2020) Pulp and paper mill wastewater: Ecotoxicological effects and bioremediation approaches for environmental safety. *Bioremediation of Industrial Waste for Environmental Safety: Volume II: Biological Agents and Methods for Industrial Waste Management* 333–356. https://doi.org/10.1007/978-981-13-3426-9_14

Haritash AK, Kaushik CP (2009) Biodegradation aspects of polycyclic aromatic hydrocarbons (PAHs): A review. *Journal of Hazardous Materials* 169: 1–15.

Hasan S (2015) Water quality of river ganga–pre and post GAP: A review. *International Journal of Advances Research in Science, Engineering and Technology* 2: 361–364.

Homas RS (2000) Microbes and urban watersheds: Concentrations, sources, & pathways. *Watershed Protection Techniques* 3: 554–565.

Hsien TY, Lin YH (2005) Biodegradation of phenolic wastewater in a fixed biofilm reactor. *Biochemical Engineering Journal* 27: 95–103.

Intasit R, Cheirsilp B, Louhasakul Y, Boonsawang P, Chaiprapat S, Yeesang J (2020) Valorization of palm biomass wastes for biodiesel feedstock and clean solid biofuel through non-sterile repeated solid-state fermentation. *Bioresource Technology* 298: 122551.

Ishak S, Malakahmad A, Isa MH (2012) Refinery wastewater biological treatment: A short review. *Journal of Scientific & Industrial Research* 71: 251–256.

Itai T, Otsuka M, Asante KA, Muto M, Opoku-Ankomah Y, Ansa-Asare OD, Tanabe S (2014) Variation and distribution of metals and metalloids in soil/ash mixtures from Agbogbloshie e-waste recycling site in Accra, Ghana. *The Science of the Total Environment* 470–471: 707–716.

Kanter DR (2018) Nitrogen pollution: a key building block for addressing climate change. *Climatic Change* 147: 11–21.

Kibler KM, Reinhart D, Hawkins C, Motlagh AM, Wright J (2018) Food waste and the food-energy-water nexus: A review of food waste management alternatives. *Waste Management* 74: 52–62.

Koptsik GN (2014) Problems and prospects concerning the phytoremediation of heavy metal polluted soils: A review. *Eurasian Soil Science* 47: 923–939.

Korcheva A (2021) Basel convention on the control of hazardous wastes. In: Idowu S, Schmidpeter R, Capaldi N, Zu L, Del Baldo M, Abreu R, editors. *Encyclopedia of Sustainable Management*. Cham: Springer.

Lakshmidevi R, Muthukumar K (2010) Enzymatic saccharification and fermentation of paper and pulp industry effluent for biohydrogen production. *International Journal of Hydrogen Energy* 35: 3389–3400.

Landberg T, Greger M (1996) Differences in uptake and tolerance to heavy metals in Salix from unpolluted and polluted areas. *Applied Geochemistry* 11:175–180.

Letchinger M (2000) Pollution and water quality, neighbourhood water quality assessment. *Project Oceanography*: 1–14.

Limmer M, Burken J (2016) Phytovolatilization of organic contaminants. *Environmental Science & Technology* 50: 6632–6643.

Martinkova L, Uhnakova B, Patek M, Nesvera J, Kren V (2009) Biodegradation potential of the genus *Rhodococcus*. *Environment International* 35:162–177.

Masson-Delmotte V, Zhai P, Pörtner HO, Roberts D, Skea J, Shukla PR, Pirani A, Moufouma Okia W, Péan C, Pidcock R, Connors S, Matthews JBR, Chen Y, Zhou X, Gomis MI, Lonnoy E, Tignor MM, Waterfield T (2018) *Global Warming of 1.5°C: An IPCC Special Report on the Impacts of Global Warming of 1.5°C Above Pre-Industrial Levels and Related Global Greenhouse Gas Emission Pathways, in the Context of Strengthening the Global Response to the Threat of Climate Change*. New York: IPCC; in press.

Metcalf L, Eddy HP, Tchobanoglous G (1991) *Wastewater Engineering: Treatment, Disposal, and Reuse* (vol. 4). New York: McGraw-Hill.

Mielke HW, Reagan PL (1998) Soil is an important pathway of human lead exposure. *Environmental Health Perspectives* 106: 217–229.

Mirsal I (2008) Soil pollution: Origin, monitoring & remediation. *Springer Science & Business Media*: 310.

Mo J, Yang Q, Zhang N, Zhang W, Zheng Y, Zhang Z (2018) A review on agro-industrial waste (AIW) derived adsorbents for water and wastewater treatment. *Journal of Environmental Management* 227: 395–405.

Mohammadi A, Sandberg M, Venkatesh G, Eskandari S, Dalgaard T, Joseph S, Granström K (2019) Environmental performance of end-of-life handling alternatives for paper-and-pulp-mill sludge: Using digestate as a source of energy or for biochar production. *Energy* 182: 594–605.

Mohammed B, Ouhammou B, El Gnaoui Y, Kerrou O, Mhamdi H, El Bari H (2021) Anaerobic digestion by biofilm digester technology treating wastewater of pulp and paper recycled: Energy variation as function of ambient environment. Available at SSRN: https://ssrn.com/abstract=3919667 or http://dx.doi.org/10.2139/ssrn.3919667

Mor RS, Sangwan KS, Singh S, Singh A, Kharub M (2021) E-waste management for environmental sustainability: An exploratory study. *Procedia CIRP* 98: 193–198.

Napoli M, Cecchi S, Grassi C, Baldi A, Zanchi CA, Orlandini S (2019) Phytoextraction of copper from a contaminated soil using arable and vegetable crops. *Chemosphere* 219: 122–129.

Panesar R, Kaur S, Panesar PS (2015) Production of microbial pigments utilizing agro-industrial waste: A review. *Current Opinion in Food Science* 1: 70–76.

Panizza CE, Shvetsov YB, Harmon BE, Wilkens LR, Le Marchand L, Haiman C, Reedy J, Boushey CJ (2018) Testing the predictive validity of the Healthy Eating Index-2015 in the multiethnic cohort: Is the score associated with a reduced risk of all-cause and cause-specific mortality? *Nutrients* 10: 452.

Patel PP, Mondal S, Ghosh KG (2020) Some respite for India's dirtiest river? Examining the Yamuna's water quality at Delhi during the COVID-19 lockdown period. *Science of the Total Environment* 744: 140851.

Perkins DN, Brune Drisse MN, Nxele T, Sly PD (2014) E-waste: A global hazard. *Annals of Global Health* 80: 286–295.

Polishchuk A, Valev D, Tarvainen M, Mishra S, Kinnunen V, Antal T, Yang B, Rintala J, Tyystjärvi E (2015) Cultivation of Nannochloropsis for eicosapentaenoic acid production in wastewaters of pulp and paper industry. *Bioresource Technology* 193: 469–476.

Preisner M (2020) Surface water pollution by untreated municipal wastewater discharge due to a sewer failure. *Environmental Processes* 7: 767–780.

Puschenreiter M, Wieczorek S, Horak O, Wenzel WW (2003) Chemical changes in the rhizosphere of metal hyperaccumulator and excluder Thlaspi species. *Journal of Plant Nutrition and Soil Science* 166(5): 579–584.

Radziemska M, Vaverková MD, Baryła A (2017) Phytostabilization—management strategy for stabilizing trace elements in contaminated soils. *International Journal of Environmental Research and Public Health* 14: 958.

Rashad FM, Saleh WD, Moselhy MA (2010) Bioconversion of rice straw and certain agro-industrial wastes to amendments for organic farming systems: 1. Composting, quality, stability and maturity indices. *Bioresource Technology* 101: 5952–5960.

Ravindran R, Hassan SS, Williams GA, Jaiswal AK (2018) A review on bioconversion of agro-industrial wastes to industrially important enzymes. *Bioengineering* 5: 93.

Rodríguez-Eugenio N, McLaughlin M, Pennock D (2018) Soil pollution: A hidden reality. *Food and Agriculture Organisation of the United Nations*: 142.

Sagar NA, Pareek S, Sharma S, Yahia EM, Lobo MG (2018) Fruit and vegetable waste: Bioactive compounds, their extraction, and possible utilization. *Comprehensive Reviews in Food Science and Food Safety* 17: 512–531.

Sajna KV, Gottumukkala LD (2019) Biosurfactants in bioremediation and soil health. *Microbes and Enzymes in Soil Health and Bioremediation*: 353–378.

Salim AA, Grbavčić S, Šekuljica N, Stefanović A, Tanasković SJ, Luković N, Knežević-Jugović Z (2017) Production of enzymes by a newly isolated *Bacillus* sp. TMF-1 in solid state fermentation on agricultural by-products: The evaluation of substrate pretreatment methods. *Bioresource Technology* 228: 193–200.

Sasikumar CS, Papinazath T (2003) Environmental management: Bioremediation of polluted environment. *Proceedings of the Third International Conference on Environment and Health, Chennai, India*: 15–17.

Schleussner CF, Ler TK, Fischer EM, Wohland J, Perrette M, Golly A, Rogelj J (2018) Differential climate impacts for policy-relevant limits to global warming. *Earth System Dynamics*: 426.

Schneiker S, Martins dos Santos VA, Bartels D, Bekel T, Brecht M, Buhrmester J (2006) Genome sequence of the ubiquitous hydrocarbon-degrading marine bacterium *Alcanivorax borkumensis*. *Nature Biotechnology* 24: 997–1004.

Schuldt JP, Konrath SH, Schwarz N (2011) Global warming or climate change? *Public Opinion Quarterly* 75: 115–124.

Schwarzenbach RP, Egli T, Hofstetter TB, Von Gunten U, Wehrli B (2010) Global water pollution and human health. *Annual Review of Environment and Resources* 35: 109–136.

Selin NE (2009) Global biogeochemical cycling of mercury: A review. *Annual Review of Environment and Resources* 34: 43–63.

Seo JS, Keum YS, Harada RM, Li QX (2007) Isolation and characterization of bacteria capable of degrading polycyclic aromatic hydrocarbons (PAHs) and organophosphorus pesticides from PAH-contaminated soil in Hilo, Hawaii. *Journal of Agriculture and Food Chemistry* 55: 5383–5389.

Sharma S, Zhang M, Gao J, Zhang H, Kota SH (2020) Effect of restricted emissions during COVID-19 on air quality in India. *Science of the Total Environment* 728:138878.

Sharrock P, Fiallo M, Nzihou A, Chkir M (2009) Hazardous animal waste carcasses transformation into slow release fertilizers. *Journal of Hazardous Materials* 167: 119–123.

Smidt H, de Vos WM (2004) Anaerobic microbial dehalogenation. *Annual Review of Microbiology* 58: 43–73.

Smith OM, Cohen AL, Rieser CJ, Davis AG, Taylor JM, Adesanya AW, Jones MS, Meier AR, Reganold JP, Orpet RJ, Northfield TD (2019) Organic farming provides reliable environmental benefits but increases variability in crop yields: A global meta-analysis. *Frontiers in Sustainable Food Systems* 3: 82.

Sokka L, Lindroos TJ, Ekholm T, Koljonen T (2020) Impacts of climate change and its mitigation in the Barents region. *Cogent Environmental Science* 6: 1805959.

Squire JN (2013) Biomedical pollutants in the urban environment and implications for public health: a case study. *International Scholarly Research Notices*: 2013.

Srivastava A, Prathna TC (2022) Yamuna action plan-III: Impact on water quality of river Yamuna, India. *Fine Chemical Engineering*: 1–10.

Stewart WM, Dibb DW, Johnston AE, Smyth TJ (2005) The contribution of commercial fertilizer nutrients to food production. *Agronomy Journal* 97: 1.

Swati, Ghosh P, Das MT, Thakur IS (2014) In vitro toxicity evaluation of organic extract of landfill soil and its detoxification by indigenous pyrene-degrading *Bacillus* sp. ISTPY1. *International Biodeterioration & Biodegradation* 90: 145–151.

Szabo K, Cătoi AF, Vodnar DC (2018) Bioactive compounds extracted from tomato processing by products as a source of valuable nutrients. *Plant Foods for Human Nutrition* 73: 268–277.

Tyagi RD, Tran FT, Chowdhury AKMM (1992) Performance of RBC coupled to a polyurethane foam to biodegrade petroleum refinery wastewater. *Environmental Pollution* 76: 61–70.

Ullah N, Ali J, Khan FA, Khurram M, Hussain A, Rahman IU, Rahman ZU, Ullah S (2012) Proximate composition, minerals content, antibacterial and antifungal activity evaluation of pomegranate (*Punica granatum* L.) peels powder. *Middle East Journal of Scientific Research* 11: 396–401.

Vandermeer KD, Daugulis AJ (2007) Enhanced degradation of a mixture of polycyclic aromatic hydrocarbons by a defined microbial consortium in a two-phase partitioning bioreactor. *Biodegradation* 18: 211–221.

Veldman JW, Putz FE (2011) Grass dominated vegetation, not species-diverse natural savanna, replaces degraded tropical forests on the southern edge of the Amazon Basin. *Biological Conservation* 144: 1419–1429.

Welz PJ (2019) Edible seed oil waste: Status quo and future perspectives. *Water Science and Technology* 80: 2107–2116.

Werner S (2002) Recent developments in international management research: A review of 20 top management journals. *Journal of Management* 28(3): 277–305.

Whitmarsh L (2009) What's in a name? Commonalities and differences in public understanding of "climate change" and "global warming". *Public Understanding of Science* 18: 401–420.

Zalas M, Cynarzewska A (2018) Application of paper industry waste materials containing TiO2 for dye-sensitized solar cells fabrication. *Optik* 158: 469–476.

Zhang, H, Luo Y, Wu L, Huang Y, Christie P (2015) Residues and potential ecological risks of veterinary antibiotics in manures and composts associated with protected vegetable farming. *Environmental Science and Pollution Research* 22: 5908–5918.

# 2 Potential Eco-Friendly Techniques for Water, Soil, and Waste Management

*Muhammad Farhan Qadir, Khuram Shehzad Khan, Muhammad Naveed, Taqi Raza, and Neal S. Eash*

## 2.1 INTRODUCTION

With the advent of the 21st century, although the exponential advancement of the agriculture and food industry occurred, every seventh person goes to bed hungry (Sunil, 2017). In fact, it has currently been proven difficult to feed 7 billion people, but it is estimated that this number will rise by 1.3% (another 2 billion) in the coming two decades. Only 11% of the land area is devoted to productive lands, which are finite yet provide food for almost > 9 billion people all over the world (MacFarquhar, 2009). This scenario undoubtedly serves as a warning about how we are looking towards food security and food safety. In short, if we are employing all feasible approaches, perhaps sustainable consumption and production could be more critical than what we currently suppose (Lu et al., 2022).

Basically, it is most important to understand the cause of global food security prior to suggested solutions. As a result, extensive exploitation of the earth's limited resources poses a serious risk to both the environment's quality and the capability to feed the population. According to studies, from 1950 to 2010, ecosystem services were reduced by 60% because of accelerated resource degradation, resulting in 0.5 billion ha deterioration in the tropics and one-third of land degradation globally (Power, 2010). A clear transformation of "industry" is reported in that period. Since that day, the population has more than doubled, and it was the first era there were more migrants in cities from the rural areas. As new sectors and innovative technologies emerged, the need for energy also increased, and trade reached six times on a worldwide scale. Similarly, six-time water consumption and river damming were reported globally, and more than two-thirds of fresh water is directly used for agriculture to meet worldwide food hunger (Hurlimann et al., 2021). So collectively, these problems contributed to food security. It's not our problem since we didn't attempt to address the issue of food safety. However, the transformation that's been currently happening since the 1960s is occurring faster than we could ever respond.

DOI: 10.1201/9781003358169-2

On this basis, we could clearly state that climate change, also referred to as global change, is not only occurring but it's also happening spontaneously faster.

The sharp next question is, What's the meaning and real cause of global change? Global warming and climate change are the substantial shifts that have been encountered during the last 65 years. Global climate change (GCC) is a multifaceted geopolitical concern with an impact on sociopolitical, biological, environmental, and socioeconomic aspects (Figure 2.1).

With the dawn of human civilization and the onset of the industrial revolution, the problem of GCC amplified manifolds, resulting in high temperatures (> 3°C), melting glaciers, changing precipitation trends, rising sea water levels globally, and shifting the industrial flow drastically (Diffenbaugh and Burke, 2019; Huang et al., 2022). So the exact interpretation of the devastating consequences of GCC is not plausible, while the immediate intentions and considerable steps might rehabilitate its threats and catastrophic effects. In general, the interdisciplinary changes result in severe consequences on the earth system, also known as global climate change. The main components of the earth system are life, water, environment, soil, and the biogeochemical cycles that are fundamental earth processes (Jansson and Hofmockel, 2020). In that global village, every component relates to each other in a dynamic equilibrium, as a consequential change in one component results in substantial effects on the other components, while mostly, the retardation is reported. However, the GCC is not a new problem, as it's happening for millions of years, but the novelty is that changes occur drastically (Gelybó et al., 2018). Industrialization resulted in high anthropogenic activities containing intensive agricultural practices, burning of fossil fuels and agricultural waste, mechanization, deforestation, and catalytic

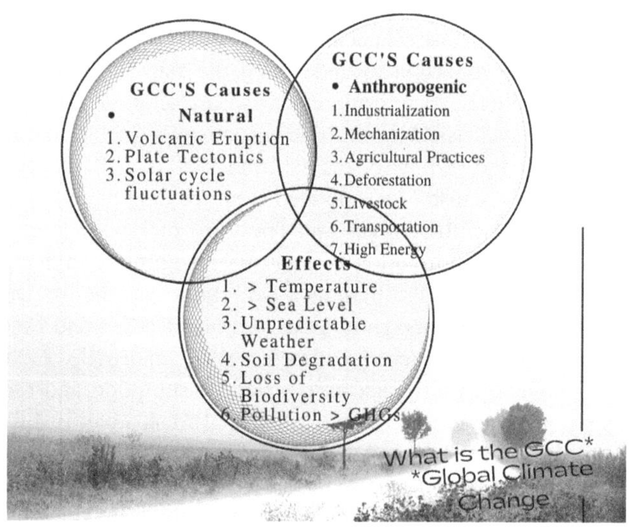


**FIGURE 2.1** Global climate change (GCC) impacts on sociopolitical, biological, and environmental aspects.

transportation (El Bilali et al., 2020). Consequently, these activities are reducing soil productivity, damaging infrastructure, retarding human health, and leading to catastrophic GCC, extreme greenhouse gases (GHGs), and warming temperatures. This demands us numerous precautionary measures, adaptation practices, and sustainable mitigation technologies to cope with the GCC effects and a firm substitute for sustaining global resources.

## 2.2 GLOBAL CLIMATE CHANGE (GCC)

Three decades ago, the devastating global climate change was considered as a projected theory while the ground realities basis. Now it becomes a well-heated discussion and drags scientists, politicians, farmers, and ultimately, every human being's attention immediately. Finally, it is considered real, and a pile of debates on how the GCC can adapt and sustain life in that change is occurring. Global climate change is directly triggered by natural changes that have been happening for millions of years, but human activities which emit heat in the form of gases into the atmosphere are increasingly causing the world to warm by contributing to the greenhouse effect (Manabe, 2019).

More than 18,000 scientific research papers were cited in the IPCC report, which 278 writers from 65 distinct countries worked on together. According to the report, carbon emissions around the world are rising (Molina and Abadal, 2021). Emissions in 2019 were 12% higher than in 2010 and more than 50% higher than in 1990, and according to Imperial College London during the media briefing, the fact is that GHGs emissions, which are responsible for global warming, are at their greatest levels in human history. "Our research indicates that controlling warming to 1.5°C is beyond reach if there are immediate and substantial emission reductions should be adopted across all sectors". The IPCC has now published three reports in the last eight months. The panel intends to publish its sixth assessment later this year, which will combine these findings on mitigation with those from reports published in August 2021 on the contribution of humans to climate change and in February 2022 on climate change adaptation and vulnerability.

Climate change has basic impact on interactions of living organisms to their environment. As there exists a minor fluctuation, the relationships with ecosystems are altered. Health impact is one of major issue that is influenced by the climate change. There are many health disasters spreading such as the African Rift Valley fever outbreaks (Martin et al., 2008), the escalating rate of fungus-caused annihilation of amphibians (Pounds et al., 2006), and affected food and water availability for humans (Anderson et al., 2004). The exact GCC risks are unpredictable, as GCC hazards will combine with natural and anthropogenic causes to enhance the mobility and transforming reactions of certain noxious pollutants in the environment (Armitage et al., 2012; Gouin et al., 2012). The persistence of organic chemicals in the environment is changed by the increase in temperature and moisture due to GCC (Boxall et al., 2009; Bloomfield et al., 2006). The continuous rise in temperature results in increasing volatilization of persistent organic molecules (Bogdal et al., 2011), causing its increased distant transportation of these toxins (Lamon et al., 2009; Armitage et al., 2012). GCC may modify the ecosystem in such a way that contaminants like mercury that have been released into the environment without treatment reside in the soil for longer periods and are mobilized or released more quickly. Increase in temperature may increase metabolic rates of many organisms to intensify, raising the possibility for bioaccumulation and biomagnification of some pollutants (Moore et al., 2007). Temperature-induced upsurge in metal uptake and bioaccumulation have been seen in a variety of marine organisms, including crustaceans, echinoderms, and mollusks (Marques et al., 2010). In a recent modeling study on methylmercury intake, increases in temperature resulted in higher amounts of the toxin in fish and mammals (Booth and Zeller, 2005); temperature rises are also predicted to expedite the conversion of mercury to methylmercury. (Downs et al., 1998). In rare cases, uptake may be reduced. Between 1994 and 2008, studies of organochlorine concentrations in fish revealed a decrease in concentrations with increasing temperature, a trend that might be explained by changes in lipid content over time (Riget et al., 2010). The cause of the shift in lipid content and its probable relationship with climate changes remain unknown (Riget et al., 2010).

Changes in the origins, destiny, and transmission of contaminants will have an impact on both the positive and negative contamination of food, drinking water supplies, and air, thus increasing susceptibility (Noyes et al., 2009). However, there have been relatively few studies to assess the potential modifications in exposure levels. In a UK study, Beulke et al. (2007) evaluated the impacts of changes in pesticide use, fate, and transport caused by GCC, either directly or indirectly, on surface water and groundwater concentrations. The study found that pesticide concentrations in surface and groundwater reservoirs are expected to rise under GCC and that maximum concentrations for several pesticides could rise by orders of magnitude. The indirect impacts of GCC (i.e., impacts on pesticide applied and timing) on surface water exposure were shown to be larger than the influence of climate change alone on chemical destiny and transport. Additional research of this type is required for various pollutant classes and geographical regions. Minor changes in exposure diversity or sensitivity to toxins or other hazards can lead to significant variations in uncertainty and danger. GCC is anticipated to boost variability and bilateral changes in human exposure to pollutants as well as other toxins. This could be due to changes in pesticide application practices and modifications in the nature and transit of those compounds caused by the GCC.

Changes in soil properties (organic carbon, dustiness) and hydrology can affect how toxins are carried throughout a terrestrial ecosystem, as well as the dilution potential of contaminants in rivers and streams. Increased prevalence of extreme weather events, such as floods and droughts, will most likely change the pollutant mobility. For example, flooding has been proven to move dioxins, heavy metals, and hydrocarbons from polluted to unpolluted regions (Lake et al., 2005; Manuel, 2006). Changes in irrigation methods in response to GCC could also transport toxins from water bodies on land in agricultural areas (Tirado et al., 2010). Variations in soil parameters (Mitchell and Likens, 2011), the persistence and severity of wildfires, fluctuations in the temperature and chemical characteristics of sea zones, and increasing sea levels are all possible outcomes of GCC (Hallegraeff, 2010). Indirect effects of GCC arise from modifications of ecological structures or processes by direct effects of GCC and can be demonstrated as combinations of direct and indirect effects such as desynchronization of forage/forager or predator/prey phonologies (Stenseth et al., 2002). The removal or dislocation of historical ranges and abundances of same organisms (Hof et al., 2011), the upsurge of native species or ailments which get advantage from dynamic climate conditions (Rohr et al., 2011), and the emergence of new habitats and biodiversity with new interspecies interactions are all possible outcomes of climate change (Parry et al., 2007). Furthermore, indirect GCC consequences can occur as a result of changes in human activities in reaction to GCC. For example, through management and adaptation measures, humans can alter resource development and agricultural techniques (Boxall et al., 2009) or, in certain situations, relocate to areas that are less influenced by the GCC.

Global climate change will have an impact not just on the amounts of harmful substances in water, food, and air but also on how humans interact with these mediums, which will have an influence on the extent of human exposure. For example, a decline in the availability of drinking water for many communities may alter exposure to waterborne toxins as human populations move to other drinking water sources (e.g., water from water-reuse and reclamation systems) (Bower, 2002). Climate change may also affect the amount of time people spend inside and outside, influencing their susceptibility to both indoor and outdoor toxins (Cecchi et al., 2010). In autotrophs, GCC directly influences the two chief types of photosynthesis process, characterized as C3 (comprises rice, wheat, soybeans, most of vegetables, and trees) and C4 (comprises mostly of tropical originated plants like maize, sorghum, sugarcane, and millet). Under increased atmospheric $CO_2$ concentrations, C3 plants are capable to carry more photosynthetic production; however, C4 plants are incapable of doing it. Tropical areas, other zones, and regions growing C4 crops may perhaps have more losses, ultimately causing lower yield and less accessibility of forage resources (Aydinalp and Cresser, 2008).

Variations in the extent and duration of ice cover may have an impact on the breakdown of legacy pollutants in specific areas (Point et al., 2011). Legacy pollutants are tenacious elements that have gathered in natural reservoirs such as surface soils, ice, sediments, and forests (McKone and MacLeod, 2004; Conner et al., 2007). The gradual and continual emissions from these reservoirs identify a long-term risk to human population and ecological health (Cowan-Ellsberry et al., 2009). Dioxins and dioxin-like compounds, PCBs, mercury discharged into the environment by mining and combustion operations, radioactive chemicals from nuclear weapons testing, DDT, lindane, and other pollutants are examples of legacy pollutants (McKone and MacLeod, 2004). The health risks of these contaminants range from cancer, reproductive issues, and altered neurodevelopment to endocrine and immune system. Contrary to other concerns identified in this study, legacy pollutant management, at particularly in industrialized countries, emphasizes tracking pollution sources and managing vulnerability instead of managing or controlling pollution from continuing economic/industrial processes (Gouin and Wania, 2007). In addition to generally lacking efficient surveillance and control of legacy pollutant dumpsites, developing countries may have populations that are significantly subjected to continuing recycling and waste-treatment operations (Linderholm et al., 2011). Since legacy pollutants retain and bioaccumulate in the atmosphere, long-term atmospheric activities associated with GCC may change their fate and transport, hence, altering human and ecosystem risks (MacLeod et al., 2005). Legacy pollutant transmission through environmental compartments influences their relative abundance in mobile (water or air) vs reserve mediums (soil and sediments). Movement of biosphere phases like air or water transports legacy contaminants from one place to another (Scheringer, 1997). Regional and worldwide migration patterns will be altered if GCC leads to hurricanes and stronger rivers, lakes, ocean, and water currents. The permanence of legacy pollutants is based on chemical transitions, a few of which are climate related. It will be crucial to comprehend the potential effects of GCC on activities, including degradation and transformation reactions, that are critical in the removal of toxins from environments like soil, water, and sediments. In addition, GCC causes variation in water cycles that cause impacts for indigenous meteorological conditions and runoff events (Frey et al., 2010).

However, these considerations highlight the importance of understanding GCC's possible impact not merely on the principles of environmental contamination and biochemistry but also on their implementations in the monitoring of pollutants in the environment. Human exposure to pollutants and other toxins should be monitored and sampled at a rate adequate to reflect changed variability caused by GCC. In both developed and developing countries, increased monitoring and action will be required to reduce any voids in legislation or regulatory actions to protect the population from excessive exposure to hazardous substances and pollutants (Balbus et al., 2012).

## 2.3  IMPACTS

Implementing appropriate adaptation methods and skills will be critical in reducing the negative effects of GCC and achieving sustainable development. The appropriate adaption strategies for the key sectors need to include agriculture, forestry, flood and drought control, coast protection, grassland management and animal husbandry, and climate prediction and forecast.

Climate change is most likely to have an impact on human health, either directly through the physical impacts of heat and cold or indirectly through greater transfer of food-borne or vector-borne infections, or effects on health from storms. There are evaluations of the worldwide consequences of GCC on health by globe region, most significantly the WHO global epidemic burden (McMichael and Lindgren, 2011). Nonetheless, while there will most likely be rise in heat-related deaths, these must be balanced even against decrease in cold-related death rates that will undoubtedly occur as a result of climate change.

Furthermore, there is some ambiguity about the net impacts for industrialized countries, as well as the division of the benefits and costs throughout more temperate geographical areas (Confalonieri et al., 2014). The increased hazards of temperature fluctuations linked with heat waves and urban heat island effects are particularly prominent at the city scale. As a result, there is a huge body of work highlighting, in qualitative or quantitative measures, the clinical implications of existing heat extremes (and cold extremes).

Climate change has the potential to impact water consumption, as well as availability and quality of water. Rise in average ambient temperature will hasten evaporation and possibly raise the requirement for cooling water in urban habitations (IPCC, 2001), thereby increasing average per capita water needs. But based on the amount of rainfall and the expected temperature fluctuations, food and water supplies may expand or decline. It will also be determined by prospective socioeconomic development and whether any further source can be obtained or generated, such as through desalination process, despite its high-power requirements.

Groundwater quality may also decline in regions where flow rate lowers. The evaluation of climate change impacts on groundwater sources involves watershed-level information, which is usually far beyond administrative boundaries of a certain city. Agricultural production is the main sector accountable for 30%–40% of all greenhouse emissions, making it a prominent sector both generating climate change and being strongly influenced by it (Grieg; Mishra et al., 2021; Ortiz-Bobea et al., 2021; Thornton and Lipper, 2014). Several agro-environmental and climatic elements have a significant impact on agricultural yield (Pautasso et al., 2012), comprising crops, forest land, and droughts.

GCC is the result of two distinct factors. The first category is natural factors, and the second is anthropogenic acts (Karami et al., 2012). It is also predicted that the earth will suffer an usual temperature increase ranging from 1°C to 3.7°C by the end of this century (Pachauri et al., 2014). Crop output around the world is also extremely susceptible to global temperature changes, as higher temperatures have a significant detrimental influence on crop development (Reidsma et al., 2009).

Global biodiversity is one of the most serious victims of climate change since it is the fastest rising source of species extinction. Several studies have shown that large-scale species transitions are strongly related to various climate occurrences. The rate and degree of CC are changing the suitable habitat range of coastal, aquatic, and terrestrial living organisms. Variations in climatological regimes have a variety of effects on ecological processes, including variations in the comparative diversity of species, habitats, changes in activities, and habitat utilization (Bates et al., 2014).

The extinction of species due to climate change has been frequently discussed in the literature (Beesley et al., 2019; Urban, 2015), and estimates of extinction till the 21st century are terrible (Abbass et al., 2019; Pereira et al., 2013). In other circumstances, species migrating towards north may be beneficial since it helps mountain-dwelling species to find optimal climates. However, due to loss of topography and habitat, migrating species may become stuck in solitary and inappropriate environments (Dullinger et al., 2012). For example, the American pika has been driven to extinction or severely reduced in several areas, owing mostly to CC-related extinction (Stewart et al., 2015).

Biodiversity is relatively susceptible to the other effects of climate change, such as rising global temperatures, shortage of water, and invading aphid attacks. For example, one research found changes in the nature of plankton populations caused by increased temperatures. Changes in such aquatic producer populations, such as diatoms and calcareous plants, may lead to variations in global carbon cycle. Furthermore, such variations are seen as a possible contribution to $CO_2$ variations in between Pleistocene glacial and interglacial periods (Kohfeld et al., 2005).

## 2.4　FUNDAMENTAL ADAPTATION STRATEGIES IN THE CONTEXT OF GLOBAL CLIMATE CHANGE

Global climate change adaptation is a broad term and contains sustainable strategies for variations and reactions to current and upcoming circumstances or their impacts. There is not a constant solution to adapt, as the actions are vulnerable to change with time and anthropogenic activities occurring, so collectively, people are doing great work for adaptation. The ongoing debates at every forum and the constant dialogues have clearly increased awareness worldwide. These conversations have helped to make "global change" a term that many people now use this vocabulary and understand the phenomenal cause of GCC. Sustainable adaptation generally includes measures to mitigate climate change's adverse effects and exploit significant novel economic possibilities. It comprises modifying strategies and behaviors based on actual or expected global climate change. The theory of adaptation is not unique; scientists have created numerous strategies in the past to deal with the very changing environment (Manes et al., 2022). For instance, the communities in the prairie provinces have been developed to survive with substantial variations in seasonal climates. However, there will be certain unexpected difficulties caused by the magnitude and intensity of future climate change. The statement "one of the most pressing environmental challenges" was used frequently to express climate change. Climate has an impact on all aspects of our lives, including our economics, social well-being, health, and way of life. All regions of the globe and almost every business market could be affected by global climate change. All economies need to deal with global climate change in such a manner, even though its effects won't be distributed equally across the globe. (da Graça Carvalho, 2012) A rising data collection portrays a collective image of a warming planet and other disturbances throughout the climatic changes according to global climate change.

Measures of regional adaptation to deal with GCC may include the following. Adaptation includes building water-saving technologies, avoiding and monitoring degradation of fertile lands, and improving regional socioeconomic sustainability. Administrators should emphasize the rational allotment of water management, the promotion of water-saving farming, the conservation and enhancement of the natural environment, and the strengthening of adaptation and mitigation of agriculture in arid regions.

Capabilities for regulating and controlling droughts, storms, and other hazards, enhancing moisture retention and outflow and tracking, and controlling disease should be developed. Climate change adaptation includes increasing forecasting and early-warning mechanisms for earthquakes and flooding, accelerating, and upgrading water and soil conservation. Because of sea level rise, the safety level of hurricane stopping dams should be increased, and the ability to assess and issue early warnings for typhoons and severe storms should be reinforced.

A schematic overview of the potential implications of climate change on agriculture, and appropriate management and adaptation techniques to prevent such impacts also discussed by Abbass et al. (2022).

Natural variations in climate have caused significant transformations throughout the earth's history. The world climate has undergone warm periods, interglacial eras, and ice ages over the past 2 million years. Climate changes also occur over narrower time frames. For instance, most areas of the state have experienced climates that have been warmer, cooler, wetter, and drier than what is currently observed at various points throughout the past 10,000 years. The only constant regarding the climate is that it is constantly changing (Louthan and Morris, 2021). The causes of climatic fluctuation are numerous. These consist of variations in the earth's orbit, solar radiation, solar cycles, volcanic eruptions, GHGs, and aerosol emissions. These forces work on a variety of time scales but, when examined collectively, successfully explain most of the climate variability during the last many 1,000 years (Mora et al., 2018). However, the rise in temperature and associated series of climatic variations recorded over the 20th century cannot be explained by these natural forces alone.

After the green revolution in the 1960s, the fertilizer consumptions increase five times to meet hunger, but fertilizer production and industrialization, particularly in agriculture, improved the terrestrial to productive land and high production from less area are all success stories of efficient use of resources. While feeding the > 7.5 billion people itself was a great challenge, that wasn't possible without the use of synthetic fertilizer. Although this mechanization produces reactive gases such as carbon dioxide ($CO_2$), nitrous oxide ($N_2O$), and methane ($CH_4$) which change the climate dramatically, at that time, these achievements were a basic need for the nation's sustainability. Every process in nature transforms the energy from one form to another, according to the "energy can't be produced nor be destroyed but it changes from one form to another". So the benefits of the green revolution, synthetic fertilizer, and mechanization outweigh the losses or effects occurring in their response and adaptation (John and Babu, 2021). As earlier stated, solutions are not constant because the climate changes day after day, so innovative and sustainable development needs time with the ongoing processing. Secondly, the question is rising whether the only novel scientific approaches and innovative engineering techniques in the farming system's sustainable goals are to meet the food demand for the future. Whereas science and engineering technology are the vast fields to umbrella all the other factors affecting the GCC, which included a wide range of management and sustainability models, policymaking, and human responses that are the major concerns of adaptations. To effectively solve the GCC and its complex problems, we should be ready to think outside the box. A good place to start is by reviewing the mentioned management concepts and techniques that were employed to enhance our resource efficiency previously.

## 2.5 SOIL WATER AND WASTE—IN THE CONTEXT OF GLOBAL CLIMATE CHANGE (GCC)

As we are considering the management of our resources, i.e., "soil water and waste", on a second, however with a serious note because life on earth will end whenever these natural resources are replenished. Basically, these three natural resources are fundamental pillars of the agricultural-based food production system (Figure 2.2). Water is the first and foremost natural resource that represents a wider range of purposes in food safety. A civil society leader quoted that "Water is life. If you are the owner of hundreds of herds or a major producer of rice or millet, it only takes one

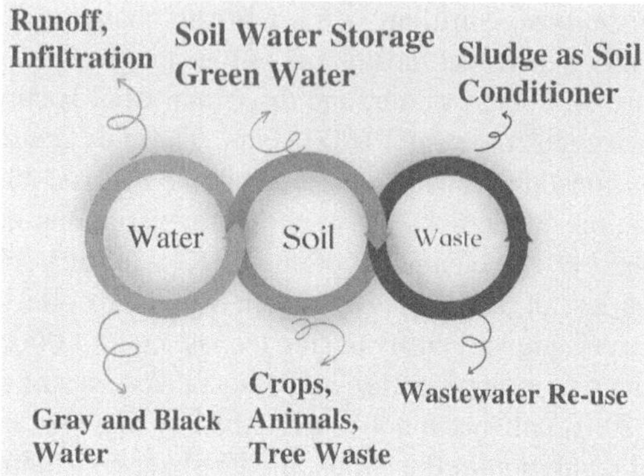


**FIGURE 2.2** Natural resources nexus.

year of drought for you to be ruined. This is happening in our farming, herding, and fishing villages now. In this insecure situation that we are experiencing today, everything is getting worse." Soil is the second most fundamental pillar and basic need of our routine life while particularly essential in food production. Human beings could not sustain themselves without soil, as the soil is the foundation of life. It is an essential natural resource that directly produces food, feed, fuel, and fiber to sustain humans on earth. It resolves the food security crisis and sustains environmental quality, which is mandatory for human existence.

Human activities generate waste, which is completely different from the aforementioned resources, and people mostly take it as a problem and nuisance. In fact, it is an anthropogenic, nutrient-enriched resource. The quality of waste really matters, which differs from place to place and based on the source of the waste, and mostly, it is mixed material with different compositions, e.g., iron ore deposits (especially low-grade material). Waste is basically concentrated in metropolitan cities due to the highly populated area. Its production accelerated in tandem with increasing population, industrialization, mechanization, urbanization, and ultimately, income of people.

As the world's population increases, the demographic values of the people are improving highly, and the soil is experiencing the highest levels of soil sealing by mechanization and infrastructure establishment, so most of this development are occurring in small and medium cities. Alternatively, the consumption of protein, vehicles, fossil fuel burning, electricity, and food products is in greater quantities by this exponential growth of the rich population (Zhai et al., 2020). However, to meet this demand, more natural resources are utilized, and as result, more waste are produced. Municipal solid waste (MSW) is categorized into organic and inorganic wastes, and there are suitable solution approaches to convert this garbage into nutritional fertilizer (Eden et al., 2017). The organic component of MSW contained value-added compounds (i.e., proteins, carbohydrates, and carbon enriched), and the chemical composition is a complete nutritional-enriched fraction that can be added to soil after compositing, acidification, and biochar production as the soil conditioner amendment.

Furthermore, integrated resource management "IRM" is a potential alternative and a comprehensive tool for the natural and man-made environmental resources management, such as soil, water, and waste, respectively. For example, the integrated water resource management (IWRM) is a well-heated discussion and debate, as news quoted the importance of water and stated that the next world war will be about a water crisis because people demand affordable and safe drinking water and basic sanitation (Nagata et al., 2021). There are many countries experiencing serious water resource issues, as pressure from the local economic sector of energy and agriculture due to the shortage of water caused by transboundary conflicts and worldwide agreements crises on the water are all open opportunities for the competent authorities of local governments to inaugurate and improve integrated water resource management projects. So the city administration is interested on how the drinking and sanitation water should be distributed and how wastewater should be collected outside that efficiently treated and reused within its borderlines. These authorities are looking forward to other local bodies to manage the industries and their water requirements outside of their boundaries due to the congested and populated areas with limited water resources. However, the national government of that country is considering dialogues with the other countries through international NGOs and governments on how they possibly share the one river for water resources to meet their agricultural requirements and energy production. Ultimately, they all need to manage the natural water resources collectively, and the idea of coordinative management and sustainable development of IWRM is crucial to minimize the losses and maximize economic and social welfare.

The IWRM management strategy has been proven to be effective. IWRM also faces significant flaws like many other integrated management methods which hamper their usefulness and popularity among policymakers. The interfaces that the primary resources with other resources are frequently ignored while managing the primary resource. Each resource management decision may have a constructive or destructive effect on another. These illustrations of exactly "how" the three proposed resources soil water and waste are interconnected and can easily be understood when considering wastewater management and application. After the wastewater provides the reuse of water for particular irrigation in agricultural systems, well-managed wastewater also yields nutrients that can be returned to the soil, e.g., phosphorus-laden biochar can be harvested from eutrophic water treated with engineered biochar and applied to soil amendment for increasing soil biodiversity, improving soil health in terms of high water holding capacity with high soil productivity and carbon sequestration.

Furthermore, if we are considering the management in terms of "technologies" and "skill resources" to take advantage of these synergies, then the answer is "no". As sewage sludge (SS) is basically a by-product of wastewater treatment units and is commonly dumped in landfills, it needs to get rid of and should be reused for reapplication to soil fertilizer, after processing and value addition as co-composting with rock phosphates, bio-activation, sludge biochar formulation, acidification, and bioleaching processes should be adapted to develop value-added SS fertilizer (Kacprzak et al., 2017).

A formal mechanism to make use of such synergies are approaches we have suggested earlier, although these may not be feasible until the management of three resources (soil, water, waste) is integrated. Certainly, a novel concept is integrated management (IM) practices at such a higher level that bypasses resource bottlenecks. We refer to it as the nexus strategy.

## 2.6 NEXUS STRATEGIES

The word "nexus" is a series of simultaneous approaches for the numerous sectors. Nexus techniques are synergies and reduce the input cost and negative outstrips with improved integrated approaches for management and governance. The term "nexus" was initially quoted in relation to environmental resource management in the 1980s, particularly in a study by the United Nations University on Food-Energy Nexus Program. In the 1980s and 1990s, the nexus idea was still widely used in resource management forums to describe the interrelation between various resources (Elmqvist et al., 2015). Water-electricity, water-energy, groundwater-electricity, water-agriculture, and lastly, water-energy-food were a few examples of these nexuses.

The nexus concept, however, did not gain popularity among international academics and policymakers up to the preliminary conference on the "Water, Energy, and Food Security Nexus" by the University of Bonn in 2011. The symposium showed a strong case for how integrating "management and governance across sectors and scales" will lead to better water, energy, and food security. It also made note of how this strategy will facilitate the transition to a green economy by lowering trade-offs, creating synergies, and promoting sustainability. The 2011 Bonn conference also emphasized the need for better coordinated and harmonized nexus methodological approaches, as well as for more consolidated decisions and policies in all primary sectors. This explains one of the nexus techniques' distinctive characteristics. The effectiveness of the nexus approach depends on how well it is accelerated by novel and advantageous strategies because it aims to combine a few disciplines into one action plan (Rasul and Sharma, 2016). As a result, the nexus model now includes fair governance and capacity development. Without them, it would not be feasible to implement innovations. These novel ways undoubtedly come at a higher cost, but it is acceptable to conclude that the predicted long-term advantages from the nexus strategy would be much higher.

We briefly discussed the challenges encountered by integrated management models in the previous section, as well as the policy circles' reluctance to accept certain outcomes carried by these management strategies and models. Preferably, it's important to briefly examine the causes and how the nexus method prevents it from becoming a problem. Returning to the IWRM example, we are aware that water distribution is a crucial factor. The equations continue to regard all other aspects of water consumers (energy, agriculture, and industry) as fixed constraints, even though in practice, they should also be considered as variables. Sometimes, failure to represent facts gives less expected outcomes. The recommendations based on such integrated modeling tools are rarely implemented by decision-makers, or they are usually unwilling of doing so.

The main difference between integrated management and the nexus method in terms of modeling is that the nexus approach substitutes the idea of linked cycle management for the conventional input-output models. As a result, when compared to models created using an integrated approach, model findings must be substantially more accurate. We agree that it is easier to assume than to execute.

Another issue to be solved in the upcoming years is the creation of nexus-strategies-based modeling devices and tools.

## 2.7 GLOBAL CLIMATE CHANGE IMPACTS AND ITS ADAPTATION

Population growth, industrialization, mechanization, urbanization, and ultimately, climate change are various scientific issues that contribute to the symposium on global climate change, as was briefly covered in the first portion of this chapter. While certain factors are a direct outcome of GCC, others help to speed up the changes happening. We highlight three of the key factors (climate change, urbanization, and population increase) that have a significant influence on the management of our natural resources, particularly soil, water, and waste. Briefly, climate change (CC) is the term used to describe unusually persistent changes in weather patterns, and these changes may occur for decades or millions of years.

There were a lot of "if" issues when the World Climate Research Program (WCRP) was founded in 1980, including if the climate was altering, whether anthropogenic activities were at least partially to blame, and whether variations can be forecast (Meehl et al., 2007). However, researchers found that climatic variations are a piece of the larger picture that is now known as worldwide change. Secondly, global climate change is thought to be brought on by biotic procedures, plate tectonics, volcanic eruptions, and changes in the amount of solar energy that land receives. Additionally, different anthropogenic activities are also accelerating these aspects of global climate change, such as global warming.

Scientists are over 95% positive that rising greenhouse gas (GHGs) concentrations and other human-caused activities are the primary causes of global warming, according to a study just released by the Intergovernmental Panel on Climate Change (IPCC). While electricity production and heat generation cause more greenhouse gas emissions globally than in the agri-food-related sectors. Forest and other land use variations put up 17%, while agriculture provides 14%. However, of all environmental resources affected, the soil and water are thought to be the most climate susceptible. Arid regions become increasingly drier due to climate change, while crops produce less due to harsh weather occurrences. The entire globe believes that adapting to climate change will be the answer to considering climate change and managing soil, water, and waste. However, it has also been shown that climate change adaptation is expensive. If irrigation is the answer to the water shortage, it should be highlighted that it is always more expensive than agriculture reliant on rainfall.

Additionally, desalination and the utilization of deep groundwater are more expensive than using regular water sources. Urbanization is a major contributor to climate change, even if it is a by-product of global change. City residents already make up more than 50% of the world's population, and by the year 2050, that number is projected to rise to 70% (Shannon et al., 2010).

Undoubtedly, resource management equations are made more difficult by the rapid expansion in urbanization, necessitating the development of fresh approaches to improve resource efficiency. With the right combination of nexus-thinking financial and technical resources, metropolitan communities can turn this risk into an opportunity. In this case, the increased amount of trash and wastewater produced because of the growing population might serve as a secondary supply of water and a source of nutrients.

Population increase is a similar force behind global climate change to urbanization. By the middle of the century, the population of the world will have increased by a definite trend, reaching 9 billion people. The potential effects on environmental resources, particularly those necessary for food production, water availability, and yield biomass to support population growth, are less clear (Molotoks et al., 2021; Godfray et al., 2010). The population of several emerging nations is increasing more quickly than their food supply. Additionally, a huge portion of agricultural land has been degraded because of population pressures.

## 2.8  FUTURE PERSPECTIVE

Nexus strategy is a novel idea that is improving accordingly. As of the moment, the nexus method does not have widespread agreement, as is usual with many other novel ideas. Many people are aware of it, appreciate it, and hold slightly diverse opinions on it. A novel project involves using the nexus method to manage our natural resources, particularly water, soil, and waste. However, the trial has positive potential.

This book's objective is to provide a forum for the discussion of various points of view about the nexus strategies, their application to the sustainability of our natural resources, and how they could improve our ability the adaptation to the dynamic rate of global change. Whereas the introductory book chapter offers a brief yet comprehensive summary, subsequent chapters include the opinions of many innovators. They talk about the potential benefits of the nexus method for waste, water, and soil management. We think this book will offer a clear, objective assessment of the nexus method's contribution to our natural resources management and adaption. We are looking that this chapter will completely contribute to developing the critical nexus strategies and modeling tools for the future.

## REFERENCES

Abbass K, Begum H, Alam AF, Awang AH, Abdelsalam MK, Egdair IMM, Wahid R. 2022. Fresh insight through a Keynesian theory approach to investigate the economic impact of the COVID-19 pandemic in Pakistan. *Sustainability* 14(3):1054.

Abbass K, Song H, Shah SM, Aziz B. 2019. Determinants of stock return for non-financial sector: Evidence from energy sector of Pakistan. *J Bus Fin Aff* 8(370):2167–0234.

Anderson PK, Cunningham AA, Patel NG, Morales FJ, Epstein PR, Daszak P. 2004. Emerging infectious diseases of plants: Pathogen pollution, climate change and agrotechnology drivers. *Trends Ecol Evol* 19:535–544.

Armitage D, Béné C, Charles AT, Johnson D, Allison EH. 2012. The interplay of well-being and resilience in applying a social-ecological perspective. *Ecol Soc* 17(4).

Aydinalp C, Cresser MS. 2008. The effects of global climate change on agriculture. *Am Eurasian J Agric Environ Sci* 3:672–676.

Balbus J, Boxall A, Fenske RA, McKone T, Zeise L. 2012. Implications of global climate change for the assessment and management of human health risks of chemicals in the natural environment. *Environ Toxicol Chem* 32:62–78.

Bates DW, Saria S, Ohno-Machado L, Shah A, Escobar G. 2014. Big data in health care: Using analytics to identify and manage high-risk and high-cost patients. *Health Aff* 33(7):1123–1131.

Beesley L, Close PG, Gwinn DC, Long M, Moroz M, Koster WM, Storer T. 2019. Flow-mediated movement of freshwater catfish, Tandanus bostocki, in a regulated semi-urban river, to inform environmental water releases. *Ecol Freshwater Fish* 28(3):434–445.

Beulke S, Boxall A, Brown C, Thomas M. 2007. *Assessing and managing the impacts of climate change on the environmental risks of agricultural pathogens and contaminants: Pathways for transport of contaminants in agricultural systems and potential effects of climate change*. Final Report to Defra for Project SD0441. Fera, Sand Hutton, York, UK.

Bloomfield JP, Williams RJ, Gooddy DC, Cape JN, Guha P. 2006. Impacts of climate change on the fate and behaviour of pesticides in surface and groundwater—a UK perspective. *Sci Total Environ* 369:163–177.

Bogdal C, Scheringer M. 2011. *Release of POPs to the environment*. UNEP/AMAP Expert Group Report—Climate Change and POPs: Predicting the Impacts. United Nations Environment Programme/Arctic Monitoring and Assessment Programme, Geneva, Switzerland.

Booth S, Zeller D. 2005. Mercury, food webs, and marine mammals: Implications of diet and climate change for human health. *Environ Health Perspect* 113:521–526.

Bower H. 2002. Integrated water management for the 21st century: Problems and solutions. *J Irrig Drain Eng* 128:193–202.

Boxall AB, Hardy A, Beulke S, Boucard T, Burgin L, Falloon PD, Haygarth PM, Hutchinson T, Kovats RS, Leonardi G, Levy LS, Nichols G, Parsons SA, Potts L, Stone D, Topp E, Turley DB, Walsh K, Wellington EM, Williams RJ. 2009. Impacts of climate change on indirect human exposure to pathogens and chemicals from agriculture. *Environ Health Perspect* 117:508–514.

Cecchi L, D'Amato G, Ayres JG, Galan C, Forastiere F, Forsberg B, Gerritsen J, Nunes C, Behrendt H, Akdis C, Dahl R, Annesi-Maesano I. 2010. Projections of the effects of climate change on allergic asthma: The contribution of aerobiology. *Allergy* 65:1073–1081.

Confalonieri UE, Lima ACL, Brito I, Quintão AF. 2014. Social, environmental and health vulnerability to climate change in the Brazilian Northeastern Region. *Clim Change* 127(1):123–137.

Conner MS, Davis JA, Leatherbarrow J, Greenfield BK, Gunther A, Hardin D, Mumley T, Oram JJ, Werne C. 2007. The slow recovery of San Francisco Bay from the legacy of organochlorine pesticides. *Environ Res* 105:87–100.

Cowan-Ellsberry C, McLachlan M, Arnot J, MacLeod M, McKone TE, Wania F. 2009. Modeling exposure to persistent chemicals in hazard and risk assessment. *Integr Environ Assess Manag* 5:662–679.

Da Graça Carvalho M. 2012. EU energy and climate change strategy. *Energy* 40(1):19–22.

Diffenbaugh NS, Burke M. 2019. Global warming has increased global economic inequality. *Proc Natl Acad Sci* 116(20):9808–9813.

Downs SG, MacLeod CL, Lester JN. 1998. Mercury in precipitation and its relation to bioaccumulation in fish: A literature review. *Water Air Soil Pollut* 108:149–187.

Dullinger S, Gattringer A, Thuiller W, Moser D, Zimmermann NE, Guisan A, Willner W, Plutzar C, Leitner M, Mang T, Caccianiga M. 2012. Extinction debt of high-mountain plants under twenty-first-century climate change. *Nat Clim Change* 2(8):619–622.

Eden M, Gerke HH, Houot S. 2017. Organic waste recycling in agriculture and related effects on soil water retention and plant available water: A review. *Agron Sustain Dev* 37(2):1–21.

El Bilali H, Bassole IHN, Dambo L, Berjan S. 2020. Climate change and food security. *Agric For/Poljoprivreda i Sumarstvo* 66(3).

Elmqvist T, Setälä H, Handel SN, Van Der Ploeg S, Aronson J, Blignaut JN, Gomez-Baggethun E, Nowak DJ, Kronenberg J, De Groot R. 2015. Benefits of restoring ecosystem services in urban areas. *Curr Opin Environ Sustain* 14:101–108.

Frey AE, Olivera F, Irish JL, Dunkin LM, Kaihatu JM, Ferreira CM, Edge BL. 2010. Potential impact of climate change on hurricane flooding inundation, population affected and property damages in Corpus Christi. *J Am Water Res Assoc* 46(5). DOI:10.1111/j.1752-1688.2010.00475.

Gelybó G, Tóth E, Farkas C, Horel Á, Kása I, Bakacsi Z. 2018. Potential impacts of climate change on soil properties. *Agrokem Talajt* 67(1):121–141.

Godfray HCJ, Beddington JR, Crute IR, Haddad L, Lawrence D, Muir JF, Pretty J, Robinson S, Thomas SM, Toulmin C. 2010. Food security: The challenge of feeding 9 billion people. *Science* 327(5967):812–818.

Gouin T, Armitage J, Cousins I, Muir DC, Ng CA, Tao S. 2012. Influence of global climate change on chemical fate and bioaccumulation: The role of multimedia models. *Environ Toxicol Chem* 32:20–31 (this issue).

Gouin T, Wania F. 2007. Time trends of Arctic contamination in relation to emission history and chemical persistence and partitioning properties. *Environ Sci Technol* 41:5986–5992.

Hallegraeff GM. 2010. Ocean climate change, phytoplankton community responses, and harmful algal blooms: A formidable predictive challenge. *J Phycol* 46:220–235.

Hof C, Levinsky I, Araujo MB, Rahbek C. 2011. Rethinking species' ability to cope with rapid climate change. *Glob Change Biol* 17:2987–2990.

Huang X, Hao L, Sun G, Yang ZL, Li W, Chen D. 2022. Urbanization aggravates effects of global warming on local atmospheric drying. *Geophys Res Lett* 49(2):e2021GL095709.

Hurlimann AC, Moosavi S, Browne GR. 2021. Climate change transformation: A definition and typology to guide decision making in urban environments. *Sustain Cities Soc* 70:102890.

Jansson JK, Hofmockel KS. 2020. Soil microbiomes and climate change. *Nat Rev Microbiol* 18(1):35–46.

John DA, Babu GR. 2021. Lessons from the aftermaths of green revolution on food system and health. *Front Sustain Food Syst* 5:644559.

Kacprzak M, Neczaj E, Fijałkowski K, Grobelak A, Grosser A, Worwag M, Rorat A, Brattebo H, Almås Å, Singh BR. 2017. Sewage sludge disposal strategies for sustainable development. *Environ Res* 156:39–46.

Karami A, Homaee M, Afzalinia S, Ruhipour H, Basirat S. 2012. Organic resource management: Impacts on soil aggregate stability and other soil physico-chemical properties. *Agric Ecosyst Environ* 148:22–28.

Kohfeld KE, Quéré CL, Harrison SP, Anderson RF. 2005. Role of marine biology in glacial-interglacial $CO_2$ cycles. *Science* 308(5718):74–78.

Lake IR, Foxall CD, Lovett AA, Fernandes A, Dowding A, White S, Rose M. 2005. Effects of river flooding on PCDD/F and PCB levels in cows' milk, soil, and grass. *Environ Sci Technol* 39:9033–9038.

Lamon L, VonWaldow H, Macleod M, Scheringer M, Marcomini A, Hungerbuhler K. 2009. Modeling the global levels and distribution of polychlorinated biphenyls in air under a climate change scenario. *Environ Sci Technol* 43:5818–5824.

Linderholm L, Jakobsson K, Lundh T, Zamir R, Shoeb M, Nahar N, Bergman A. 2011. Environmental exposure to POPs and heavy metals in urban children from Dhaka, Bangladesh. *J Environ Monit* 13: 2728–2734.

Louthan AM, Morris W. 2021. Climate change impacts on population growth across a species' range differ due to nonlinear responses of populations to climate and variation in rates of climate change. *PLOS One* 16(3):e0247290.

Lu Y, Zhang Y, Hong Y, He L, Chen Y. 2022. Experiences and lessons from agri-food system transformation for sustainable food security: A review of China's practices. *Foods* 11(2):137.

MacFarquhar N. 2009. Experts worry as population and hunger grow. *New York Times*.

MacLeod MJ, Riley WJ, McKone TE. 2005. Assessing the influence of climate variability on atmospheric concentrations of persistent organic pollutants using a global-scale mass balance model (BETR-Global). *Environ Sci Technol* 39:6749–6756.

Manabe S. 2019. Role of greenhouse gas in climate change. *Tellus A Dyn Meteorol Oceanogr* 71(1):1620078.

Manes S, Vale MM, Malecha A, Pires AP. 2022. Nature-based solutions promote climate change adaptation safeguarding ecosystem services. *Ecosyst Serv* 55:101439.

Manuel M. 2006. In Katrina's wake. *Environ Health Perspect* 114: A32–A39.

Marques A, Nunes ML, Moore SK, Strom MS. 2010. Climate change and seafood safety: Human health implications. *Food Res Int* 43:1766–1779.

Martin V, Chevalier V, Ceccato P, Anyamba A, De Simone L, Lubroth J, de La Rocque S, Domenech J. 2008. The impact of climate change on the epidemiology and control of Rift Valley fever. *Rev Sci Tech* 27:413–426.

McKone TE, MacLeod M. 2004. Tracking multiple pathways of human exposure to persistent multimedia pollutants: Regional, continental, and global scale models. *Annu Rev Environ Resour* 28:463–492.

McMichael AJ, Lindgren E. 2011. Climate change: Present and future risks to health, and necessary responses. *J Int Med* 270(5):401–413.

Meehl GA, Covey C, Delworth T, Latif M, McAvaney B, Mitchell JF, Stouffer RJ, Taylor KE. 2007. The WCRP CMIP3 multimodel dataset: A new era in climate change research. *Bull Am Meterol Soc* 88(9):1383–1394.

Mishra SN, Gupta HS, Kulkarni N. 2021. Impact of climate change on the distribution of Sal species. *Ecol Inform* 61:101244.

Mitchell MJ, Likens GE. 2011. Watershed sulfur biogeochemistry: Shift from atmospheric deposition dominance to climatic regulation. *Environ Sci Technol* 45:5267–5271.

Molina T, Abadal E. 2021. The evolution of communicating the uncertainty of climate change to policymakers: A study of IPCC synthesis reports. *Sustainability* 13(5):2466.

Molotoks A, Smith P, Dawson TP. 2021. Impacts of land use, population, and climate change on global food security. *Food Energy Secur* 10(1):e261.

Moore MJ, Goldstein D, Hamm J, Figer A, Hecht JR, Gallinger S, Au HJ, Murawa P, Walde D, Wolff RA, Campos D. 2007. Erlotinib plus gemcitabine compared with gemcitabine alone in patients with advanced pancreatic cancer: A phase III trial of the National Cancer Institute of Canada Clinical Trials Group. *J Clin Oncol* 25(15):1960–1966.

Mora C, Spirandelli D, Franklin EC, Lynham J, Kantar MB, Miles W, Smith CZ, Freel K, Moy J, Louis LV, Barba EW. 2018. Broad threat to humanity from cumulative climate hazards intensified by greenhouse gas emissions. *Nat Clim Change* 8(12):1062–1071.

Nagata K, Shoji I, Arima T, Otsuka T, Kato K, Matsubayashi M, Omura M. 2021. Practicality of integrated water resources management (IWRM) in different contexts. *Int J Water Resour Dev*:1–23.

Noyes PD, McElwee MK, Miller HD, Clark BW, Van Tiem LA, Walcott KC, Erwin KN, Levin ED. 2009. The toxicology of climate change: Environmental contaminants in a warming world. *Environ Int* 35:971–986.

Ortiz-Bobea A, Ault TR, Carrillo CM, Chambers RG, Lobell DB. 2021. Anthropogenic climate change has slowed global agricultural productivity growth. *Nat Clim Change* 11(4):306–312.

Pachauri RK, Allen MR, Barros VR, Broome J, Cramer W, Christ R, Church JA, Clarke L, Dahe Q, Dasgupta P, Dubash NK. 2014. Climate change 2014: Synthesis report. Contribution of Working Groups I, II and III to the fifth assessment report of the Intergovernmental Panel on Climate Change (p. 151). IPCC.

Parry ML, Canziani OF, Palutikof JP, van der Linden PJ, Hanson CE. 2007. Cross-chapter case study. In *Climate change 2007: Impacts, adaptation and vulnerability*. Contribution of Working Group II to the Fourth Assessment Report of the Intergovernmental Panel on Climate Change. Cambridge University Press, Cambridge, UK, pp. 843–868.

Pautasso M, Döring TF, Garbelotto M, Pellis L, Jeger MJ. 2012. Impacts of climate change on plant diseases—Opinions and trends. *Eur J Plant Pathol* 133:295–313.

Pereira HM, Ferrier S, Walters M, Geller GN, Jongman R, Scholes RJ, Cardoso A. 2013. Essential biodiversity variables. *Science* 339(6117):277–278.

Point D, Sonke JE, Day RD, Roseneau DG, Hobson KA, Vander Poll SS, Moors AJ, Pugh RS, Donard OFX, Becker PR. 2011. Methylmercury photodegradation influenced by sea-ice cover in Arctic marine ecosystems. *Nat Geosci* 4:188–194.

Pounds AJ, Bustamonte MR, Coloma LA, Consuegra JA, Fogden MPL, Foster PN, La Marca E, Masters KL, Merino-Viteri A, Puschendorf R, Ron SR, Sanchez-Azofeifa GA, Still CJ, Young BE. 2006. Widespread amphibian extinctions from epidemic disease driven by global warming. *Nature* 439:161–167.

Power AG. 2010. Ecosystem services and agriculture: Tradeoffs and synergies. *Philos Trans R Soc B Biol Sci* 365(1554):2959–2971.

Rasul G, Sharma B. 2016. The nexus approach to water-energy-food security: An option for adaptation to climate change. *Clim Policy* 16(6):682–702.

Reidsma P, Ewert F, Boogaard H, van Diepen K. 2009. Regional crop modelling in Europe: The impact of climatic conditions and farm characteristics on maize yields. *Agric Syst* 100(1–3):51–60.

Riget F, Vorkamp K, Muir D. 2010. Temporaltrends of contaminants in Arctic (char Salvelinus alpinus) from a smalllake, Southwest Greenland during a warming climate. *J Environ Monit* 12:2252–2258.

Rohr JR, Dobson AP, Johnson PT, Kilpatrick AM, Paull SH, Raffel TR, Ruiz-Moreno D, Thomas MB. 2011. Frontiers in climate change-disease research. *Trends Ecol Evol* 26:270–277.

Scheringer M. 1997. Characterization of the environmental distribution behavior of organic chemicals by means of persistence and spatial range. *Environ Sci Technol* 31:2891–2897.

Shannon MA, Bohn PW, Elimelech M, Georgiadis JG, Marinas BJ, Mayes AM. 2010. Science and technology for water purification in the coming decades. *Nat Nanotechnol Collect Rev Nature J*:337–346.

Stenseth NC, Mysterud A, Ottersen G, Hurrell JW, Chan K-S, Lima M. 2002. Ecological effects of climate fluctuations. *Science* 297:1292–1296.

Stewart JA, Perrine JD, Nichols LB, Thorne JH, Millar CI, Goehring KE, Massing CP, Wright DH. 2015. Revisiting the past to foretell the future: Summer temperature and habitat area predict pika extirpations in California. *J Biogeogr* 42(5):880–890.

Sunil P. 2017. *Novel postharvest treatments of fresh produce*. CRC Press, Boca Raton, FL.

Thornton PK, Lipper L. 2014. How does climate change alter agricultural strategies to support food security? *Intl Food Policy Res Inst* 1340. http://dx.doi.org/10.2139/ssrn.2423763

Tirado MC, Clarke R, Jaykus LA, McQuatters-Gollop A, Frank JM. 2010. Climate change and food safety: A review. *Food Res Int* 43:1745–1765.

Urban MC. 2015. Accelerating extinction risk from climate change. *Science* 348(6234):571–573.

Zhai Z, Martínez JF, Beltran V, Martínez NL. 2020. Decision support systems for agriculture 4.0: Survey and challenges. *Comput Electron Agric* 170:105256.

# 3 Potential Role of Healthy Soil for Global Environment

*Muhammad Farhan Qadir, Khuram Shehzad Khan,*
*Muhammad Naveed, Noman Younas,*
*Muhammad Yousuf, Muhammad Hamza Khalil,*
*Azmat Qadir Maria, Taqi Raza, and Neal S. Eash*

## 3.1  INTRODUCTION: FUNCTIONS OF SOIL WITH SDGS

The basic question is whether nature plays a vital role in accomplishing each of the 17 Sustainable Development Goals (SDGs) proposed by the UN General Assembly. Open working questions also arise from a study of the key aspects of the goals. Nature should not be allowed to exist without interference from human needs, while human existence is a component of nature and, hence, intimately connected to how it functions.

The SDGs aim to improve human survival, sustainability, and overall welfare, while SDGs also protect the land, environment, and its natural occurring resources, such as the climate, oceans, and terrestrial ecosystems (SDGs 13, 14, and 15, respectively) (Keesstra et al. 2016). Many SDGs are intended to maintain and improve human existence, which cannot be fulfilled without reverence and the protection of natural resources because human beings are an integral part of nature and must abide by its basic laws. The subsequent discussion will explain both risks to soils that threaten the sustainable usage of land and soil resources, as well as the role that soil function plays in attaining the SDGs, while without them, many of the SDGs cannot be attained. The second goal, which mostly relates to soil and its role in the natural environment, aims to "end hunger, establish food security and improved nutrition, and stimulate sustainable agriculture" from the 17 SDGs. Article 6's No. 6 and No. 7's "Guarantee access to cheap, dependable, sustainable and contemporary energy for all", which touches on soil and land in part, are two additional goals. We shall now talk about goal numbers because of this—regarding the availability of water resources, their sustainable management, and Goal No. 6. Under Goal No. 7, bioenergy is given special consideration in terms of the supply of energy. Aim for sustainable agriculture and food security.

The role of soils and how land and soils supply the world's requirements for food, water, and energy are discussed in the subsequent sections. As healthy food and clean, pure, and fresh water are two essentials for the survival of human beings, we shall primarily concentrate on these two areas.

## 3.2  LAND VS SOIL

Land refers to the surface of the earth's crust with soil serving as a primary constituent of land. As shown in Figure 3.1, soils are three-dimensional surface objects that are composed of solid, liquid, and gaseous components. They contain both inorganic and organic substances and a wide diversity and abundance of living creatures. Soil is a non-renewable resource since it develops at an incredibly slow rate. Soils serve six key purposes that are significant for humans and the environment (Lehmann 2010). For further information, see the findings of the Commission of the European Communities (2002) regarding SDGs: Cultivation of crops for plant biomass occurred by adapting the physical, chemical, and biological factors that affect how plants grow to provide food, feed, and renewable energy. As this vital role is the basis for all existence, both human and animal, as this is the primary role of SDG no. 2. In addition, the foundation for producing bioenergy, see SDG no. 7. Cultivation of crops for plant biomass occurred by adapting the physical, chemical, and biological factors that affect how plants grow to provide food, feed, and renewable energy. This vital role is the basis for all existence, both human and animal. This is the primary role of SDG no. 2. In addition, for the foundation for producing bioenergy, see SDG no. 7. The nutrients required for plant growth are restored and maintained by filtration and buffering the form of adsorption and desorption of all inorganic and organic ingredients and nutrients from the internal surface of the soil. By filtering and holding onto contaminants arising from environmental pollution or by bio-transforming organic matter to detoxify it, they prevent the pollution of the food chain and groundwater. SDG no. 6 includes the processes that enable the generation of hygienic groundwater resources designed for human use and the filtering of precipitation. Three to four times the biomass and species count of all the biota are reported on the soil surface. Soil biodiversity represents the biggest genetic stockpile in the world and must be preserved and protected for the sustainability of life (Steffan 2018).

Land and soil serve three additional technical, economic, and social purposes in addition to these three ecological ones: (1) Buildings, factories, roads, landfills, sports and leisure facilities, other technical infrastructure essential to human life, and the production and trading of goods and services all began with the development of soil. Since the soil is sealed for these uses, all ecological services are lost. (2) Land and soil serve as a supply of water, mostly groundwater reservoirs, as well as mineral raw materials like clay, sand, gravel, or stone for the construction of the infrastructures. Due to the clay and silt concentration in places lacking stone or gravel, fertile soil is dug up and utilized to make bricks and tiles, which compromises food security in many places. Finally, the soil is a biological and cultural legacy that shapes as well as conceals cultural landscapes, protects paleontological and archaeological artifacts, and acts as a historical memory that enables us to comprehend the evolution of people and their relationship with the natural world. While the three natural and ecological soil functions are the most significant in terms of SDG fulfilment (Blum and Swaran 2004).

**FIGURE 3.1**  Soil-covered land mass constituting the bulk of the earth's surface and constituents.

## 3.3   RESOURCE MANAGEMENT REGARDING SDGS

The SDGs can just be accomplished through sustainable agriculture management (SagM), particularly agricultural land management. So the problem arises: What is sustainable agricultural land management?

The goods and services that SagM may offer are shown in Figure 3.2. This figure makes it abundantly clear that agricultural land can also produce enough clean groundwater below the surface of the soil, as every droplet of rain that falls on the earth must first pass through the topsoil before it can become underground water, which is frequently the only source of drinking water, along with agricultural soil preferably can produce biomass in the form of food (life nutrition), fiber, and bioenergy on top of the land. This demands quantitative and qualitative coordination between agricultural management of topsoil and plant production (particularly the use of agrochemicals) and groundwater production. Additionally, to prevent eutrophication and obstruction of waterways caused by chemical compounds (e.g., fertilizer application and sediments), soil management should be adopted in the way to reduce or completely prevent surface runoff of land material into free waters like lakes, ponds, and rivers (Robert et al. 2018).

However, all agricultural practices must control and reduce the exhausting amount of greenhouse gases (GHGs) to stop climate change according to SDG no. 13 because soils also release climate-relevant toxic gases such as carbon dioxide ($CO_2$), methane

**FIGURE 3.2** Services and products that can be directly attributed to the earth.

($CH_4$), and nitrous oxide ($N_2O$), while an exponential surge was noticed with anthropogenic activities after advents of 21st century (Barnosky 2012). All the soil activities associated with sustainable food production and the preservation of groundwater supplies also depend on soil biodiversity. While the aboveground biodiversity, pests, and pathogenic diseases that damage food production and storage are also closely tied to soil biodiversity. Depending on how the soil is managed, even human health may be directly or indirectly impacted by the chemical and biotic makeup of the soil, as stated by Lal (2002) "healthy soils produce the healthy plants that produce the healthy humans". However, the quality and precise distribution of the world's earth resources influence the provision of goods and services, particularly the supportable production of food (SDG 2) through SAM practices.

## 3.4 FOOD SECURITY THREAT IN THE CONTEXT OF SOIL DISTRIBUTIONS

SDG no. 2 is related to soils and soil functions in its roles in promoting food security. Food security was defined as "the situation in which people have physical, social, and economic access at all times to adequate safe and nutritious food to suit their dietary requirements and food choices, for energetic life and healthful living" according to the first World Summit Initiative Seto (2016). The objectives of this definition are difficult since it encompasses access to food under all circumstances through sustainable production through available natural resources, including land surface, terrain, soil, water, biota, and climatic conditions. As technical dangers, availability (production) also involves environmental, social, and economic opportunities. Whereas the availability of instruments, especially agricultural machinery and agrochemicals (plant protection products, fertilizers, etc.), agricultural systems, land ownership, and food policy frameworks, are also crucial. Food accessibility is quite different due to several restrictions, including physical availability (e.g., via storing, transportation, market place disturbance, quality, security rules, etc.) and economic and social availability, from food production. Most people must buy food at national and international food policy, so price structures and revenue generation all also play a significant role. Different eating habits depending on worldviews, mythologies, religions, and general cultural beliefs are most noticeable and quite diverse internationally. Finally, because people have varied genetic makeups and health conditions, different food kinds have variable physiological accessibility over the world. The more cases of allergies throughout the world are one picture of the food web (Oliver 2015).

Seminars about soil-related food security emphasize accessibility (i.e., sustainable food production through SAM practices), and it is dependent on the availability of natural resources, which are accessible land surface, topography, soil distribution, and quality, as well as water resources, natural biota, and livestock flora and fauna (such as seed material) and climatic conditions (Doran 2018).

## 3.5 SOIL DISTRIBUTION AND ITS ROLE IN THE SUSTAINABLE POPULATION RESOURCES

The quality and geographic spreading of the global land and soil resources, involving the population distribution and soil capability to sustain life, are covered here. It is important to understand how vastly different land and soil resources are distributed around the world and how they affect the quality of food production, which will clarify the actual concerns with global food security (Mueller 2010).

In terms of the eleven soil orders of the soil taxonomic system and related populations, the distribution of soil and land resources meeting the worldwide distribution has been described (Blum and Swaran 2004). According to 2014 estimates of the world population, population statistics have been modified (7.2 billion), and the expected population for each soil class as well as the land area that each occupies is increasing accordingly (Baillie 2001).

More than 70% of the world's population is sustained by the following soil orders distribution—Ultisols, Alfisols, Inceptisols, and Entisols—while it is a considerable population proportion. Despite only making up 41% of the land area, these groups of soils offer the ideal conditions for agriculture. Preferably, communities began to settle on plains and small-top, rolling terrain that required little effort to manage. Later, technical advancements can also be leveraged to employ soil types more intensively like Vertisols and Aridisols that were previously restricted by agricultural management. Alfisols and Mollisols have dense populations in temperate parts of the planet. The Mollisols soil makes up around 6.6% of the earth's population and occupies about 6.4% of the total land area. Although these two soil orders are among the world's most productive, they are mainly found in temperate climates. A substantial portion of the population in the tropics is connected to the Entisol and Inceptisol river terraces and Ultisol. Significant issues have been brought about by ongoing low-input agricultural production by Ultisols and Oxisols. The Andisols, which form from volcanic material, has the greatest population density of over 129 persons/km$^2$, while the Gelisols in the north have the lowest resident density of approximately 2 persons/km$^2$. Ultisol and Vertisol are two tropical locations where agricultural production is severely exploited, with population densities of more than 100 and 120 people per km². The land area dominated by Histosols and Aridisols has densities of 21 and 24 per km², respectively, and these soils are also considered fragile systems. Although the densities of these soils are low in comparison to other soil orders, they are endangering the sustainability of several regional systems. The landless poor people have been pushed to move into urban areas and vulnerable marginal habitats or to destroy their nation's resources via excessive or improper usage in many tropical regions due to rapid population expansion (Braimoh 2002).

## 3.6  SOIL USE AND ITS MANAGEMENT IN THE CONTEXT OF LAND QUALITY

The capacity of the land to perform a specific soil function is measured by its quality. It is typical to categorize land based on how well it can accomplish the services, including the performance of biomass production and its agricultural component, which are the main subjects of quality evaluation in most systems (Liu 2020). Lands and soils are categorized following their capacity to conduct the functions, distinguishing between two diverse intrinsic parameters: performance, which refers to their capacity to produce biomass, and resilience, which denotes their capacity to return to a new equilibrium after a disturbance. So soils are further divided into a nine-level land categorization with an emphasis on agricultural food production based on the two intrinsic soil aspects: "performance" and "resilience" (Beinroth 2019).

The Class I land category is only 2.38% of the world's land-living area but is accounting for an extra 40% of food production. Lands belonging to this category have ideal soils and are situated in ideal climates for crop production (such as Mollisols). They are characterized by high production and resilience, so these soils are highly responsive to management and have few limitations. The usage of 9.53% of Class II and III land resources is not permanently restricted by small limitations (like Alfisols) that can be quickly fixed. Most of these areas are found in temperate portions of the world, where the weather is pleasant and there are seldom extremes in temperature or precipitation. These soils are adaptable to treatment, and the advantages of good management can endure for a remarkably prolonged period. More than 50% of the world's population lives in areas with land quality classifications IV, V, and VI (such as Ultisol, Oxisols, and Vertisols), which make up around 34% of the global land area and are primarily found in the tropics. These soils and the sites in which they are located are susceptible to a range of restrictions, from elevated ambient temperatures that lower germination rates to poor nutrient accessibility and limiting annual crop biomass output. These lands must be carefully maintained since many of these categories are found in tropical areas, where production is frequently predominately low input, with big populations and greater growth rates than elsewhere. If these lands are not managed correctly, they are susceptible to deterioration. In the earth's surface, > 9% is classified as Class VII land, which includes shallow soils like Entisols and Inceptisols, salty soils like Aridisols, and soils with a lot of organic matter (such as Histosols and Spodosols). Saline soils can be used for specific adapted crops and planting techniques, while shallow soils are typically thought to be unsuitable for crops. Peatlands with prominent levels of organic matter are more prone to deterioration and long-term loss from drainage. Almost 17% of the land is classified as Class VIII land. They are typically regarded as unfavorable for agriculture since they have minimal temperatures (like Gelisols) and/or vertical slopes. More than 28% of the land is classified as Category IX, and the amount of soil moisture is insufficient to enable the cultivation of annual crops (e.g., dry soil, quicksand, rocky soil, etc.) (Blum 2008).

Around 12% of the global land surface is suitable to produce foodstuff and fiber, 24% is available for grazing, 31% is productive forest, and 33% is unsuitable for any sustainable usage, primarily due to climatical and topographic restrictions, according to the Food and Agriculture Organization Farsund (2015). Overall, the 12% of world land and land surfaces, along with soil quality classes I to III, where around 25% of the world's population lives and all food and fiber sold for global markets are produced, are the primary sources of food security (Buringh 1985).

## 3.7 SAM PRACTICES PROVIDE GOODS AND SERVICES

SAM (sustainable agricultural management) practices involve long-term biomass provision, food, feed, and renewable energy, as well as provide fresh and clean water in the form of groundwater and biodiversity preservation. Land management also should control trace gas exchanges between soil and atmosphere that are relevant to climate change, safeguard human health from pollution in the food chain, and preserve free water resources like lakes and rivers from pollution. SAM practices are undoubtedly a challenging project because there are so many competing interests. For instance, agricultural production through the application of excessive fertilizer and plant protection compounds on top of the soil, while the protection of underground groundwater is in direct competition with one another because every raindrop that falls on the soil must first pass through the soil earlier than it turns into groundwater. Additionally, the kinds and quantities of trace gases released by soils, such as $(CO_2)$, $(CH_4)$, and $(N_2O)$ are wedged by the physical and chemical effects of agricultural technology and fertilizer application on soils, which also directly contribute to climate change. The quantity and quality of groundwater resources, another essential requirement for human being life that is partially addressed in SDG no. 6, are, therefore, impacted by the management of agricultural soils. In this way, SDGs 2 and 6 are connected through the management of land and soil.

Land management includes the production of biomass, which serves as the foundation for producing bioenergy, such as ethanol from carbohydrates like grains, diesel from vegetable oils like rapeseed and palm oil, and biogas from the fermentation of organic waste. To a lesser extent, land and soil management, food, or solid fuels (such as wood and straw) are also related to SDG no. 7. The sustainability of biomass production will also assure the sustainability of clean underground water and surface water sources (SDG 6); hence, we will exclusively examine sustainable native land and soil management for biomass production in the sections that follow. Additionally, sustainable forestry and agriculture may produce renewable energy for humans (SDG 7) (Zhang 2021).

## 3.8   RISK ASSESSMENT OF SOIL FOR GLOBAL FOOD SATISFACTION

Although soils adapt to a variety of environmental factors, there are serious barriers that global climate change (GCC) is endangering the inherent functioning of soils (Bonan and Doney 2018). Often, these variations are brought about by or affected by humans through anthropogenic activities (Figure 3.3, SDG no. 15). GCC is significantly affected by agricultural operations, and a variety of potential challenges to the ability of soils to carry out their functions have been highlighted by current national and international approaches to soil conservation. For instance, the CeuC drew several risks to sustainable soil use in its explanation for the European Soil Conservation Strategy, including (a) soil surface sealing, (b) erosion, (c) sedimentation (compaction), (d) loss of organic matter, (e) loss of biodiversity, (f) pollution, (g) salinization, and (h) floods and landslides. Although the CEuC commission has focused on these risks concerning Europe, these threats exist worldwide, and the proportional importance of each depends on the location and the usage of land. These dangers are partially brought on by the fact that soils are required to carry out a variety of management strategies, often many at once according to the damage (Crist 2017).

We frequently produce an unbalanced system, as soils are less resilient and additionally vulnerable to these risks because of increased demands on soils for production activities. As a result, soils' capacity to conduct their typical tasks frequently suffers greatly (Pretty 2008). In some circumstances, too much emphasis on the execution of one function may endanger the capacity to manage other soil functions. These risks are increasingly seen to be especially related to the practices that soils produce biomass, involving the provision of biofuels and the production of freshwater resources.

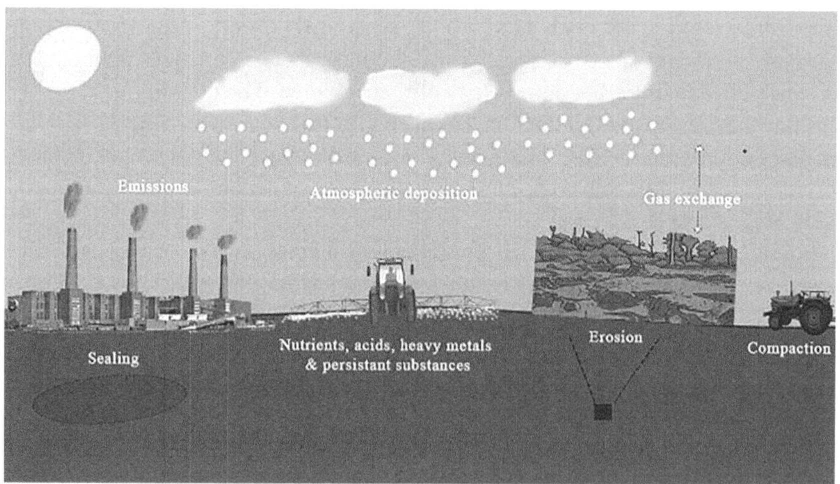


**FIGURE 3.3** Human activities that degrade soil quality.

### 3.8.1 Urbanization, Industrialization, and Transportation Resulted in Soil Sealing

Soil sealing refers to the process of covering soil surfaces to build infrastructure for modern life, homes, roads, or other properties. Most of the functions of soil, such as collecting precipitation to penetrate and filter pollutants, are hindered when land is covered. Additionally, sealed spaces change the patterns of surface water flow, which causes floods and more biodiversity fragmentation (Scalenghe 2009). Governments and environmental authorities are worried about this persistent soil loss and the resulting loss of ecosystem function due to soil sealing, which is irreversible. The real amount of global soil sealing could be checked by viewing the land from space at nighttime, which shows that most of the prime earth, according to the wide distribution of cropland, is related to areas with high-level urbanization (Pielke 2005). Elevated stages of urbanization may be seen in America, Europe, and Japan, while these places are related to agricultural productive lands (Class I–III). This feature is increasingly visible as China and the Indian subcontinent become more urbanized. Our ancestors selected the best soils to settle on to guarantee that fundamental necessities like food, fiber, and water were satisfied, which is highly prone to soil sealing. These early communities have developed into huge metropolitan areas over time. Therefore, any more urban expansion will result in the closure of more desirable places. The second effect of the seal is a change in lifestyle from rural to urban, which results in less biomass production and compromises food security. According to estimates, the current daily soil loss in developed countries (total surface area 4,324,782 km$^2$) caused by urbanization, industrialization, and transportation is about 15–20 km$^2$/day. Around 250–300 km$^2$/day of soil is thought to be lost on average worldwide. Therefore, the greatest danger to sustainable agriculture and forest management is sealing (Prokop 2011).

### 3.8.2 Soil Damages in the Context of Land Excavation

There is not any solid stone accessible for the construction of homes, factories, or other types of infrastructures in frequent sedimentary locations or vast alluvial plains, including in developing countries and various other parts of the world. To produce bricks, tiles, and other construction materials, agricultural topsoil with greater clay and silt content and lacking carbonates is dug up to a depth of 3–5 m, sometimes even beyond. Most of the time, they are created close to where the movable furnace building was excavated. The excavation holes are often several hectares in size, thus forming ponds or lakes. There are typically no additional soil resources accessible for agricultural cultivation in the same location, although excavation is irreversible. Food shortages are likely to happen or to arise in the coming days in these areas since they are among the most inhabited and increasing at one of the quickest rates in the world.

## 3.9 IMPROPER SOIL MANAGEMENT LEADS TO SOIL DEGRADATION

In addition to soil sealing and excavation, include other types of soil degradation, such as erosion, compaction, loss of biodiversity, pollution, salinization, loss of organic matter, desertification, nutrient extraction, floods, and landslides (Stolte 2016).

### 3.9.1 Erosion

Steep slopes, climate (severe weather), inappropriate land use, inadequate vegetation covering, and occasionally, tectonic activity or ecological calamities like deforestation all contribute to erosion (Gibbs and Salmon 2015). Erosion is a natural geographical occurrence that arises due to the combination of these conditions. Silty texture and a lack of organic matter are two inherent qualities of soils that render them prone to erosion. Although erosion is a natural procedure, some human actions, particularly the inappropriate use of agricultural land, can cause erosion rates to dramatically rise. Loss of topsoil function, along with severe circumstances, and the loss of the entire topsoil are consequences of soil erosion. The ability of the soil to support elevated levels of biomass production can be adversely affected by even a few millimeters (mm) of topsoil erosion. There is an emphasis on a strong connection between topsoil loss (erosion) and potential effects on the world's carbon budget. Additionally, as dirt is moved into streams, it has negative effects on the ecosystem through downstream sedimentation, the addition of contaminants, and an increase in water turbidity (Verheijen et al. 2009).

### 3.9.2 Compaction

When soil is physically pressured by using heavyweight machinery or overgrazing, particularly under humid soil conditions, soil compaction develops in agriculture (Horn 2011). It has been seen as an issue in agricultural systems for a long time. Deteriorations in soil organic matter levels have been noticed, in addition to the requirement to prevent compaction by reducing traffic flow when soils are moist and more susceptible to damage. These were increasing the susceptibility of compaction and are the possible promoting causes. Skiing and hiking can also be problematic in sensitive locations, and compaction occurs often on building sites. The amount of pore space between soil particles is reduced during compaction, and as a result, the soil loses some or all its capacity to soak up water. Reversing compaction in deeper soil layers is the most complicated (Pagliai 2000). Compaction causes the soil's general structure to deteriorate, which in turn affects fertility and biomass production and restricts root development, water storage capacity, biological activity, and stability. Additionally, when it rains heavily, water cannot effortlessly permeate the soil which can result in surface water runoff, erosion, and even floods.

### 3.9.3 Decline in Organic Matter

The three components of organic matter (OM) include live creatures (bacteria, fungus, earthworms, and other soil biotas), humus, and organic matter (plant root remnants, leaves, and other potted plant matter, dead animals, and feces). Organic matter is continually introduced to the soil and further breaks down in natural processes. Plant nutrients are produced and promptly absorbed by plants' roots during this breakdown process, resulting in a closed active nutrient cycle. In addition, carbon is freed as $CO_2$ into the atmosphere and then is reabsorbed during photosynthesis. Key soil processes are maintained by organic matter, which also significantly affects soil fertility and erosion resistance (Schuman 2002). It helps to agglomerate soil particles and increase the soil's ability to function as a physical and chemical buffer, which reduces the amount of pollution that is transferred from the soil to the water. The influence of agroforestry methods on soil organic matter is well acknowledged. Although the significance of sustaining soil organic matter content is acknowledged, there is proof that over the past 50 years, especially in temperate regions, the shift to greater specialization and cereal monoculture has frequently prevented or completely replaced soil organic matter loss due to natural decomposition (Schulp 2008).

Crop rotation methods, which used to be crucial for sustaining soil organic matter content, are frequently no longer a characteristic of agricultural systems because of the partition of livestock from the production of arable land brought about by agricultural specialization (Beinroth 2001). Concerns are also raised about the possibility of additional alterations in soil organic matter levels because of current climate change. The reduction in the organic matter has been highly reported because of inadequate management; the formation of organic matter into the soil is a slow procedure. However, by using land management strategies like conservation tillage, no-till cropping, organic farming, permanent grassland, cover crops, manuring and mulching with mungbeans, farmyard fertilizer, and compost application, these OM decreases can be stopped or even reversed. Numerous writers have highlighted that the use of conservation tillage techniques results in considerable carbon accumulation in soils. In the main highlands of Mexico, scientists evaluated the soil's organic carbon content after 16 years of conventional and zero tillage and discovered noticeably greater amounts of organic matter in zero tillage locations. They also observed that the amount of soil carbon was significantly affected by the preservation of organic residues at the location. Research in Spain noted that under low or no-till systems, there is an improvement in soil carbon content and a rehabilitation of soil structural solidity. Additionally, they mentioned how to cultivate crops, crop rotation, and the quantity and quality of crop remains left on the soil surface all have an impact on soil carbon concentration. These methods have been proven to improve soil biodiversity, prevent erosion, and increase fertility in addition to tackling difficulties with soil organic matter (Ismaili 2015).

The provision of nutrients for crop production by soil organic matter is Irucial in numerous smallholder farming systems in the tropics and subtropics. Previously, the traditional farming methods sometimes include fallow seasons, during which no crops are grown, as well as soil amendments made using crop and plant leftovers and fertilizers, which raise or at the very least sustain levels of organic matter. Land pressure increases due to population development. After the 20th century, land loss through sealing and erosion frequently shortens fallow periods and lowers levels of organic recycling. Erosion is also frequently caused by declining soil organic matter levels, reducing ability to produce biomass, and increasing susceptibility and compaction. The global carbon cycle also heavily depends on soil organic carbon. Recent research has improved understanding of the role of land use management in upholding soil carbon stocks and practices that may increase carbon sequestration in soil pools and has highlighted the significance of soil organic carbon pools in the context of global carbon fluxes (Laganiere 2010).

### 3.9.4 Diminution in Soil Microbiome

The world's biggest genetic reserve is found in soil, which also serves as a habitat for a diverse range of creatures. The characteristics of soils and soil organisms vary greatly because of the interaction of soil-establishing elements such as geology, climate, plant cover, and topography (Wall 2005). Our understanding of the relative abundance of soil biota, particularly the nature and role of microorganisms, is quite restricted. The physical, chemical, and particularly biological characteristics necessary for soil productivity are maintained in large part by the bacteria, fungus, protozoa, and other microorganisms that live in the soil (Brussaard 2012). The size of organic leftovers is reduced by larger creatures like worms, snails, and tiny arthropods before they are moved to deeper strata of more stable soil and further destroyed by microbes. Additionally, soil organisms themselves can serve as nutrient reservoirs, suppress outside diseases, and degrade organic contaminants into less dangerous subcomponents (Brussaard 2012). Complicated interactions and interdependence exist between species. Because of this interrelation, the disappearance of one species can have a domino effect that results in the extinction of other species and related services. Reduced soil biodiversity makes soils extra susceptible to other processes that cause degradation and frequently reduces their capacity to conduct numerous ecological services (Hunter 1999). Because soil biodiversity and aboveground biodiversity are intimately linked, both have an impact on several soils and additional environmental functions which are frequently employed as indicators of soil health or the condition of larger environmental systems (Isbell et al. 2017).

### 3.9.5 Soil Contamination

In addition to contaminated groundwater and surface water reserves, the introduction of pollutants into the soil also causes the impairment or loss of one or more of the soil's functions. If contaminants are present in soil over a specific level, the food chain may be negatively impacted on many distinct levels, including human health, various ecosystems, and other natural resources. It is crucial to consider not only the concentrations of soil contaminants but also their environmental behavior and exposure pathways to human and ecosystem well-being that determine their potential effects. It is common to distinguish between soil contamination from diffuse sources and soil pollution from clearly defined sources (also known as localized or point-source pollution) (Baran 2008). Mining, industrial sites, landfills, and other facilities are frequently linked to localized (or point-source) pollution, either while they are in use or after they have been shut down. Water and soil may be in danger because of these actions. The handling or removal of tailings, acid mine drainage, and the employment of some chemical agents all carry dangers in the mining industry. Another significant local pollution source might be industrial facilities. Contamination can sometimes last for years. Diffuse pollution is frequently linked to agricultural activities, air deposition, and insufficient wastewater and waste recycling and treatment. There are several causes of diffuse air pollution, including emissions from industry, transportation, and agriculture. Acidifying pollutants (such as $sO_2$ and NOx), metals (such as cadmium, lead, arsenic, and mercury), and a variety of organic substances (such as dioxins, PCBs, and PAHs) are released into the soil because of air pollution deposition.

There is a relationship between soil and the environment. Soil also has a strong relationship with the hydrosphere. There is a need to be concerned about the consequences of dispersed contamination, which may result from the use of agrochemicals on water resources. Because the connections between soil and the underlying hydrological systems are frequently complex, it may take some time before the effects of this pollution are seen in the quality of the groundwater. Therefore, while attempting to manage or stop additional pollution and repair damaged soil and water resources, it is vital to consider the nature of these links. There are two ways that soil pollution might affect food production. The first immediate result is a decrease in soil production due to pollution's growth-inhibiting effects. The possibility that food produced for human consumption is unsuitable owing to contaminant absorption during growth is a second influence on food production. Without proper planning and management, the increases in urbanization and industrialization might lead to soil degradation and negatively affect the world's food security (Van Beek 2012).

### 3.9.6 Soil Salinization

Salinization, which causes a drastic decrease in crop yields, is the build-up of soluble salts in the soil, primarily sodium, magnesium, and calcium. This process is frequently linked to improper irrigation water application because irrigation water often contains salt in varying amounts. This is especially true in regions with low precipitation, high evapotranspiration rates, or soil textures (> clay contents) that prevent the salt from being washed away. The salt is then trapped in the soil to form a surface layer. In coastal places, groundwater overuse may also contribute to salinization by lowering groundwater levels, which can lead to saltwater intrusion (Arora 2017). Marginal areas are frequently employed for agricultural production due to the always rising need for food from a populating world and dietary changes brought on by rising prosperity (high wealth). Irrigation is frequently the most straightforward way to utilize the land for food production when a lack of water (good quality) storage is the root problem. Soil salinization problem will be high in the future due to the salty water sources utilized for irrigation and the soils irrigated in the arid and semiarid climatic and ecological conditions, particularly considering the rapid depletion of freshwater resources globally and the real expansion of irrigated land on a global scale to meet the ever-increasing demands of food of a rising world population. Due to the lack of high-quality water supplies and the contradictory needs of human consumption, the only resource accessible for irrigating crops will be lower-quality, frequently brackish water (Bernstein 2019).

### 3.9.7 Soil Desertification

A complicated method of land degradation brought on by both natural and human-caused factors is called desertification (e.g., because of the environmental effects of global climate change) (Sheikh 2006). Increased drought and excessive use of natural sources, particularly plant cover, graze, or firewood collection, are its first symptoms. These are followed by soil deterioration and soil loss, including salinization. However, they are always a significant component of agricultural production regions and are frequently utilized as pastures, even by wanderers. Desertification-prone lands are typically not employed as farmland for food production. Desertification's loss of these places puts more pressure on the still-productive lands and soils utilized for food production and may even affect societal instability. 3,592 billion hectares of land are thought to be threatened by desertification, reported in the last decade. According to current estimation, "desertification stress zones" impact a total of around 4.23 billion hectares of land, of which 1.17 billion hectares are found in regions with a high resident density (41 people/km$^2$), around 24% of the earth's land area, and are affected by land degradation. To address the challenges of desertification, offering another sustainable method for using drylands worldwide is an emerging well-heated discussion (Bai et al. 2008).

### 3.9.8   SOIL NUTRIENT MINING

Soil nutrient extraction is undoubtedly one of the biggest risks to food production in most of the tropics, even though it has not been recognized in Europe or any other newly formed soil conservation strategy. Nutrient extraction threatens the ability of most of Africa's agriculture to produce food (De Jager 2001). A substantial portion of the African continent's agricultural production is dependent on delicate ecosystems, naturally poor soil richness, and little usage of contemporary inputs such as inorganic fertilizers, plant protection agents, and better crop types. The soil is left uncultivated to allow for "healing" under the fallow system, which has been used traditionally in Africa, particularly in sub-Saharan Africa. In several nations, the removal of indigenous communities from certain portions of the land because of land seizing has led to shorter and, in other cases, longer fallow periods. Rising populations are placing greater demand on the land, the period being cut shorter. This pattern frequently results in the reduction or "mining" of soil nutrient reserves, especially when fertilization is absent or extremely low. To sustain overall levels of food production when nutrient levels drop, soil productivity declines (Henao and Baanante 2006). Marginal lands, which frequently have lower intrinsic fertility and other crop production restrictions, are sought out at the expense of ecologically significant fields, reduced food security, reduced crop yields, and decreased per capita food production because of nutrient extraction from the soil in Africa may be major medium- to long-term causes of widespread land degradation. Estimates of the size and scope of nutrient extraction have been made using nutrient balances that account for system inputs and outputs. Fertilizers, organic waste, manure, nitrogen losses, phosphorus fixation, and sediments are examples of inputs. Crop consumption and uptake, erosion, leaching, and volatilization all are methods of the output of the nutrient cycle. This deterioration process is likely to continue unless regulations are put in place to restrict this "mining". For African farmers to adopt a component of forwarding planning and bring considerable progress to their goods, support for agricultural inputs and related conservation measures, as well as improvements to the competitive environment, are required. The introduction of farmers' land tenure rights for the land they cultivate is a significant development in some areas. There is no motivation for sustainable and long-term soil management in the absence of land tenure rights. Nutrient extraction will be encouraged via short-term management.

### 3.9.9 Soil Floods and Landslides

Although their impacts are sometimes worsened by unexpected weather circumstances, floods and landslides are natural hazards that are strongly tied to soil and land management activities. Landslides mostly affect regional food production, while they may briefly impair communication networks and impede food supply. Flooding can occur locally, impacting a few hectares, or nationally, covering hundreds of square kilometers, in severe circumstances. Flooding may cause soil erosion, which can destroy soil, destroy seeds, and in severe situations, destroy crops and contaminate them with debris. Along with harm to the soil and the ecosystem, there are frequently serious repercussions on human activities and lives, including the destruction of buildings and infrastructure and the loss of agricultural land. Other than the dangers already mentioned, floods and landslides constitute a hazard to the soil. However, in rare instances, compaction or blockage of the soil may prevent it from playing its normal duty in regulating the water cycle, contributing to floods. Flooding and landslides, which are frequently brought on by deforestation, abandoned land, and climatic conditions, result in the erosion of the soil's fertile layer.

## 3.10 FOOD SAFETY AND SECURITY VS GLOBAL CLIMATIC CHANGES

### 3.10.1 An Introduction

Global processes that threaten food and water security include pollution, salinization, desertification, nutrient extraction, floods, and landslides. The erosion, compaction, and loss of soil organic matter that occur because of local or regional processes, including sealing and excavation, as well as soil loss and degradation, pose threats to soils. These include issues like climate change, population expansion, movement from rural to urban regions, and modifications to eating and lifestyle patterns, all of which enhance the demand for food and space while simultaneously negatively affecting the capacity to provide it. These things are intimately linked to the products and services of the soil. Rising energy demand and consumption, as well as the loss of natural resources due to increasing food and fiber production, have other consequences on climate change that worsen the problems already associated with the use of fossil fuels. New economic practices for food production and marketing, such as land gripping, protecting, and the use of goods in the food industry, are also growing alongside the current economic issues. Under GCC, these global changes and processes will be examined.

## 3.10.2 GLOBAL CHANGES

The worldwide population distribution is quite unequal, and the global population is increasing by around 80 million people a year (Dyson 2015). The number of undernourished persons worldwide has grown by almost 9% since 1990, despite a more than 12% increase in food production (Swinnen 2012). The number of people facing food insecurity grew from 854 million in 2007 to around 1.02 billion in 2009, owing to changes in food prices and market circumstances (see also FAO 2011; IFPRI 2008 findings). According to FAO figures, a decade ago, there were around 926 million people who were food insecure FAO statistics (2011). All these statistics only account for people who are undernourished in terms of protein and calories; they do not account for those who are undernourished in terms of iron (about 2 billion), iodine (roughly 750 million), or any other nutritional shortages (WHO 2000). From approximately 2.5 billion in 1950 to approximately 6 billion in 2000, the world's food production increased by nearly 250%, maintaining population, largely due to "the success of the green revolution" and the input of significant amounts of fossil energy in the shape of fertilizers and plant defensive compounds (Godfray et al. 2010). But as the 20th century ended, population expansion started to exceed production increase. As a result, grain output per person decreased significantly throughout this time. Furthermore, it was noted that the lack of fertilizers, plant protection products, and agricultural tools in some areas (such as Africa) contributed to the green revolution's limited impact. Between 1990 and 2030, the world's need for food is estimated to increase three- to fivefold. In addition, because of population expansion, increased soil loss from urbanization, and land and soil deprivation, the per capita land area for food production—presently around 0.27 hectares—continues to decline (Armar-Klemesu 2000). Whereas 97% of worldwide food insecurity is concentrated, Africa and Asia are the primary focus of efforts to reduce poverty (Hejazi et al. 2014). Food shortages were made worse by price hikes in 2008, which left a lot of people in Africa and Asia without the financial means to buy food. The global climatic change is predicted by the Intergovernmental Panel on Climate Change (IPCC), and the increase in cereal yields is estimated to have slowed over the majority of the second half of the 20th century (Bai et al. 2008).

The production of cereals might decrease by a further 20%–40% because of climate change, mainly in Africa and Asia. By 2050, there will be an additional 2.3 billion people, which means that to end the poverty crisis and feed the growing population, grain production must expand by 70% and productivity in emerging nations must increase by at least double. Over the previous 50 years, food production has increased by more than 200%, yet the amount of land used for farming has only expanded by less than 10% (Godfray et al. 2010). We require 70%–100% more food by 2050 to feed an anticipated 9 billion people worldwide (Chartres 2015). Will there be another acreage for the farming of food? This is becoming less and less possible due to competition for land from other purposes, particularly the production of biofuels. It seems doubtful that the amount of land utilized for agriculture would significantly rise given the recent knowledge of the need to safeguard biodiversity and preserve the products and essential services offered by natural ecosystems. In addition to the growth in global population, the migration of 8 million to 100 million people annually from rural to urban regions is accelerating urbanization and causing more productive lands to be closed or degraded, resulting in many nearby slum neighborhoods. Furthermore, the use of grains in the livestock sector is putting food security at risk due to the rising demand for animal products throughout the world, particularly meat and protein in the shape of eggs and milk. Approximately 2–3 kilograms of grain are needed to produce 1 kg of chicken, 4–6 kg of food grain for 1 kg of pig, and 7–10 kg of cereal for 1 kg of beef (McClelland et al. 2018). Because they are utilized as animal feed instead of considering extra environmental implications, substantial volumes of plant-based foods are lost to human consumption.

### 3.10.3 Food Safety, Security, and Economics in the Context of Global Climate Change (GCC)

All biological procedures in terrestrial and aquatic ecosystems are being impacted by global climate change, which poses various threats to the food supply. In turn, more common extreme weather events impact soil loss and degradation by increasing water and wind erosion, floods and landslides, desertification, and salinization. Extreme shifts between dry and wet circumstances are stressful to plant development, especially during prolonged droughts. Additionally, there is a rising shortage of clean water for irrigation and increased rivalry for water for the household, industrial, and agricultural usage as well as the generation of biofuels. Additionally, the production of food and biofuels accelerates urbanization, industrialization, transportation, and deforestation for agricultural land, which reduces soil carbon absorption and exacerbates global climate change. Methane ($CH_4$) and nitrous oxide ($N_2O$) emissions rise because of intensive agricultural production of food and biofuels, and these gases are significantly more responsible for climate change than soil and animal $CO_2$. On the other hand, warmer weather, particularly in the Americas, Asia, and Europe where freshwater resources are plentiful, may partially make up for declines in food production in water-scarce areas and places with high rates of soil erosion and deprivation.

When the global economic crisis started in the first few years of the 21st century, new economic models in food production and marketing either evolved or were accomplished (Wahl 2009). For the first time, food was the focus of extensive speculative activity through derivative trading in international and local food markets, leading to irrational price volatility and issues with the availability of food. Land grasping the practice of purchasing or leasing vast parcels of land, particularly in developing nations, to produce food or biofuels has also considerably risen. Derivative trading is a type of economic speculating on agricultural goods, especially food, as explained for Africa, whereas land grabs may fundamentally alter a country's rural and agricultural environment. Additionally, it is anticipated that immediate issues with soil fertility would occur, particularly when food and biofuel production is increased to maximize earnings from property that has been bought or leased. Land gripping is a blatant indication that several industrialized nations and multinational corporations have realized that the amount of available land cannot be increased and that there is a region sustaining food and biofuel due to the constant loss of productive land from their environments and the increasing demand for food and energy at both the local and worldwide levels (Rulli 2014).

### 3.10.4 BIOFUEL: A BRIEF INTRODUCTION

The production of liquid biofuels (like biodiesel or ethanol), solid biofuels (like straw and wood), and gaseous biofuels (like biogas) have become economically interesting, particularly in the places where economic subsidies are given, considering the use of biofuels' contribution to mitigating global climate change is considered. Natural vegetation, such as forests, is destroyed in many regions of the world, particularly in Asia and Latin America, to obtain access to new surfaces to produce biofuels (mostly sugarcane and oil). In the medium and long run, biofuel production competes with food production and, consequently, food security, owing to the ever-decreasing area of food production caused by urbanization, industrialization, transportation, as well as soil erosion and other types of soil loss and degradation (Stewart 2009). Additionally, there are obvious warnings that the production of agricultural biofuels is harming soil quality since organic matter is totally extracted and then restored to the soil to preserve biodiversity and the soil's organic matter status. There is even more competition for water. Already, there is not enough water available in many regions of the world to produce enough food (Gerbens-Leenes et al. 2009).

## 3.11 FINDINGS AND WAY FORWARD

Nearly a billion people suffer from hunger today, primarily due to the inability to obtain food, despite the significant rise in food production in the second half of the 20th century, primarily due to the onset of the green revolution in the 1960s and other optimistic developments, particularly the use of fertilizers and plant protection products. A total of 4 billion individuals are estimated to be malnourished because of other vitamin deficiencies. Sustainable production and the accessibility of food are also in increasing danger from the effects of human activity, notably changes in land use patterns on a local and global level. The greatest danger to food security comes from soil loss brought by urbanization, industrialization, and transportation, as well as water and wind erosion and more serious types of soil degradation, such as loss of organic matter, pollution, and soil biodiversity. Food security is threatened by hardness, salinization, nutrient extraction, deserts, and floods. Additionally, there are other dangers to food security besides the shrinking agriculturally productive area and the deterioration of the residual productive surfaces. The population of the globe is increasing by 80 million people a year, and 8 million to 100 million people move from rural to urban regions every year, leaving these communities food insecure and reliant on food markets. Food security is also directly threatened by climate change, mostly due to increased soil erosion and degradation brought on by extreme weather. Rainfed and irrigated agriculture is being threatened in many areas by dwindling freshwater supplies. At the same time, there are severe conflicts for land, energy, and water resources to produce biofuels, and there is also a rising requirement for food and fiber in both the domestic and global food markets. High price volatility in the food market may be a result of new economic theories and speculative activity in the food sector (such as trading in derivatives). Land seizing (compaction and sealing processes), mostly in emerging nations in Africa, Asia, and Latin America, has resulted from the loss of agriculturally productive land, endangering the quality of the land and the soil. All of this demonstrates that we have a long way to go before attaining SDGs 1 and 2, which call for eradicating hunger, attaining food security and enhanced nutrition, and indorsing sustainable agriculture, as well as SDGs 6 and 7, which call for ensuring that everyone has access to energy. One must foresee in the upcoming one to two decades, without innovative approaches to biomass production, particularly food production locally and globally while safeguarding groundwater resources from contamination and generating enough biomass to satisfy local and global energy demands, millions of people will be gravely threatened this year by food and water shortages, which will further worsen hunger, particularly in emerging nations.

## REFERENCES

Armar-Klemesu, M. (2000). Urban agriculture and food security, nutrition and health. *Growing Cities, Growing Food: Urban Agriculture on the Policy Agenda. A Reader on Urban Agriculture*, 99–117.

Arora, S., Singh, A. K., & Singh, Y. P. (Eds.). (2017). *Bioremediation of salt affected soils: An Indian perspective*. Springer.

Bai, Z. G., Dent, D. L., Olsson, L., & Schaepman, M. E. (2008). Proxy global assessment of land degradation. *Soil Use and Management*, *24*(3), 223–234.

Baillie, I. C. (2001). *Soil survey staff 1999, soil taxonomy: A basic system of soil classification for making and interpreting soil surveys, agricultural handbook 436* (p. 869). Natural Resources Conservation Service, USDA.

Baran, N., Lepiller, M., & Mouvet, C. (2008). Agricultural diffuse pollution in a chalk aquifer (Trois Fontaines, France): Influence of pesticide properties and hydrodynamic constraints. *Journal of Hydrology*, *358*(1–2), 56–69.

Barnosky, A. D., Hadly, E. A., Bascompte, J., Berlow, E. L., Brown, J. H., Fortelius, M., . . . & Smith, A. B. (2012). Approaching a state shift in Earth's biosphere. *Nature*, *486*(7401), 52–58.

Beinroth, F. H., Eswaran, H., & Reich, P. F. (2001). Global assessment of land quality. *Sustaining the Global Farm*, 569–574.

Beinroth, F. H., Eswaran, H., & Reich, P. F. (2019). Land quality and food security in Asia. In *Response to land degradation* (pp. 83–100). CRC Press.

Bernstein, N. (2019). Plants and salt: Plant response and adaptations to salinity. In *Model ecosystems in extreme environments* (pp. 101–112). Academic Press.

Blum, W. E. (2008). Characterisation of soil degradation risk: An overview. *Threats to Soil Quality in Europe*, *23438*, 5–10.

Blum, W. E., & Swaran, H. (2004). Soils for sustaining global food production. *Journal of Food Science*, *69*(2), crh37–crh42.

Bonan, G. B., & Doney, S. C. (2018). Climate, ecosystems, and planetary futures: The challenge to predict life in Earth system models. *Science*, *359*(6375), eaam8328.

Braimoh, A. K. (2002). Integrating indigenous knowledge and soil science to develop a national soil classification system for Nigeria. *Agriculture and Human Values*, *19*(1), 75–80.

Brussaard, L. (2012). Ecosystem services provided by the soil biota. *Soil Ecology and Ecosystem Services*, *1995*.

Buringh, P. (1985). The land resource for agriculture. *Philosophical Transactions of the Royal Society of London. B, Biological Sciences*, *310*(1144), 151–159.

Chartres, C. J., & Noble, A. (2015). Sustainable intensification: Overcoming land and water constraints on food production. *Food Security, 7*(2), 235–245.

Crist, E., Mora, C., & Engelman, R. (2017). The interaction of human population, food production, and biodiversity protection. *Science, 356*(6335), 260–264.

De Jager, A., Onduru, D., Van Wijk, M. S., Vlaming, J., & Gachini, G. N. (2001). Assessing sustainability of low-external-input farm management systems with the nutrient monitoring approach: A case study in Kenya. *Agricultural Systems, 69*(1–2), 99–118.

Doran, J. W., Jones, A. J., Arshad, M. A., & Gilley, J. E. (2018). Determinants of soil quality and health. In *Soil quality and soil erosion* (pp. 17–36). CRC Press.

Dyson, T. (2015). *World population and human capital in the twenty-first century*. London School of Economics.

European Commission, Directorate-General for Employment, European Institute for Social Security, & Agence pour le Développement et la Coordination des Relations Internationales. (2002). *MISSCEEC II: Mutual information system on social protection in the Central and Eastern European Countries: Bulgaria, Czech Republic, Estonia, Hungary, Latvia, Lithuania, Poland, Romania, Slovak Republic and Slovenia*. European Commission.

FAO. (2011). *The State of 2001 Food insecurity in the world*. https://www.fao.org/3/y1500e/y1500e.pdf.

FAOSTAT. (2011). *FAOSTAT database on agriculture*. http://faostat.fao.org/. Verified 3 January 2012.

Farsund, A. A., Daugbjerg, C., & Langhelle, O. (2015). Food security and trade: Reconciling discourses in the food and agriculture organization and the world trade organization. *Food Security, 7*(2), 383–391.

Gerbens-Leenes, P. W., Hoekstra, A. Y., & Van der Meer, T. H. (2009). The water footprint of energy from biomass: A quantitative assessment and consequences of an increasing share of bio-energy in energy supply. *Ecological Economics, 68*(4), 1052–1060.

Gibbs, H. K., & Salmon, J. M. (2015). Mapping the world's degraded lands. *Applied Geography, 57*, 12–21.

Godfray, H. C. J., Beddington, J. R., Crute, I. R., Haddad, L., Lawrence, D., Muir, J. F., . . . & Toulmin, C. (2010). Food security: The challenge of feeding 9 billion people. *Science, 327*(5967), 812–818.

Hejazi, M. I., Edmonds, J., Clarke, L., Kyle, P., Davies, E., Chaturvedi, V., Wise, M., Patel, P., Eom, J., & Calvin, K. (2014). Integrated assessment of global water scarcity over the 21st century under multiple climate change mitigation policies. *Hydrology and Earth System Sciences, 18*, 2859–2883. https://doi.org/10.5194/hess-18-2859-2014.

Henao, J., & Baanante, C. (2006). *Agricultural production and soil nutrient mining in Africa: Implications for resource conservation and policy development*. IFDC, An International Center for Soil Fertility and Agricultural Development. www.ifdc.org https://vtechworks.lib.vt.edu/server/api/core/bitstreams/b610d19d-c439-4475-9a1a-13347a993872/content.

Horn, R., & Peth, S. (2011). Mechanics of unsaturated soils for agricultural applications, chapter 3. In *Handbook of soil sciences* (2nd ed., p. 31). Taylor & Francis.

Hunter, M. L., & Hunter, M. L., Jr. (Eds.). (1999). *Maintaining biodiversity in forest ecosystems*. Cambridge University Press.

IFPRI. (2008). *Reaching sustainable food security for all by 2020*. http://www.ifpri.org/2020/books/actionlong.pdf

Isbell, F., Adler, P. R., Eisenhauer, N., Fornara, D., Kimmel, K., Kremen, C., Letourneau, D. K., Liebman, M., Polley, H. W., Quijas, S., & Scherer-Lorenzen, M. (2017). Benefits of increasing plant diversity in sustainable agroecosystems. *Journal of Ecology, 105*(4), 871–879.

Ismaili, K., Ismaili, M., & Ibijbijen, J. (2015). The use of 13C and 15N based isotopic techniques for assessing soil C and N changes under conservation agriculture. *European Journal of Agronomy*, *64*, 1–7.

Keesstra, S., Pereira, P., Novara, A., Brevik, E. C., Azorin-Molina, C., Parras-Alcántara, L., Jordán, A., & Cerdà, A. (2016). Effects of soil management techniques on soil water erosion in apricot orchards. *Science of the Total Environment*, *551*, 357–366.

Laganiere, J., Angers, D. A., & Pare, D. (2010). Carbon accumulation in agricultural soils after afforestation: A meta-analysis. *Global Change Biology*, *16*(1), 439–453.

Lal, R. (2002). Soil carbon dynamics in cropland and rangeland. *Environmental Pollution*, *116*(3), 353–362.

Lehmann, A., & Stahr, K. (2010). The potential of soil functions and planner-oriented soil evaluation to achieve sustainable land use. *Journal of Soils and Sediments*, *10*(6), 1092–1102.

Liu, K. L., Han, T. F., Huang, J., Zhang, S. Q., Gao, H. J., Zhang, L., . . . & Zhang, H. M. (2020). Change of soil productivity in three different soils after long-term field fertilization treatments. *Journal of Integrative Agriculture*, *19*(3), 848–858.

McClelland, S. C., Arndt, C., Gordon, D. R., & Thoma, G. (2018). Type and number of environmental impact categories used in livestock life cycle assessment: A systematic review. *Livestock Science*, *209*, 39–45.

Mueller, L., Schindler, U., Mirschel, W., Shepherd, T. G., Ball, B. C., Helming, K., . . . & Wiggering, H. (2010). Assessing the productivity function of soils. A review. *Agronomy for Sustainable Development*, *30*(3), 601–614.

Oliver, M. A., & Gregory, P. J. (2015). Soil, food security and human health: A review. *European Journal of Soil Science*, *66*(2), 257–276.

Pagliai, M., Pellegrini, S., Vignozzi, N., Rousseva, S., & Grasselli, O. (2000). The quantification of the effect of subsoil compaction on soil porosity and related physical properties under conventional to reduced management practices. *Advances in Geo Ecology*, *32*, 305–313.

Pielke, R. A., Sr. (2005). Land use and climate change. *Science*, *310*(5754), 1625–1626.

Pretty, J. (2008). Agricultural sustainability: concepts, principles and evidence. *Philosophical Transactions of the Royal Society B: Biological Sciences*, *363*(1491), 447–465.

Prokop, G., Jobstmann, H., & Schönbauer, A. (2011). *Overview of best practices for limiting soil sealing or mitigating its effects in EU-27: Final report*. Technical Report 2011-050. European Commission. http://ec.europa.eu/environment/soil/pdf/sealing/Soil%20seal ing%20-%20Final%20Report.pdf.

Robert, F. C., Sisodia, G. S., & Gopalan, S. (2018). A critical review on the utilization of storage and demand response for the implementation of renewable energy microgrids. *Sustainable Cities and Society*, *40*, 735–745.

Rulli, M. C., & D'Odorico, P. (2014). Food appropriation through large scale land acquisitions. *Environmental Research Letters*, *9*(6), 064030.

Scalenghe, R., & Marsan, F. A. (2009). The anthropogenic sealing of soils in urban areas. *Landscape and Urban Planning*, *90*(1–2), 1–10.

Schulp, C. J., & Veldkamp, A. (2008). Long-term landscape-land use interactions as explaining factor for soil organic matter variability in Dutch agricultural landscapes. *Geoderma*, *146*(3–4), 457–465.

Schuman, G. E., Janzen, H. H., & Herrick, J. E. (2002). Soil carbon dynamics and potential carbon sequestration by rangelands. *Environmental Pollution*, *116*(3), 391–396.

Seto, K. C., & Ramankutty, N. (2016). Hidden linkages between urbanization and food systems. *Science*, *352*(6288), 943–945.

Sheikh, B. A., & Soomro, G. H. (2006). Desertification: causes, consequences and remedies. *Pakistan Journal of Agriculture, Agricultural Engineering, Veterinary Sciences*, *22*(1), 44–51.

Steffan, J. J., Brevik, E. C., Burgess, L. C., & Cerdà, A. (2018). The effect of soil on human health: an overview. *European Journal of Soil Science, 69*(1), 159–171.

Stewart, B. A. (2009). Economic balance: Competition between food production and biofuels expansion. In *Soil quality and biofuel production* (pp. 163–194). CRC Press.

Stolte, J., Tesfai, M., Oygarden, L., Kvaerno, S., Keizer, J., Verheijen, F., . . . & Hessel, R. (2016). *Soil threats in Europe: Status, methods, drivers and effects on ecosystem services: Deliverable 2.1 RECARE project.* European Commission DG Joint Research Centre.

Swinnen, J., & Squicciarini, P. (2012). Mixed messages on prices and food security. *Science, 335*(6067), 405–406.

Van Beek, C. L., & Tóth, G. (2012). Risk assessment methodologies of soil threats in Europe. *JRC Scientific and Policy Reports EUR, 24097.*

Verheijen, F. G., Jones, R. J., Rickson, R. J., & Smith, C. J. (2009). Tolerable versus actual soil erosion rates in Europe. *Earth-Science Reviews, 94*(1–4), 23–38.

Wahl, P. (2009). *Food speculation: The main factor of the price bubble in 2008.* Briefing Paper. World Economy, Ecology & Development, WEED – Weltwirtschaft, Ökologie & Entwicklung Eldenaer Straße 60 D-10247, Berlin.

Wall, D. H., Fitter, A. H., & Paul, E. A. (2005). Developing new perspectives from advances in soil biodiversity research. *Biological Diversity and Function in Soils,* 3–27.

World Health Organization. (2000). *Nutrition for health and development: A global agenda for combating malnutrition* (No. WHO/NHD/00.6). World Health Organization.

Zhang, B., Xu, J., Lin, Z., Lin, T., &Faaij, A. P. (2021). Spatially explicit analyses of sustainable agricultural residue potential for bioenergy in China under various soil and land management scenarios. *Renewable and Sustainable Energy Reviews, 137,* 110614.

# 4 Soil-Based Implication Approach for Environmental Nexus

*Sonal Sharma, Anuj Saraswat, Neha Khardia,*
*Rukoo Chawla, Shri Ram, Rajeew Kumar,*
*Md Basit Raza, Kumar Gaurav,*
*Biswaranjan Behera, Amit K. Dash, and*
*Abdel Rahman Al Tawaha*

## 4.1 INTRODUCTION

Sustainability concerns are associated with the interdependence between water, food, and energy because these needs deplete natural resources to ensure a secure future (Majhi et al. 2022). The ability of agricultural lands to produce more food at an unacceptably high cost to the environment has been put to the test by the increasing pressure to do so. Poor quality soils have steadily decreased yields on the lands, while environmental sensitivity to farming has increased (Saraswat et al. 2023). Discussions about the future needs for water, food, and energy to feed the world's growing population haven't taken into account the available soil resource, which is the most important part of our ability to grow food, retain water, and generate energy. Soils are critical in producing food and feed. The greatest threat to food security comes from soil losses brought on by urbanization, industrialization, and transportation (Behera et al. 2021). Although salinization, desertification, nutrient mining, flooding, pollution, organic matter depletion, compaction, soil biodiversity decline, and other severe types of soil degradation also pose a danger to food security. Further, soil degradation has been accelerated in regions where flood and drought situations are becoming increasingly prevalent due to climate and increased warming. Increasing food production by 70%–100% will be necessary by 2050 to fulfill the needs of the growing population (Das et al. 2022). The rising demand for food requires the development of soil management strategies. Soil lacks recognition for its crucial role in ecosystem functioning, including providing food, water, and energy. It seems as if we have forgotten the important role of soil and operating under the assumption that its capacity to supply ecosystem functions will always be there. For us to address the difficulties of the future that we will confront regarding food, water, and energy, we must understand the importance of our soil and how it influences functions in terms of the food, water, and energy nexus. Soils are strongly interconnected with provisioning, regulatory, and supporting services, as well as our

DOI: 10.1201/9781003358169-4

capacity to create and retain a viable future. Soils are the foundation upon which our future will be built. Increasing energy and water availability and ensuring food security will require the integration of soil into mainstream discourse.

## 4.2 SOIL AND SUSTAINABLE DEVELOPMENT GOALS

The Sustainable Development Goals (SDGs) replaced the Millennium Development Goals (MDGs) in September 2015 after the political agreement was reached on the definition of 17 goals to improve human lives and safeguard the environment on a global scale (UN 2016). The eradication of hunger, better education and health, sustainable cities, climate change mitigation, conserving terrestrial and aquatic ecosystems, and many other aims are included in the 169 targets that make up these 17 goals. These 17 Sustainable Development Goals (SDGs) represent two intertwined but distinct priorities: the earth's well-being and its inhabitants (Griggs et al. 2013). Among these SDGs, the SDG numbers 2, 6, and 7 mainly represent the soil, water, and energy, respectively. Objective 2 of the 17 SDGs is to "end hunger, achieve food security and better nutrition, and promote sustainable agriculture". This goal represents healthy soils and the natural ecosystems in which they are located. Goal 6 addresses water availability and sustainable management, while Goal 7 addresses the provision of energy (Blum 2016). SDG 2 aims to establish food security and a world free of hunger by 2030 by implementing long-term solutions. This goal ensures that everyone has access to nutritious foods to achieve a healthy lifestyle. It is very necessary, in order to accomplish this objective, to increase both the scope of sustainable agriculture and the availability of food. Soil health directly impacts SDG 2 (zero hunger). In a human time frame, soil is a non-renewable resource on which many ecosystem services rely. However, the 21st-century global concerns highlight the need for restoration and sustainable management of soil health. The food's nutritional content depends on the soil's ecological health and its management. Soil organic carbon (SOC) concentration is crucial for managing soil health (Lal 2016). Soil health is critical for ensuring the supply of diverse ecosystem services and environmental protection.

Sustainable water resource management and access to sanitized and safe water are critical for unleashing productivity and economic growth and providing a substantial return on investment for ongoing systems in education and health care. SDG 6, which wheels around ensuring access to water and sanitation for all, works in this regard. The natural environment, such as forests, soils, and wetlands, aids in the management and regulates water quantity and quality by bolstering watershed resilience and complementing investment opportunities in physical infrastructure and institutional and legislative arrangements for contingency planning, water use, and its access. In addition to potable water, sanitation, and hygiene, SDG 6 seeks to increase the sustainability and quality of water resources in order to benefit both individuals and the environment. SDG 7 assures universal access to reliable, economical, sustainable, and smart energy. The environment offers various sources of energy (renewable and non-renewable), including solar power, hydroelectric, geothermal, wind, natural gas, bioenergy, coal, petroleum, etc. An increase in fossil fuel consumption without accompanying measures to reduce greenhouse gas emissions will have far-reaching consequences for global climate. Saving money on energy costs and promoting the utilization of renewable energy aid to limit climate change and, as a result, the risk of natural catastrophes. Also, soils have direct as well as indirect effects on available energy. Burning peat yields around 0.02% of global energy supply while emitting approximately 0.7%–0.8% of carbon losses due to land use change and forestry (LUCF). Comprehensive rules and guidelines for management are needed to avoid peats, convert permanent land uses, and restore energy conversion waste to the soil (Smith et al. 2021).

## 4.3  SOIL AND ITS FUNCTIONS

Soil is a three-dimensional entity beneath the earth's surface. Weathering of rocks contributes to the formation of soil. It forms a part of many ecosystems, as it includes biotic and abiotic components that interact and influence each other to maintain the dynamics of life on earth. The biotic part includes earthworms, bacteria, algae, nematodes, fungi, actinomycetes, and many other living organisms, whereas the abiotic component encompasses mineral matter which forms the bulk of soil solids. Soil performs many functions: It acts as a reservoir for nutrients and water necessary for plant growth and development, provides space for aeration, gives physical support to roots, enables plants to stand erect, provides shelter to micro and macro flora and fauna, etc. In addition to nutrient cycling, buffering and filtering, water dynamics, supporting plant ecosystems and human structures, and fostering biodiversity and habitat are all biophysical soil functions. Figure 4.1 shows six key uses of soil and ecosystem services that benefit are beneficial to the people (Blum 2005; Commission of the European Communities 2006). Soil is the source of vital nutrients and the life support required for plants. Healthy soil is capable of reducing climate change by preserving or increasing its carbon content. It possesses the essential microbes required for nutrient cycling. Soils are inextricably tied to agricultural production. According to estimates, humans get 98.8% of their daily energy from soil (i.e., 2849 kcal per person) and about 1.2% from aquatic sources (i.e., 35 kcal per person) (FAOSTAT 2018). Figure 4.2 discusses the different perspectives on the importance of soil contributions to food security.

## 4.4  LAND QUALITY FOR SOIL MANAGEMENT AND USE

The capacity of land to perform a particular task is measured by its quality. Land use or management decisions at any scale, from a single farm family to a nation, require knowledge of the soil system and its relationships with the environment and management options. As with land use, these linkages are complicated, with feedback generating diverse metastable conditions and nonlinear, often delayed, responses to change. This suggests that some changes in external conditions may not result in a discernible change in the soil's condition. In contrast, others may result in a rapid or gradual decline or improvement. Changes in soil parameters are affected by how land is used and managed (Zhang et al. 2018). Soil degradation is a result of indiscriminate usage of land due to anthropogenic activities, removing natural vegetation. Converting natural vegetation to farming decreases soil nutrient and organic carbon content while increasing the content of sand and bulk density (Biro et al. 2013). Land

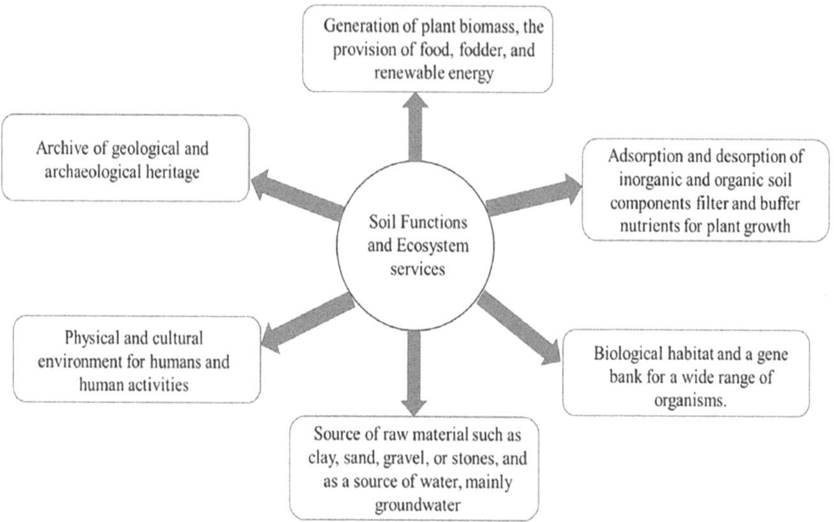


**FIGURE 4.1**   Functions of soil and its ecosystem services.

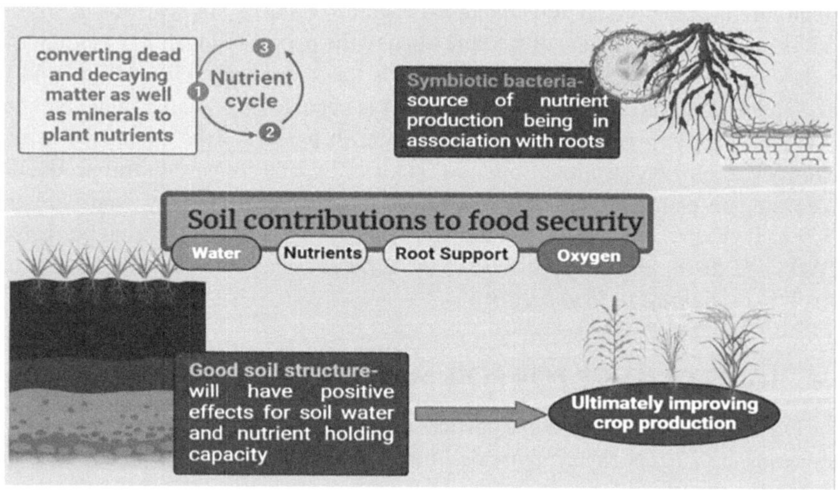


**FIGURE 4.2**   Different perspectives of role of soil contributions to food security.

quality must be monitored in order to assess the impact of long-term agronomic practices on soil characteristics, determine the economic advantages of alternate till-age and cover crop, and evaluate the efficacy of policies designed to improve soil quality. Land's economic and environmental value should be considered together when assessing its economic potential. For measuring the soil and land quality (soil

**TABLE 4.1**

**Classification of Land Quality and its Properties**

| Land quality classes | Properties |
| --- | --- |
| I | This is primary land. Few management-related constraints limit the soil's productivity. The temperature and moisture conditions of the soil are optimal for annual plants. Sustainable crop production generally entails less than 20% risk. |
| II and III | These soils have few problems associated with sustainable agriculture. Though their output is generally good, care must be taken to limit degradation. In general, 20%–40% of sustainable agriculture production is at risk, although the risk can be minimized through proper conservation practices. |
| IV, V, and VI | These soils are not suitable for grain crops without substantial conservation management efforts. A key limitation is the insufficiency of plant nutrients. The degradation of soil should be tracked regularly. In other words, productivity is low, and there is a 40%–60% chance of failure in growing grains sustainably. |
| VII | Crops with few inputs do not grow well on these soils. They are prone to degradation because of their low resilience. In either case, they should be left as forest or rangeland. 60%–80% of grain crop production is at risk of not being sustainable. |
| VIII and IX | These areas either have extremely fragile ecosystems or are unfavorable for agricultural usage. They should be kept in their natural condition. The risk for sustainable agricultural production is greater than 80%. |

*Source:* Eswaran et al. (1999), Beinroth et al. (2001), Kettler (2019)

and land are used synonymously) (Eswaran et al. 1999; Beinroth et al. 2001; Kettler 2019), two intrinsic soil parameters are defined: "soil resilience" as the capacity to reinstate to equilibrium after disturbance and "soil performance" as the capacity to produce biomass. This divided soils into nine land quality classes. Table 4.1 explains nine land quality classes and their properties.

## 4.5 DEGRADATION OF LAND AND SOIL

Land degradation is deterioration in soil quality and a decline in its capacity to be a multipurpose resource due to various natural and anthropogenic activities (Lal 1989). Land degradation reflects the unsustainability and low potential productivity of land use systems. It can no longer execute its environmental regulating responsibilities of storing, receiving, and recycling nutrients, water, and energy (Eswaran et al. 2001). The term "land degradation" is more specific than "soil degradation". This is a serious worldwide issue since it affects the environment and food production. The bare hillsides, water and wind eroded sites, cracked and parched plains, nutrient-deficient bleak land, and massive waste dumps are all signs of land degradation. Soil degradation is the loss of structural and functional properties of soil (Osman 2013). There are many types and causes of land and soil degradation, which are discussed next.

### 4.5.1 Types of Soil Degradation

The global population is projected to rise to as high as 9.5 billion by 2050, thus necessitating an increase in food production to cater to the rising food demand. Soil degradation has resulted in deterioration of soil quality and ecosystem services, especially in the tropics and subtropics. Globally, there is an increase in degraded soil and a decrease in arable land, which raises severe concerns for agricultural productivity and human food and nutritional security. The term "soil degradation" is associated with the decline in soil functions and ecosystem services. Four conceptual categories of soil deterioration exist: (a) physical, (b) chemical, (c) biological, and (d) ecological aspects (Figure 4.3) (Acosta et al. 2009). Physical degradation can be seen with disturbed soil structure, including the disturbance in pore space or bulk density that makes soil susceptible to crusting. Compaction, poor water infiltration, surface sealing, water erosion, wind erosion, high soil temperature variations, surface runoff, and chances for desertification are also increased (Lal 2015). Chemically degrading soil includes the soil under the process of acidification, alkalization, and salinization, as well as reduced cation exchange capacity (CEC), nutrient depletion, volatilization of $NH_4^+$-N, leaching loss $NO_3^-$-N or other nutrients, and eutrophication of water bodies due to heavy use of fertilizers or contamination by industrial waste (Edelstein and Ben-Hur 2018).

"Soil biological degradation" refers to alterations in soil biodiversity, loss of SOC pool, decrease in soil C sink ability, and increase in greenhouse gases emission ($CO_2$ and $CH_4$) from the soil into the atmosphere (Oertel et al. 2016). Ecological degradation is the outcome of the amalgamation of three forms of degradation: physical, chemical, and biological. Environmental management strategies generally result in

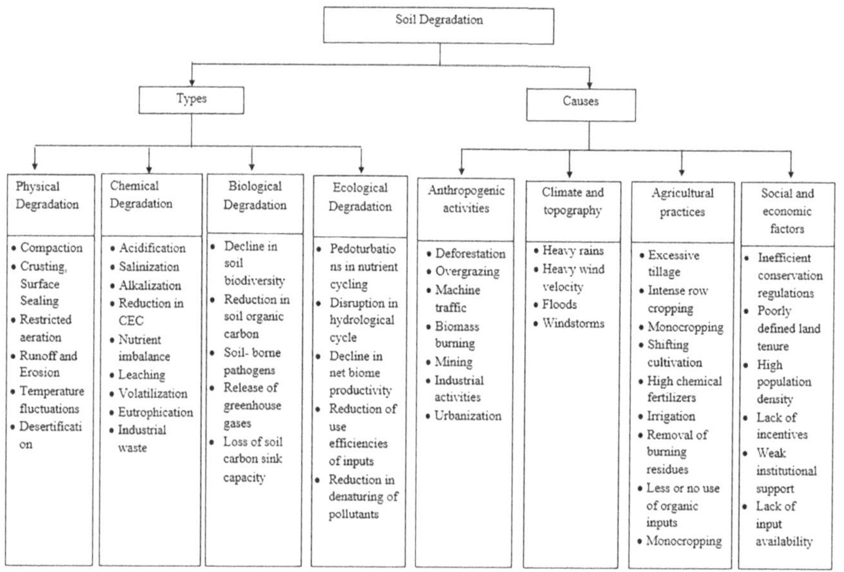


**FIGURE 4.3**    Types and causes of soil degradation.

ecological damage, involve disruption of ecosystem services like elemental cycling, disturb the hydrological cycle, create low use efficiency of input, and result in low net biome productivity (Barrios 2007).

### 4.5.2 Causes of Soil Degradation

Soil degradation is caused due to factors such as changing weather patterns, variable topography, changing land use patterns, cultivation practices, and various social, political, and economic factors, as shown in Figure 4.3 (Lal 2015; Ayub et al. 2020; Blanco-Canqui and Lal 2010). Deforestation, overgrazing, heavy load of farm machinery, biomass burning, mining, industrial activities, urbanization, and construction of roads are examples of anthropogenic activities that are involved in soil degradation. Climatic and topography factors such as heavy rains, windstorms or high-velocity winds, and floods cause wind and water erosion (Montgomery 2007). The soil deteriorates due to agricultural methods such as excessive plough, tillage, intense row cropping, monoculture, shifting cultivation, high chemical input, irrigation, and residue clearance. Social and economic factors contributing to soil erosion include inefficient conservation regulations, lack of incentives, ill-defined land tenure, high population density, poor institutional support, minimal income, and the lack of labour. Soil erosion is influenced by social and economic conditions like poorly defined land tenure, ineffective conservation policies, weak institutional support and lack of incentives, low income, high population density, and non-availability of input (Lal 2007).

## 4.6 EFFECT OF CLIMATE CHANGE ON WATER AND FOOD SECURITY

The most recent IPCC report reiterates the key findings of earlier IPCC reports about climate change and its key physical impacts, such as alterations in land and ocean temperatures, increasing sea levels, and acidification of the oceans (IPCC 2021). Extreme weather patterns exacerbate soil degradation by causing water and wind erosion, landslides, flood, salinization, and desertification (Haron and Dragovich 2010; Blum 2016). In many locations, climate change causes great uncertainty regarding future water availability. It will impact runoff, precipitation, hydrological systems, snowmelt, temperature, water quality, and groundwater recharge (FAO 2016). Sea level rise will increase the salinity of both groundwater and surface in coastal locations. Food security is one of the key issues caused by climate change. The effects of climate change on food security are complex. It influences agriculture, cattle, fisheries, forests, and aquaculture and may have serious economic and social implications such as damaged livelihoods, revenue losses, adverse health effects, and trade disruption. A new multi-model study utilizing the IPCC's greatest warming scenario indicated a minus 17% global effect on yields of four crop groupings (oil seeds, coarse grains, rice, and wheat accounting for around 70% of global crop harvested land) by 2050 compared to an unchanging climate scenario. Chakrabarty (2016) discussed the intricate effects of climatic changes on India's food security. Flavelle (2019) reported that climate change threatens the world's food supply, while many researchers investigated climate change's global and regional impacts on wheat and maize yield (FAO 2016).

Using fossil fuels increases the concentrations of greenhouse gases in the atmosphere, such as di-nitrous oxide ($N_2O$) and methane ($CH_4$), which significantly impacts climate change and reduces the nutritional quality of food. In contrast, food production is significantly impacted as temperature rises in areas with water stress and soil degradation.

## 4.7 SOIL DEGRADATION PROCESS: A BARRIER TO ATTAINING FOOD SECURITY

Both directly and indirectly, the population has impacted the health of the soil. The demands of people have disrupted nature (Dietz et al. 2004). The growing population has compelled urbanization. Urban areas are expected to occupy increasing land, rising from 213 million hectares in 2000 (1.6% of ice-free land) to 621 million hectares (4.7%) by 2040 (Van Vliet et al. 2017). Resurfacing of the soil has led to civilization which ultimately causes soil productivity to decline. Industrialization in the agriculture sector has enhanced the production of commercial need-based crop production to support the sustained provision of raw inputs to agro-industries. Monocropping has thus increased and is responsible for reducing soil organic matter and decreasing useful microorganisms (SARE 2019; Zhao et al. 2018; Meena et al. 2022). Soil pollution has been caused by the use of chemical pesticides, herbicides, insecticides, etc. (Miao et al. 2003). Soil type and its makeup, as well as the kind of pesticide, have a significant impact on pesticide residues in soil and their long-term persistence in the soil (Natural Resources Conservation Service 1998). Agriculture has focused predominantly on growing productivity, but the emphasis has changed to include achieving nutritional security over time. Around 17.3% of the population worldwide is in danger of undernutrition (Liu et al. 2018). More than two billion people suffer from micro nutrient deficiencies worldwide (McGuire 2015). In addition, over 41 million people are overweight, 155 million are stunted, and 52 million children under five are wasted (WHO 2022). Micro nutrient deficiency has been a major cause of malnutrition. Zn and Fe deficits are thought to impact around 60% and 30% of the world's population, respectively (WHO 2021; Yang et al. 2021). In order to mitigate the scenario mentioned earlier, soil management becomes an ideal solution. The intrinsic nutritional availability of soils and, consequently, their capacity to provide food with nourishment vary substantially. Analyzing regional and worldwide trends in the incidence of malnutrition reveals this. A variety of climatic factors for the production of agricultural commodities is governed by the inherent soil qualities and climate, which also influence worldwide patterns in crop production, the nutritional value of the food and feed produced, and the soil's overall impact.

## 4.8 SUSTAINABLE LAND MANAGEMENT: MEETING GLOBAL DEMANDS FOR GOODS AND SERVICES

Sustainable land management embraces land use systems by using suitable management techniques that aid in maximizing the social and financial advantages of

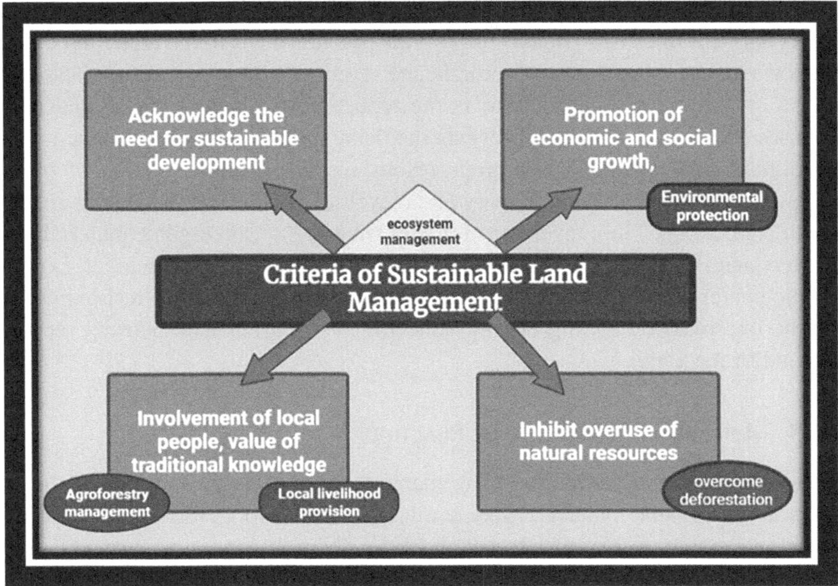


**FIGURE 4.4**  Different criteria taken into consideration for sustainable land management.

the land. It also preserves or improves the natural support of the land resources. Figure 4.4 shows different criteria taken into consideration for sustainable land management. The practices for sustainable land operations are as follows:

### 4.8.1 Agroforestry Management

This involves wide practices of planting on contours and cultivating woody species and trees. It reduces erosion, increases soil fertility, and provides greater infiltration (Garrity 2004). Some legume trees (e.g., *Leucaena* sp., *Acacia* sp., etc.) increase soil fertility by biological N fixation. Planting trees ensures better nutrient availability (Sánchez et al. 1994). In many agroecosystems, agroforestry systems are essential for improving soil health and amelioration. Agroforestry may also lessen erosive carbon sequestration losses, enhancing the organic carbon store (Verchot et al. 2007). Different studies have reported upliftment of soil health status; for instance, after five years, *Acacia nilotica* + *Saccharum munja* and *Acacia nilotica* + *Eulaliopsis binata* demonstrated an increase in soil organic carbon status from 0.39% to 0.52% and 0.44% to 0.55%, respectively (Samra and Singh 2000). Based on research into how effectively nutrients are transferred from trash to plants in natural ecosystems, agroforestry improves soil (Vitousek and Sanford 1986).

### 4.8.2 Conservation Agriculture

Future farming methods that are more sustainable include conservation agriculture, which is indicated by minimum soil disturbance. Crop rotations decrease the danger of pests and diseases caused by pathogenic organisms, as the biological variety checks harmful microbes (Leake 2003). This practice helps in the reduction of wind and soil erosion. It also controls weeds and requires tillage operations. The major agenda of conservative agriculture is the reduce costs, such as diesel and labour expenses (Raza et al. 2023). It prevents the delay in operations, thus reducing time expenditure. Inclusion of cover crops results in surface accumulation of organic carbon-stimulating microbial activity and recycling of nutrients (Roldán et al. 2003; Alvear et al. 2005). The conversion of organic material to soil organic material in the leftovers and the storage of SOC depend on careful nutrient management. Low soil organic content results from low input and subsistence agriculture techniques; this may be improved by adding organic amendments that enhance nutrient recycling (Lal and Bruce 1999).

### 4.8.3 APPROPRIATE MANAGEMENT PRACTICES

The root biomass may be increased by adequate water management, raising the concentration of organic material. More available water in roots improves soil organic carbon sequestration (Kimmelshue et al. 1995). The right kind of planning takes into account both regional planning that takes into account both on- and off-site interactions, as well as local engagement by the participants. Production of livestock is of utmost importance. Managing livestock lowers hazards. Livestock is an important part of sustainable land management. It is a source of provision of good services if well managed.

## 4.9 CONCLUSIONS

Climate change, population growth shifts in diet, and soil degradation pose challenges to the future production of food and feed worldwide. Inadequate nutrition causes malnutrition in approximately 4 billion people globally. Human activities, particularly shifting patterns of land use on both local and global dimensions, also threaten the availability and sustainability of food production. With the expansion of international agricultural trade, soil nutrients are being shipped from their original location to areas with vastly different geology, climate, fauna, and soil properties. In addition, increasing losses and soil degradation due to climate change directly threaten food security. At the same time, making biofuels puts much pressure on space, energy, and water. There is a continuous increase in food and fiber demand in local and global food markets. Many ecological services that are necessary to sustain human life depend on healthy soil, yet this neglected component of the food, water, and energy nexus is often overlooked. To attain global food security, crop production practices will have to undergo a radical shift to address soil degradation, polluted and limited water supplies, biodiversity loss, limited fossil fuels, climate change, and variable weather conditions during the planting season. Soil scientists must become more active in the agricultural food production, energy generation, and water resources policy talks. To achieve water, food, and energy security, we need to make more people aware of how soil properties, functions, and ecosystem services are linked.

## REFERENCES

Acosta, J.A., Faz, Á. and Martínez Martínez, S., 2009. Evaluation of land degradation in an area affected by different management systems, SE Spain. In *Land degradation and rehabilitation: Dryland ecosystems. Papers presented at the Fourth International Conference on Land Degradation, Cartagena, Murcia, Spain, 12–17 September 2004* (pp. 81–96). Catena Verlag, Reiskirchen.

Alvear, M., Rosas, A., Rouanet, J.L. and Borie, F., 2005. Effects of three soil tillage systems on some biological activities in an Ultisol from Southern Chile. *Soil Tillage Res* 82(2), pp. 195–202.

Ayub, M.A., Usman, M., Faiz, T., Umair, M., Rizwan, M., Ali, S. and Zia ur Rehman, M., 2020. Restoration of degraded soil for sustainable agriculture. In *Soil health restoration and management* (pp. 31–81). Springer, Singapore.

Barrios, E., 2007. Soil biota, ecosystem services and land productivity. *Ecol Econ* 64(2), pp. 269–285.

Beinroth, F.H., Eswaran, H. and Reich, P.F., 2001. Global assessment of land quality. *J Sustain Agric Sci* pp. 569–574.

Behera, B., Das, T.K., Raj, R., Ghosh, S., Raza, M.B. and Sen, S. 2021. Microbial consortia for sustaining productivity of non-legume crops: Prospects and challenges. *Agric Sci* 10, pp. 1–14.

Biro, K., Pradhan, B., Buchroithner, M. and Makeschin, F., 2013. Land use/land cover change analysis and its impact on soil properties in the Northern part of Gadarif region, Sudan. *Land Degrad Dev* 24(1), pp. 90–102.

Blanco-Canqui, H. and Lal, R., 2010. Soil and water conservation. In *Principles of soil conservation and management* (pp. 1–19). Springer, Dordrecht.

Blum, W.E., 2005. Functions of soil for society and the environment. *Rev Environ Sci Biotechnol* 4(3), pp. 75–79.

Blum, W.E., 2016. Role of soils for satisfying global demands for food, water, and bioenergy. In *Environmental resource management and the nexus approach* (pp. 143–177). Springer, Cham.

Chakrabarty, M., 2016. *Climate change and food security in India.* Observer Research Foundation (ORF), New Delhi, India.

Commission of the European Communities, 2006. *Proposal for a directive of the European Parliament and of the council establishing a framework for the protection of soil and amending directive 2004/35/EC. COM (2006) 232 final.* European Commission, Brussels. Available at https://eur-lex.europa.cu/LexUriServ/LexUriServ.do?uri=COM:2006:0232:FIN:EN:PDF

Das, D., Sahoo, J., Raza, M. B., Barman, M. and Das, R. 2022. Ongoing soil potassium depletion under intensive cropping in India and probable mitigation strategies: A review. *Agron Sustain Dev* 42(1), p. 4.

Dietz, T., Millar, D., Dittoh, S., Obeng, F. and Ofori-Sarpong, E., 2004. Climate and livelihood change in North East Ghana. In *The impact of climate change on drylands* (pp. 149–172). Springer, Dordrecht.

Edelstein, M. and Ben-Hur, M., 2018. Heavy metals and metalloids: Sources, risks and strategies to reduce their accumulation in horticultural crops. *Sci Hortic* 234, pp. 431–444.

Eswaran, H., Beinroth, F. and Reich, P., 1999. Global land resources and population-supporting capacity. *Am J Agric Res* 14(3), pp. 129–136.

Eswaran, H., Lal, R. and Reich, P.F., 2001. Land degradation: An overview. Responses to land degradation. In E. Bridges, I. Hannam, L. Oldeman, F. Penning de Vries, S. Scherr and S. Sompatpanit (Eds.), *Proceedings of 2nd international conference on land degradation and desertification. KhonKaen, Thailand.* Oxford University Press, New Delhi.

FAO, 2016. *Climate change and food security: Risks and responses.* Food and Agriculture Organization, Rome. Available at www.fao.org/3/a i5188e.pdf

FAOSTAT, 2018. *Statistical database.* Food and Agriculture Organization of the United Nations, Rome. Available at www.fao.org/faostat/en/#home

Flavelle, C., 2019. *Climate change threatens the world's food supply, United Nations warns.* Available at www.nytimes.com/2019/08/08/climate/climate-change-food-supply.html

Garrity, D.P., 2004. Agroforestry and the achievement of the millennium development goals. *Agrofor Syst* 61(1), pp. 5–17.

Griggs, D., Stafford-Smith, M., Gaffney, O., Rockström, J., Öhman, M.C., Shyamsundar, P., Steffen, W., Glaser, G., Kanie, N. and Noble, I., 2013. Sustainable development goals for people and planet. *Nature* 495(7441), pp. 305–307.

Haron, M. and Dragovich, D., 2010. Climatic influences on dryland salinity in Central West New South Wales, Australia. *J Arid Environ* 74(10), pp. 1216–1224.

IPCC, 2021. Summary for policymakers. In Masson-Delmotte, V., Zhai, P., Pirani, A., Connors, S. L., Péan, C., Berger, S., Caud, N., Chen, Y., Goldfarb, L., Gomis, M.I., Huang, M., Leitzell, K., Lonnooy, E., Matthews, J.B.R., Maycock, T.K., Waterfield, T., Yelekci, O., Yu, R., and Zhou, B. (eds.), *Climate change 2021: The physical science basis* (pp. 3–32). Contribution of Working Group I to the Sixth Assessment Report of the Intergovernmental Panel on Climate Change. Cambridge University Press, Cambridge, UK and New York, NY. Available at www.ipcc.ch/report/ar6/wg1/downloads/report/IPCC_AR6_WGI_SPM_final.pdf

Kettler, T., 2019. *Soil genesis and development, lesson 6-global soil resources and distribution.* Department of Agronomy and Horticulture, University of Nebraska-Lincoln.

Kimmelshue, J.E., Gilliam, J.W. and Volk, R.J., 1995. Water management effects on mineralization of soil organic matter and corn residue. *Soil Sci Soc Am J* 59(4), pp. 1156–1162.

Lal, R., 1989. *Land degradation and its impact on food and other resources* (Vol. 85). Academic Press, San Diego.

Lal, R., 2007. Anthropogenic influences on world soils and implications to global food security. *Adv Agron* 93, pp. 69–93.

Lal, R., 2015. Restoring soil quality to mitigate soil degradation. *Sustainability* 7(5), pp. 5875–5895.

Lal, R., 2016. Soil health and carbon management. *Food Energy Secur* 5(4), pp. 212–222.

Lal, R. and Bruce, J.P., 1999. The potential of world cropland soils to sequester C and mitigate the greenhouse effect. *Environ Sci Policy* 2(2), pp. 177–185.

Leake, A.R., 2003. Integrated pest management for conservation agriculture. In *Conservation agriculture* (pp. 271–279). Springer, Dordrecht.

Liu, E., Pimpin, L., Shulkin, M., Kranz, S., Duggan, C.P., Mozaffarian, D. and Fawzi, W.W., 2018. Effect of zinc supplementation on growth outcomes in children under 5 years of age. *Nutrients* 10(3), p. 377.

Majhi, P.K., Raza, B., Behera, P.P., Singh, S.K., Shiv, A., Mogali, S.C., Tanmaya Kumar, B., Biswaranjan, P. and Behera, B., 2022. Future-proofing plants against climate change: A path to ensure sustainable food systems. In *Biodiversity, functional ecosystems and sustainable food production* (pp. 73–116). Springer International Publishing, Cham.

McGuire, S., 2015. FAO, IFAD, and WFP, the state of food insecurity in the world 2015: meeting the 2015 international hunger targets: Taking stock of uneven progress. Rome: FAO, 2015. *Adv Nutr* 6(5), pp. 623–624.

Meena, O.P., Sammauria, R., Gupta, A.K., Gupta, K.C., Behera, B., Saxena, R., Yadav, M.L., Singh, P., Meena, R.K., Raza, M.B. and Lal, M.K., 2022. Energy-carbon footprint vis-à-vis system productivity and profitability of diversified crop rotations in semi-arid plains of North-West India. *J Soil Sci Plant Nut* 22(2), pp. 2026–2041.

Miao, Z., Padovani, L., Riparbelli, C., Ritter, A.M., Trevisan, M. and Capri, E., 2003. Prediction of the environmental concentration of pesticide in paddy field and surrounding surface water bodies. *Paddy Water Environ* 1(3), pp. 121–132.

Montgomery, D.R., 2007. Soil erosion and agricultural sustainability. *Proc Natl Acad Sci* 104(33), pp. 13268–13272.

Natural Resources Conservation Service, 1998. Soil quality concerns: Pesticides. *Soil Quality Information Sheet*. United States Department of Agriculture. Available at www.nrcs. usda.gov/Internet/FSE_DOCUMENTS/nrcs142p2_052821.pdf

Oertel, C., Matschullat, J., Zurba, K., Zimmermann, F. and Erasmi, S., 2016. Greenhouse gas emissions from soils—a review. *Geochem* 76(3), pp. 327–352.

Osman, K.T., 2013. Soil resources and soil degradation. In *Soils* (pp. 175–213). Springer, Dordrecht.

Raza, M.B., Hombegowda, H.C., Kumar, P. and Kumar, S., 2023. Effect of conservation agriculture on soil health. *A Train Manual Conserv Agric* 41.

Roldán, A., Caravaca, F., Hernández, M.T., Garcıa, C., Sánchez-Brito, C., Velásquez, M. and Tiscareno, M., 2003. No-tillage, crop residue additions, and legume cover cropping effects on soil quality characteristics under maize in Patzcuaro watershed (Mexico). *Soil Tillage Res* 72(1), pp. 65–73.

Samra, J.S. and Singh, S.C., 2000. Silvopasture systems for soil, water and nutrient conservation on degraded land of Shiwalik foothills (sub-tropical northern India). *J Soil Water Conserv* 28(1), pp. 35–42.

Sánchez, P.A., Woomer, P.L. and Palm, C.A., 1994. Agroforestry approaches for rehabilitating degraded lands after tropical deforestation. In *JIRCAS International Symposium Series* (Vol. 1, pp. 108–119). Technical Approach, Tsukuba, Ibaraki, Japan, 17 September 1992, held by Tropical Agriculture Research Center (TARC). https://www.jircas.go.jp/ sites/default/files/publication/intlsymp/intlsymp-1_108-119.pdf

Saraswat, A., Ram, S., Abdel Rahman, M.A., Raza, M.B., Golui, D., Hombegowda, H.C., Lawate, P., Sharma, S., Dash A.K., Scopa, A. and Rahman, M.M., 2023. Combining fuzzy, multicriteria and mapping techniques to assess soil fertility for agricultural development: A case study of Firozabad District, Uttar Pradesh, India. *Land* 12(4), p. 860.

SARE, 2019. *Building soils for better crops: Ecological management for healthy soils*. Sustainable Agriculture Research and Education, National Institute of Food and Agriculture (NIFA), U.S. Available at www.sare.org/wp-content/uploads/Building-Soils-for-Better-Crops.pdf

Smith, J., Farmer, J., Smith, P. and Nayak, D., 2021. The role of soils in provision of energy. *Philos Trans R Soc* 376(1834), p. 20200180.

U.N, 2016. *The sustainable development goals report 2016*. United Nation. Available at https:// unstats.un.org/sdgs/report/2016/

Van Vliet, J., Eitelberg, D.A. and Verburg, P.H., 2017. A global analysis of land take in cropland areas and production displacement from urbanization. *Glob Environ Change* 43, pp. 107–115.

Verchot, L.V., Van Noordwijk, M., Kandji, S., Tomich, T., Ong, C., Albrecht, A., Mackensen, J., Bantilan, C., Anupama, K.V. and Palm, C., 2007. Climate change: linking adaptation and mitigation through agroforestry. *Mitig Adapt Strateg Glob Chang* 12(5), pp. 901–918.

Vitousek, P.M. and Sanford, R.L., 1986. Nutrient cycling in moist tropical forest. *Annu Rev Ecol Evo Syst* pp. 137–167.

WHO. 2021. *Levels and trends in child malnutrition: UNICEF/WHO/The World Bank Group joint child malnutrition estimates*. World Health Organization, Geneva. Available at www.who.int/publications/i/item/9789240025257

WHO. 2022. World Health Organization, Geneva. Available at www.who.int/news-room/ fact-sheets/detail/malnutrition

Yang, Q.Q., Yu, W.H., Wu, H.Y., Zhang, C.Q., Sun, S.S.M. and Liu, Q.Q., 2021. Lysine biofortification in rice by modulating feedback inhibition of aspartate kinase and dihydrodipicolinate synthase. *Plant Biotechnol J* 19(3), pp. 490–501.

Zhang, Y., Wei, L., Wei, X., Liu, X. and Shao, M., 2018. Long-term afforestation significantly improves the fertility of abandoned farmland along a soil clay gradient on the Chinese Loess Plateau. *Land Degrad Dev* 29(10), pp. 3521–3534.

Zhao, Q., Xiong, W., Xing, Y., Sun, Y., Lin, X. and Dong, Y., 2018. Long-term coffee monoculture alters soil chemical properties and microbial communities. *Sci Rep* 8(1), pp. 1–11.

# 5 Soil Security to Address Potential Global Issues

*Hina Fatima, Man Park, Muhammad Ameen,
Iqra Aslam, Tabinda Athar, Syed Shabbar
Hussain Shah, Ghulam Hassan Abbasi,
Muhammad Ali, Aisha Abdul Waris,
Muhammad Naveed Arshad, and
Muhammad Ashar Ayub*

## 5.1 INTRODUCTION

Soil security is defined as the improvement and maintenance of soil resources to produce food, fiber, and fresh water along with increasing energy demand, preventing climate change and maintaining soil biodiversity (Koch et al., 2013). In 2015, McBratney and Field defined soil security as the homology of food, water, and energy security. Despite its importance, it has remained unnoticed in the past. Recently, people's attention has been drawn to the significance of soil health, the millions of organisms that live there, and their connections to both food and climate change (D'hondt et al., 2021; Yang et al., 2021).

Soils deliver various ecosystem services and are essential for the existence of human and animal life. For instance, soils not only ensure sustainable food, fiber, and bioenergy production but also filter and buffer inorganic and organic components by adsorption and desorption. Doing so protects the food chain from contamination and filters groundwater, providing safe and clean drinking water. Additionally, soils retain and protect soil biodiversity and provide the basis for the development of infrastructure, including buildings, industries, and roads (Blum, 2016). In 2015, the General Assembly of the United Nations (UN) proposed 17 Sustainable Development Goals (SDGs) for the betterment of human life and planet earth. 5 out of 17 SDGs are directly and indirectly linked to soil security, i.e., to end hunger and achieve food security (SDG#2), to ensure clean water and sanitation (SDG#6), to ensure reliable and cheap energy for all (SDG#7), to combat climate change (SDG#13), and to guard and reestablish terrestrial ecosystems by managing forests and restoring degraded soils (Horn et al., 2018; Lal et al., 2021).

However, soils are being degraded rapidly by direct and indirect human actions, leading to unimaginable devastating consequences. In 2021, Food and Agriculture Organization (FAO) reported that around 34% of planet earth's agricultural land is degraded. And if soil security is not taken at serious note, around 90% of the earth's area could be degraded by 2050 (Scholes et al., 2018). FAO Global Soil Partnership report released in 2017 stated that around 75 billion metric tonnes of fertile soils

DOI: 10.1201/9781003358169-5

were eroded from arable lands (Pimentel et al., 1995). Loss of fertile layer impedes farming and farming activities, makes it difficult to achieve UN SDGs, and results in far-reaching climatic disasters. Therefore, it's imperative to shed light on the importance of soil, its conservation, and its relationship with associated global challenges to prevent our soils from both anthropogenic and naturally induced disasters and preserve this non-renewable resource for future generations.

## 5.2 MAJOR THREATS PREVENTING SOILS FROM ATTAINING SDGS

Sustainable development (SD) has converted a prevalent slogan into a present advanced dialogue. SD has developed a universal growth pattern for international assistance organizations along with motto of growth with ecological campaigners (Ukaga et al., 2011). This idea fascinates broad-based consideration that other upgrading perceptions lack and seems dignified to endure the persistent progress model for an extended time (Scopelliti et al., 2018).

Nevertheless, the dominance and status of sustainable development goals evoke those who are interested about their implication or meaning and what it needs as well as infers for expansion theory and training (Montaldo, 2013; Shahzalal and Hassan, 2019). SD, therefore, deals with risk alleviation regarding health, climate, and food. Many disciplines offer solutions to achieve sustainable development goals, but no one appears to describe them with accuracy and precision to avoid food security in the future (Mensah and Enu-Kwesi, 2018). In the effort to transfer past the stability oratory and follow a more evocative schedule for workable growth, a clear description of this thought and elucidation of its crucial measurements are desirable (Gray, 2010). Soil plays various essential roles regarding food security, evaluated by sustainable development goals. The main threats regarding soil security include soil erosion, soil degradation, contamination, and salinization. All the facts discussed above can be indirectly observed during urbanization and industrialization (Figure 5.1). Sustainable development goals especially 2, 11, 14, and 15 are directly relevant to soil conservation.

### 5.2.1 URBANIZATION AND INDUSTRIALIZATION

The urban conversions over the previous two decades have enhanced local financial progress and improved the living standards of the citizens. Nevertheless, it's prevalent that the tradition of the town changeover is imperfect, and leftovers are essentially unmaintainable. The urban residents of numerous towns in any area are usually adults, presenting > 60% of the world's youth population. The atmospheric contaminations are the main threat upsetting many of the towns. Worldwide Asia and the Pacific have 19 of the most atmosphere-polluted cities out of 20. In accordance to WHO, open-air contamination caused 3.7 million causalities worldwide in 2012. Urban zones in Asia and the Pacific produce nearly 960,000 tonnes of public solid waste per day, and by 2025, this number will rise to 2.4 million tonnes. The amount of urban population living with extreme numerous threats is presently about

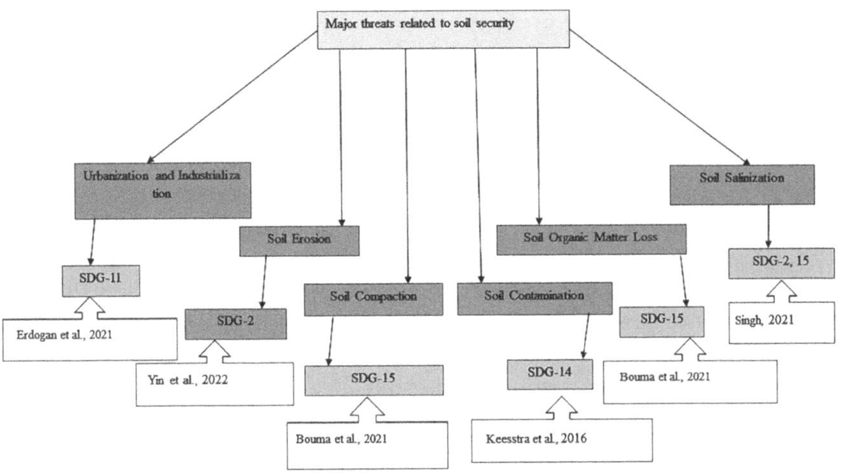


**FIGURE 5.1** Relationship between major threats of soil security and sustainable development goals.

742 million and could increase up to 1 billion by 2030. In the 2030 schedule, special importance has been devoted to cities and their stability (SDG#11). The alteration of factory work is one of the immense tasks on the way to SD. Industrialized procedures are implemented by dispersed, logical, and partly self-directed machines, with self-organized skills. For instance, industry 4.0 workshops are extremely elastic, allowing the self-adjusted industrialized developments that aid mass production according to our desires (Lasi et al., 2014). In spite of this scientific progress, industry 4.0 compromises numerous tasks for all three-way measurements of sustainability (Renn et al., 2021).

The prospects of the input of industry 4.0 to SD and the accomplishment of the SDGs of the United Nation's 2030 schedule are relatively extraordinary, particularly between strategy makers and legislatures of the factories (Kunkel and Matthess, 2020; Nara et al., 2021). The Global e-Sustainability Initiative (GeSI), which is an enterprise of ICT corporations, announces that ICT has "the spark to produce influential ecological, financial and communal benefits", fluctuating from decreasing $CO_2$ discharges (20%), creating added incomes (over $11 trillion) to extensive social welfares by 2030 (GeSI, 2015).

The urbanization and environmental change are joining in hazardous behaviors. The world's population by this time is > 50% urban, and this is predicted to increase to two-thirds. The urban environment by now has endured tragedies like flooding and hot storms. Besides, the coastlines, seashores, rivers, and flood plans, including a lot of the world's biggest towns and cities, are supremely susceptible to natural catastrophic strikes. It is expected that in the upcoming years, climate variation may reduce hundreds of millions of city population due to water overflows, mudslides, life-threatening climate, and unusual calamities.

## 5.2.2　Soil Erosion

The world population is increasing incessantly, as it was 1 billion in 1820, 7 billion in 2012, and would be 10 billion in 2056 as a result of rational proliferation. To feed such a huge population, a lot of assets including diet, vigor, water, soil, and air should be addressed (Ferreira et al., 2018). Man-made activities (overgrazing, afforestation, improper agricultural activities, and tillage) and associated earth utility fluctuations are the chief causes of enhanced soil degradation, which has considerable inferences for the C-cycle and nutrients (Borrelli et al., 2018). Water-eroded soil is the main threat for Europe according to their thematic strategy commission. It decreases soil output and hints to afforestation in susceptible zones (European Commission, 2006; Kirkby et al., 2008). The main strategic reaction is compulsory to tackle the influences of soil erosion in demolished zones, predominantly in the vision of present environment fluctuations and the upcoming water catastrophe (Panagos et al., 2016). The scientific knowledge of the new evaluation of soil injury by soil water erosion in Europe has generated a systematic debate demonstrating the conclusions, ways of their improvement, as well as efficient utilization within the policy sequence (Panagos et al., 2019). The resultant soil erosion atlases are utilized to recognize zones where comprehensive revisions and remedial activities are desirable. In addition, it permits for comparative four-dimensional and sequential judgements, described as situation examination and incorporation of the erosion scheme with the carbon cycle. The United Nations SDGs clearly recognize soils being vital for SD and endorse the security of soil assets in order to attain the motivating objective of zero land deterioration by 2030 (Keesstra et al., 2016). SDG#2 and SDG#15 are dependent mainly on soil erosion by water.

## 5.2.3　Soil Sealing/Soil Compaction

Compaction of the layer beneath the upper soil layer is a main ongoing soil deterioration process. Soil compaction is described as "the process in which total air-filled pores are decreased due to any natural and man-made activity, causing decline or damage to one or more than one soil roles" (Van den Akker, 2008). Although upper soil compaction is flexible, subsoil compaction is successfully tenacious, accumulative, and undistinguishable externally. Therefore, the compaction of subsoil should be prohibited due to unavailability of efficient reclamation choices (Schjønning et al., 2013). The danger of subsoil compaction has improved, following the mechanical and technical expansion of current farming, owing to the rise in the extent of equipment used on the farm (Schjønning et al., 2015). An intense propagation in wheel burden since the imposition of machines in agriculture has been observed and exceeds the load capacity of the soils in many areas (van den Akker and Schjønning, 2004). Soil compaction below the upper layer is a hazardous for all the soils, but it is perhaps more common in humid zones with automatic agriculture. The outcomes of soil compaction include the reduction in infiltration and buffer capacity of the soil, resulting in no filtration of pollutants by soil. It also increases the release of greenhouse gases such as nitrous oxides. Compaction of subsoil layer is a challenge in a way to achieve sustainable development goals and for the authorities due to several reasons. Firstly, it is a huge problem in the sense that the type of problem is uncertain, when different benefits and perceptions support a query but have no optimum answers (Alrøe and

Noe, 2014). Secondly, straight fluctuations are indistinguishable as they take place within the earth, and variations are steady and collective (Horn et al., 1995). Thirdly, after ending the flexible soil limit, it becomes compacted which remains tenacious (Schjønning et al., 2013). Fourthly, the remedial actions are not rewarding. In spite of the tenacious adverse possessions on plant's yield, it may be financially feasible for growers to use heavy machines and compress the subsoils compared to accepting precautionary measures. Fifthly, the compaction is forceful and location-based. Threat varies by soil category and hydrological status. Therefore, it is tough to make standard protective endorsements regarding field traffic. As a final point, the continuing organizational advances in European agriculture suggest an extra growth in equipment size that creates lower soil compaction and is hazardous for soil assets.

### 5.2.4 Soil Contamination

An increase in pesticides and agrochemicals results in the plant uptake. The remaining are leached down into the soil, water, and atmosphere, ultimately becoming a reason for pollution or may donate to overall changes in climate. This supplementation by agro toxins may have secondary and antagonistic consequences on flora and fauna in cultivated lands and within contiguous ecologies. If SDGs are reviewed thoroughly, SDG#3 emphasizes on good health and well-being and SDG#2 stresses on production of ample amount of food for the increasing population with an aim of zero hunger. Unfortunately, the emphasis of majority of the farmers around the world is on SDG#2 and not on SDG#3.

The soils deliver water, nutrients, and anaerobic atmosphere to plants, and soil construction allows the root—according to their genetics—to travel for nutrients. As a medium for plant growth, it should be free of chemicals/pollutants and contain enough nutrients and organic matter. Soil can be described with respect to its health parameters which are supported by clear visual observations. These are about the root and shoot proliferation (Veerman et al., 2020). Soil health can be determined by observing various factors: (i) lack of pollutants and osmosis, (ii) sufficient C storage, (iii) promising soil build-up, (iv) advantageous soil community, and (v) sufficient nutrients. In comparison to other soil health plans, an inadequate indicator has been suggested for operative causes (Norris et al., 2020). Soils on a large scale are being converted to polluted soils which not only have aftereffects on the environment but also on the food chain (Steffan et al., 2017).

Plant growth should be improved by the collaborating work of soil scientists and hydrologists for soil nutrient and soil moisture categorization. Extensively accessible and well-verified simulation models of the interactive approach between water, soil, plant, and atmosphere system (White et al., 2013) are perfect ways to recognize interdisciplinary collaboration, motivated by significant social matters like water supervision (Hackten Broeke et al., 2019).

### 5.2.5 Soil Organic Matter Loss

The effects of climate variability are well managed by the demonstration method, which indicates the significance of fluctuation of earth utility, soil carbon accumulation, soil wetness, as well as heat stress with respect to chemistry, geology, and biology (Denman et al., 2007). There is a need to increase soil organic carbon by adding more organic matter by inventive administration which has reasonably diverse possibilities in diverse soils and is significant for climate change mitigation. Soil health and fertility are mainly dependent on two important constituents: soil organic matter and soil nutritional status. The reduction in soil health relates to loss of soil organic matter, organic carbon, and C/N ratio. In 2050, 70% more food production would be a crucial challenge to feed 9.6 billion people (Lal, 2013). To achieve that 2050 goal, the soil should be preserved from various losses like soil erosion, organic matter loss, contamination, biodiversity loss, and lack of sufficient nutrition since these all eventually lead to soil degradation which is a major threat in a way to achieve SD.

### 5.2.6 Soil Salinization

The improvement of irrigated cultivation is essential to nourish the growing worldwide population, which is anticipated to exceed 9.9 billion in the coming 30 years (UN, 2017). The increased population creates more problems due to enhanced anthropogenic activities. The ultimate outcome of the advancement would be improper drainage, which brings salinization and water logging problems in cultivated regions (Zhang et al., 2019; Jiang et al., 2019). These concerns about farming lands are tenacious in irrigated areas throughout the world (Mosaffa and Sepaskhah, 2019). The salt stress has posed severe threats to the sustainability of irrigated agriculture (Manasa et al., 2020). More than 3% of total soil assets are salt-stressed globally by now (FAO, 2022). In addition, this number is obstinately increasing at a degree of 2 million hectares (mha) per annum (Singh, 2018). In arid and semi-arid climates, global crop production exhibits 18% to 43% losses due to salinity (Chang et al., 2019). Hossain et al. (2011) described that salinity stress in root zone aggravated by water logging widely limit the field crop production. Similar conclusions were drawn by Al-Muaini et al. (2019) and Zheng et al. (2009). Further, Chang et al. (2019) described that globally, > 20% of the total cultivated area is adversely deteriorated by soil salinity and water logging. If it remains unattended, salt-affected land could increase to > 50% of the total irrigated land worldwide by 2050 (Wang et al., 2020a).

## 5.3 GLOBAL CHANGE ENDANGERING AGRICULTURAL SUSTAINABILITY

The demand for food is rising as a result of changing diets and a growing worldwide population. As crop yields level off globally, ocean health deteriorates, and natural resources like soils, water, and biodiversity are severely depleted, production is finding it difficult to keep up. According to a 2020 report, the number of starving population is increasing at the rate of 60 million in every five years and an estimated 690 million, or 8.9% of the global population, are starving. Since the globe would have to produce nearly 70% more food by 2050 in order to feed an estimated population of 9 billion, the issue of food security will only get worse. So it's crucial to understand what influences agricultural sustainability. We will discuss a few key elements in this chapter that could seriously jeopardize the viability of agriculture in the future.

### 5.3.1 CLIMATE CHANGE

World is confronting one of the most pressing issues in the form of climate change. Global warming, unpredictable weather, the melting of ice glaciers, and rising sea levels are some examples of its effects. It is defined as significant changes in the average values of meteorological elements, such as precipitation and temperature (Lipczynska-Kochany, 2018). The emissions of greenhouse gases (GHGs), especially $CO_2$, $CH_4$, and $N_2O$, are the deriving factors for climate change owing to their bulk presence in nature and significant effect on climate. The "underestimation" of climate change is a result of the slow rise in harmful atmospheric events brought on by global warming. Environmental deterioration is primarily caused by natural events such as solar cycles, earthquake activity, and volcanic eruptions in addition to human activity (Roy, 2018). This issue becomes more severe by agriculture's increased sensitivity to climate change. Terrible consequences because of climate change are already being felt, in the form of growing temperatures, climate variability, shifting agroecosystem boundaries, invasive plants and pests, and uncommon extreme weather events. On farms, climate change is diminishing animal outputs, the nutritional value of main cereals, and crop yields (Udara Willhelm Abeydeera, 2019). Effects of climate change on crop productivity of numerous cereal plants in the form of yield variations have been listed in Table 5.2. An exceptionally high rate of land degradation due to climate change is producing accelerated desertification and nutrient-deficient soils. Land degradation is a serious global concern that is said to be getting worse every day. A quarter of the world's land area can now be classified as deteriorated, according to the Global Assessment of Land Degradation and Improvement (GLADA). An estimated 1.5 billion people are claimed to be affected by land deterioration. Anthropogenic activities and climate change result in the annual loss of 15 billion tonnes of productive soil. Extreme drought conditions, which are becoming more common as a result of climate change, reduce crop output by immobilizing nutrients and allowing salt to build up in the soil, which makes it unhealthy, saline, and finally, unproductive. Farmers eventually forsake such barren fields since they are no longer fertile, which causes economic losses and social problems. Depending on the geography and irrigation technique, different crops are affected differently by climate change. Expanding irrigated regions can boost crop production but may have a negative impact on the environment (Kang et al., 2009). By shrinking their growing seasons, many crops are likely to have lower yields (Mahato, 2014). If both the temperate and tropical regions suffer warming of 2°C, it is anticipated that production of wheat, rice, and maize will decline (Challinor et al., 2014). Flooding has become more frequent and sporadic in recent years as a result of climate change. If remedial and preventive remediation techniques are not implemented, the topsoil and nutrients in it will be lost, leading to low productivity for several years to come. Agricultural areas in coastal locations may decline as a result of rising sea levels or the frequency of heavy rains. This has caused crops in coastal areas to experience challenges such as decreased respiration, photosynthesis, and transpiration, which in turn puts the

**TABLE 5.1**

**Direct and Indirect Impacts of Climate Change on Agriculture**

| Factors | Effects | References |
|---|---|---|
| Change in mean climate | Temperature getting higher within growing seasons impact the productivity of agriculture, income of farms, and security of food | (Battisti and Naylor, 2009) |
| | Change in precipitation enhances the water availability which eventually raises the temperature and increases the time span of the growing season | (Döll, 2002; Fischer et al., 2007) |
| Climate variability and extreme weather events | Variation in the possibility of heat waves | (Ciais et al., 2005) |
| Extreme temperature | Widespread disruption in cereal market and food security | (Wollenweber et al., 2003; Vara Prasad et al., 2003; Battisti and Naylor, 2009) |
| | Effect on enzyme activity and gene expression | |
| | Impact on carbon assimilation leading to yield reduction | |
| Metrological drought | Low precipitation | (Holton et al., 2003) |
| Agricultural drought | Lack in soil moisture, escalation in plant water stress | (Burke et al., 2006) |
| Hydrological drought | Reduced stream flow | (Li et al., 2009) |
| Socioeconomic drought | Disrupt balance between supply and demand | (Alcamo et al., 2007) |
| Heavy rainfall and flooding | Increased water logging, reduced plant growth, delayed farming operation | (Kettlewell et al., 1999) |
| Tropical storms | Flooding | (Webster, 2008) |
| Pests and diseases | Increased pathogen by mutation under stress | (Gregory et al., 2009) |
| Changes in water availability | High altitude leading to warming | (Barnett et al., 2005) |

supply and security of food in such areas in danger. Few additional effects of climate alternate have been enumerated in Table 5.1.

### 5.3.2 Competition for Limited Freshwater Resources

Fresh water constitutes only 2.5% of the total amount of water on earth, and less than 1% of it is consumable, despite the fact that it is vital for human survival. Drinking water, farming, manufacturing, and the creation of electricity all depend on fresh water. Additionally, 10% of all animal species on the planet, many of which are endangered, only exist in freshwater settings. Over two-thirds of all human withdrawals, or 2,600 km$^3$ of water, are used to irrigate crops each year (AQUASTAT, 2021). Numerous river basins around the world are on the verge of disappearing as

**TABLE 5.2**

**Yield Variations of Various Crops Due to Climate Change and their Possible Management Strategies**

| Crops | Yield variations | Causes | Smart Technologies | References |
|---|---|---|---|---|
| Cotton | 30%–46% | Slow warming scenario, rapid warming scenario | Management of pest attack | (Schlenker and Roberts, 2009; Wagan et al., 2015) |
| Wheat | −6% | Medium-high and low GHG emissions | Laser land leveling | (Lee et al., 2011; Latif et al., 2013) |
| Rainfed corn | −24%–34% | Increased temperature and precipitation variability | Stress-tolerant crop varieties | (Zhao et al., 2017; Ogada et al., 2020) |
| Maize | +1.6% | Extreme weather events and global warming | Drought-tolerant varieties | (Cai et al., 2009; Aryal et al., 2015) |
| Cassava | +1.7% | Up surge in mean growing season temperature by 1°C | Stress-tolerant crop varieties | (Singh et al., 2015) |
| Rice | −3.7% | Increase in world's mean temperature | Site-specific nutrient management | (Tao et al., 2006; Khatri-Chhetri et al., 2016) |
| Sorghum | −2.2% | Increased temperature | Zero tillage | (Carleton and Hsiang, 2016; Liu and Basso, 2020) |
| Soyabean | −0.5% | Increased temperature | Bed furrow | (Kukal and Irmak, 2018) (Birthal et al., 2012) |

water shortages worsen (Smakhtin, 2004). More and more water is being diverted away from agriculture to meet other needs, such as electricity production or expanding urban populations (Pimentel et al., 1997). Global groundwater extraction has dramatically expanded between 1950 and 2000 to provide municipal, industrial, and agricultural applications. As a result, aquifer salinization has occurred, land has subsided, and yields have drastically decreased in many parts of the world where groundwater reserves have decreased (Konikow and Kendy, 2005). Vorosmarty and colleagues asserted in 2000 that the world's water supplies are already under stress at the current population levels and that this stress will only worsen as populations continue to grow. What could make the situation worse in the future is the wages rise that lead to diets, including more water-intensive agricultural goods, eventually causing water service levels to rise. Together, these factors are driving the per capita water demand in developing countries to increase dramatically. Agriculture is simultaneously expanding into new areas and becoming more productive, both of which are quickly raising the need for water in order to meet increased food demands for a growing population. Industrial water use is rising as a result of rising energy consumption and other industrial activity in many regions.

### 5.3.3 Increased Population and Changing Lifestyle

Population of the world is increasing day by day which eventually increases the need and demand for food, clothes, water, and other needs which are required to be there for survival. By 2050, it is projected to grow to 9.7 billion people, which will increase the burden on agricultural lands to satisfy an increased global food demand already being impacted by the effects of climate change. According to statistics from the Food and Agriculture Organization (FAO) released in 2016, if the current state of GHG emissions and climate change persists, major cereal crop yields (20%–45% for maize, 5%–50% for wheat, and 20%–30% for rice) will fall by the year 2100 (FAO, 2016). Therefore, if the current trends continue, crop losses may rise at an unprecedented rate in the very near future, which will significantly contribute to decreased production and raised food prices and make it more challenging to meet the increasing demands of the expanding population. Due to the population's constant growth, intense agricultural methods have been adopted, including the unprecedented use of agrochemicals, the production of livestock (for meat and other forms of income), and the exploitation of water resources (Arora, 2019). This has exacerbated the problem by causing the contamination of natural resources and the release of GHG (due to agricultural operations). Though forests serve as a sink for the rising $CO_2$ levels, the unregulated rate of deforestation has unbalanced the carbon cycle. This has led to an increase in carbon emissions and an uneven pattern of the climate, which have a negative influence on a number of factors, including agricultural production.

### 5.3.4 Escalated Energy Demand and its Rising Cost

A large amount of energy is used in agricultural production, either directly through the burning of fossil fuels or indirectly through the use of energy-intensive inputs, particularly fertilizer. For instance, total production cost from direct energy usage reached to an average of 6.7% during 2005–2008. A large amount of energy is used in agricultural production, either directly through the burning of fossil fuels or indirectly through the use of energy-intensive inputs, particularly fertilizer. The average cost of direct energy usage from 2005 to 2008 was roughly 6.7% of overall production cost. Therefore, whether the changes in energy prices are brought on by global oil markets, policies to meet environmental goals, or policies to improve energy security, agricultural production is always sensitive to those changes. Producers of livestock are similarly impacted by energy-related costs. Despite the fact that their direct energy costs are lower than those of crop production, livestock producers will bear increased feed costs under both the lower and higher energy price change scenarios. However, poultry is the animal species that converts feed the most effectively into meat. Its production would be less impacted than that of beef and pork (Sands et al., 2011). Consumer food prices are impacted by higher agricultural commodity prices, but energy costs associated with food processing, distribution, and marketing have a greater impact on retail food prices than do costs associated with the production of agricultural commodities.

## 5.4 ROLE OF SOIL SECURITY IN ADDRESSING GLOBAL CHALLENGES

### 5.4.1 FOOD SECURITY

The world population is accelerating at an unprecedented level which is alarming in many ways. In order to face the existing challenges, we should comprehend earth systems more deeply along with minimizing the detrimental impacts humans are causing to the planet. Since soil security is directly and indirectly related to the most of the existential challenges, such as food security, water security, climate change, biodiversity protection, and human health protection, therefore, instead of emphasizing each challenge separately, there is a need to collectively understand these challenges. Without soils, it's impossible to imagine food security, clean water, sustainable water supply, biodiversity, and various ecosystem services (Bouma and McBratney, 2013; Holt et al., 2016; Pozza and Field, 2020).

Similar to soil security, food security has its own certain dimensions, i.e., availability, access, utilization, and stability. Availability has been defined as the physical presence of a sufficient amount of good quality food which is directly correlated with agricultural land management and its availability (Hwalla et al., 2016). Access is when a person has enough resources to obtain safe and healthy food which could be acquired from the market (indirect) and home production (direct) (Leroy et al., 2015). Utilization ranges from the way food is consumed and absorbed in the human body to food choices, preparation, and hygiene (Gross et al., 2000). The fourth dimension of food security is "stability" which means the guaranteed availability, access, and utilization of food and its sustainable availability despite environmental changes, political changes, and conflicts.

Since 1990, experts have joined their heads to ponder whether there is enough land available to feed the future generations. According to an estimate, if food consumption patterns kept changing and the food consumption rate kept increasing at the same rate, then the production level of 2005 needs to be increased up to 70% to 110% in order to meet food demands in 2050 (Tilman et al., 2011; Bruinsma, 2011; Alexandratos and Bruinsma, 2012; Gomiero, 2016). In order to satisfy wheat, rice, and maize demands in upcoming decades, an estimated 1% to 1.5% increase in the yield of cereal crops is required (Cassman, 1999). In 2013, Lambin and coworkers suggested that in comparison with the year 2000, an additional 81 to 147 mha needs to be cultivated to meet the production demands of 2030. However, keeping in view the incessant upsurge in population, rapid urbanization, forest plantation, and conversion of fertile lands into industries, houses, roads, and bioenergy policy mandates, which further reduce the available croplands, it is required to reserve an additional 285 to 792 mha to croplands to reach production benchmarks of 2030. It's axiomatic that soil is much more than dirt. It performs various functions supporting human, animal, and plant lives. One of its imperative functions is to provide a medium to sustain plant life along with storing nutrients and water for their growth (Koch et al., 2013; Pozza and Field, 2020). Hence, it's crucial to properly manage soils and protect them from degradation and erosion in order to feed our future generations.

### 5.4.2 WATER SECURITY

Soil is a non-renewable natural resource that plays role in moving, storing, filtering, and transferring water. Soil affects the quantity and quality of water, which is relevant to SDG#6 (clean water and sanitation). Water is not only required for quenching human and animal's thirst and performing various household activities but is also essential for food production, poverty reduction, performing various ecosystem services, and education. 70% of global water is being consumed solely by the agriculture sector. An incessant upsurge in the global population, shifts in human lifestyle, and continuously changing global trends would exert more pressure on the available water supply, ultimately creating water scarcity issues (Ki-Moon, 2014; Lal et al., 2021). An increase has been observed in the global safe drinking water consumption from 61% in 2000 to 71% in 2017. Despite this increase, 4.2 billion people lack access to properly managed sanitation, and 2.2 billion lack access to safe drinking water (UN, 2020).

In 2050, it's anticipated that there will be 9 billion people on earth, and food demand for a population of 9 billion will put more strain on water resources causing the two-thirds population to face water shortage. Besides, around two-thirds of the world's population is likely to dwell in cities in 2050 which would put more pressure on the sustainable supply of urban water resources. (Rosegrant et al., 2009; Jägerskog and Jønch Clausen, 2012; Falkenmark, 2013). Climate change, natural hazards, and contamination of surface and groundwater hinder the chances of attaining SDG#6. In order to preserve this valuable resource proper land management strategies should be followed. Since agriculture is the primary sector utilizing 70% of the fresh water where the most of the water of is utilized by irrigation. Following model irrigation techniques would improve irrigation efficiency and prevent water loss by runoff (Garrick and Hall, 2014). In 2019 Visser and coworkers reported few cases of integrated land management approaches that not only helped in achieving SDG#6 but also played a major role in acquiring other SDGs i.e., organic farming and conservation of infiltrating rainwater which in turn assisted in preserving groundwater against drought. Besides, practicing tillage, mulching and surface residue management have shown promising results (Hatfield et al., 2001; Lal et al., 2021).

### 5.4.3 SUSTAINABLE BIOENERGY

Fossil fuels are the major source of global energy, contributing greater than 80% to global consumption. However, owing to constant demand and excessive consumption fossil fuel reserves are diminishing and there is a need to seek alternative, eco-friendly and sustainable sources of energy to fulfill the increasing demands for energy and to reduce greenhouse gases (GHGs) emission (Singh et al., 2018). Bioenergy has emerged as a potential alternative to fossil fuels which not only could provide sustainable energy by reducing the consumption of fossil fuels but also reduces GHGs emission and sequesters carbon (George and Cowie, 2011). Currently, half of the global renewable energy (wind, water and solar) demand is being satisfied by bioenergy (Leirpoll et al., 2021; Shorabeh et al., 2021).

Bioenergy is derived from biomass and its production is dependent on the amount of biomass (Umar et al., 2021). For sustainable bioenergy production uninterrupted supply of biomass is essential (Nwozor et al., 2021; Malode et al., 2021). However, excessive biomass production and modifications to land use and management may have an impact on the biological, chemical, and properties of soil as well as soil carbon stocks. Additionally, bioenergy production could affect water quality and may develop land competition between food and fuel eventually leading to a loss in carbon stocks. Bioenergy compared to fossil fuel sources is relatively expensive. Following proper management strategies can reduce these negative outcomes and ensure several benefits, including sustainable energy, mitigation of climate change, rural development, and improved human health (Duer and Christensen, 2010; George and Cowie, 2011).

Cultivation of biofuel feedstock crops like switch grass, willows, and poplars on marginal and degraded lands has shown promising results. Many studies have reported promising yields of switchgrass on marginal lands and degraded lands (abandoned coal mines) for bioenergy production (Marra et al., 2013; Brown et al., 2016; Chen et al., 2016). Production of perennial biofuel crops in the filter strips and multifunctional riparian buffers not only improves water quality but also ensures sustainable bioenergy and additional income for the farmers (Ferrarini et al., 2017). Growing cover crops for biofuel feedstocks is another promising approach with minimal-to-no effects on the food systems (Jones et al., 2018). Besides, the implementation of optimum land use planning i.e., employing a functional land management framework and spatial land use planning could be employed for sustainable bioenergy production (Ale et al., 2019).

### 5.4.4 Climate Change Abatement

Global warming and climate change are causing both direct and indirect adverse effects on all life sectors throughout the globe. Most commonly it is causing human health disorders, temperature increase, wildfires, floods, drought, changes in the patterns of disease vectors, air pollution, and ecosystem disruption (Simpson et al., 2021). It is also causing negative effects on the welfare and health of livestock by disrupting their metabolic activities, causing oxidative stresses, reduced immunity, and more infections. Soil has gained significant importance for global climate change mitigation and carbon reduction agenda (Nave et al., 2021). The storehouse of carbon is the soil, and soil security is greatly helpful to reduce the adverse effects of climate change, global warming, and food insecurity. Carbon sequestration has been advocated as an effective measure for the abatement of climate change (Hunt et al., 2020). Soil security is concerned with improving soil health, biodiversity, and its physical, chemical, and biological properties. Whereas soil health and soil security are being adversely affected due to urbanization, global warming, and industrialization. Thus, the capacity and functioning of soil for nutrient cycling, gaseous regulation, supporting biodiversity, and ecosystem functioning, and mitigating the adversities of global warming and climate change is reduced (Zhang et al., 2020).

Naturally, the soil is the storehouse of carbon and holds the significant capacity to sequester carbon from the surroundings. Grasslands, forests, and wetlands have more carbon sequestration capacity and help to reduce the adverse effects of climate change. These are also known as carbon saving accounts and help to stabilize and regulate the global climate (Nunes et al., 2020). While the carbon sequestering capacity of agricultural soils can also be enhanced by using organic cultivation practices and reducing the use of chemical-based inputs. Enhancing organic matter in the soils is also important to improve carbon sequestration potential. Soils with good microbial communities, organic matter, and carbon function much better to reduce the adverse effect of climate change, global warming, and GHG emissions (Jansson and Hofmockel, 2020). Healthy soils are significantly important to improve resilience against climate change by offering regulation of flow and drainage and storing more water. It also causes the regulation of dissolved solids, nutrients, and contaminants. Soil security assures healthy and functional soils having different sections for agricultural and microbial activities, soil macrobiota, leaching, and groundwater recharge (Keesstra et al., 2021). Compacted and poorly drained soils are deprived of soil biodiversity and vegetation and favor the phenomenon of climate change and global warming. Healthy soils also improve the resilience towards wet periods by acting as reservoirs and filters, improving percolation and infiltration through the soil, reducing sedimentation and soil erosion, flooding mitigation, and improving water quality. Similarly, healthy soils are also helpful during heating periods due to their ability to hold more water and regulate soil temperature (Chang et al., 2021).

### 5.4.5 HUMAN HEALTH PROTECTION

Soil security is directly related to the production of safe and nutritional food, fiber, biodiversity, and fuel. The soil security is not only related to the agriculture and environment, but it also supports human life and ecosystem functioning on the earth (Beerling et al., 2018). Soils naturally contain essential elements and support various functions which eventually play a role in providing human health protection, clean water, and air. Healthy soils help to get more forests, diversify wildlife, improve the productivity of grazing lands, and produce nutrient-rich foods having better quality (Ruppel, 2022). Soil security also helps to maximize the production of ornamental plants in public places, homes, and hospitals that exert positive effects on the mental state and wellness of people (Vibholm et al., 2022). Thus, their work potential and productivity are enhanced due to mental satisfaction. Moreover, soil security also favors the better uptake of nutrients from the soil and their translocation and transformation in the edible plants (O'Connor, et al., 2022).

Thus, it is significantly helpful to improve social justice and equality by ensuring soil security all over the globe. The linkage between soil and human beings is complex, yet it can be simplified by the use of right and target-oriented approaches. Soil is the host to numerous organisms, and moderation of their interaction for soil security can be used for the betterment of human life (Ling et al., 2022). There is the assimilation of beneficial and harmful microorganisms by human beings either through direct or indirect routes. There have been extensive studies to check the effects of healthy and contaminated soils on human health. There has been significant degradation and deterioration of soil health and quality due to urbanization, industrialization, climate change, and global warming. Overuse of fertilizers, pesticides, synthetic chemicals, and other inputs has also threatened and affected soil security because of which the whole world is facing far-reaching consequences (Riyaz et al., 2022).

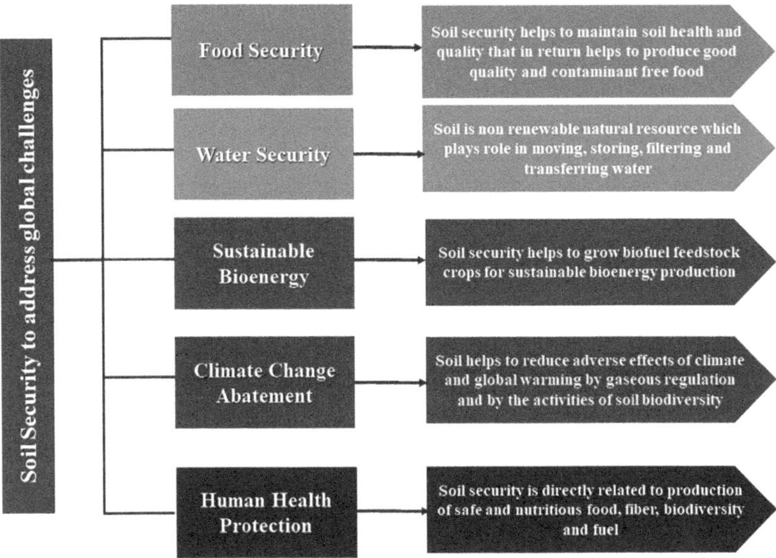


**FIGURE 5.2** Role of soil security in addressing prevalent global challenges.

The presence of pharmaceutical compounds, organic pollutants, and potentially toxic elements in the soil and the use of agricultural chemicals is causing an enhanced risk of various diseases, including asthma, cancer, diabetics, headaches, eyes and skin irritation, nausea, and dizziness (Parra-Arroyo et al., 2022). Millions of agricultural workers suffer from pesticide poisoning, and the adverse effect can be greatly minimized by soil security measures. Soil security also minimizes the chances of airborne dust and erosion (Plutino et al., 2022). The suspension and free movement of soil particles in the air causes respiratory diseases and lungs cancer. Airborne dust also contains additional contaminants, such as radioactive materials, pollens, insects, heavy metals, organic chemicals, harmful chemicals, and pathogens. Breathing of these toxicants promotes their entry into the bloodstream and causes adverse effects on health and wellness. Soil security measures significantly reduce the soilborne particles and protect human beings from their toxic effects (Steffan et al., 2018). A recapitulation of soil security and its role in resolving prevalent global challenges has been given in Figure 5.2.

## 5.5  VARIOUS LAND MANAGEMENT STRATEGIES

Food security is closely linked with the ability of the land to sustain the population. The deterioration of soil structure, decline of soil fertility, salinization, alkalization, water and wind erosion, loss of biodiversity, and organic matter are all factors that contribute to land degradation. Therefore, improvement of soil health and rehabilitation of degraded lands to achieve sustainable development should be our first priority. In order to achieve targeted food security, practicing proper land management and conservation strategies is important.

### 5.5.1 MANAGEMENT OF SOIL BIODIVERSITY

Soil biodiversity is defined as a diverse range of organisms that interact with plants and small animals to form a web of life and perform biological activities in the soil system. It improves plant nutrition and water storage, providing recycling of organic matter, controlling soil pests and diseases and resistance to soil erosion (Chandra, 2021). Soil biodiversity is a critical component of healthy soil for sustainable soil and crop productivity (Fan et al., 2021).

Soil biodiversity loss is directly associated with the degradation of soil and change in ecosystem functions (Bach et al., 2020). It includes physical (crusting, compaction, deterioration of soil structure), chemical (salinization, acidification, decline in nitrogen mineralization, elemental imbalance, nutrient depletion), and biological (loss of soil organic matter, soil species and activity of soil microorganisms) deterioration (Lal, 2004). Changes in belowground diversity are assumed as the reflection of aboveground (Wall and Nielsen, 2012). The biodiversity aboveground started to decrease as land conversion and agricultural intensification took place, having an impact on the associated ecosystem both above and belowground (Tibbett et al., 2020).

Soil biodiversity conservation and utilization of ecosystem functions are convenient for both agriculture and its consumers (Robertson and Swinton, 2005; Delgado-Baquerizo et al., 2020). Different management strategies, such as minimum tillage, crop rotation, and incorporation of manure and crop residues, can contribute to the conservation of soil biodiversity and improve the capacity of soil to perform its functions. The application of manure can increase nematode-trapping fungi (Wachira et al., 2009). It is suggested that frequent bio inoculant applications should be encouraged to increase their numbers in soil (Okoth et al., 2011). Another management practice that can be used for soil biodiversity is crop rotation. Crop rotation is a well-known approach for preventing one group of organisms from accumulating at the expense of another. It can also break down the life cycle of successive plant pathogens and pests in the soil (Barbieri et al., 2019). Intercropping or relay cropping system helps to promote several species to grow. Cabbage/cowpea, spider plant/eggplant, and maize/bean are examples of intercrop systems that are beneficial for increasing soil biodiversity and reducing hazardous species.

### 5.5.2 SOIL ORGANIC MATTER RESTORATION

Soil organic matter (SOM) has an important role in sustainable agriculture and the environment and in preventing soil degradation (Table 5.3). SOM is considered one of the key measures of soil quality and health for centuries (Darwin, 1840; Jenny, 1941; Rodale, 1945; Allison, 1973; Jenny, 1994; Howard, 2006). Soil fertility is directly linked with SOM. Both microorganisms and organic matter are considered reservoirs of plant nutrients (Imran et al., 2019). Organic matter acts as a nutrient reservoir and soil conditioner, preventing nutrient loss due to leaching and erosion (Imran, 2018). Soil organic matter also has a key role to improve structural stability and water retention (Lal, 2020).

The soil organic matter is a non-replaceable component with a variety of important qualities, in both well-developed and severely degraded soils, which may change

**TABLE 5.3**

**Organic Matter and Its Role in the Betterment of the Ecosystem**

| Functions | Explanations | References |
| --- | --- | --- |
| Erosion control (aggregation) | OM plays role in aggregation with mineral surface | (Hernandez-Soriano et al., 2018; Totsche et al., 2018; Bucka et al., 2019) |
| | Protein- and carbohydrate-rich litters support faunal activity that help aggregation | (Mizuta et al., 2015; Guhra et al., 2022) |
| | Persistent aggregation through constant growth of a microbial community that produces binding agents on recalcitrant material | (Miltner et al., 2012; Sarker et al., 2018) |
| | OM hydrophobicity reduces aggregate wetting and increases aggregate stability | (Kimura et al., 2017) |
| | Through humic and fulvic acids make good aggregates by binding to mineral surface | (Polláková et al., 2018) |
| | Dissolved OM can cause binding through rapid bacterial production | (Krause et al., 2019) |
| Erosion control (water retention) | OM provide numerous pore network to retain water | (Eden et al., 2017) |
| Climate regulation (carbon sequestration) | Promote production of microbes that use carbon energy | (Cotrufo et al., 2013) |
| | Production of microbes and necro mass to stabilize fast decomposition, recalcitrant OM property can reduce decomposition | (Wang et al., 2020b) |
| Improve soil properties | SOM can contribute to aromaticity or aliphaticity in soil | (Dos Santos et al., 2020) |
| | Persistent OM contribute to thermal stability and oxidation resistance in soil | (Han et al., 2018) |
| | Solubility of OM helps in stabilization of mineral surfaces | (Ge et al., 2020) |
| | OM has a number of functional groups which bind to phyllosilicate (amide groups) and metal oxides (carboxylate and aromatic groups) | (Tipping et al., 2011; Janzen, 2019; Kleber and Lehmann, 2019; Olk et al., 2019) |
| | OM improve wetting properties through amphiphilic molecules | (Hoffland et al., 2020) |
| Soil biodiversity | Through soil aggregation provides diverse microenvironment for microbes | (George et al., 2017; Bach et al., 2018; Feng et al., 2022) |
| | Decomposition of SOM produces different types of faunal and microbial guilds | (Frouz, 2018; de Graaff et al., 2019) |
| Plant growth | SOM produces hormones that stimulate root and shoot growth | (Hijbeek et al., 2017; Olaetxea et al., 2018; Oldfield et al., 2019) |

(*Continued*)

**TABLE 5.3** *(Continued)*
**Organic Matter and Its Role in the Betterment of the Ecosystem**

| Functions | Explanations | References |
| --- | --- | --- |
| | Increased carbon use efficiency and nitrogen use efficiency | (Hijbeek et al., 2018; Bonanomi et al., 2019) |
| | SOM provides fresh litter that declines C/nutrient ratios and increases mineralization | (Kleber et al., 2007) |
| | OM decomposition produces saprotrophic competitors against pathogens | (Bonanomi et al., 2018) |
| | With high carbon use efficiency, co-metabolic degradation of organo-pollutants is enabled | (Straathof, 2015) |
| | Aggregates provide more pore spaces for water retention that are eventually used by plants | (Eden et al., 2017) |
| | Aggregation also provides pore spaces for good aeration | (Reichert et al., 2022) |
| Better water quality | Through a variety of charges and functional groups, OM produces good quality water | (Wang et al., 2019) |

the physical, chemical, and biological conditions (Li et al., 2017; Orgill et al., 2017). As a result, long-term and sustainable management systems should improve and maintain SOM content, but there is no one-size-fits-all solution for improving and sustaining SOM content. Conservation agriculture (such as no-till, cover cropping, residue retention, and integrated nutrient management), composite farming systems, and crops with tree and livestock integration are some of the approaches that attain these conditions of improving soil health by restoration of SOM (Smith and Kucera, 2018). Compost can be considered to enhance SOM if correctly applied (Adiyah et al., 2021). Short legume fallows are effective for maintaining SOM levels (Muzangwa et al., 2020). The effects of grassland revegetation on soil's physicochemical parameters and surface electrochemical properties can be significant. As the grassland restoration period grows, the aggregate stability, water content, and SOM content increases (Liu et al., 2020).

SOM content depends on the amount of plant biomass produced and the amount of dead organic matter introduced into the soil to supply energy and nourishment to soil organisms (Hairiah et al., 2006). Microorganism quantity and composition are influenced by litter and root inputs from trees (Błońska et al., 2016; Lladó et al., 2017). Soil microorganisms, which are key contributors to the formation of SOM and chemical stability, may also vary in the chemical stability of SOM due to differences in biochemical decomposability and transformation of residues (Zhu et al., 2020). Organic carbon derived from plants changes into SOM, and higher organic components in combination with soil microbial residues contribute to the restoration and stabilization of SOM (Li et al., 2021).

### 5.5.3 MANAGEMENT OF SOIL NUTRITION

Soil nutrient depletion is a major problem causing soil degradation and is directly connected to food insecurity and agriculture sustainability. Intensive use of land for agricultural production and improper management practices cause low fertility and nutrient depletion in soils (Nayakekorale, 2020). During the 20th century, advancements in genetics and the development of high yield verities and high-cost fertilizer usage resulted in soil fertility depletion (Ganguly et al., 2021; Jayne and Sanchez, 2021). There are many other reasons responsible for fertility depletion, such as soil erosion, water logging, crusting, compaction, the decline in organic matter, less crop rotation, and soil salinization (Lal, 2015; Aleminew and Alemayehu, 2020).

Soil productivity is directly linked with soil fertility, so it's very important to pay attention to soil fertility management. Integrated nutrient management involves using organic and inorganic resources wisely. Nutrient cycling is optimized through integrated nutrient management, as it ensures fitted cycling with synchronization between crop nutrient demand and soil nutrient release while reducing losses due to leaching, runoff, volatilization, and immobilization (Geta et al., 2013; Shah and Wu, 2019). Increasing organic matter, using cover crops, crop rotation, and crop diversification helps in maintaining soil fertility (Silveira and Kohmann, 2020; Shah et al., 2021). The addition of crop residues (straw, roots, and stems) and manures and improving the management of fertilizer in the right amount at the right time with the right place and the right method are significant measures to protect soil losses and sustain production (Johnston and Bruulsema, 2014; Zhao et al., 2020). Conservation agriculture practices are important in degraded soils with nutrient depletion for sustainable agriculture.

### 5.5.4 MULCHING AND COVER CROPS

Mulching is the technique of putting various covering materials to the soil's surface in order to enhance crop production and halt moisture loss (Kader et al., 2019). Mulching has significant environmental advantages, i.e., regulates soil and plant root temperature, lowers nutrient losses, prevents soil erosion and compaction, and enhances the physical characteristics of the soil (Ngouajio and McGiffen, 2002; Lamont, 2005). Cover crops are another effective way of soil conservation to reduce runoff and water erosion and improve soil productivity (Haruna et al., 2020). The advantages of the cover crops in land management are well recognized. Cover crops not only give shelter to unprotected land from water and wind erosion but they also reduce evaporation and help soil moisture conversation (Dabney et al., 2001). It reduces water runoff and improves infiltration by declining surface seal formation and enhances soil porosity and earthworm activity (Ruan et al., 2001). Cover crops reduce soil erosion through splash detachment by lowering the kinetic energy of raindrops (Haramoto and Gallandt, 2005).

Mostly, cover crops are cultivated not only for the economic benefits but also for the ecosystem benefits. It provides conditioning for successive crops. Mostly, cover crops are planted as green manures in the fields and then incorporated as mulch and soil amendments to improve fertility (Kaspar et al., 2011).

Cover crops also contribute to environmental sustainability, as soil organic carbon has shown an increase in lands cultivated with cover crops compared to non-cover crops management (Haruna et al., 2017; Haruna, 2019). Cover crops are also effective in reducing soil and nutrient loss into streams and rivers (Aryal et al., 2018; Adler et al., 2020). There are three main categories of cover crops that vary according to their properties and uses: leguminous, grasses, and broadleaf non-leguminous.

In conservation agriculture (CA), cover crops residues are kept on the soil surface as dead or live mulch which protect the soil from erosion and also improve the physical, chemical, and biological features of soil (Calegari et al., 2013). Globally, various researchers and farmers agreed on the fact that soil mulch cover is a good management strategy for soil protection in various cropping systems and different agro-ecological zones (Sharma et al., 2018).

## 5.6  CONCLUSIONS/SUMMARY

Research in the past decade including the SDGs of the UN has emphasized and assured the importance of soil conservation, owing to its association with existential global challenges of food, water, energy, climate change, and human health. However, soils are exposed to various threats coming from both natural and anthropogenic sources, which are the biggest hurdles in a way of achieving the SDGs of the UN. Therefore, there is a need to conserve soil from degradation and restore the degraded soils by following proper land management strategies. Research without practice is nothing more than a piece of random information. Despite much attention given to soil security, lots of research questions and knowledge gaps are there requiring the attention of researchers and policymakers. There is a need for a strategic relationship between soil scientists and policymakers to put that research into action, to make them realize the importance of soil security and support policy delivery, and to achieve SD by 2030. There is a need to follow proper land management strategies, i.e., management of soil biodiversity, restoration of soil organic matter, and nutrition and protection of soils from land degradation. Besides, since agriculture is directly and indirectly linked with various SDGs, especially SDG#1, SDG#2, SDG#3, SDG#6 SDG#7, and SDG#13; therefore, there is also a need to increase the area under cultivation in order to cope with the food, water, and energy demands for the upcoming decades.

## REFERENCES

Adiyah, F., Fuchs, M., Mich, Eli, E., & Dawoe, E. (2021). Cocoa farmers perceptions of soil organic carbon effects on fertility, management and climate change in the Ashanti region of Ghana. *African Journal of Agricultural Research*, *17*(5), 714–725.

Adler, R., Singh, G., Nelson, K., Weirich, J., Motavalli, P., & Miles, R. (2020). Cover crop impact on crop production and nutrient loss in a no-till terrace topography. *Journal of Soil and Water Conservation*, *75*(2), 153–165.

Alcamo, J., Dronin, N., Endejan, M., Golubev, G., & Kirilenko, A. (2007). A new assessment of climate change impacts on food production shortfalls and water availability in Russia. *Global Environmental Change*, *17*(3–4), 429–444.

Ale, S., Femeena, P. V., Mehan, S., & Cibin, R. (2019). Environmental impacts of bioenergy crop production and benefits of multifunctional bioenergy systems. In *Bioenergy with carbon capture and storage* (pp. 195–217). Academic Press.

Aleminew, À., & Alemayehu, M. (2020). Soil fertility depletion and its management options under crop production perspectives in Ethiopia. *Agricultural Reviews*, *41*, 91–105.

Alexandratos, N., & Bruinsma, J. (2012). *World agriculture towards 2030/2050: The 2012 revision*. ESA working, agricultural development economics division food and agriculture organization of the united nations. www.fao.org/economic/esa

Allison, F. E. (1973). *Soil organic matter and its role in crop production*. Elsevier.

Al-Muaini, A., Green, S., Dakheel, A., Abdullah, A. H., Abdelwahid Abou Dahr, W., Dixon, S., Kemp, P., & Clothier, B. (2019). Irrigation management with saline groundwater of a date palm cultivar in the hyper-arid United Arab Emirates. *Agricultural Water Management*, *211*, 123–131.

Alrøe, H. F., & Noe, E. B. (2014). Second-order science of interdisciplinary research: A polyocular framework for wicked problems. *Constructivist Foundations*, *10*(1), 65–76.

AQUASTAT FAO. (2021). *AQUASTAT-FAO's global information system on water and agriculture*. Retrieved from www.fao.org/nr/water/aquastat/main/index.stm.

Arora, N. K. (2019). Impact of climate change on agriculture production and its sustainable solutions. *Environmental Sustainability*, *2*(2), 95–96.

Aryal, J. P., Mehrotra, M. B., Jat, M. L., & Sidhu, H. S. (2015). Impacts of laser land leveling in rice–wheat systems of the north-western indo-gangetic plains of India. *Food Security*, *7*(3), 725–738.

Aryal, N., Reba, M., Straitt, N., Teague, T., Bouldin, J., & Dabney, S. (2018). Impact of cover crop and season on nutrients and sediment in runoff water measured at the edge of fields in the Mississippi Delta of Arkansas. *Journal of Soil and Water Conservation*, *73*(1), 24–34.

Bach, E. M., Ramirez, K. S., Fraser, T. D., & Wall, D. H. (2020). Soil biodiversity integrates solutions for a sustainable future. *Sustainability*, *12*(7), 2662.

Bach, E. M., Williams, R. J., Hargreaves, S. K., Yang, F., & Hofmockel, K. S. (2018). Greatest soil microbial diversity found in micro-habitats. *Soil Biology and Biochemistry*, *118*, 217–226.

Barbieri, P., Pellerin, S., Seufert, V., & Nesme, T. (2019). Changes in crop rotations would impact food production in an organically farmed world. *Nature Sustainability*, *2*(5), 378–385.

Barnett, T. P., Adam, J. C., & Lettenmaier, D. P. (2005). Potential impacts of a warming climate on water availability in snow-dominated regions. *Nature*, *438*(7066), 303–309.

Battisti, D. S., & Naylor, R. L. (2009). Historical warnings of future food insecurity with unprecedented seasonal heat. *Science*, *323*(5911), 240–244.

Beerling, D. J., Leake, J. R., Long, S. P., Scholes, J. D., Ton, J., Nelson, P. N., . . . & Hansen, J. (2018). Farming with crops and rocks to address global climate, food and soil security. *Nature Plants*, *4*(3), 138–147.

Birthal, P. S., Nigam, S. N., Narayanan, A. V., & Kareem, K. A. (2012). Potential economic benefits from adoption of improved drought-tolerant groundnut in India. *Agricultural Economics Research Review*, *25*(347–2016–16900), 1–14.

Błońska, E., Lasota, J., & Gruba, P. (2016). Effect of temperate forest tree species on soil dehydrogenase and urease activities in relation to other properties of soil derived from loess and glaciofluvial sand. *Ecological Research*, *31*(5), 655–664.

Blum, W. E. (2016). Role of soils for satisfying global demands for food, water, and bioenergy. In *Environmental resource management and the nexus approach* (pp. 143–177). Springer.

Bonanomi, G., Incerti, G., Abd El-Gawad, A. M., Cesarano, G., Sarker, T. C., Saulino, L., . . . & Mazzoleni, S. (2018). Comparing chemistry and bioactivity of burned vs. decomposed plant litter: different pathways but same result? *Ecology*, *99*(1), 158–171.

Bonanomi, G., Sarker, T. C., Zotti, M., Cesarano, G., Allevato, E., & Mazzoleni, S. (2019). Predicting nitrogen mineralization from organic amendments: beyond C/N ratio by 13C-CPMAS NMR approach. *Plant and Soil, 441*(1), 129–146.

Borrelli, P., Van Oost, K., Meusburger, K., Alewell, C., Lugato, E., & Panagos, P. (2018). A step towards a holistic assessment of soil degradation in Europe: Coupling on-site erosion with sediment transfer and carbon fluxes. *Environmental Research, 161*, 291–298.

Bouma, J., & McBratney, A. (2013). Framing soils as an actor when dealing with wicked environmental problems. *Geoderma, 200*, 130–139.

Brown, C., Griggs, T., Keene, T., Marra, M., & Skousen, J. (2016). Switchgrass biofuel production on reclaimed surface mines: I. Soil quality and dry matter yield. *BioEnergy Research, 9*(1), 31–39.

Bruinsma, J. (2011). The resources outlook: By how much do land, water and crop yields need to increase by 2050? *Looking Ahead in World Food and Agriculture: Perspectives to 2050*, 233–278.

Bucka, F. B., Kölbl, A., Uteau, D., Peth, S., & Kögel-Knabner, I. (2019). Organic matter input determines structure development and aggregate formation in artificial soils. *Geoderma, 354*, 113881.

Burke, E. J., Brown, S. J., & Christidis, N. (2006). Modeling the recent evolution of global drought and projections for the twenty-first century with the Hadley Centre climate model. *Journal of Hydrometeorology, 7*(5), 1113–1125.

Cai, X., Wang, D., & Laurent, R. (2009). Impact of climate change on crop yield: A case study of rainfed corn in central Illinois. *Journal of Applied Meteorology and Climatology, 48*(9), 1868–1881.

Calegari, A., Tourdonnet, S., Tessier, D., Rheinheimer, D., Ralisch, R., Hargrove, W., Guimarães, M., & Filho, J. (2013). Influence of soil management and crop rotation on physical properties in a long-term experiment in Paraná, Brazil. *Communications in Soil Science and Plant Analysis, 44*(13), 2019–2031.

Carleton, T. A., & Hsiang, S. M. (2016). Social and economic impacts of climate. *Science, 353*(6304), aad9837.

Cassman, K. G. (1999). Ecological intensification of cereal production systems: yield potential, soil quality, and precision agriculture. *Proceedings of the National Academy of Sciences, 96*(11), 5952–5959.

Challinor, A. J., Watson, J., Lobell, D. B., Howden, S. M., Smith, D. R., & Chhetri, N. (2014). A meta-analysis of crop yield under climate change and adaptation. *Nature Climate Change, 4*(4), 287–291.

Chandra, R. (2021). Soil biodiversity and community composition for ecosystem services. In *Soil science: Fundamentals to recent advances* (pp. 69–84). Springer.

Chang, X., Gao, Z. Wang, S., & Chen, H. (2019). Modelling long-term soil salinity dynamics using SaltMod in Hetao Irrigation District, China. *Computers and Electronics in Agriculture, 156*, 447–458.

Chang, Y., Rossi, L., Zotarelli, L., Gao, B., Shahid, M. A., & Sarkhosh, A. (2021). Biochar improves soil physical characteristics and strengthens root architecture in Muscadine grape (Vitis rotundifolia L.). *Chemical and Biological Technologies in Agriculture, 8*(1), 1–11.

Chen, Y., Ale, S., & Rajan, N. (2016). Spatial variability of biofuel production potential and hydrologic fluxes of land use change from cotton (Gossypium hirsutum L.) to Alamo switchgrass (Panicum virgatum L.) in the Texas High Plains. *BioEnergy Research, 9*(4), 1126–1141.

Ciais, P., Reichstein, M., Viovy, N., Granier, A., Ogée, J., Allard, V., . . . & Valentini, R. (2005). Europe-wide reduction in primary productivity caused by the heat and drought in 2003. *Nature, 437*(7058), 529–533.

Cotrufo, M. F., Wallenstein, M. D., Boot, C. M., Denef, K., & Paul, E. (2013). The microbial efficiency-matrix stabilization (MEMS) framework integrates plant litter decomposition with soil organic matter stabilization: Do labile plant inputs form stable soil organic matter? *Global Change Biology, 19*(4), 988–995.

Dabney, S. M., Delgado, J. A., & Reeves, D. W. (2001). Using winter cover crops to improve soil and water quality. *Communications in Soil Science and Plant Analysis, 32*(7–8), 1221–1250.

Darwin, C. (1840). XXXV—on the formation of mould. *Transactions of the Geological Society of London, 2*(3), 505–509.

de Graaff, M. A., Hornslein, N., Throop, H. L., Kardol, P., & van Diepen, L. T. (2019). Effects of agricultural intensification on soil biodiversity and implications for ecosystem functioning: A meta-analysis. *Advances in Agronomy, 155*, 1–44.

Delgado-Baquerizo, M., Reich, P. B., Trivedi, C., Eldridge, D. J., Abades, S., Alfaro, F. D., Bastida, F., Berhe, A. A., Cutler, N. A., & Gallardo, A. (2020). Multiple elements of soil biodiversity drive ecosystem functions across biomes. *Nature Ecology & Evolution, 4*(2), 210–220.

Denman, K. L., Brasseur, G., Chidthaisong, A., Ciais, P., Cox, P. M., Dickinson, R. E., Hauglustaine, D., Heinze, C., Holland, E., Jacob, D., Lohmann, U., Ramachandran, S., da Silva Dias, P. L., Wofsy, S. C., & Zhang, X. (2007). Couplings between changes in the climate system and biogeochemistry. In S. Solomon, D. Qin, M. Manning, Z. Chen, M. Marquis, K. B. Averyt, M. Tignor, & H. L. Miller (Eds.), *Climate change 2007: The physical science basis. Contribution of working group I to the fourth assessment report of the intergovernmental panel on climate change* (pp. 499–587). Cambridge University Press.

D'hondt, K., Kostic, T., McDowell, R., Eudes, F., Singh, B. K., Sarkar, S., . . . & Sessitsch, A. (2021). Microbiome innovations for a sustainable future. *Nature Microbiology, 6*(2), 138–142.

Döll, P. (2002). Impact of climate change and variability on irrigation requirements: A global perspective. *Climatic Change, 54*(3), 269–293.

Dos Santos, O. A. Q., Tavares, O. C. H., García, A. C., Rossi, C. Q., de Moura, O. V. T., Pereira, W., . . . & Pereira, M. G. (2020). Fire lead to disturbance on organic carbon under sugarcane cultivation but is recovered by amendment with vinasse. *Science of the Total Environment, 739*, 140063.

Duer, H., & Christensen, P. O. (2010). Socio-economic aspects of different biofuel development pathways. *Biomass and Bioenergy, 34*(2), 237–243.

Eden, M., Gerke, H. H., & Houot, S. (2017). Organic waste recycling in agriculture and related effects on soil water retention and plant available water: A review. *Agronomy for Sustainable Development, 37*(2), 1–21.

European Commission. (2006). *Communication from the commission to the council, the European parliament, the European economic and social committee and the committee of regions.* EC.

Falkenmark, M. (2013). Growing water scarcity in agriculture: future challenge to global water security. *Philosophical Transactions of the Royal Society A: Mathematical, Physical and Engineering Sciences, 371*(2002), 20120410.

Fan, K., Delgado-Baquerizo, M., Guo, X., Wang, D., Zhu, Y.-G., & Chu, H. (2021). Biodiversity of key-stone phylotypes determines crop production in a 4-decade fertilization experiment. *The ISME Journal, 15*(2), 550–561.

FAO. (2016). *The state of food and agriculture–climate change, agriculture and food security.* Retrieved from www.fao.org/3/i6030e/i6030e.pdf.

FAO. (2017). *Global soil partnership endorses guidelines on sustainable soil management.* Retrieved from www.fao.org/global-soil-partnership/resources/highlights/detail/en/ 416516/.

FAO. (2021). *The state of the world's land and water resources for food and agriculture— systems at breaking point*. Synthesis Report. FAO.

FAO. (2022). *Management of salt affected soils*. Retrieved fromwww.fao.org/soils-portal/ soil-management/management-of-some-problem-soils/salt-affected-soils/en/.

Feng, H., Pan, H., Li, C., &Zhuge, Y. (2022). Microscale heterogeneity of soil bacterial communities under long-term fertilizations in fluvo-aquic soils. *Soil Ecology Letters*, 1–11.

Ferrarini, A., Serra, P., Almagro, M., Trevisan, M., & Amaducci, S. (2017). Multiple ecosystem services provision and biomass logistics management in bioenergy buffers: A state-of-the-art review. *Renewable and Sustainable Energy Reviews, 73*, 277–290.

Ferreira, C. S., Pereira, P., & Kalantari, Z. (2018). Human impacts on soil. *Science of the Total Environment, 644*, 830–834.

Fischer, G., Tubiello, F. N., Van Velthuizen, H., & Wiberg, D. A. (2007). Climate change impacts on irrigation water requirements: Effects of mitigation, 1990–2080. *Technological Forecasting and Social Change, 74*(7), 1083–1107.

Frouz, J. (2018). Effects of soil macro-and mesofauna on litter decomposition and soil organic matter stabilization. *Geoderma, 332*, 161–172.

Ganguly, R. K., Mukherjee, A., Chakraborty, S. K., & Verma, J. P. (2021). Impact of agrochemical application in sustainable agriculture. In *New and future developments in microbial biotechnology and bioengineering* (pp. 15–24). Elsevier.

Garrick, D., & Hall, J. W. (2014). Water security and society: risks, metrics, and pathways. *Annual Review of Environment and Resources, 39*, 611–639.

Ge, S., Pan, Y., Zheng, L., & Xie, X. (2020). Effects of organic matter components and incubation on the cement-based stabilization/solidification characteristics of lead-contaminated soil. *Chemosphere, 260*, 127646.

George, B. H., & Cowie, A. L. (2011). Bioenergy systems, soil health and climate change. In *Soil health and climate change* (pp. 369–397). Springer.

George, P. B., Keith, A. M., Creer, S., Barrett, G. L., Lebron, I., Emmett, B. A., . . . & Jones, D. L. (2017). Evaluation of mesofauna communities as soil quality indicators in a national-level monitoring programme. *Soil Biology and Biochemistry, 115*, 537–546.

Geta, E., Bogale, A., Kassa, B., & Elias, E. (2013). Determinants of farmers' decision on soil fertility management options for maize production in Southern Ethiopia. *American Journal of Experimental Agriculture, 3*(1), 226.

Global e-GeSI. (2015). *SMARTer2030: ICT solutions for 21st century challenges*. Global e-Sustainability Initiative.

Gomiero, T. (2016). Soil degradation, land scarcity and food security: Reviewing a complex challenge. *Sustainability, 8*(3), 281.

Gray, R. (2010). Is accounting for sustainability actually accounting for sustainability and how would we know? An exploration of narratives of organizations and the planet. *Accounting, Organizations and Society, 35*(1), 47–62.

Gregory, P. J., Johnson, S. N., Newton, A. C., & Ingram, J. S. (2009). Integrating pests and pathogens into the climate change/food security debate. *Journal of Experimental Botany, 60*(10), 2827–2838.

Gross, R., Schoeneberger, H., Pfeifer, H., & Preuss, H. J. (2000). The four dimensions of food and nutrition security: definitions and concepts. *SCN News, 20*(20), 20–25.

Guhra, T., Stolze, K., &Totsche, K. U. (2022). Pathways of biogenically excreted organic matter into soil aggregates. *Soil Biology and Biochemistry, 164*, 108483.

Hackten Broeke, M. J. D., Mulder, H. M., Bartholomeus, J. G., van Brakel, P. T. G., Supit, I., de Wit, A. J. W., & Ruijtenberg, R. (2019). Quantitative land evaluation implemented in Dutch water management. *Geoderma, 338*, 536–545.

Hairiah, K., Sulistyani, H., Suprayogo, D., Purnomosidhi, P., Widodo, R. H., & Van Noordwijk, M. (2006). Litter layer residence time in forest and coffee agroforestry systems in Sumberjaya, West Lampung. *Forest Ecology and Management, 224*(1–2), 45–57.

Han, L., Ro, K. S., Wang, Y., Sun, K., Sun, H., Libra, J. A., & Xing, B. (2018). Oxidation resistance of biochars as a function of feedstock and pyrolysis condition. *Science of the Total Environment, 616*, 335–344.

Haramoto, E. R., & Gallandt, E. R. (2005). Brassica cover cropping: I. Effects on weed and crop establishment. *Weed Science, 53*(5), 695–701.

Haruna, S. I. (2019). Influence of winter wheat on soil thermal properties of a Paleudalf. *International Agrophysics, 33*(3).

Haruna, S. I., Anderson, S. H., Nkongolo, N. V., Reinbott, T., & Zaibon, S. (2017). Soil thermal properties influenced by perennial biofuel and cover crop management. *Soil Science Society of America Journal, 81*(5), 1147–1156.

Haruna, S. I., Anderson, S. H., Udawatta, R. P., Gantzer, C. J., Phillips, N. C., Cui, S., & Gao, Y. (2020). Improving soil physical properties through the use of cover crops: A review. *Agrosystems, Geosciences & Environment, 3*(1), e20105.

Hatfield, J. L., Sauer, T. J., & Prueger, J. H. (2001). Managing soils to achieve greater water use efficiency: A review. *Agronomy Journal, 93*, 271–280.

Hernandez-Soriano, M. C., Dalal, R. C., Warren, F. J., Wang, P., Green, K., Tobin, M. J., . . . & Kopittke, P. M. (2018). Soil organic carbon stabilization: Mapping carbon speciation from intact microaggregates. *Environmental Science & Technology, 52*(21), 12275–12284.

Hijbeek, R. H. F. M., Ten Berge, H. F. M., Whitmore, A. P., Barkusky, D., Schröder, J. J., & Van Ittersum, M. K. (2018). Nitrogen fertiliser replacement values for organic amendments appear to increase with N application rates. *Nutrient Cycling in Agroecosystems, 110*(1), 105–115.

Hijbeek, R., van Ittersum, M. K., ten Berge, H. F., Gort, G., Spiegel, H., & Whitmore, A. P. (2017). Do organic inputs matter–a meta-analysis of additional yield effects for arable crops in Europe. *Plant and Soil, 411*(1), 293–303.

Hoffland, E., Kuyper, T. W., Comans, R. N., & Creamer, R. E. (2020). Eco-functionality of organic matter in soils. *Plant and Soil, 455*(1), 1–22.

Holt, A. R., Alix, A., Thompson, A., & Maltby, L. (2016). Food production, ecosystem services and biodiversity: We can't have it all everywhere. *Science of the Total Environment, 573*, 1422–1429.

Holton, J. R., Curry, J. A., & Pyle, J. A. (Eds.). (2003). *Encyclopedia of atmospheric sciences* (Vol. 1). Academic Press.

Horn, R., Domżżał, H., Słowińska-Jurkiewicz, A., & Van Ouwerkerk, C. (1995). Soil compaction processes and their effects on the structure of arable soils and the environment. *Soil and Tillage Research, 35*(1–2), 23–36.

Horn, R., Fleige, H., Lal, R., & Zimmermann, I. (2018). Soil health and functions as a basic requirement for advancing the SDGs. *Soils and Sustainable Development Goals. GeoEcology Essay, Catena Soil Sciences, Stuttgart, 978*(3).

Hossain, M. A., Araki, H., & Takahashi, T. (2011). Poor grain filling induced by waterlogging is similar to that in abnormal early ripening in wheat in Western Japan. *Field Crops Research, 123*, 100–108.

Howard, A. (2006). *The soil and health: A study of organic agriculture.* University Press of Kentucky.

Hunt, J. R., Celestina, C., & Kirkegaard, J. A. (2020). The realities of climate change, conservation agriculture and soil carbon sequestration. *Global Change Biology, 26*(6), 3188–3189.

Hwalla, N., El Labban, S., & Bahn, R. A. (2016). Nutrition security is an integral component of food security. *Frontiers in Life Science, 9*(3), 167–172.

Imran, A. (2018). Global impact of climate change on water, soil resources and threat towards food security: Evidence from Pakistan. *Advances in Plants & Agriculture Research, 8*(5), 350–355.

Imran, A., Khan, A., Bari, A., Ud Din, R., Ali, R., Ahmed, N., Ali, A., Maula, F., Ahmad, F., & Naveed, S. (2019). Production statistics and modern technology of maize cultivation in Khyber Pakhtunkhwa Pakistan. *Plants Science of Arc, 2*, 1–12.

Jägerskog, A., & Jønch Clausen, T. (2012). *Feeding a thirsty world: challenges and opportunities for a water and food secure future.* Stockholm International Water Institute.

Jansson, J. K., & Hofmockel, K. S. (2020). Soil microbiomes and climate change. *Nature Reviews Microbiology, 18*(1), 35–46.

Janzen, H. (2019). The future of humic substances research: Preface to a debate. *Journal of Environmental Quality, 48*(2), 205–206.

Jayne, T. S., & Sanchez, P. A. (2021). Agricultural productivity must improve in sub-Saharan Africa. *Science, 372*(6546), 1045–1047.

Jenny, H. (1941). *Factors of soil formation; a system of quantitative pedology* (No. 631.4 J45). Courier Corporation.

Jenny, H. (1994). *Factors of soil formation: a system of quantitative pedology.* Courier Corporation.

Jiang, Q., Peng, J., Biswas, A., Hu, J., Zhao, R., He, K., & Shi, Z. (2019). Characterizing dryland salinity in three dimensions. *Science of the Total Environment, 682*, 190–199.

Johnston, A. M., & Bruulsema, T. W. (2014). 4R nutrient stewardship for improved nutrient use efficiency. *Procedia Engineering, 83*, 365–370.

Jones, C. D., Oates, L. G., Robertson, G. P., & Izaurralde, R. C. (2018). Perennialization and cover cropping mitigate soil carbon loss from residue harvesting. *Journal of Environmental Quality, 47*(4), 710–717.

Kader, M. A., Singha, A., Begum, M. A., Jewel, A., Khan, F. H., & Khan, N. I. (2019). Mulching as water-saving technique in dryland agriculture. *Bulletin of the National Research Centre, 43*(1), 1–6.

Kang, Y., Khan, S., & Ma, X. (2009). Climate change impacts on crop yield, crop water productivity and food security: A review. *Progress in Natural Science, 19*(12), 1665–1674.

Kaspar, T., Singer, J., Hatfield, J., & Sauer, T. (2011). *The use of cover crops to manage soil* (Vol. 409). American Society of Agronomy and Soil Science Society of America.

Keesstra, S. D., Bouma, J., Wallinga, J., Tittonell, P., Smith, P., Cerdà, A., Montanarella, L., Quinton, J. N., Pachepsky, Y., Putten, W. H. V. D., Bardgett, R. D., Moolenaar, S., Mol, G., Jansen, B., & Fresco, L. O. (2016). The significance of soils and soil science towards realization of the United Nations sustainable development goals. *Soil, 2*, 111–128.

Keesstra, S. D., Sannigrahi, S., López-Vicente, M., Pulido, M., Novara, A., Visser, S., & Kalantari, Z. (2021). The role of soils in regulation and provision of blue and green water. *Philosophical Transactions of the Royal Society B, 376*(1834), 20200175.

Kettlewell, P. S., Sothern, R. B., & Koukkari, W. L. (1999). UK wheat quality and economic value are dependent on the North Atlantic Oscillation. *Journal of Cereal Science, 29*(3), 205–209.

Khatri-Chhetri, A., Aryal, J. P., Sapkota, T. B., & Khurana, R. (2016). Economic benefits of climate-smart agricultural practices to smallholder farmers in the Indo-Gangetic plains of India. *Current Science*, 1251–1256.

Ki-Moon, B. (2014). *The road to dignity by 2030: Ending poverty, transforming all lives and protecting the planet.* Synthesis Report of the Secretary-General on the Post-2015 Sustainable Development Agenda. United Nations.

Kimura, A., Baptista, M. B., & Scotti, M. R. (2017). Soil humic acid and aggregation as restoration indicators of a seasonally flooded riparian forest under buffer zone system. *Ecological Engineering, 98*, 146–156.

Kirkby, M. J., Irvine, B. J., Jones, R. J., Govers, G., & Team, P. (2008). The PESERA coarse scale erosion model for Europe. I. Model rationale and implementation. *European Journal of Soil Science, 59*, 1293–1306.

Kleber, M., & Lehmann, J. (2019). Humic substances extracted by alkali are invalid proxies for the dynamics and functions of organic matter in terrestrial and aquatic ecosystems. *Journal of Environmental Quality, 48*(2), 207–216.

Kleber, M., Sollins, P., & Sutton, R. (2007). A conceptual model of organo-mineral interactions in soils: self-assembly of organic molecular fragments into zonal structures on mineral surfaces. *Biogeochemistry, 85*(1), 9–24.

Koch, A., McBratney, A., Adams, M., Field, D., Hill, R., Crawford, J., . . . & Zimmermann, M. (2013). Soil security: Solving the global soil crisis. *Global Policy, 4*(4), 434–441.

Konikow, L. F., & Kendy, E. (2005). Groundwater depletion: A global problem. *Hydrogeology Journal, 13*(1), 317–320.

Krause, L., Biesgen, D., Treder, A., Schweizer, S. A., Klumpp, E., Knief, C., &Siebers, N. (2019). Initial microaggregate formation: Association of microorganisms to montmorillonite-goethite aggregates under wetting and drying cycles. *Geoderma, 351*, 250–260.

Kukal, M. S., & Irmak, S. (2018). Climate-driven crop yield and yield variability and climate change impacts on the US Great Plains agricultural production. *Scientific Reports, 8*(1), 1–18.

Kunkel, S., & Matthess, M. (2020). Digital transformation and environmental sustainability in industry: Putting expectations in Asian and African policies into perspective. *Environmental Science & Policy, 112*, 318–329.

Lal, R. (2004). Soil carbon sequestration to mitigate climate change. *Geoderma, 123*(1–2), 1–22.

Lal, R. (2013). Climate-strategic agriculture and the water-soil-waste nexus. *Journal of Plant Nutrition and Soil Science, 176*, 479–493.

Lal, R. (2015). Restoring soil quality to mitigate soil degradation. *Sustainability, 7*(5), 5875–5895.

Lal, R. (2020). Soil organic matter and water retention. *Agronomy Journal, 112*(5), 3265–3277.

Lal, R., Bouma, J., Brevik, E., Dawson, L., Field, D. J., Glaser, B., . . . & Zhang, J. (2021). Soils and sustainable development goals of the United Nations: An international union of soil sciences perspective. *Geoderma Regional, 25*, e00398.

Lambin, E. F., Gibbs, H. K., Ferreira, L., Grau, R., Mayaux, P., Meyfroidt, P., . . . & Munger, J. (2013). Estimating the world's potentially available cropland using a bottom-up approach. *Global Environmental Change, 23*(5), 892–901.

Lamont, W. J. (2005). Plastics: Modifying the microclimate for the production of vegetable crops. *Hort Technology, 15*(3), 477–481.

Lasi, H., Fettke, P., Kemper, H. G., Feld, T., & Hoffmann, M. (2014). Industry 4.0. *Business & Information Systems Engineering, 6*, 239–242.

Latif, A., Shakir, A. S., & Rashid, M. U. (2013). Appraisal of economic impact of zero tillage, laser land levelling and bed-furrow interventions in Punjab, Pakistan. *Pakistan Journal of Engineering and Applied Sciences, 13*, 65–81.

Lee, J., De Gryze, S., & Six, J. (2011). Effect of climate change on field crop production in California's Central Valley. *Climatic Change, 109*(1), 335–353.

Leirpoll, M. E., Næss, J. S., Cavalett, O., Dorber, M., Hu, X., & Cherubini, F. (2021). Optimal combination of bioenergy and solar photovoltaic for renewable energy production on abandoned cropland. *Renewable Energy, 168*, 45–56.

Leroy, J. L., Ruel, M., Frongillo, E. A., Harris, J., & Ballard, T. J. (2015). Measuring the food access dimension of food security: A critical review and mapping of indicators. *Food and Nutrition Bulletin, 36*(2), 167–195.

Li, N., Lei, W., Long, J., & Han, X. (2021). Restoration of chemical structure of soil organic matter under different agricultural practices from a severely degraded Mollisol. *Journal of Soil Science and Plant Nutrition, 21*(4), 3132–3145.

Li, N., You, M.-Y., Zhang, B., Han, X.-Z., Panakoulia, S., Yuan, Y.-R., Liu, K., Qiao, Y.-F., Zou, W.-X., &Nikolaidis, N. (2017). Modeling soil aggregation at the early pedogenesis stage from the parent material of a Mollisol under different agricultural practices. In *Advances in agronomy* (Vol. 142, pp. 181–214). Elsevier.

Li, Y., Ye, W., Wang, M., & Yan, X. (2009). Climate change and drought: A risk assessment of crop-yield impacts. *Climate Research, 39*(1), 31–46.

Ling, N., Wang, T., & Kuzyakov, Y. (2022). Rhizosphere bacteriome structure and functions. *Nature Communications, 13*(1), 1–13.

Lipczynska-Kochany, E. (2018). Effect of climate change on humic substances and associated impacts on the quality of surface water and groundwater: A review. *Science of the Total Environment, 640*, 1548–1565.

Liu, J., Wang, Z., Hu, F., Xu, C., Ma, R., & Zhao, S. (2020). Soil organic matter and silt contents determine soil particle surface electrochemical properties across a long-term natural restoration grassland. *Catena, 190*, 104526.

Liu, L., & Basso, B. (2020). Impacts of climate variability and adaptation strategies on crop yields and soil organic carbon in the US Midwest. *PLOS One, 15*(1), e0225433.

Lladó, S., López-Mondéjar, R., &Baldrian, P. (2017). Forest soil bacteria: Diversity, involvement in ecosystem processes, and response to global change. *Microbiology and Molecular Biology Reviews, 81*(2), e00063–00016.

Mahato, A. (2014). Climate change and its impact on agriculture. *International Journal of Scientific and Research Publications, 4*(4), 1–6.

Malode, S. J., Prabhu, K. K., Mascarenhas, R. J., Shetti, N. P., & Aminabhavi, T. M. (2021). Recent advances and viability in biofuel production. *Energy Conversion and Management: X, 10*, 100070.

Manasa, M. R. K., Katukuri, N. R., & Nair, S. S. D. (2020). Role of biochar and organic substrates in enhancing the functional characteristics and microbial community in a saline soil. *Journal of Environmental Management, 269*, 110737.

Marra, M., Keene, T., Skousen, J., & Griggs, T. (2013). Switchgrass yield on reclaimed surface mines for bioenergy production. *Journal of Environmental Quality, 42*(3), 696–703.

Mensah, J., & Enu-Kwesi, F. (2018). Implication of environmental sanitation management in the catchment area of Benya Lagoon, Ghana. *Journal of Integrative Environmental Sciences, 16*, 23–43.

Miltner, A., Bombach, P., Schmidt-Brücken, B., & Kästner, M. (2012). SOM genesis: Microbial biomass as a significant source. *Biogeochemistry, 111*(1), 41–55.

Mizuta, K., Taguchi, S., & Sato, S. (2015). Soil aggregate formation and stability induced by starch and cellulose. *Soil Biology and Biochemistry, 87*, 90–96.

Montaldo, C. R. B. (2013). Sustainable development approaches for rural development and poverty alleviation & community capacity building for rural development and poverty alleviation. Yonsei University. https://sustainabledevelopment.un.org/content/documents/877LR%20Sustainable%20Development%20v2.pdf

Mosaffa, H. R., & Sepaskhah, A. R. (2019). Performance of irrigation regimes and water salinity on winter wheat as influenced by planting methods. *Agricultural Water Management, 216*, 444–456.

Muzangwa, L., Mnkeni, P. N. S., & Chiduza, C. (2020). The use of residue retention and inclusion of legumes to improve soil biological activity in maize-based no-till systems of the Eastern Cape Province, South Africa. *Agricultural Research, 9*(1), 66–76.

Nara, E. O. B., da-Costa, M. B., Baierle, I. C., Schaefer, J. L., Benitez, G. B., do Santos, L. M. A. L., & Benitez, L. B. (2021). Expected impact of industry 4.0 technologies on sustainable development: A study in the context of Brazil's plastic industry. *Sustainable Production and Consumption, 25*, 102–122.

Nave, L. E., Bowman, M., Gallo, A., Hatten, J. A., Heckman, K. A., Matosziuk, L., . . . & Swanston, C. W. (2021). Patterns and predictors of soil organic carbon storage across a continental-scale network. *Biogeochemistry, 156*(1), 75–96.

Nayakekorale, H. (2020). Soil degradation. In *The soils of Sri Lanka* (pp. 103–118). Springer.

Ngouajio, M., & McGiffen, M. E. (2002, August). Sustainable vegetable production: Effects of cropping systems on weed and insect population dynamics. *XXVI International Horticultural Congress: Sustainability of Horticultural Systems in the 21st Century, 638*, 77–83.

Norris, C. E., Bean, G. M., Cappellazzi, S. B., Cope, M., Greub, K. L., Liptzin, D., . . . & Honeycutt, C. W. (2020). Introducing the North American project to evaluate soil health measurements. *Agronomy Journal, 112*(4), 3195–3215.

Nunes, L. J., Meireles, C. I., Pinto Gomes, C. J., & Almeida Ribeiro, N. (2020). Forest contribution to climate change mitigation: Management oriented to carbon capture and storage. *Climate, 8*(2), 21.

Nwozor, A., Owoeye, G., Olowojolu, O., Ake, M., Adedire, S., & Ogundele, O. (2021, February). Nigeria's quest for alternative clean energy through biofuels: An assessment. In *IOP conference series: Earth and environmental science* (Vol. 655, No. 1, p. 012054). IOP Publishing.

O'Connor, J., Hoang, S. A., Bradney, L., Rinklebe, J., Kirkham, M. B., & Bolan, N. S. (2022). Value of dehydrated food waste fertiliser products in increasing soil health and crop productivity. *Environmental Research, 204*, 111927.

Ogada, M. J., Rao, E. J., Radeny, M., Recha, J. W., & Solomon, D. (2020). Climate-smart agriculture, household income and asset accumulation among smallholder farmers in the Nyando basin of Kenya. *World Development Perspectives, 18*, 100203.

Okoth, S. A., Otadoh, J. A., & Ochanda, J. O. (2011). Improved seedling emergence and growth of maize and beans by Trichoderma harziunum. *Tropical and Subtropical Agroecosystems, 13*(1), 65–71.

Olaetxea, M., De Hita, D., Garcia, C. A., Fuentes, M., Baigorri, R., Mora, V., . . . & Garcia-Mina, J. M. (2018). Hypothetical framework integrating the main mechanisms involved in the promoting action of rhizospherichumic substances on plant root-and shoot-growth. *Applied Soil Ecology, 123*, 521–537.

Oldfield, E. E., Bradford, M. A., & Wood, S. A. (2019). Global meta-analysis of the relationship between soil organic matter and crop yields. *Soil, 5*(1), 15–32.

Olk, D. C., Bloom, P. R., Perdue, E. M., McKnight, D. M., Chen, Y., Farenhorst, A., . . . & Harir, M. (2019). Environmental and agricultural relevance of humic fractions extracted by alkali from soils and natural waters. *Journal of Environmental Quality, 48*(2), 217–232.

Orgill, S., Condon, J., Kirkby, C., Orchard, B., Conyers, M., Greene, R., & Murphy, B. (2017). Soil with high organic carbon concentration continues to sequester carbon with increasing carbon inputs. *Geoderma, 285*, 151–163.

Panagos, P., Borrelli, P., & Poesen, J. (2019). Soil loss due to crop harvesting in the European Union: A first estimation of an underrated geomorphic process. *Science of the Total Environment, 664*, 487–498.

Panagos, P., Imeson, A., Meusburger, K., Borrelli, P., Poesen, J., & Alewell, C. (2016). Soil conservation in Europe: Wish or reality? *Land Degradation & Development, 27*, 1547–1551.

Parra-Arroyo, L., González-González, R. B., Castillo-Zacarías, C., Martínez, E. M. M., Sosa-Hernández, J. E., Bilal, M., . . . & Parra-Saldívar, R. (2022). Highly hazardous pesticides and related pollutants: Toxicological, regulatory, and analytical aspects. *Science of the Total Environment, 807*, 151879.

Pimentel, D., Harvey, C., Resosudarmo, P., Sinclair, K., Kurz, D., McNair, M., . . . & Blair, R. (1995). Environmental and economic costs of soil erosion and conservation benefits. *Science, 267*(5201), 1117–1123.

Pimentel, D., Huang, X., Cordova, A., & Pimentel, M. (1997). Impact of population growth on food supplies and environment. *Population and Environment*, 9–14.

Plutino, M., Bianchetto, E., Durazzo, A., Lucarini, M., Lucini, L., & Negri, I. (2022). Rethinking the connections between ecosystem services, pollinators, pollution, and health:

Focus on air pollution and its impacts. *International Journal of Environmental Research and Public Health, 19*(5), 2997.

Polláková, N., Šimanský, V., & Kravka, M. (2018). The influence of soil organic matter fractions on aggregates stabilization in agricultural and forest soils of selected Slovak and Czech hilly lands. *Journal of Soils and Sediments, 18*(8), 2790–2800.

Pozza, L. E., & Field, D. J. (2020). The science of soil security and food security. *Soil Security, 1*, 100002.

Reichert, J. M., Corcini, A. L., Awe, G. O., Reinert, D. J., Albuquerque, J. A., Gallarreta, C. C. G., & Docampo, R. (2022). Onion-forage cropping systems on a Vertic Argiudoll in Uruguay: Onion yield and soil organic matter, aggregation, porosity and permeability. *Soil and Tillage Research, 216*, 105229.

Renn, O., Beier, G., & Schweizer P.-J. (2021). The opportunities and risks of digitalisation for sustainable development: A systemic perspective. *GAIA Ecological Perspectives for Science and Society, 30*, 23–28.

Riyaz, M., Mathew, P., Zuber, S. M., & Rather, G. A. (2022). Botanical pesticides for an eco-friendly and sustainable agriculture: New challenges and prospects. In *Sustainable agriculture* (pp. 69–96). Springer.

Robertson, G. P., & Swinton, S. M. (2005). Reconciling agricultural productivity and environmental integrity: A grand challenge for agriculture. *Frontiers in Ecology and the Environment, 3*(1), 38–46.

Rodale, J. I. (1945). *Pay dirt; farming & gardening with composts* (No. 632.86 R685p). Devin-Adair Company.

Rosegrant, M. W., Ringler, C., & Zhu, T. (2009). Water for agriculture: Maintaining food security under growing scarcity. *Annual Review of Environment and Resources, 34*, 205–222.

Roy, I. (2018). Addressing on abrupt global warming, warming trend slowdown and related features in recent decades. *Frontiers in Earth Science, 6*, 136.

Ruan, H., Ahuja, L. R., Green, T. R., & Benjamin, J. G. (2001). Residue cover and surface-sealing effects on infiltration: Numerical simulations for field applications. *Soil Science Society of America Journal, 65*(3), 853–861.

Ruppel, O. C. (2022). Soil protection and legal aspects of international trade in agriculture in times of climate change: The WTO dimension. *Soil Security*, 100038.

Sands, R., Westcott, P., Price, J. M., Beckman, J., Leibtag, E., Lucier, G., . . . & Wojan, T. R. (2011). *Impacts of higher energy prices on agriculture and rural economies* (No. 1477–2017–4002). Economic Research Service, Economic Research Report, United States Department of Agriculture.

Sarker, T. C., Incerti, G., Spaccini, R., Piccolo, A., Mazzoleni, S., & Bonanomi, G. (2018). Linking organic matter chemistry with soil aggregate stability: Insight from 13C NMR spectroscopy. *Soil Biology and Biochemistry, 117*, 175–184.

Schjønning, P., Lamandé, M., Berisso, F. E., Simojoki, A., Alakukku, L., & Andreasen, R. R. (2013). *Gas diffusion, non-darcy air permeability and CT-scans for a traffic-affected clay subsoil.* Paper Presented at the ASA, CSSA, and SSSA 2013 International Annual Meetings.

Schjønning, P., Stettler, M., Keller, T., Lassen, P., & Lamandé, M. (2015). Predicted tyre–soil interface area and vertical stress distribution based on loading characteristics. *Soil and Tillage Research, 152*, 52–66.

Schlenker, W., & Roberts, M. J. (2009). Nonlinear temperature effects indicate severe damages to US crop yields under climate change. *Proceedings of the National Academy of sciences, 106*(37), 15594–15598.

Scholes, R. J., Montanarella, L., Brainich, E., Barger, N., Ten Brink, B., Cantele, M., . . . & Willemen, L. (2018). *IPBES (2018): Summary for policymakers of the assessment report on land degradation and restoration of the intergovernmental science-policy platform on biodiversity and ecosystem services.* IPBES.

Scopelliti, M., Molinario, E., Bonaiuto, F., Bonnes, M., Cicero, L., De Dominicis, S., & Bonaiuto, M. (2018). What makes you a "hero" for nature? Socio-psychologicalprofiling of leaders committed to nature and biodiversity protection across seven; EU countries. *Journal of Environmental Planning and Management, 61*, 970–993.

Shah, F., & Wu, W. (2019). Soil and crop management strategies to ensure higher crop productivity within sustainable environments. *Sustainability, 11*(5), 1485.

Shah, K. K., Modi, B., Pandey, H. P., Subedi, A., Aryal, G., Pandey, M., & Shrestha, J. (2021). Diversified crop rotation: An approach for sustainable agriculture production. *Advances in Agriculture, 2021.*

Shahzalal, M. D., & Hassan, A. (2019). Communicating sustainability: Using community media to influence rural people's intention to adopt sustainable behavior. *Sustainability, 11*(3), 812.

Sharma, P., Singh, A., Kahlon, C. S., Brar, A. S., Grover, K. K., Dia, M., & Steiner, R. L. (2018). The role of cover crops towards sustainable soil health and agriculture—a review paper. *American Journal of Plant Sciences, 9*(9), 1935–1951.

Shorabeh, S. N., Argany, M., Rabiei, J., Firozjaei, H. K., & Nematollahi, O. (2021). Potential assessment of multi-renewable energy farms establishment using spatial multi-criteria decision analysis: A case study and mapping in Iran. *Journal of Cleaner Production, 295*, 126318.

Silveira, M. L., & Kohmann, M. M. (2020). Maintaining soil fertility and health for sustainable pastures. In *Management strategies for sustainable cattle production in southern pastures* (pp. 35–58). Elsevier.

Simpson, N. P., Mach, K. J., Constable, A., Hess, J., Hogarth, R., Howden, M., . . . & Trisos, C. H. (2021). A framework for complex climate change risk assessment. *One Earth, 4*(4), 489–501.

Singh, A. (2018). Alternative management options for irrigation-induced salinization and waterlogging under different climatic conditions. *Ecological Indicators, 90*, 184–192.

Singh, A. P., Agarwal, A. K., Agarwal, R. A., Dhar, A., & Shukla, M. K. (2018). Introduction of alternative fuels. In *Prospects of alternative transportation fuels* (pp. 3–6). Springer.

Singh, V. K., Shukla, A. K., Singh, M. P., Majumdar, K., Mishra, R. P., Rani, M., & Singh, S. K. (2015, January). *Effect of site-specific nutrient management on yield, profit and apparent nutrient balance under pre-dominant cropping systems of Upper Gangetic Plains*. ICAR.

Smakhtin, V., Revenga, C., & Döll, P. (2004). A pilot global assessment of environmental water requirements and scarcity. *Water International, 29*(3), 307–317.

Smith, C., & Kucera, M. (2018). *Effects on soil water holding capacity and soil water retention resulting from soil health management practices implementation*. USDA Natural Resources Conservation Service.

Steffan, J. J., Brevik, E. C., Burgess, L. C., & Cerdà, A. (2017). The effect of soil on human health: An overview. *European Journal of Soil Science, 69*, 159–171.

Steffan, J. J., Brevik, E. C., Burgess, L. C., & Cerdà, A. (2018). The effect of soil on human health: An overview. *European Journal of Soil Science, 69*(1), 159–171.

Straathof, A. L. (2015). *Explorations of soil microbial processes driven by dissolved organic carbon*. Wageningen University and Research.

Tao, F., Yokozawa, M., Xu, Y., Hayashi, Y., & Zhang, Z. (2006). Climate changes and trends in phenology and yields of field crops in China, 1981–2000. *Agricultural and Forest Meteorology, 138*(1–4), 82–92.

Tibbett, M., Fraser, T. D., & Duddigan, S. (2020). Identifying potential threats to soil biodiversity. *Peer J, 8*, e9271.

Tilman, D., Balzer, C., Hill, J., & Befort, B. L. (2011). Global food demand and the sustainable intensification of agriculture. *Proceedings of the National Academy of Sciences, 108*(50), 20260–20264.

Tipping, E., Lofts, S., & Sonke, J. E. (2011). Humic ion-binding model VII: A revised parameterisation of cation-binding by humic substances. *Environmental Chemistry, 8*(3), 225–235.

Totsche, K. U., Amelung, W., Gerzabek, M. H., Guggenberger, G., Klumpp, E., Knief, C., . . . & Kögel-Knabner, I. (2018). Microaggregates in soils. *Journal of Plant Nutrition and Soil Science, 181*(1), 104–136.

Udara Willhelm Abeydeera, L. H., Wadu Mesthrige, J., & Samarasinghalage, T. I. (2019). Global research on carbon emissions: A scientometric review. *Sustainability, 11*(14), 3972.

Ukaga, U., Maser, C., & Reichenbach, M. (2011). Sustainable development: Principles, frameworks, and case studies. *International Journal of Sustainability in Higher Education, 12*(2), Emerald.

Umar, M., Ji, X., Kirikkaleli, D., & Alola, A. A. (2021). The imperativeness of environmental quality in the United States transportation sector amidst biomass-fossil energy consumption and growth. *Journal of Cleaner Production, 285*, 124863.

UN. (2017). *World population prospects: 2017 revision population.* Retrieved from www.un.org/development/.

UN. (2020). *The sustainable development goals report.* United Nations.

Van den Akker, J. J. (2008). Soil compaction. In S. Huber, G. Prokop, D. Arrouays, G. Banko, A. Bispo, R. Jones, M. Kibblewhite, W. Lexer, A. Möller, & R. Rickson (Eds.), *Environmental assessment of soil for monitoring: Volume I, indicators & criteria EUR 23490 EN/1.* Office for the Official Publications of the European Communities.

Van den Akker, J. J., & Schjønning, P. (2004). Subsoil compaction and ways to prevent it. In P. Schjønning, S. Elmholt, & B. T. Christensen (Eds.), *Managing soil quality: Challenges in modern agriculture* (pp. 163–184). CABI Publishing.

Vara Prasad, P. V., Boote, K. J., Hartwell Allen, L., Jr., & Thomas, J. M. (2003). Super-optimal temperatures are detrimental to peanut (Arachis hypogaea L.) reproductive processes and yield at both ambient and elevated carbon dioxide. *Global Change Biology, 9*(12), 1775–1787.

Veerman, C., Pinto Correia, T., Bastioli, C., Biro, B., Bouma, J., & Cienciela, E. (2020). *Caring for soil is caring for life—ensure 75% of soils are healthy by 2030 for food, people, nature and climate, independent expert report.* EU Soil Health and Food Mission Board.

Vibholm, A. P., Christensen, J. R., & Pallesen, H. (2022). Occupational therapists and physiotherapists experiences of using nature-based rehabilitation. *Physiotherapy Theory and Practice*, 1–11.

Visser, S., Keesstra, S., Maas, G., De Cleen, M., & Molenaar, C. (2019). Soil as a basis to create enabling conditions for transitions towards sustainable land management as a key to achieve the SDGs by 2030. *Sustainability, 11*(23), 6792.

Vorosmarty, C. J., Green, P., Salisbury, J., & Lammers, R. B. (2000). Global water resources: Vulnerability from climate change and population growth. *Science, 289*(5477), 284–288.

Wachira, P., Kimenju, J., Okoth, S., & Mibey, R. (2009). Stimulation of nematode-destroying fungi by organic amendments applied in management of plant parasitic nematode. *Asian Journal of Plant Sciences, 8*, 153–159.

Wagan, S. A., Memon, Q. U. A., Wagan, T. A., Memon, I. H., & Wagan, A. (2015). Economic analysis of laser land leveling technology water use efficiency and crop productivity of wheat crop in Sindh. *Pakistan Journal of Earth Sciences and Environment, 5*, 21–25.

Wall, D. H., & Nielsen, U. N. (2012). Biodiversity and ecosystem services: Is it the same below ground. *Nature Education Knowledge, 3*(12), 8.

Wang, C., Wang, X. U., Pei, G., Xia, Z., Peng, B. O., Sun, L., . . . & Bai, E. (2020b). Stabilization of microbial residues in soil organic matter after two years of decomposition. *Soil Biology and Biochemistry, 141*, 107687.

Wang, F., Zhou, S., Biswas, A., Yang, S., & Ding, J. (2020a). Multi-algorithm comparison for predicting soil salinity. *Geoderma, 365*, 114211.

Wang, S., Zhao, M., Zhou, M., Li, Y. C., Wang, J., Gao, B., . . . & Ok, Y. S. (2019). Biochar-supported nZVI (nZVI/BC) for contaminant removal from soil and water: A critical review. *Journal of Hazardous Materials, 373*, 820–834.

Webster, P. J. (2008). Myanmar's deadly daffodil. *Nature Geoscience, 1*(8), 488–490.

White, J. W., Hunt, L., Boote, K. J., Jones, J. W., Koo, J., Kim, S., Porter, C. H., Wilkens, P. W., & Hoogenboom, G. (2013). Integrated description of agricultural field experiments and production: The ICASA version 2.0 data standards. *Computers and Electronics in Agriculture, 96*, 1–12.

Wollenweber, B., Porter, J. R., & Schellberg, J. (2003). Lack of interaction between extreme high-temperature events at vegetative and reproductive growth stages in wheat. *Journal of Agronomy and Crop Science, 189*(3), 142–150.

Yang, T., Lupwayi, N., Marc, S. A., Siddique, K. H., & Bainard, L. D. (2021). Anthropogenic drivers of soil microbial communities and impacts on soil biological functions in agroecosystems. *Global Ecology and Conservation, 27*, e01521.

Zhang, F., Guo, S., Zhang, C., & Guo, P. (2019). An interval multiobjective approach considering irrigation canal system conditions for managing irrigation water. *Journal of Cleaner Production, 211*, 293–302.

Zhang, J., Van Der Heijden, M. G., Zhang, F., & Bender, S. F. (2020). Soil biodiversity and crop diversification are vital components of healthy soils and agricultural sustainability. *Frontiers of Agricultural Science and Engineering, 7*(3), 236.

Zhao, C., Liu, B., Piao, S., Wang, X., Lobell, D. B., Huang, Y., . . . & Asseng, S. (2017). Temperature increase reduces global yields of major crops in four independent estimates. *Proceedings of the National Academy of Sciences, 114*(35), 9326–9331.

Zhao, X., Liu, B. Y., Liu, S. L., Qi, J. Y., Wang, X., Pu, C., Li, S. S., Zhang, X. Z., Yang, X. G., & Lal, R. (2020). Sustaining crop production in China's cropland by crop residue retention: A meta-analysis. *Land Degradation & Development, 31*(6), 694–709.

Zheng, C., Jiang, D., Liu, F., Dai, T., Jing, Q., & Cao, W. (2009). Effects of salt and waterlogging stresses and their combination on leaf photosynthesis, chloroplast ATP synthesis, and antioxidant capacity in wheat. *Plant Science, 176*, 575–582.

Zhu, X., Jackson, R. D., DeLucia, E. H., Tiedje, J. M., & Liang, C. (2020). The soil microbial carbon pump: From conceptual insights to empirical assessments. *Global Change Biology, 26*(11), 6032–6039.

# 6 Making the Water-Soil-Waste Nexus Wor
## *Framing the Boundries of Resource Flows*

*Laila Shahzad, Faiza Sharif, Arifa Tahir, Umair Raiz, Maryam Ali, Masooma Batool, and Ahmad Mahmood*

## 6.1 INTRODUCTION

Water resources are fundamental to support all life and productivity on land. Earth has 1,386 million $km^3$ water reserves of which 35 million $km^3$ (2.5%) accounts for freshwater resources. The fresh water supports 7.5 billion human population, globally (Meran et al. 2021). Developing countries rely on groundwater and different surface water sources, including rainfall, lakes, rivers, and ponds. Approximately, 85% of water resources is used for irrigation, 10% by domestic consumers, and 5% by the industry sector. Water bodies serve as a source of water supply and as a sink for runoff effluents. Watershed activities contribute to heavy pollution loads in case of untreated water discharge to water bodies (Benchbani et al., 2022). Water bodies are generally perceived as public goods, and user activities lack accountability. Flooding during the monsoon season and drought conditions prevail in dry seasons; thus, water supply also does not remain homogenous throughout the year (Sibly, 2020). Additionally, overexploitation, infrastructure development, and urbanization in metropolitan cities are reducing water infiltration rates through soil and restraining the groundwater recharge (Aitkadi & Ziyad, 2018). Bangladesh once had 4,238 water bodies (Rajshahi) in 1961; the number was reduced to 214 by 2019. The leading cause was the increase of built-up area from 10 $km^2$ to 17.7 $km^2$ during 1989–2019 (Kafy et al., 2021).

Diverse water consumption and effluent discharge sectors necessitate the development of boundaries for resource flow regulation across the sectors. Integrated water resources management (IWRM) is a holistic approach to water sustainable management (Alamanos et al., 2018). Dublin (1992) and Bonn (2011) conferences featured the IWRM approach as an integrative framework and sustainable solution for water shortage mitigation measures. The IWRM nexus approach is activated since World Summit 1992 (Ibisch et al., 2016). Nexus approach is out of water box and links other domains of terrestrial and aquatic systems. Ecological



DOI: 10.1201/9781003358169-6

integrity, equity, and efficiency are pillars of IWRM. The implementation of a strategic plan involves the appropriate allocation of scarce freshwater resource and development of participatory strategy for the stakeholder to ensure water conservation and reliable water distribution, considering economic growth factors (Benchbani et al., 2022). IWRM addresses the Sustainable Development Goal (SDG) 6 which stated, 'water availability and hygiene for all' (Galvez & Rojas, 2019). To address equity issues, IWRM combines upcoming and existing problems to address area and population needs and frames the targets under an all-inclusive approach.

IWRM framework strategy has multiple design phases. During planning phase, resources inter-relative boundaries are defined under nexus approach. Development phase has model finalization and testing of input parameters. Finally, the application or execution phase begins which is followed by the perpetuation phase. IWRM models reveal in-depth information facilitating rigorous decision-making for investigated site (Benn, 2016). Hydro-geomorphological information systems account for flow accumulation (FA) and flow direction (FD). Channelizing stream network, flow direction, and flow accumulation are essential estimations for IWRM plan (Kafy et al., 2021). Remote sensing (RS) and geographic information system (GIS) tools enable the precise implementation of the IWRM plan through spatiotemporal modeling. Generally, factorial dominance-based solutions and comprehensive modeling tools are preferred. Through modeling approaches such as support vector machine (SVM) algorithms and regression methods, a 95% accurate classification of site factors (such as stormwater runoff and winter steam flow) is achievable (Davis, 2007). In a study, optimized groundwater flow (MODFLOW, 2005) models were developed to assess basin (Himren dam system) water storage and to control water losses. Multiple solutions with weightage were selected as alternative options. For irrigation, open and drip irrigation mechanisms were recommended which caused aquifer conservation potential for three decades by minimizing 45% reliance on raw water supply (Jawad et al., 2019). Study revealed catchment area runoff, evaporation, and exploitation as major challenges.

Tennessee Valley Authority (TVA) in 1933 first adopted the IWM approach which had flood control, navigation, power production, public health, erosion control, and recreation elements (Meran et al., 2021). Now 80% of countries are adopting the IWRM principles globally. Israel, USA, and UK have adopted the IWRM approach for tackling the issues related to food and energy scarcities and to increase environment sustainability (Ngene et al., 2021). The research efforts are directed towards the production of genetically modified crops which are less water intensive. Furthermore, desalination, renewable energy, and reliable water infrastructure development are other research interests (Hwang et al., 2020).

IWRM approaches usually lack realities and legitimacy when reforms are to be implemented practically. IWRM's instruments can successfully be applied by rigorous institutional reforms. IWRM enabling conditions first require institutional integrity and stakeholder involvement in setting priorities (Saidi, 2017). Pollution prevention, improved wastewater management, technological advancement, and controlled reliance on freshwater resources are possible through sectoral collaboration.

As a matter of fact, water use is not limited to one sector; therefore, an integrated approach towards wastewater management is mandatory (Ingold & Tosun, 2020).

The current chapter is aimed at framing the boundaries of resource flow across multiple water use areas to illustrate the IWRM significance. IWRM theory will be seen under water and land integrated management systems to establish pathways for sustainable development. Additionally, after analyzing the status of IWRM approach in different countries, potential reductions in water consumption and wastewater production by different interrelating mechanism have been described.

## 6.2  INTEGRATED SOLID WASTE MANAGEMENT

ISWM is a strategic approach which basically covers all the waste collecting and sorting parameters for the sustainable management of the solid waste, and the resource utilization is the main point of consideration. Different areas are covered by the ISWM in which the geographical and administrative boundaries are considered as the top priority. Other areas for coverage are jurisdiction and administrative based. Different sectors and subsectors are also covered by the ISWM. Waste streams and types (hazardous and non-hazardous) are also considered in the ISWM (Marshall & Farahbakhsh, 2013).

### 6.2.1  CHALLENGES AND OPPORTUNITIES OF ISWM

- Cities with the major economic growth produces different types of wastes
- Composition and variety of wastes is changing with the changing habits of people
- Pessimistic impacts of wastes on the different aspects of environment
- To develop the management plan, there is a need for different stakeholders at different levels of waste generation
- This management plan is beneficial for the government which gains valuable resources from the waste and uses it in different purposes

### 6.2.2  BENEFITS OF ISWM

There are many benefits of the ISWM which are described later.

Cleaner and safe environment is the first benefit of ISWM. Higher resource use and reuse efficiency is also related to the ISWM. It also has economic benefits, like less cost, which is needed to dispose the final management of waste in case of resource reuse. Local bodies can also avail from this ISWM approach (Sridevi et al., 2012).

### 6.2.3  PLANNING OF ISWM

An ISWM plan consists of a management system, including the policies which are the main points of planning. Besides policies, technology (basic equipment and operational aspects) also plays an important role in the planning of ISWM. In the last section, the different voluntary measures (awareness raising, self-regulations, and

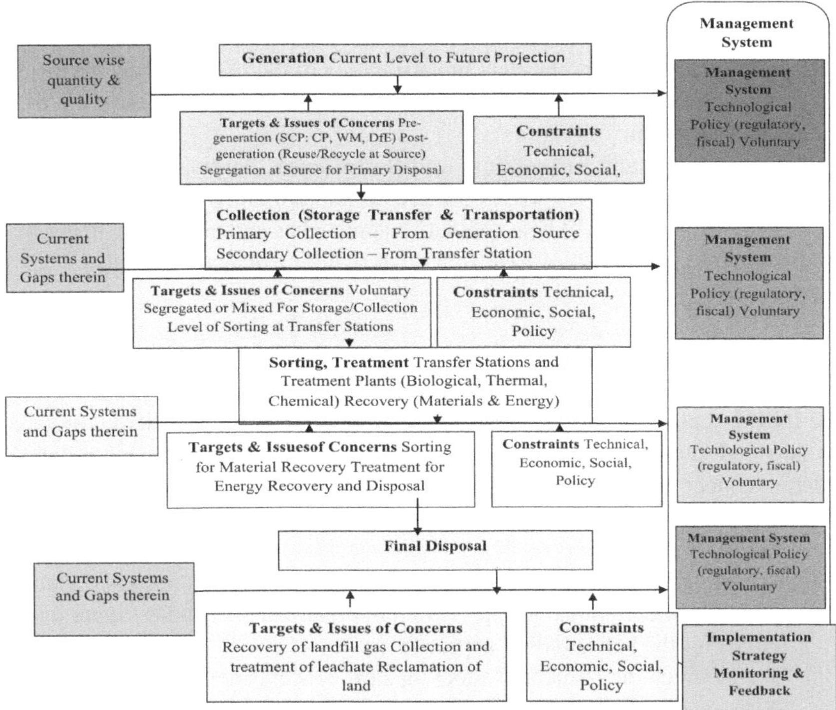



**FIGURE 6.1**    Integrated solid waste management (ISWM) plan.

other different parameters) also play their roles in the proper planning of ISWM. The planning of ISWM is shown in Figure 6.1.

There are different points of the hierarchy of ISWM which are described next.

- **Source reduction and reuse:** In this step, the minimum waste is generated and reuse of different items is encouraged.
- **Recycling:** In this step, different non-biodegradable items, like plastics, are recycled so that less waste is generated.
- **Composting:** In this step, different biodegradable materials are processed through composting in which the vermi composting is mainly used.
- **Waste to energy:** Waste is converted into energy by different methods, like RFD, bio-methanation, and incineration.
- **Landfills:** This is the last step which is not as encouraged as the other steps. In this step, waste is disposed in the sanitary landfills (Mushtaq et al., 2020).

## 6.3   INTEGRATED NATURAL RESOURCES MANAGEMENT

The INRM is a new concept which mainly focuses on the sustainable management of the all-natural resources like land, forest, water, biodiversity, and the other parameters of life (CGIAR Center Directors Committee, 1999). In operational practices, this concept plays an important role. Different conceptual and methodological frameworks are needed for the clear understanding of the INRM—how the different efforts are collaborated to attain the desired results and impacts in the form of sustainable development.

This complexity and 'integration' at different levels develop difficult organizational steps in a setup of the INRM where tasks are mainly focused on the use of the technology. A variety of new alternative approaches and models are being developed, and some were rediscovered from other scientific data and modified to INRM (Korten, 1980).

### 6.3.1 Main Elements of INRM Approach

The INRM approach emerges out of the experiences in the value-driven process of the new and latest approach in which people and different shareholders share the different concepts of the INRM. Both types of people can avail from this concept. The different levels of INRM approach are as lows.

   i. To reinforce the different policymaking concepts, there is need of collaboration of different local groups, different organizations, and different institutions for the better INRM approach
  ii. To encourage the farmers to adopt the new concepts developed by this new INRM approach and enhance their knowledge and leaning capacities and mixing them with new concepts to develop an action-based approach
 iii. To increase the collective learning through learning and action-based approach and facilitate self-reflection and sharing of concepts and knowledge
  iv. To make the NRM and the other associated policies and concepts coincide with the help of the stakeholder platforms of service providers and the different communities (Hagmann, 1999). The five capital assets of the INRM approach are shown in Figure 6.2

### 6.3.2 Key Features of INRM Research

There are different key features of the INRM approach which are described next:

* Follow a systems approach that is developed by different steps
* Be focused on the process
* Work at multiple levels and scales and involve multiple stakeholders and shareholders
* Address the different issues, especially the issue of trade-offs
* Employ and develop the new tools, methods, and the models to achieve aim
* Be agreeable to different points of the scaling up and out
* Do and perform a complement research on the improvement of the different steps of approach
* Lead and focus on the desirable and measurable impacts

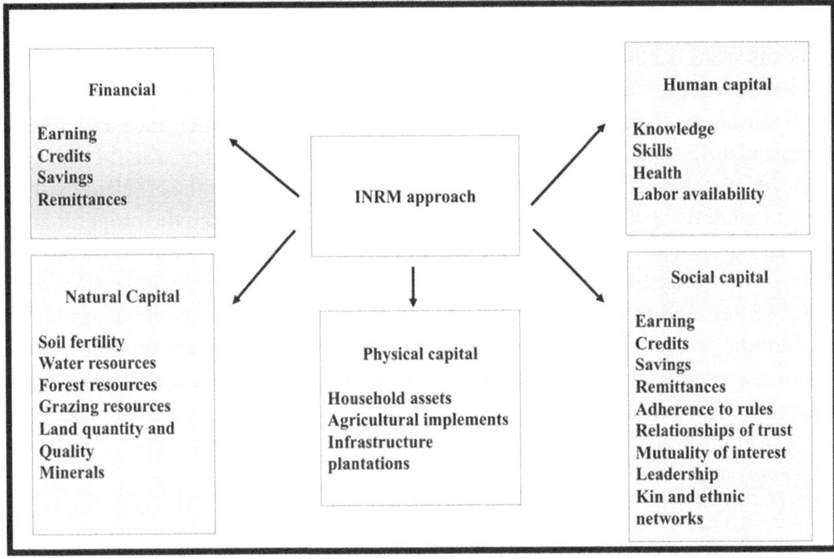



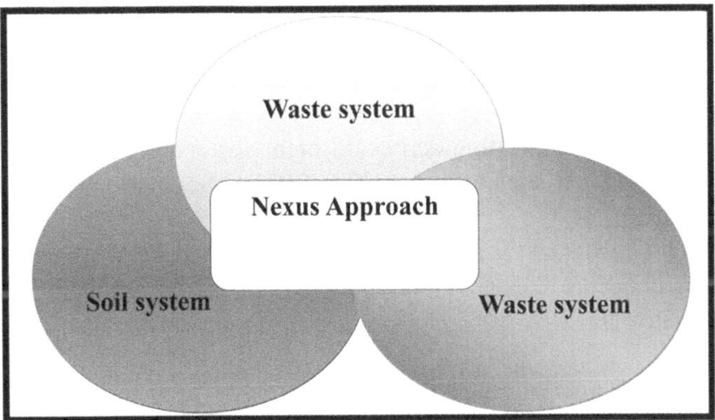



**FIGURE 6.2**   The five capital assets of INRM approach.

**Source: Carney (1998)**

### 6.3.3 Benefits of INRM

Benefits which are associated with the INRM are described as the payments and money assets for water regulation, soil conservation, carbon uses, and biodiversity-related works, which offset and gain many advantages from this. Income generation from sustainable use and management of the natural resources and many value added of forest products can be gained through this INRM approach. Other main benefits are the improved natural resource productivity and climate resilience of the environment (Lovell et al., 2002).

## 6.4  THE WATER-SOIL-WASTE NEXUS APPROACH

The nexus word is basically a Latin word meaning to combine or connect something. This word was first used in the 1980s in the environmental resource management term (food-energy nexus program). This word gained more importance and attention in the academics and policy departments as a major role on the 'water, energy, and food security nexus' program. The nexus approach can enhance the efficient use of the environmental resources like water, food, and energy by applying the proper management and governance involvement, leading towards sustainability and green economy.

The nexus concept has been increasingly discussed in the engineering sector, natural sector, and social sciences sector to make sense in the context of manifold understandings and uses. While a basic realization of the concept present is widely used in research studies and different other practices, the concept is rarely operationalized. This make hurdles in the analysis of the conditions and parameters (e.g., participation, governance) and effects or results (e.g., sustainability) of implement a nexus approach (Avellan et al., 2017).

UNU-FLORES's different initiatives on the nexus approach are described next:

- A good and basic understanding of the nexus resources among the different scientists and the people of the community
- Integration of the different concepts of the nexus approach into the policy and framework designs
- To gain practice by cutting across different sectors like academia, policy, practice, and civil society sectors for sustainable environmental resource management
- Not only focusing on the nexus parameter concepts but also paying attention in future concepts like energy use, geothermal resources, and the others like biodiversity

The WSW (water-soil-waste) nexus has a close association with the WEF (water-energy-food) nexus. WEF nexus mainly focused on the sectors; the WSW nexus defines how the resources are arranged in a manner to gain the sustainable management. In the WEF approach, the use of waste was omitted, as it requires more management. This approach mainly depended on the sectors, so the WSM approach basically used this waste dimension in different uses. In this regard, the materials and the energy flow can be seen and understood by the natural sciences processes (Schwarz & Bless, 2013). The WSM approach was shown in Figure 6.3.

Different socioeconomic benefits can be drawn by using this approach. At the end, different policymaking concepts and frameworks can be designed by using this approach. The WSW nexus approach is defined as follows.

The nexus approach to environmental resources management examines the following:

- The interconnections and interdependencies of environmental resources
- Their transitions and fluxes across spatial scales and between compartments
- Instead of just looking at separate components, the functioning, productivity, and management of a complex system is taken into consideration

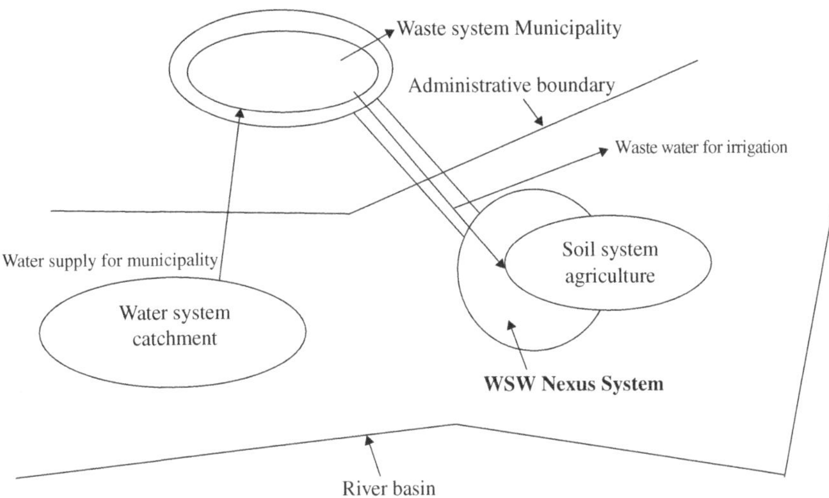


**FIGURE 6.3**  WSW nexus approach.

## 6.4.1  WSW Nexus Boundaries

Combining the inter-linkages of the resources describes the different benefits of agriculture which can improve the quality of soil and enhance the productivity. To clearly understand the concepts of the WSW nexus approach concept, then three types of boundaries are used to describe it.

1. Water: In this case, the wastewater is collected from the different human-based uses in a defined basin, and this basin boundary is partly or nor partly applicable
2. Soil: In this case, the wastewater can be collected into the desired ecosystem to fulfill the needs, and this is at an agro system level
3. Waste: In this case, wastewater can be transported from one point to another, and this is a municipal-based administrative boundary (not applicable) (UNU-FLORES, 2017)

All described approaches like INRM, ISWM, and IWRM are lacking to correct usage of wastewater for the different environmental uses. The WSW nexus system is basically a geographical part in which basically two systems are merging and the wastewater can be collected in the fields like the soil system. This nexus system gives advantages because of the reuse efficiency of the wastewater, and also, the productivity of the plants and the soil can be improved. Figure 6.3 describes the WSW nexus system in an efficient way which is described under. A small framework was described which is used to clearly understand the WSW nexus approach which says the following:

- Considering the level of wastewater that was entering the overall water-wastewater system is about 81.9 m³/s

- The soil condition can be improved by the addition of wastewater. In wastewater, different things are included, like the different nutrients and minerals and primary treatment of wastewater (quality)
- This basically decreases the complexity of different systems analysis and to make it able to mainly focus on bonding the different environmental resources to gain collective benefits
- The boundaries of the different resources like the water, soil, and waste systems are clear by their scholars and departments (Kurian & Ardakanian, 2015)

## 6.5  INTEGRATED WATER RESOURCE MANAGEMENT (IWRM)

IWRM framework has majorly emphasized components of water quality and ecological management, cross sectors integration, and stakeholder participation to guarantee the efficient use and supply of water resources (Badham et al., 2019). IWRM approaches can be evaluated by modeling for any site to categorize problems and to execute the IWRM-prioritized components. Hydrological modeling and water balance analyses brought by the IWRM tools facilitate the measure of social, economic, and environmental sustainability targets. The following table indicates the IWRM global implementation perspective as comparative scenario analyses.

## 6.6  POTENTIAL REDUCTIONS IN WATER CONSUMPTION AND WASTEWATER PRODUCTION BY INTEGRATED MECHANISMS

Hydrological cycle circulates water between water bodies, land, and the atmosphere. Water controls atmospheric conditions such as temperature and precipitation. Floods, riverbank erosion, water pollution, overexploitation, and management issues are the challenges that need attention in order to manage finite water reserves (Mees et al., 2017). Water is shared by multiple sectors; thus, the IWRM framework would bring coordinated development across all economic spheres. The following inter-relative mechanisms are worth considering to potentially reduce water consumption and wastewater production rates.

### 6.6.1  EFFICIENT USE OF WATER RESOURCES

Efficiency refers to enhanced users' productivities within the resource limits. However, productivity, if described in the context of agriculture, would also depend on other site-specific characteristics such as climatic and technological differences. Efficiency must be linked with fairness or equity as an allocation assessment criteria. Without IWRM variables modeling and simulation, equitable resource management is not possible (Momm et al., 2022). The equity application tools would be water pricing, water metering, polluter pay principle, resource allocation, and water resource-use permits. IWRM legal framework must incorporate the equity and obligatory fundamentals in infrastructure management programs. Depending on the multi-user perspective of water, integrated management strategy adoption by energy, sanitation, agriculture, and industrial as well as environmental sectors is necessary (Satriani

**TABLE 6.1**

**Integrated Water Resource Management (IWRM) Global Implementation Perspective for Proposed and In-Practice Scenarios**

| Country | Water availability per capita | Pressure points on water consumption | IWRM status | IWRM nexus approach | IWRM tools/components | Parameters studied | Comments/outcomes | References |
|---|---|---|---|---|---|---|---|---|
| Rajshahi, Bangladesh | 7,568 cubic meters/year | Irrigation, land use, urbanization | Proposed study | Water-agriculture land | GPS*, SVM site classification, Land Cover Classification System (LCCS), ERDAS Imagine (2015) tool for image assessment, flow accumulation (FA) and flow direction (FD) measuring tool (ArcGIS) to track water transformation level | Baseline data for water flow direction, watershed topography | Farming, technology, institutional, and water management need to be prioritized in IWRM | (Kafy et al., 2021) |
| Shihmen Reservoir at Tahan River, northern Taiwan | 1188 cubic meters/year | High water demand | Proposed study | Reservoir operation-community-agriculture | Multicriteria decision support (integrated scenario) system (SMC-DSS); multitierinteractive genetic algorithm (MIGA) analytics to establish aquifer storage hydrographs | Water demand, water inflow, and reservoir storage capacity, population size | Water discharge from agricultural and public sectors must be reduced by 50% and 20%, respectively | (Wang et al., 2011) |
| Tocantins-Araguaia River basin, Brazil | 43,027 cubic meters/year | High water demand | Proposed study | Environmental-political-administrative-socioeconomic | Water Resources Planning and Management (WARPLAM) decision support system (DSS); GIS for spatial analyses | Hydrographic basin limits, hydrologic data, and number of hydropower generation units | WARPLAM revealed socioeconomic, environmental, and political aspects of integration; DSS adjusted priorities for IWRM framework | (Coelho et al., 2012) |
| Upper Thames River basin, Southwestern Ontario, Canada | 144 cubic meters/year | Groundwater overexploitation at the sub-basin level, insufficient groundwater recharge | Proposed study | Water-land-socioeconomic environment | Integrated socio-hydrologic-economic model (multi-method modeling criteria) and spatial database and hydrologic model; multidisciplinary approach towards water resource management | Data about registered water users (municipal, agricultural, commercial), climatic condition (precipitation and temperature), and site characteristics | Local aquifers were under threat due to overexploitation and changing climatic conditions; urbanization, climate variations, and higher runoff volumes were major threats to river basin water | (Nikolic & Simonovic, 2015) |

(Continued)

## TABLE 6.1  *(Continued)*

## Integrated Water Resource Management (IWRM) Global Implementation Perspective for Proposed and In-Practice Scenarios

| Country | Water availability per capita | Pressure points on water consumption | IWRM status | IWRM nexus approach | IWRM tools/components | Parameters studied | Comments/outcomes | References |
|---|---|---|---|---|---|---|---|---|
| Zayandehrud river basin, Iran | 1,688 cubic meters/year | Frequent droughts and surface water shortage, high consumption rates | Proposed study | Water demand management, supply management, meta-supply management-based scenarios and analyses for basin water management | Zayandehrud planning model (ANFIS) forganiza resources sustainability index incorporated temperature, precipitation, and runoff variability factor and baseline scenario to find the rate of climate change; Fuzzy method applied to determine performance criteria and sustainability index and resilience-vulnerability factors in the system | Water demand-supply scenarios temporal analyses, volumetric factors, vulnerability, resilience, and maximum deficit | Meta scenario demonstrated demand-and-supply scenarios application to avoid resource degradation; study pointed sustainable resource management policies related to technical, educational, tariff, economic, and cultural areas for reduction (15%) in water demand | (Safavi et al., 2016) |
| Diyala River basin, Iraq | 2,348 cubic meters/year | Porrganiza economic, environmental (high salinity, climate changes, and high water requirement), and social factors | Proposed study | Water-energy-agriculture | ε-DSEA optimization algorithm for optimized integrated water resource management (OP-IWRM) | Water quality, river morphology, river bed level alterations, groundwater storage, hydropower production potential, etc. | Dam water release control and power production alternations would be by approx. 50% and 5.1 GW respectively; river water volumes will reduce to approx. 66% and water delivery to decline to approx. 16%; energy and food sectoral reforms were recommended | (Jawad et al., 2019) |
| Morocco (Souss region) | 804.9 cubic meters/year | Water scarcity, uneven rainfall (140 billion cubic meters), frequent droughts, urbanization | Execution phase | Water-land | Legislative amendments, irrigation water charges, water quality assessment plan at national-level incorporated water resources quality diagnosis, water quality management and remedial measures plan | Gap analyses (scenario evaluation) | National Plan for Water Quality Improvements; land erosion control, water-saving measures, sea water desalination, wastewater treatment and reuse, and energy sector reforms were adopted | (Kadi & Ziyad, 2018) |

| Cuvelai-Etosha Basin (CEB), Namibia, Sub-Saharan Africa | — | High fresh water dependence, saline groundwater, high population growth, increasing urbanization | Execution (2006) | Water-technology-research | Multi-resource-mix approach (rainwater, wastewater, floodwater, and groundwater) and multi-use approach was developed based on water quality. Multi-mechanisms for water collection, storage, treatment, and reuse; multiple technologies for flood and rainwater harvesting, groundwater desalination, etc.; sectoral collaborations prioritization and stakeholders' involvement | Ongoing activities impacts evaluation, such as treatment plants test results assessment | Rain and floodwater harvesting, groundwater desalination, and wastewater treatment and reuse technologies upgradation were demonstrated from pilot to commercial scale after results validation | (Liehr et al., 2016) |
|---|---|---|---|---|---|---|---|---|
| Africa | — | Water shortage, poor water quality, and poor water resources management | Execution phase | Water | Water resources management legislative mandate implementation tool, regional developmental priorities, municipal infrastructure, water quality, and pollution management areas prioritization | Performance evaluation indicators such as resources range, consumption rate, and resource quality | Forestry permits, catchment area management, technical capacity, fund for farmers, community-based rainwater harvesting, departmental collaboration, and equity share based recommendations | (van Koppen & Schreiner, 2014) |
| Nepal | 7,482 cubic meters/year | Flooding and droughts, uneven rainfall | Execution phase | Water-land | River basin management, water resources development | Water consumption rate for agriculture, hydropower and drinking water as mass balance | Lack of integration among sectors was a challenge; recommendations for sectoral decision-making processes, benefit sharing, public participation, and conflict resolution | (Suhardiman et al., 2015) |
| The Rio Grande, Mexico, basin | 3,660 cubic meters/year | Groundwater and surface water shortages | Proposed study | Water | General circulation models (GCMs) for Intercomparison Project Phases. Multi-model ID 020 Morelia-Quererndaro for aquifer based on demands for irrigation; environmental flows determination through Mexican Standard NMX-AA-159-SCFI-2012; climate change scenarios and flow volume ratio analyses, climate models (CICESE, IMTA y Corganizatissemblage | Indices, reliability, resilience factors, environmental flow; vulnerability, sustainability index | Aquifer recharge of 284.1 hm$^3$/year was expected to reduce by −11.4%; irrigation water demand increase from 88.3 hm$^3$/year to 95.6 (8.3%) hm$^3$/year and urban water demand increase from 82.1 hm$^3$/year to 108 hm$^3$/year estimated. Low aquifer recharge, decreasing precipitation, and lower water availability were estimated, and policy approaches were suggested | (Bedolla et al., 2017) |

(*Continued*)

**TABLE 6.1** *(Continued)*

**Integrated Water Resource Management (IWRM) Global Implementation Perspective for Proposed and In-Practice Scenarios**

| Country | Water availability per capita | Pressure points on water consumption | IWRM status | IWRM nexus approach | IWRM tools/components | Parameters studied | Comments/outcomes | References |
|---|---|---|---|---|---|---|---|---|
| Bangladesh | 7,568 cubic meters/year | Intense cropping and irrigation activity | Execution phase | Water-land | Ecosystem restoration, livelihood improvement, natural disasters mitigation, and public consultations during the IWRM planning phase; evaluation of institutional transition | Water resource analyses (quantification, numbers, usage, output) | Afforestation, rebuilding of ponds and cross dams, volumetric irrigation fee, metering system, flood control, water conservation, river management, sectoral coordination, and institutional transition and population (76%) access to clean water were the outcomes | (Gain et al., 2017) |
| Indonesia | 7,648 cubic meters/year | Drought, flood, lands subsidence, water pollution, erosion, sedimentation, uneven rainfall, riverbank encroachment, and deforestation | Execution phase | Water | Capacity development for flood management to reduce soil erosion and surface runoff; spatial planning; water management structures, pollution prevention, and reliable water supply and distribution Water resources functionality, cost-and-benefit analyses at catchment scale | | Land rehabilitation projects execution reduced runoff, illegal logging, and intensive land use. However, proposed projects did not meet execution to full potential; program delays and cancellations reduced IWRM-projects-related negotiations and collaborations across sectors | (Fulazzaky, 2014) |
| Lake Albert basin, Uganda | 1,407 cubic meters/year | Population growth, water and land demand, water resources encroachment, and infrastructure development | Execution phase (2006) | Land-water | Stakeholder's assessment for resources management; implementation and assessment of catchment interventions such as research, advocacy, data generation, service delivery, and capacity build Data evaluation for fishing, agriculture, deforestation, grazing, flooding and water quality, tourism, and peri-urbanism prospects | | Check and balance for activities, local stakeholders' indulgence | (Katusiime & Schutt, 2020) |

*Note:*   * GPS: Global positioning system

et al., 2015). Through sectoral collaboration and stakeholders' involvement, demand and supply of water can be controlled through the adoption of equity principles and water reuse approaches. For instance, the water can be reused to augment the diminishing groundwater sources. Furthermore, through the flood-zone mapping, mitigative measures can be considered such as flooded water harvesting and storage for future use. Through water audits and smart water metering techniques, the water input and effluent discharge (mass balance) can be determined for any utility or area (Sibly, 2020). The water use areas and stressors under variable climatic, financial, and socioeconomic aspects can be evaluated to device effective policy options. Usually, system boundaries, goals, and associated risks are stakeholder-interest specific. Through interdisciplinary IWRM modeling, problems can better be negotiated and water management options can be put forward for mutual collaboration.

## 6.6.2 Technical Aspects of IWRM

The IWRM framework implementation needs monitoring technologies and assessment tools to establish the baseline data and simulated projections over time. IWRM is a cross-sectoral water management approach. The IWRM plan can effectively be implemented by incorporating stakeholder views from all the sectors into one umbrella policy agenda. Sectors (electricity, food sector, etc.) have specified social, economic, and environmental activities and issues that can be looked into by the IWRM paradigm. Nexus approach works well for integrated solutions (Kashem & Mondal, 2022). IWRM's 'technical' aspects can be fulfilled by context-based measures which may incorporate hydrology, meteorology, geohydrology, and other sciences such as political, social, economic, and engineering prospects. Mathematical models are devised to illustrate the complexity of hydro-economic simulations and optimizing the networks for spatial allocation of water share. Surface water geographical zone and runoff water flow routes can be traced by GIS-, RS-, and GPS-based modeling technologies (Gooch & Baggett, 2013). The computation of multi-objective problems can be performed by the latest auto-adaptive constraints approach for criteria assessment to attain feasible site-specific solutions. For instance, the optimum IWRM (OP-IWRM) method of the decision support systems (DSSs) multi-objectives algorithm approach was employed to implement IWRM for Iraqi's Diyala River basin. Optimization models encountered governance, climatic factors, user demand, and alternate use options as criteria factors. Case 1 was taken as conventional while Case 2 as IWRM for comparison for the period of 1981 to 2013. Results suggested that the river water level, groundwater storage, and river discharge would reduce to 84% (Jawad et al., 2019). To hamper the losses, improved farming practices such as drip system, crops rotation, crop type, and crop pattern modifications would reduce the uncertainty level to 50%. Similarly, saline water treatment would ensure the continual water supply in the long term. Major policy forms were emphasized in food and energy sectors. Furthermore, 50% uncertainty associated with alternatives can be overcome by appropriate system maintenance and government policy reform to check water overexploitation. Nikolic and Simonovic (2015) mentioned the system dynamic simulation (SDS) tool for decision support for IWRM implementation, which processes spatial data analyses features. Allende et al. (2009) employed GIS along with cluster analysis for Watershed Cuitzeo Lake, Central Mexico. Through multicriteria decision, clustered geographical and ecological features were ranked to find hydrometeorological priority features in the context of political, socioeconomic, and environmental aspects. However, the major challenges were cross-sector variables optimization under a holistic approach. Decision uncertainty would be minimized through trade-offs clarification, capacity building, transparency, and inclusiveness across the sectors for a successive IWRM accomplishment (Coelho et al., 2012).

### 6.6.3 Treatment of Wastewater

Water is not an isolated commodity; rather, it's an ecosystem of resource flow across spheres and sectors. Through hydrogeological cycle, water becomes accessible to all life forms. However, the runoff water is not simply a water but a mix of toxic and hazardous substances. Instead of the direct washout of effluent to water bodies and agrifields, alternatives are considered based on spatial (range) and temporal (weather) scales (Ioannou & Laspidou, 2022). IWRM models give manipulation of simulated stressors and conditions for which alternatives can be considered. Through hydrological modeling, surface water runoff, water flow, and chemical sources origin, water recharge rate, water abstraction, pollution level, and water demand quantification can be taken into account. There is again the need for data generation and accessibility to monitor socioeconomic, meteorological, and hydrological regimes on temporal and spatial scales. Through IWRM planning, demolishing factors would be avoided for surface water body (Jawad et al., 2019).

Water quality assessment for water resources quality diagnosis suggests to opt the remedial measures plan. The municipal wastewater (grey water + black water) can be treated to reuse in agriculture, gardening, aquifer recharge, and other secondary purposes. IWRM refers aquifers artificial recharge and rainwater harvesting. Eliraz et al. (2017) used desalinated sea water with natural groundwater for aquifer recharge in Israel. Column experiments revealed the magnesium and calcium presence; magnesium enrichment in water enhanced water desalination. The combination enhanced the water temporal storage and also enhanced water treatment. IWRM is a water conservation approach whereby the increased groundwater recharge maintains water supply during dry season, ensures food production, and controls intense weather events (Friesen et al., 2016). In agriculture, drip irrigation was beneficial for better yield and quality in comparison to flood irrigation. It brought high water-saving potential. Similarly, the energy sector can be a potential reuse sector for treated wastewater.

### 6.6.4 IWRM and Water Conservation

To manage hydrological issues of deteriorating water quality, IWRM integrates distributional patterns, rules enforceability, and ethical dimensions for safe water use and management. All efforts of water management are oriented to water conservation. Water bodies and watersheds give any place an economic worth based on irrigation and other use perspectives. Arid regions due to less vegetation and more bare land are exposed to solar energy which results in high rates of evapotranspiration. Land-water has nutrient fluxes at catchments subject to agriculture activities (Haileslassie, 2019). Northern Mongolia has severe water problems and population expansion and infrastructure development. In Mongolia, agricultural soil analyses revealed nutrient deficiency that was attributed to nutrient washout and deficient input of fertilizer use. There, the modeling of water-solute fluxes at the catchment scale would precisely determine the nutrient concentration gradient and the severity of the runoff problem (Ibisch et al., 2016). Water management for improved livestock and agricultural production and ecological conservation is mandatory. Forests promote groundwater recharge, hold runoff water, reduce erosion, and regulate climate by being the evapotranspiration source. In this perspective, community integrated assessment modeling (CIAM) determines temperature and rainfall measures, while the hydrological model makes the simulation of the hydrological cycle and determines the rainfall runoff for the selected area (Bedolla et al., 2017).

### 6.6.5 IWRM Economic and Legal Perspectives

The IWRM core pillars are effective legislation, regulatory framework, and social welfare with the inclusion of every individual of society. The IWRM approach emphasizes on the decentralization of responsibilities, integration of hydrological principles, and economic prospects. IWRM works with the balance of water demand and economic apportionment of limited water sources (Safavi et al., 2016). Further, it addresses the economic valuation of multiple and alternate water uses to ensure water-efficient management. Households, agriculture, and industrial areas need to incorporate use-based water share allocation based on efficiency, equity, social fairness, and eco-sustainability, particularly in water-scarce areas (Gooch & Baggett, 2013). The implementation of water rights and water tariffs is not possible without strong institutional frameworks to maintain water services sustainably. If water is extracted from a groundwater source or reservoir, then it can be considered a private good based on rules. Innate endowment concerns can be overcome by tax-based systems. So strict equality is recommended instead of innate entitlements to avoid unequal water distribution and resource overexploitation under the IWRM framework (Grigg, 2019). In a case study, climate modeling and remote sensing tools were used to evaluate the Souss region, Morrocco, aquifer management. To improve water productivity, the government recommended to restrict the access to boreholes and wells digging and control for irrigated areas expansion. Legislative amendments were observed as farmers were paying for irrigation water (Aitkadi & Ziyad, 2018). Ostrom (1990) mentioned the Philippines and Spain monitoring systems and conflict-resolution schemes regarding water usage that were supportive to ecosystem conservation functions.

In another study, the Rhine River (60th largest basin of world) basin IWRM plan considered climate impacts management (Ludwig et al., 2014). European Union has the Water Framework Directive for the River Rhine adaptation to climate change. Ecology, hydrology, water quality, economy, and land use were the major focus areas. Under the framework, the efforts were directive to floodplains restoration and controlled effluent runoff to water body (Ludwig et al., 2014). To manage the

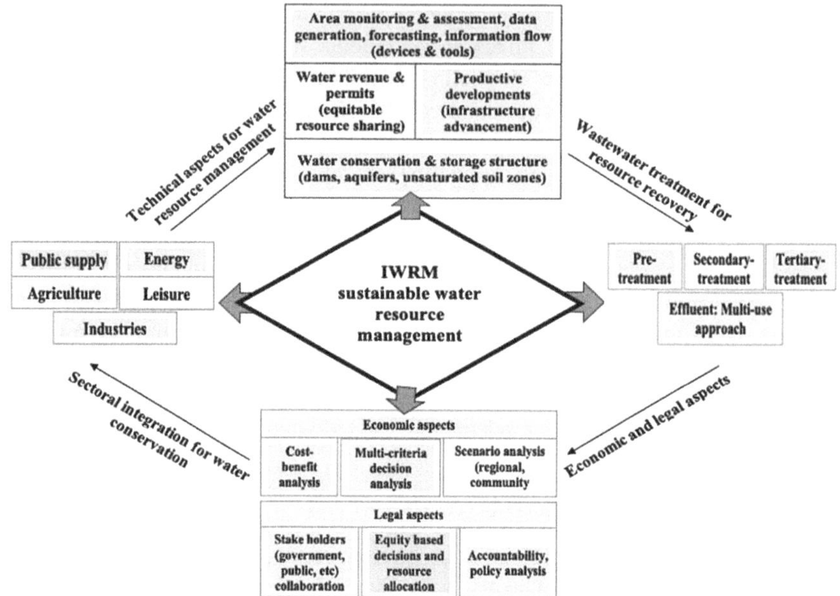



**FIGURE 6.4** IWRM frameworks for sustainable water resource management.

watershed activities, industrial, agricultural, and navigation sectors were incorporated in the decision-making. Project execution also required large-scale negotiations with other stakeholders of flood control, ecological infrastructure, landscape planning, nature conservation, and groundwater and land management sectors. Overall, the infrastructural planning must consider future projections to bring long-term effectiveness.

## 6.7 CONCLUSION

IWRM elucidates the factors of governance and socioeconomic and environmental aspects for water resource effective management. IWRM promotes water resources conservation through the stakeholders' participation, polluter-pays principle, and public–private sector coordination for economic mutualization and transparency. We tried exploring the boundaries of different systems like ISWM, INRM, and the WSW nexus in this chapter. We derived different criteria and management planning of these systems and also defined the system boundaries. The WSW nexus system must be designed in a way to enhance resources circulation and waste management sustainably. The WSW reduces the complexity and problems at each level of the nexus without affecting the structure integrity which is basically inherent in each of these underlying systems. Policy shifts and institutional transitions for successful IWRM implementation are mandatory, as the framework opts for a decentralized approach for decision-making. Multiple decision-makers and stakeholders can mutually collaborate to address challenges of resources shortage and uncertainty.

## REFERENCES

AitKadi, M., & Ziyad, A. (2018). *Integrated Water Resources Management in Morocco*. General Council of Agriculture Development, Ministry of Agriculture, Fishery, Rural Development and Forestry, Rabat, Morocco. https://doi.org/10.1007/978-981-10-7913-9_6.

Alamanos, A., Mylopoulos, N., Loukas, A., & Gaitanaros, D. (2018). An Integrated Multicriteria Analysis Tool for Evaluating Water Resource Management Strategies. *Water*, *10*(12), 1795. https://doi.org/10.3390/w10121795.

Allende, C., Mendoza, M., Granados, E., & Morales, M.L. (2009). Hydro-Geographical Regionalization: An Approach for Evaluating the Effects of Land Cover Change in Watersheds: A Case Study in the Cuitzeo Lake Watershed, Central Mexico. *Water Resources Management*, 23, 2587–2603. https://doi.org/10.1007/s11269-008-9398-6.

Avellan, T., Roidt, M., Emmer, A., Von Koerber, J., Schneider, P., & Raber, W. (2017). Making the water–soil–waste nexus work: Framing the boundaries of resource flows. *Sustainability*, 9(10), 1881. https://doi.org/10.3390/su9101881.

Badham, J., Elsawah, S., Guillaume, J., Hamilton, S., Hunt, R., Jakeman, A., et al. (2019). Effective Modeling for Integrated Water Resource Management: A Guide to Contextual Practices by Phases and Steps and Future Opportunities. *Environmental Modelling & Software*, *116*, 40–56. https://doi.org/10.1016/j.envsoft.2019.02.013.

Bedolla, H.J., Solera, A., Arquiola, P.J., Monzonís, P.M., Andreu, J., & Quispe, S.T. (2017). The Assessment of Sustainability Indexes and Climate Change Impacts on Integrated Water Resource Management. *Water*, 9, 213. https://doi.org/10.3390/w9030213.

Benchbani, I., Sebari, K., & Zemzami, M. (2022). Integrated Water Resources Management in the Loukkos Basin (Morocco): An Approach to Improve Resilience Under Climate Change Impact. *E3S Web of Conferences*, *346*, 03024. https://doi.org/10.1051/e3sconf/20223460302.

Benn, K. (2016). Integrated Water Resource Management. *Australasian Journal of Environmental Management*, *24*(1), 84–85. https://doi.org/10.1080/14486563.2016.1220666.

Carney, M. (1998). The Competitiveness of Networked Production: The Role of Trust and Asset Specificity. *Journal of Management Studies*, *35*(4), 457–479.

CGIAR Center Directors Committee. (1999). *Integrated Natural Resources Management: The Bilderberg Consensus*. International Centers, Washington, DC. https://cgspace.cgiar.org/server/api/core/bitstreams/2d92f0ab-46ad-4ad6-bbf5-1dc90cc15903/content

Coelho, A.C., Labadie, J.W., & Fontane, D.G. (2012). Multicriteria decision support system for regionalization of integrated water resources management. *Water Resources Management*, *26*, 1325–1346. https://doi.org/10.1007/s11269-011-9961-4.

Davis, M. (2007). Integrated Water Resource Management and Water Sharing. *Journal of Water Resources Planning and Management*, *133*(5), 427–445. https://doi.org/10.1061/(asce)0733-9496(2007)133:5(427)

Eliraz, R.G., Russak, A., Nitzan, I., Guttman, J., & Kurtzman, D. (2017). Investigating Geochemical Aspects of Managed Aquifer Recharge by Column Experiments with Alternating Desalinated Water and Groundwater. *Science of the Total Environment, 1*, 574, 1174–1181. https://doi.org/10.1016/j.scitotenv.2016.09.075.

Friesen, J., Rodriguez, S.L., Laura, F., & Ralf, L. (2017). Environmental and Socio-Economic Methodologies and Solutions Towards Integrated Water Resources Management. *Science of the Total Environment*, 581–582, 906–908. https://doi.org/doi:10.1016/j.scitotenv.2016.12.051.

Fulazzaky, M.A. (2014). Challenges of Integrated Water Resources Management in Indonesia. *Water*, *6*, 2000–2020. https://doi.org/10.3390/w6072000.

Gain, A.K., Mondal, M.S., & Rahman, R. (2017). From Flood Control to Water Management: A Journey of Bangladesh towards Integrated Water Resources Management. *Water*, *9*, 55. https://doi.org/10.3390/w9010055.

Galvez, V., &Rojas, R. (2019). Collaboration and Integrated Water Resources Management: A Literature Review. *World Water Policy*, 5, 179–191. https://doi.org/10.1002/wwp2.12013

Gooch, G., & Baggett, S. (2013). IWRM in the Swedish Context: A Voluntary Move to IWRM Principles or a Legal Necessity to Comply with the European Union Water Framework Directive? *International Journal of Water Governance*, 1(3), 361–378. https://doi.org/10.7564/13-ijwg10.

Grigg, N. (2019). IWRM and the Nexus Approach: Versatile Concepts for Water Resources Education. *Journal of Contemporary Water Research & Education*, 166(1), 24–34. https://doi.org/10.1111/j.1936-704x.2019.03299.x.

Hagmann, J. (1999). *Learning Together for a Change: Facilitating Innovation in Natural Resource Management Through Learning Process Approaches in Rural Livelihoods in Zimbabwe*. Weikersheim: Margraf.

Haileslassie, Z. (2019). Promoting Federalism, IWRM, and Functional Approach to Water Governance under Ethiopian Water Laws. *Mizan Law Review*, 13(3), 384–418. https://doi.org/10.4314/mlr.v13i3.3.

Hwang, J., Park, S., & Song, C. (2020). A Study on an Integrated Water Quantity and Water Quality Evaluation Method for the Implementation of Integrated Water Resource Management Policies in the Republic of Korea. *Water*, 12(9), 2346. https://doi.org/10.3390/w12092346.

Ibisch, R.B., Bogardi, J.J., & Borchardt, D. (2016). *Integrated Water Resources Management: Concept, Research and Implementation*. Springer, Cham. https://doi.org/10.1007/978-3-319-25071-7_1.

Ingold, K., & Tosun, J. (2020). Special Issue 'Public Policy Analysis of Integrated Water Resource Management'. *Water*, 12(9), 2321. https://doi.org/10.3390/w12092321.

Ioannou, A., & Laspidou, C. (2022). Resilience Analysis Framework for a Water–Energy–Food Nexus System Under Climate Change. *Frontiers in Environmental Science*, 10. https://doi.org/10.3389/fenvs.2022.820125.

Jawad, J.Y., Alsaffar, H.M., Bertram, D., & Kalin, R.M. (2019). A Comprehensive Optimum Integrated Water Resources Management Approach for Multidisciplinary Water Resources Management Problems. *Journal of Environmental Management*, 1, 239, 211–224. https://doi.org/10.1016/j.jenvman.2019.03.045.

Kafy, A., Faisal, A., Raikwar, V., Rakib, A., Kona, M.A., & Ferdousi, J. (2021). Geospatial Approach for Developing an Integrated Water Resource Management Plan in Rajshahi, Bangladesh. *Environmental Challenges*, 4. https://doi.org/10.1016/j.envc.2021.100139.

Kashem, S., & Mondal, M. (2022). Development of a Water-Pricing Model for Domestic Water Uses in Dhaka City Using an IWRM Framework. *Water*, 14(9), 1328. https://doi.org/10.3390/w14091328.

Katusiime, J., & Schutt, B. (2020). Integrated Water Resources Management Approaches to Improve Water Resources Governance. *Water*, 12, 3424. https://doi.org/10.3390/w12123424.

Korten, D. C. (1980). Community Organization and Rural Development: A Learning Process Approach. *Public Administration Review*, 480–511. https://doi.org/10.2307/3110204

Kurian, M., & Ardakanian, R. (2015). The Nexus Approach to Governance of Environmental Resources Considering Global Change. In *Governing the Nexus* (pp. 3–13). Springer, Cham. https://doi.org/10.1007/978-3-319-05747-7_1.

Liehr, S., Brenda, M., Cornel, M.P., Deffner, J., Felmeden, J., Bjokisch, A., et al. (2016). *From the Concept to the Tap—Integrated Water Resources Management in Northern Namibia*. Springer, Cham. https://doi.org/10.1007/978-3-319-25071-7_26.

Lovell, C., Mandondo, A., & Moriarty, P. (2002). The Question of Scale in Integrated Natural Resource Management. *Conservation Ecology*, 5(2). www.consecol.org/vol5/iss2/art25/

Ludwig, F., Slobbe, E., & Cofino, W. (2014). Climate Change Adaptation and Integrated Water Resource Management in the Water Sector. *Journal of Hydrology, 518*, 235–242. https://doi.org/10.1016/j.jhydrol.2013.08.010.

Marshall, R. E., & Farahbakhsh, K. (2013). Systems Approaches to Integrated Solid Waste Management in Developing Countries. *Waste Management, 33*(4), 988–1003. doi:10.1016/j.wasman.2012.12.023

Mees, H., Suykens, C., & Crabbé, A. (2017). Evaluating Conditions for Integrated Water Resource Management at Sub-Basin Scale: A Comparison of the Flemish Sub-Basin Boards and Walloon River Contracts. *Environmental Policy and Governance, 27*(1), 59–73. https://doi.org/10.1002/eet.1736.

Meran, G., Siehlow, M., & Hirschhausen, C. (2021). *The Economics of Water Rules and Institutions. The Economics of Water.* Springer Nature Switzerland AG, Springer Cham. https://doi.org/10.1007/978-3-030-48485-9.

MODFLOW. (2005). *The U.S. Geological Survey Modular Ground-Water Model – The Ground-Water Flow Process: U.S. Geological Survey Techniques and Methods 6-A16.* https://www.usgs.gov/software/modflow-2005-usgs-three-dimensional-finite-difference-ground-water-model.

Momm, H., Bingner, R., Moore, K., & Herring, G. (2022). Integrated Surface and Groundwater Modeling to Enhance Water Resource Sustainability in Agricultural Watersheds. *Agricultural Water Management, 269*, 107692. https://doi.org/10.1016/j.agwat.2022.107692.

Mushtaq, J., Dar, A.Q., & Ahsan, N. (2020). Spatial-Temporal Variations and Forecasting Analysis of Municipal Solid Waste in the Mountainous City of North-Western Himalayas. *SN Applied Sciences, 2*(7), 1–18. https://doi.org/10.1007/s42452-020-2975-x.

Ngene, B., Nwafor, C., Bamigboye, G., Ogbiye, A., Ogundare, J., & Akpan, V. (2021). Assessment of Water Resources Development and Exploitation in Nigeria: A Review of Integrated Water Resources Management Approach. *Heliyon, 7*(1), 05955, 2405–8440. https://doi.org/10.1016/j.heliyon.2021.e05955.

Nikolic, V.V., & Simonovic, S.P. (2015). Multi-Method Modeling Framework for Support of Integrated Water Resources Management. *Environmental Processes, 2*, 461–483. https://doi.org/10.1007/s40710-015-0082-6

Ostrom, E. (1990). *Governing the Commons: The Evolution of Institution for Collective Action. The Political Economy of Institutions and Decisions.* Cambridge University Press, Cambridge.

Safavi, H.R., Golmohammadi, M.H., & Solis, S.S. (2016). Scenario Analysis for Integrated Water Resources Planning and Management Under Uncertainty in the Zayandehrud River Basin. *Journal of Hydrology, 539*, 625–639. https://doi.org/10.1016/j.jhydrol.2016.05.073.

Saidi, M. (2017). Conflicts and Security in Integrated Water Resources Management. *Environmental Science & Policy, 73*, 38–44. https://doi.org/10.1016/j.envsci.2017.03.015.

Satriani, A., Loperte, A., & Soldovieri, F. (2015). Integrated Geophysical Techniques for Sustainable Management of Water Resource: A Case Study of Local Dry Bean Versus Commercial Common Bean Cultivars. *Agricultural Water Management, 162*, 57–66. https://doi.org/10.1016/j.agwat.2015.08.010.

Schwarz, N., & Bless, H. (2013). Constructing Reality and Its Alternatives: An Inclusion/Exclusion Model of Assimilation and Contrast Effects in Social Judgment. In *The Construction of Social Judgments* (pp. 217–245). Hillsdale, NJ: Erlbaum, Psychology Press.

Sibly, H. (2020). Urban Water Policy When Environment Inflows Are Uncertain. *Water Resources and Economics, 30*, 100149. https://doi.org/10.1016/j.wre.2019.100149.

Sridevi, V., Modi, M., Ch, M.V.V., Lakshmi, A., & Kesavarao, L. (2012). *A Review on Integrated Solid Waste Management.* https://citeseerx.ist.psu.edu/viewdoc/summary?doi=10.1.1.300.9119.

Suhardiman, D., Clement, F., & Bharati, L. (2015). Integrated Water Resources Management in Nepal: Key Stakeholders' Perceptions and Lessons Learned. *International Journal of Water Resources Development, 31*. https://doi.org/10.1080/07900627.2015.1020999.

United Nations University Institute for Integrated Management of Material Fluxes and of Resources (UNU-FLORES). (2017). National Research Institute for Rural Engineering, Water, and Forestry (INRGREF), Technical University of Denmark, Institution of Agricultural Research and Higher Education (IRESA).

van Koppen, B., & Schreiner, B. (2014). Moving Beyond Integrated Water Resource Management: Developmental Water Management in South Africa. *International Journal of Water Resources Development, 30*(3), 543–558.

Wang, K., Chang, L., & Chang, F. (2011). Multi-Tier Interactive Genetic Algorithms for the Optimization of Long-Term Reservoir Operation. *Advances in Water Resources, 34*(10), 1343–1351. https://doi.org/10.1016/j.advwatres.2011.07.004.

# 7 Nexus Approach
## *Resource Management for Soil Productivity*

*Rajneesh Thakur*

## 7.1 INTRODUCTION

By 2050, there will likely be more than 9 billion people on the planet, making it difficult to provide their needs while minimising environmental harm. The majority of people rely heavily on the land, water, and forest systems as their primary sources of nourishment. Inadequate farming and management methods, as well as population increase, are placing great strain on these assets. The intense farming practices, overgrazing and deforestation, soil erosion and declining soil fertility, water scarcity, and livestock feed and fuel wood shortage pose significant challenges for small-scale farmers who rely on these resources. A downward trend of decreasing livestock and crop output, food insecurity, rapid population expansion, and environmental pollution is frequently brought on by these factors interacting with one another (referred to as the nexus problem).

Overall, a vicious cycle of poverty, food shortages, and resource deterioration is created, trapping an increasing number of rural residents. Since local producers make up the majority of people living below the poverty line, enhancing the natural resource base is crucial to any attempt to stop this "vicious cycle" and increase their production. The parts of the project effort, however, is based on the "intensification package strategy" and is particularly based on accelerating production utilising fertilizer and upgraded seed (mostly hybrid maize), regardless of farmers' capabilities and agro-ecological zones. The majority of farmers who lack the resources have found this to be both unproductive and insufficient to address the fundamental problems they are facing. We will require safe and reliable sources of nutrition energy and water to sustain this population. One of the most complicated yet crucial concerns that civilisation must address is the connection between food, energy, and water. There isn't any more available land to exploit, and in some parts of the world, there is a shortage of fresh water, which restricts the use of land for agriculture. The additional issue of global warming must be addressed by all approaches. To meet the needs of the current and future populations, a reliable supply of food, energy, and water will be needed. The International Union for Conservation of Nature (IUCN) recently established worldwide criteria for nature-based sorganizatiophasising ecosystem-based adaptability among other things (IUCN 2020). The connections between water, energy, food, and ecosystems are depicted in Figure 7.1.

DOI: 10.1201/9781003358169-7

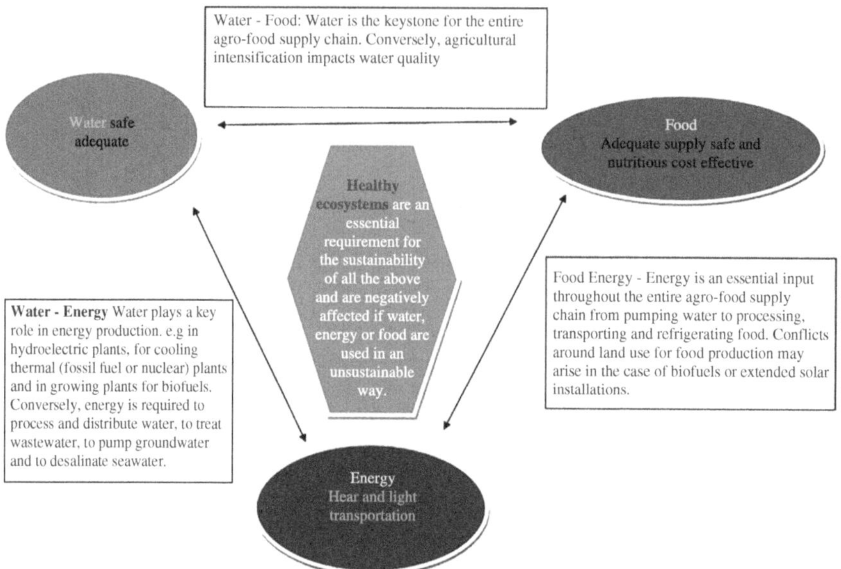



**FIGURE 7.1** Inter-linkages between water, food, and energy.

A nexus strategy is needed to boost food, energy, and water management while balancing the management of scarce resources and moving towards a more "green" civilisation that provides enough food, energy, and water for the expanding human population. Due to the concept's comprehensive resource utilisation approach and the problems the modern world is dealing with as a result of the effects of climate change on water, energy, and food sources, international organisations have taken note of it (Wichelns 2017). Nexus believes that management of natural or modified resources, such as land, water, energy, and biodiversity, is very successful at lowering devastation risk and assisting impoverished countries in adapting to and lessening the effects of climate change (Rasul and Sharma 2016; Krchnak et al. 2011). The Nexus approach was developed in response to the realisation that water, energy, agriculture, and natural ecosystems exhibit strong interconnections and that with a conventional spectrum of applications, merely attempting to achieve resource security independently frequently endangers sustainable development and safety in some or all of the other sectors. The nexus approach looks at linkages, synergies, and trade-offs to develop solutions, advance the security and efficiency of water, food, and energy, and minimise the effects and risks on ecosystems.

## 7.2 WHY A "NEXUS" APPROACH TO MANAGING SOIL AND LAND IS NECESSARY

Globally, soils are rapidly degrading, which reduces ecosystem diversity and various essential services, jeopardises food security and other human rights, and increases people's susceptibility to climate change. This leads to even more unsustainable land use practices, creating a vicious cycle. It can be made into a beneficial cycle with the help of integrated (or "nexus") management of the soil, land, water, and ecosystems.

Deteriorating environmental conditions consequently have a negative impact on farming capacity. Around 1.5 billion rural residents rely on degraded land. According to a 2008 assessment from the Global Assessment of Land Degradation and Improvement (GLADA) project, the issue is particularly severe in parts of the boreal forest in Siberia and North America, Southeast Asia, southern China, north-central Australia, and Africa south of the equator. Surprisingly, 95% of the land in Swaziland is degraded, as are 66% of the lands in Angola, 64% of the lands in Gabon, 60% of the lands in Thailand, and 60% of the lands in Zambia. Land degradation affects 457 million people in China.

In the context of planetary boundaries related to water (reducing "green water" stored in the soil and atmospheric moisture recycling), nitrogen, and phosphorus (not only in terms of water pollution but also regarding depletion of global phosphorus resources), soil and land degradation is also a serious issue. Soil and land deterioration must consequently be slowed down and reversed in order to effectively manage the world's commons.

## 7.2.1 CAUSES

Numerous traditional although unsustainable practices, such as excessive tillage, extensive monocultures, insufficient crop rotation and fallow periods, overgrazing, cultivation of steep slopes, removal of vegetation, excessive chemical use, and others, are to blame for land degradation. Agricultural yields are impacted by the loss of soil organic matter, which reduces the soil's capacity to hold onto water and nutrients. As their livelihoods are threatened, farmers may exacerbate the problems by utilising resources excessively even more, creating a vicious cycle. The consequences spread to the surrounding landscapes, impacting ecosystems' ability to operate, for instance, by water pollution and shifting regional climate.

Pollination, natural pest control, and climatic resilience may all suffer as a result of declining biodiversity, which can potentially impair groundwater resources and reservoir storage downstream. Floods also occur more frequently. Deteriorating environmental conditions also have a negative impact on possibilities for agricultural biodiversity and soil and moisture absorption recycling (Rockström et al. 2009). Soil and land deterioration must consequently be slowed down and rectified in order to effectively manage the world's resources.

## 7.2.2 NEXUS' APPROACHES FOR MANAGEMENT OF RESOURCES

The nexus approach to natural resources planning investigates how environmental resources are related and dependent upon one another, as well as how they move across various geographical scales and divisions. Instead of only considering individual components, a complex system's administration, productivity, and efficiency are taken into consideration. The Latin term "nexus" means "that which connects" or "that which binds together." In the 1980s, the term "nexus" was first applied to the management of environmental resources, particularly in a study by the UN

University (Sachs and Silk 1990). However, it wasn't until the lead-up to the Bonn 2011 meeting on the "Water, Energy, and Food Security Nexus" that the nexus solution gained widespread recognition in worldwide academic and policy circles. The conference asserted that this method may enhance water, energy, and food security while simultaneously supporting sustainability and the transition to a green economy by integrating "management and governance across sectors and scales," minimising trade-offs, and establishing synergies (Hoff 2011). UNU-FLORES strives to manage water, soil, waste, and other resources in an integrated way as a direct response to the nexus initiative (closely related to the water-energy-food security nexus).

Institutional barriers can be overcome by organisations that establish broad, long-term goals and consider trade-offs. Institutions with a more sector-focused mission should be improved compared to those with a "nexus mandate," such as ministries of the environment (e.g., ministries of water or energy). By directing benefits towards the ecosystem services towards integration at the landscape or regional scale, investments can help with scaling up, for example. As with the upcoming TEEB (The Economics of Ecosystems and Biodiversity) agriculture and food studies, this requires understanding the financial benefits of a nexus strategy. A growing amount of information supports the assertion that "agricultural output depends on services given by healthy ecological systems," as stated in a concept note for the study from July 2013. The note draws attention to a critical knowledge gap that, if filled, could strengthen the case for integrated approaches: the value of various ecological systems "are essentially invisible in marketplaces and are therefore not reflected in national accounting and statistics nor in land use and management decisions."

Such comprehensive approaches will safeguard the environment, the weather, and ultimately, the political system while also enhancing human security. This is made sure of using the so-called "nexus approach" (Hoff 2011; Allan et al. 2015; Al-Saidi and Elagib 2017). The main challenge for trying to implement (context-specific) nexus approaches to soil and land management will be trying to engage performers at all levels, from growers who must switch to environmentally friendly practices, agro-businesses that will need new business models, to policymakers who must coordinate across sectors.

## 7.3  WHAT POTENTIAL NEXUS-BASED CHANGES MIGHT FARMING EXPERIENCE?

It will probably become less labour-intensive (in terms of energy, irrigation, agrochemicals, and other non-renewable inputs). By reducing tillage, for instance, and increasing crop diversity and rotation, it will place a higher priority on soil and water conservation. Additionally, it supports agro-ecological methods, including waste product recycling, integrated pest management, water harvesting, and "green manuring," which involves growing cover crops during fallow periods to improve the soil's organic matter and nutritional content. All of these strategies can support agricultural development, albeit more sustainably, and offer additional benefits like greater terrestrial carbon storage and decreased water pollution from agricultural runoff.

Nexus or integrated techniques have been acknowledged by science for a while as offering chances to increase source use efficiencies across multiple resources while limiting overexploitation and environmental degradation. The potential advantages of nexus techniques have been shown in a number of early evaluations (see, e.g., Howells et al. 2013). Additionally, some smallholder farmers already use nexus strategies out of necessity. They rely heavily on all natural resources and frequently cannot afford extra inputs like energy or agrochemicals. Implementing nexus ideas at bigger sizes, like as across regions or land masses, is still quite difficult (Hoff et al., 2013)

It can be useful to develop a broad framework that can be used to arrange studies on the water-soil-waste nexus because the nexus concept is all about connection. Such a plan can serve to both more effectively explain the findings of prior investigations as well as serve to direct future research. The nexus approach seeks to standardize the relationships and offer instruments to gauge how all resources are being used. It takes a system-wide approach, acknowledges the natural interdependencies between the food, water, and energy sectors in terms of resource consumption, aims to maximize trade-offs and synergies, and considers the effects on the environment and society. Understanding the connections between the food, energy, and water nexus can open up opportunities to improve the efficiency of resource usage as well as cooperation and policy coherence between the three sectors (Rasul and Sharma 2016).

1. Concentrating on the relationships and trade-offs among the three constituent parts;
2. Serving as a resource for cross- and interdisciplinary research; and
3. Enhancing communication.

## 7.4  WHY DO WE CURRENTLY NEED A "NEXUS" STRATEGY AND WHAT DOES THE TERM "NEXUS" MEAN?

Food, energy, and water security are all interconnected. Industries and nations compete for resource sharing in the absence of management.

- Sustainable Development Goals of Agenda 2030: Sectoral goals and trade-offs
- Need for efficient and effective investment, spreading costs, and maximising profits in the context of climate change and post–COVID-19 recovery
- The "nexus" method strives to balance and co-optimise the interests of the various sectors while considering the needs of the environment and human rights and to enhance governance
  Nexus is not a novel concept; rather, it is a new inference of earlier system-wide ideas and combinations of tactics, such as integrated water resource management, landscape approaches supported by the World Bank, and multifunctional production systems like agroforestry, crop-livestock-biofuels, ecological sanitation, and ecosystem approaches introduced by the Convention on Biological Diversity (CBD) (IWRM). The identification of new "win-win" chances will be a key strategy. International organisations like the UN Food and Agriculture Organization (FAO), as well as other

donors and networks like the Global Soil Partnership, can offer pertinent information and a crucial venue for discussion in this area.

### 7.4.1 The Convention on Biological Diversity Is a Crucial International Law for Sustainable Development

A legal framework for "the protection of biological diversity, the sustainable use of its components, and the fair and equitable sharing of the benefits deriving from the exploitation of genetic resources," the Convention on Biological Variety, has been ratified by 196 countries (CBD). Other aspects of biodiversity that are covered by the Convention on Biological Diversity include species, ecosystems, and genetic resources. The Cartagena Protocol on Biosafety is a factor in its discussion of biotechnology. Actually, it covers every area that has to do with biodiversity and how it affects development, including business, culture, education, politics, and even agriculture and the environment. The convention's three primary goals are as follows.

**The following principles must be followed:** A proper and equitable distribution of the benefits attained from the utilisation of these resources and related traditional knowledge is necessary for the preservation of biological diversity, sustainable resource use, and sustainable resource use.

## 7.5 THE CARTAGENA PROTOCOL AND THE NAGOYA PROTOCOL ARE TWO PROTOCOLS

### 7.5.1 Cartagena Protocol

The safe handling, transportation, and application of live modified organisms is the goal of a global accord known as the Cartagena Protocol on Biosafety to the Convention on Biological Diversity (LMOs). On January 29, 2000, it was accepted, and on September 11, 2003, it became effective. There are 171 parties to this protocol as of right now.

#### 7.5.1.1 Nagoya Protocol

To ensure a fair and equitable distribution of the advantages derived from the use of genetic resources, a global agreement known as the Nagoya Protocol on Access to Genetic Resources and the Fair and Equitable Sharing of Advantages deriving from their Utilization to the Convention on Biological Diversity was created. It was ratified on October 12, 2014, after being accepted on October 29, 2010, in Nagoya, Japan.

### 7.5.2 Integrated Water Resources Management (IWRM)

In order to increase social and economic advantages in an equitable manner while threaten the integrity of critical habitats, integrated water resources management (IWRM) is a technique that promotes the integrated regulation of water, land, and related resources. UNEP assists the Department of Irrigation and Waterways in managing its water supplies and putting IWRM strategies into practice. This comprises the following:

- Promoting participatory methods that incorporate all categories of water consumers; incorporating residential, agricultural, economic, and ecological demands into water harvesting administration; emphasising the function of striking a balance between social equality, ecological preservation, and economical effectiveness.
- This includes funding studies on the efficient utilisation surface and groundwater resources, pushing contributors and the global humanitarian organization to do climate impact analyses of water infrastructure investments, and creating drought and flood contingencies and preventative plans.

### 7.5.3 Natural Resource Nexuses in the Uneven Region Pressure on Natural Resources Continues to Increase

Land and resources are under pressure from growing demands, altering climates and technology, urbanisation, rising population, socioeconomic demand and inequality, globalisation, and other global trends, the majority of which are non-renewable.

## 7.6 CLIMATE CHANGE

The rising demand for food may have a negative impact on agricultural output and a substantial knock-on impact on both energy and water use. There are threats to public health as well as consequences for economic growth, worldwide food security, and environmental biodiversity and ecosystems.

### 7.6.1 Calling for the Sustainable and Coordinated Use of Natural Resources

A nexus approach is required and offers potential for integrated planning, management, and control of land and resources. Natural resource nexus hotspots have been identified. The UNECE nexus work's framework is shown in Figure 7.2 and is described next.

### 7.6.2 Four Main Approaches to Sustainable Development

#### 7.6.2.1 Appraisal of the Environment

The environmental impact of any sustainability effort must first be assessed. It is crucial to take into account how the local ecosystem, social structure, and economic situation of the area in question interact in a symbiotic way. Any region's biophysical conditions are closely linked to its economic structure, which in turn affects socioeconomic development. An extensive biophysical examination aids in the evaluation process.

#### 7.6.2.2 Estimation of the Environmental Impact

Environmental and smart use of it is important components of sustainability. Unplanned developer's detrimental impacts on the environment could threaten

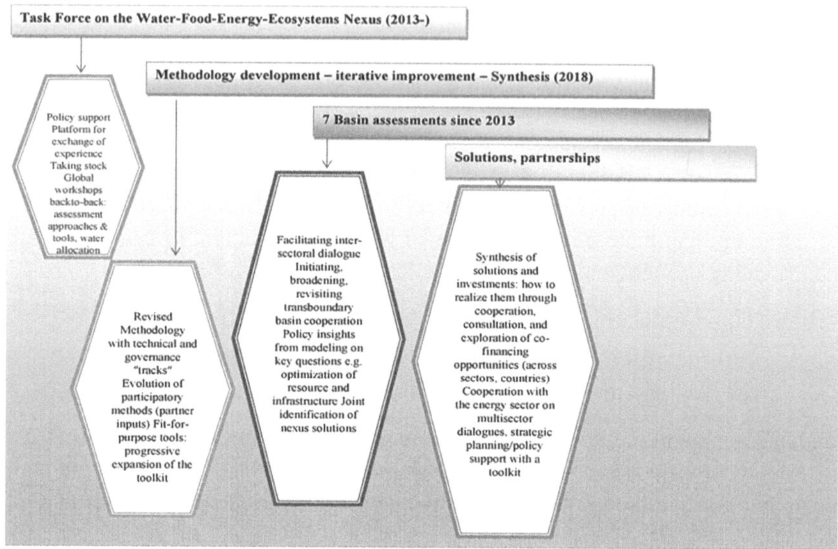


**FIGURE 7.2** Backbone of UNECE nexus work.

sustainability as a whole. Sustainable growth is defined as any expansion that perfectly balances social and ecological needs. To do this, it is necessary to recognise the essential positive traits and to ensure their efficient use and supervision.

It is important to carefully analyze how development would affect various ecological inputs, climate, soil, vegetation, and drainage resources, as well as how these resources will be assessed and monitored. As a result, the study of the interactions between human development and natural systems is known as estimated environmental influence.

### 7.6.2.3 Natural Resource Accounting

Wind, water, and land are assets that aren't capitalised in business or governmental accounts. All countries, rich or poor, must fully take into consideration changes in the stock of natural resources when measuring economic growth. Deterioration and exhaustion of natural resources such as fertile land, fresh water, forests, marine resources, and livestock national productivity, resource generation, and national wealth are all directly impacted by natural resource decline. Ecology and economy are fundamentally linked. The Organization for Economic Co-operation and Development (OECD) stated in its "Declaration on Environment: Resources for the Future" treaty that the climate should be given proper account when pursuing economic development.

### 7.6.2.4 Government Policies and Economic Outlook

Sustainability in growth is hampered by the growing disparity between rich and poor nations, individual inequality, massive trade, money flow, and inconsistent government policies. Sustainable development, which is prevalent in the majority of

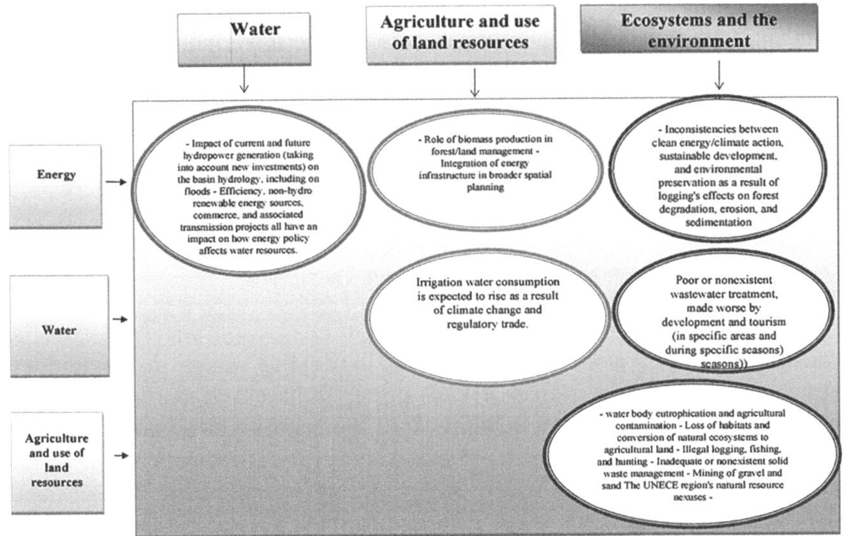



**FIGURE 7.3**  Illustration: Problems with nexus in the Drin River basin.

undeveloped countries, may be hampered by integrated, cohesive development plans, a lack of pricing incentives, awareness, and an absence of information campaigns. Stopping the unsustainable use of natural resources should involve taxes, decreased or removed subsidies, and other actions. Government policy may thus be crucial to the process of sustainable development. The Drin River basin has nexus issues, as seen in Figure 7.3.

## 7.7  WATER-ENERGY-FOOD (WEF)

The nexus idea has attracted attention on a global scale in response to climate change. The Food and Agriculture Organization's (FAO) WEF nexus concept can involve a variety of stakeholders (FAO 2014). In order to address the growing demand for finite resources without endangering their sustainability, it represents a novel approach to examining how water, energy, and food interact. The WEF nexus is an approach to sustainable livelihoods that seek to balance a variety of goals, profits, human needs, and the environment. Future sustainable development depends on thorough studies that can help decision-makers evaluate various decisions' effects by giving more precise information for other sectors' policy, planning, monitoring, and evaluation (Giampietro 2013).

Water-energy nexus (Hamiche et al. 2016), water-food nexus (Mortada et al. 2018), and water-energy-food nexus are examples of nexus approaches (El-Gafy 2017). Investigating the nexus of climate-energy-water-land (Kraucunas et al. 2015), modeling water-energy-food-land use-climate nexus, integrated water-energy-land nexus (Cremades 2019), and nexus bridging water energy-food-land requirements (Ringler et al. 2013; Laspidou et al. 2018). Additionally, it has been suggested that water harvesting be included as a crucial stage in the WEF nexus (Kumari and Eslamian 2021). It would be advantageous to include more industries in the WEF nexus, but doing so would take a lot of coordination, money, and industry expertise. When tackling environmental issues, it is necessary to take regional studies into account because resource usage efficiency, sustainable consumption patterns, product revenues, and resource restrictions vary depending on the various facilities in each location.

### 7.7.1 THE NEXUS APPROACH IS AN EFFECTIVE TACTIC FOR OVERCOMING RESOURCE LIMITATIONS

In this case, the WEF nexus strategy can help in deciding what to do to address the absence of appropriate resources in each place. To have a deeper understanding, the study must be centered on quantity since the majority of research employing the WEF nexus approach to date has focused on quality. If one merely concentrates on one of the interconnected water, energy, or food sectors, there is a substantial risk of overlooking the connections between them. As a result, balancing the many essential biomass components in a WEF correlation approach is a fundamental principle of water resources management.

### 7.7.2 ENCOURAGE SUSTAINABLE GROWTH AND RAISE THE STANDARD OF LIVING

This plan can help communities within the watershed live more comfortably, maintain natural and social capital, and support the long-term sustainability of water resources. Use of the WEF nexus index (WEFNI) is advocated in this regard (Sadeghi et al. 2020). It may be used annually to manage food, energy, and water usage as well as how they interact, which will cut down on water and energy use and increase productivity using the best crop patterning techniques. A crop pattern based on economic factors and resources that provide essential support in satisfying human needs and nature conservation objectives can considerably enhance the management of agriculture in a particular area.

For this, it is necessary to determine the best farming pattern for the region using the optimisation techniques described in some studies, such as a graphical method for optimising the water-energy nexus (Tsolas et al. 2018), developed an optimisation model for the best resource allocation towards sustainable water and food security (Mortada et al. 2018), optimisation of the water-food-energy nexus in response to urbanisation (Uen et al. 2018), and land use (Nie et al. 2019). The optimisation process requires several choice variables and complex calculation procedures in practice; therefore, a computer model might be useful if the variables and functions can be properly described in computer code (Wicaksono et al. 2019). Rare studies have examined the WFE nexus's best advantages in light of $CO_2$ emissions. The AWEFSM model, which aims to maximise system profit while protecting the environment, has made attempts to take $CO_2$ emissions into account by examining trade-offs between economic, environmental, and carbon-abatement objectives (Li et al. 2019).

Alternately, you may optimise the WEFNI (watershed scale) in the agricultural sector but without combining it with $CO_2$ emissions (Sadeghi et al. 2020). This study set out to model short-term joint operations for a multi-objective problem in order to optimise the balance between $CO_2$ emissions, water usage, power consumption, agricultural production, cost, and advantages during the cropping season and keep improving the strategic advantages of the WFE nexus in the coming years. The importance of the WEF nexus methodology in highlighting the connections between water, energy, and food as well as in identifying methods for reducing $CO_2$ emissions to generate the optimal crop pattern was the subject of a second objective.

## 7.8  CASE STUDY

1. Using the nexus concept, multipurpose sustainable wastewater treatment and water reuse in agriculture "Safe Use of Wastewater in Agriculture: Good Practice, Examples" is the source for the information (UNU-FLORES et al. 2016).
2. Integrated reservoir management, a WEF nexus approach application, in the Durance-Verdon basin, France. Based on the material by Hülsmann, Rinke, Paul, and Diez Santos (in press).
3. Benin's pilot program for a drip irrigation system driven by solar energy. (In 2007, Burney and her colleagues collaborated with the nonprofit Solar Electric Light Fund [SELF] on a trial irrigation project in rural Benin, with funding from Stanford's Woods Institute for the Environment.)

The possibility for wastewater treatment facilities to generate electricity from biogas and provide purified water for agricultural or environmental uses, the installation of floating solar panels on reservoirs, and the use of renewable energy for desalination facilities are typical examples.

## 7.9  ADVANTAGES OF THE NEXUS STRATEGY

Using a nexus approach can lead to a number of cross-sector benefits, including the following:

1. **Economic benefits**, including improved medium- and long-term viability of commercial development, resilience to climate change effects, reduced risks and costs from flooding and droughts, higher value added in the agriculture and travel sectors, increased efficiencies in the use of resources and infrastructure, and optimised trade and innovation.
2. **Advantages on the social and environmental fronts**, such as advancing the 2030 Agenda for Sustainable Development, improving population health, creating jobs, improving water and sewage systems, enhancing conservation, and restoring habitat and ecosystems.
3. **Regional cooperation and geopolitical advantages**, such as rising transnational investments, adoption of more transnational agreements, common standards and norms, expansion of regional markets for goods, services, and labour, and greater cooperation in the management of resource sharing.

### 7.9.1   Projects Tackling the "Food, Water, and Energy Nexus"

| S. No. | Projects |
|---|---|
| 1. | CITYFOOD: smart, integrated, multitrophic city food production systems offer a water and energy-saving strategy for addressing the effects of urbanisation on the world |
| 2. | Creating Interfaces: enhancing integrated governance capabilities at the food-water-energy nexus in coastal cities |
| 3. | CRUNCH: food-water-energy operationalised in climate-resilient urban nexus choices |
| 4. | ENLARGE: allowing for the large-scale integration of technology hubs via DDS in diverse urban FWE nexuses to improve community resilience |
| 5. | FEW-meter: allowing for the DDS-enabled, large-scale integration of technology hubs to improve community resilience in diverse urban FWE nexuses |
| 6. | FUSE: For Urban Sustainable Environments: food, water, and energy |
| 7. | GLOCULL: sustainable food, water, and energy innovation in urban living labs on a global and local scale |
| 8. | IFWEN: understanding new approaches to urban food, water, and energy governance |
| 9. | IN-SOURCE: management of susInable urban FWE resources through integrated analysis and modeling |
| 11. | M-NEX: an FWE nexus-based approach to intelligent urban metabolic systems for future green cities |
| 12. | SUNEX: creating a sustainable urban food, water, and energy (FWE) plan through maximising system synergy |
| 13. | Urbanising in place: from the ground up, creating the food-water-energy link |
| 14. | Vertical green 2.0: vertical greening for liveable cities, using innovation to speed up the adoption of an established idea |
| 15. | WASTE FEW ULL: food-energy-water waste mapping and reducing waste in the food-energy-water nexus—urban living lab |

*Source:*   JPI Urban Europe/Belmont Forum (2018)

## 7.10   CONCLUSION

A few of the many factors that contribute to land degradation include excessive tillage, large monocultures, insufficient crop rotation and fallow seasons, overgrazing, farming steep slopes, removing vegetation, excessive chemical usage, and other widespread but unsustainable practices. This contributes to a vicious cycle that encourages even more irresponsible land use practices. The loss of organic matter reduces the soil's ability to hold onto water and nutrients, and declining agricultural yields also have an adverse effect on soil fertility and productivity. A beneficial cycle can be created by managing the soil, land, water, and ecosystems as a "nexus."

## REFERENCES

Allan, T., Keulertz, M., and Woertz, E. 2015. The water-food-energy nexus: an introduction to nexus concepts and some conceptual and operational problems. *Int J Water Res Dev.* 31, 301–311. doi: 10.1080/07900627.2015.1029118

Al-Saidi, M., and Elagib, N. A. 2017. Towards understanding the integrative approach of the water, energy and food nexus. *Sci Total Environ.* 574, 1131–1139. doi: 10.1016/j.scitotenv.2016.09.046

Cremades, R., Mitter, H., Tudose, N. C., Sanchez-Plaza, A., Graves, A., and Broekman, A. 2019. Ten principles to integrate the water-energy-land nexus with climate services for co-producing local and regional integrated assessments. *Sci Total Environ.* 693, 133–662. doi: 10.1016/j.scitotenv.2019.133662.

El-Gafy, I. 2017. Water-food-energy nexus index: analysis of water-energy-food nexus of crop's production system applying the indicators approach. *Appl Water Sci.* 7(6), 2857–2868.

FAO. 2014. *The Water-Energy-Food Nexus: A New Approach in Support of Food Security and Sustainable Agriculture.* FAO, Rome.

Giampietro, M. 2013. Innovative accounting framework for the food-energy-water nexus. *Environ Nat Resour Series.* 56.

Hamiche, A. M., Stambouli, A. B., and Flazi, S. 2016. A review of the water-energy nexus. *Renew Sustain Energy Rev*, 65, 319–331.

Hoff, H. 2011. Understanding the nexus. In *Background Paper for the Bonn2011 Conference: The Water, Energy and Food Security Nexus.* Stockholm Environment Institute, Stockholm.

Hoff, H., Fielding, M., and Davis, M. 2013. A 'nexus' approach to soil and land management: turning vicious cycles into virtuous ones. *Rural.* 21(47), 10–12.

Howells, M., Hermann, S., Welsch, M., Bazilian, M., Segerström, R., Alfstad, T., Gielen, D., Rogner, H., Fischer, G., Van Velthuizen, H., and Wiberg, D. 2013. Integrated analysis of climate change, land-use, energy and water strategies. *Nat Clim Change*, 3(7), 621–626.

IUCN. 2020. *Global Standard for Nature-Based Solutions, a User-Friendly Framework for the Verification, Design and Scaling Up of NbS*, 1st ed. IUCN, Gland, Switzerland.

Kraucunas, I., Clarke, L., Dirks, J., Hathaway, J., Hejazi, M., and Hibbard, K. 2015. Investigating the nexus of climate, energy, water, and land at decision-relevant scales: the platform for regional integrated modeling and analysis (PRIMA). *Clim Change*, 129(3–4), 573–588.

Krchnak, K. M., Smith, D. M., and Deutz, A. 2011. *Putting Nature in the Nexus: Investing in Natural Infrastructure to Advance Water-Energy-Food Security.* IUCN and The Nature Conservancy, Gland, Switzerland.

Kumari, R., and Eslamian, S. 2021. Water harvesting in forests: an important step in water-food-energy nexus. In Eslamian, S, editor. *Handbook of Water Harvesting and Conservation: Basic Concepts and Fundamentals.* Wiley, New York, 337–353. doi: 10.1002/9781119478911.ch22.

Laspidou, C. S., Kofinas, D. T., Mellios, N. K., and Witmer, M. 2018. Modelling the water-energy-food-land use-climate nexus: the nexus tree approach. *Multidiscip Digi Publish Inst Proceed.* 2(11), 617. doi: 10.3390/proceedings2110617.

Li, M., Fu, Q., Singh, V. P., Ji, Yi., Liu, D., and Zhang, C. 2019. An optimal modelling approach for managing agricultural water-energy-food nexus under uncertainty. *Sci Total Environ.* 651, 1416–1434.

Mortada, S., Abou Najm, M., Yassine, A., El Fadel, M., and Alamiddine, I. 2018. Towards sustainable water-food nexus: an optimization approach. *J Cleaner Prod.* 178, 408–418.

Nie, Y., Avraamidou, S., Xiao, X., Pistikopoulos, E. N., Li, J., and Zeng, Y. 2019. A food-energy-water nexus approach for land use optimization. *Sci Total Environ.* 659, 7–19.

Rasul, G., and Sharma, B. 2016. The nexus approach to water-energy-food security: an option for adaptation to climate change. *Clim Policy.* 6, 682–702.

Ringler, C., Bhaduri, A., and Lawford, R. 2013. The nexus across water, energy, land and food (WELF): potential for improved resource use efficiency? *Curr Opin Environ Sustain.* 5(6), 617–264.

Rockström, J., Steffen, W., Noone, K., Persson, Å., Chapin, F. S., Lambin, E. F., Lenton, T. M., Scheffer, M., Folke, C., Schellnhuber, H. J., and Nykvist, B. 2009. A safe operating space for humanity. *Nature.* 461(7263), 472–475.

Sachs, I., and Silk, D. 1990. *Food and Energy: Strategies for Sustainable Development.* United Nations University Press, Tokyo.

Sadeghi, S. H., Sharifi, M. E., Delavar, M., and Zarghami, M. 2020. Application of water-energy-food nexus approach for designating optimal agricultural management pattern at a watershed scale. *Agric Water Manag.* 233, 106071. doi: 10.1016/j.agwat.2020.106071.

Tsolas, S. D., Karim, M. N., and Hasan, M. M. F. 2018. Optimization of water-energy nexus: a network representation-based graphical approach. *Appl Energy.* 224, 230–250.

Uen, T. S., Chang, F. J., Zhou, Y., Tsai, W. P. 2018. Exploring synergistic benefits of water-food-energy nexus through multi-objective reservoir optimization schemes. *Sci Total Environ.* 633, 341–351.

UNU-FLORES, Hettiarachchi, H., & Ardakanian, R. (2016). *Safe Use of Wastewater in Agriculture: Good Practice Examples.* United Nations University Institute for Integrated Management of Material Fluxes and of Resources (UNU-FLORES), Dresden.

Urban Europe/Belmont Forum. 2018. *The 15 Projects That Will Take on the Food-Water-Energy Nexus.* https://jpi-urbaneurope.eu/news/the-15-projects-that-will-take-on-the-food-water-energy-nexus/

Wicaksono, A., Jeong, G., and Kang, D. 2019. Water-energy-food nexus simulation: an optimization approach for resource security. *Water.* 11(4), 667. doi: 10.3390/w11040667.

Wichelns, D. 2017. The water-energy-food nexus: is the increasing attention warranted, from either a research or policy perspective? *Environ Sci Policy.* 69, 113–123.

# 8 Climate Change, Profligacy, Poverty and Destruction

## *All Things Are Interconnected*

*Laila Shahzad, Umair Raiz, Sadia Yousaf,
Nimra Riaz, Rabiya Nasir, and Qaiser Fareed Khan*

## 8.1 CLIMATE CHANGE AND DEPLETION OF RESOURCES

Climate change and the increasing human population has always put pressure on natural resources, causing them to deplete and definitely more exclusive to use. The subsequent complications of natural resource depletion affect people from various sides. On one side, loss of species is continuously leading to mass extinction, whereas conservation of natural ecosystems is vulnerable due to the introduction of invasive species, urbanization and the added demand for the natural resources. The increased stress on climate change is also altering the weather patterns and considerable changes on agricultural productivity. The increased ratio of hurricanes and floods have always seen too often now, causing destruction and exploiting the natural resources such as land degradation and economic loss. Moreover, droughts as well as wildfires are persisting and intensifying with every passing day, causing loss of crop productivity and having a direct impact on the growing economy of a country. Economic growth and poverty decline strongly vary in refining natural resources. The innovative tools of economic and transparency evaluation disclose the factual value of natural capital and supportable management for poverty decline and sustainable management of natural resources (IPCC 2007; MEA 2005).

Soil erosion, salinization, nutrient reduction, as well as desertification have disintegrated more than a quarter of the world's land area (Bai et al. 2008). In the last 50 years, water extractions have tripled, resulting in water shortage and groundwater depletion. By the end of 2030, extraction of water in developing nations are expected to climb by another 50%, putting two-thirds of the world's population in places with moderate to severe water stress. Ecosystems have also been pressured as a result of growth. Approximately 60% of the world's ecosystem services are now of lesser quality than they were 50 years ago, and the current rate of species loss is 100–1,000 times that of prehuman times (WMO 2013).

DOI: 10.1201/9781003358169-8

Many societies and economies tend to cope the situation of extreme climatic conditions and other hazards. Many temporary solutions are seen as an escape route to face the climatic challenges such as trade, migration, as well as storage of food. The competence to face the climatic change and severe weather events is favorably reliant on the economic growth. Generally, income sources of the poor are generally narrower and are exposed to more climate delicacy. Severe weather events that have the capability to cause limited destruction and few fatalities in a developed country can cause massive damage in a developing country. Underprivileged people are intensely susceptible to alteration from average climatic conditions such as extended drought period and natural calamities such as floods as well as droughts. Among the underprivileged, vulnerability differs, as most of the groups are financially, socially and politically unstable; they are more subjected to risk than others.

### 8.1.1 WATER SCARCITY

Water scarcity is one of the leading issues especially seen in developing countries. It has been estimated globally that people influenced by water shortage is expected to increase from 1.7 billion to 5 billion by the year of 2025 (IPCC 2001a). Climate change is expected to further decrease water accessibility in many water-scarce areas especially in the subtropics due to increasing rates of droughts, evaporation and alterations in the average pattern of rainfall. On the other hand, precipitation tends to increase in the regions of equatorials and areas having high latitude that face less water shortage. As the events of rainfall are anticipated to become more intense, the frequency of floods increases threatening communities and infrastructure. Elevation in temperatures and changes in the pattern of rainfall are expected to increase the retreat and melting of glaciers (IPCC 2001). The related changes are more likely to have downstream impacts on agricultural activities. In the areas of the Himalayas, melting of glaciers has become a life-threatening concern due to increasing risk of glacial lake causing outburst floods (UNEP/ICIMOD 2002), while the underground water is also polluted and depleted day by day by many factors in different areas (Riaz et al. 2018).

### 8.1.2 AGRICULTURE AND FOOD SECURITY

Agriculture and food security are the most considerable sectors for developing countries, as the effect of agricultural progress on poverty alleviation is likely to excel the effect of growth on other sectors other than agriculture (ODI 2002). Food security is seen as a task of many interrelated factors such as food production as well as purchasing. Climate change has dramatically deteriorated the incidence of hunger having direct negative consequences and indirect consequences on buying. Land deprivation, price fluctuations and population explosions are major problems for sustainable agricultural productivity. Alterations in climatic patterns such as increasing temperature and precipitation put a stress on agricultural reservoirs relatively in progressing countries. These stressors compromise the quality of land areas which resultantly cause serious problems such as increasing periods of droughts, land desertification and floods. The accessibility of land to poor community can reduce rural poverty. Low-lying coastal groups are more exposed to combat sea level rise and the consequence of climate change on marine reservoirs. The rise in sea level may direct salinization and make agricultural areas unproductive. Poor communities, where fish make a significant source of food (protein), are diminishing due to fluctuations in climatic patterns and related changes in aquatic environment. The effect of climatic change on food security is a major concern in Africa. Due to water scarcity, the sub-Saharan countries have extreme shortage of food as the length of growing season decreases. In Asia and Latin America, due to unprecedented livestock activities, the productivity of crop is expected to decrease in the coming years. The possible impacts of the change in climate are likely to enhance vulnerability and minimizing opportunities to human health as well as by meddling with education and the capability to work. Any effort at forecasting and evaluating climate change on human health is a strenuous talk, as climate change has both direct as well as indirect impact on human health.

Agricultural production can be affected by changes in water supplies, as well as other climate change effects. Droughts and other severe events can reduce agricultural productivity and quality, and as temperatures rise, production declines are expected for numerous important crop species. Changes in the length of the growing season, on the other hand, can have both good and pose negative effects on crop productivity and prices (Gowda et al. 2018).

Fish and invertebrate harvesting in marine environments generates $212 billion in yearly sales in the United States, and climate change is impacting the availability, distribution and quality of commercially significant species. Rising ocean temperatures also diminish oxygen levels, which might result in a 14%–24% reduction in average fish body size by 2050. Rising stream temperatures will have a deleterious impact on some harvested species in freshwater systems. Except for the Arctic, future warming is predicted to limit the discover the potential of all US locations (Lam et al. 2016).

A direct impact is an increase in morbidity and mortality rates due to elevation in temperature. Prolonged heat waves with humidity can increase these rates, especially in urban areas. Another direct impact which can be seen is the increase in death and injury rates due to severe weather events such as flooding, storms as well as landslides. It has been estimated that about 96% of disaster-linked deaths have occurred in developing countries (World Development Report 2000/2001). Fluctuations in temperature and changes in the pattern of rainfall may alter the rate of vector-borne ailments such as dengue and malaria. Extreme weather events such as droughts as well as floods induced by climate change can damage and minimize drinkable water supplies and upsurge the rate of waterborne diseases such as cholera and diarrhea. In undeveloped areas, inadequate access to clean water combine with bad hygienic measures are considered to cause life-threatening diseases. It has been estimated that rate of spread of these diseases has killed almost 2.21 million people in underdeveloped countries, and 90% among them are children (Prüss et al. 2002).

Economy-wide effects due to climate change can have a net negative impact on poor countries, hence hindering the economic growth. Poor adaption to climate change can enhance the rate of extreme weather events, raising the cost of rehabilitation and averting funds from long-standing development purposes. Sudden climate calamities are presently taken toll on the economies of the developing countries that gradually direct loss to human as well as economic wealth. Areas where climate change aggravates climatic severeness cause restricted adaptive capacity in the developmental scenarios due to deprivation of life and private properties and decreased productivity of economic sectors as well as damage of infrastructure. This is especially true for underdeveloped countries with little economic diversity.

Changes in provisioning services or the material products that people acquire from ecosystems as well as biodiversity are results of climate change, which can have a significant impact on human economies and well-being. Increased temperatures, variations in the pattern of rainfall and ecosystem instability such as wildfires, for example, are affecting freshwater supplies for communities, agriculture and power generation in wooded watersheds. In some areas, surface water scarcity is likely during dry years. Water quality is also affected by rising stream temperatures. Wildfires can also increase sediment and debris deposition in streams, ponds and lakes (Luce et al. 2012). These changes will put a strain on water supplies, potentially driving up the cost of water treatment (Warziniack et al. 2018).

## 8.2   IMPACTS OF CLIMATE CHANGE ON POVERTY AND NATURAL RESOURCES

The impact of climate change on poverty can be seen through direct as well as indirect channels that deprive poor people from basic necessities. Figure 8.1 tries to develop a link between the three significant pillars.

### 8.2.1   DIRECT IMPACTS

Direct channels are deep-rooted in customary impact assessment network and propose direct links between biophysical changes and market reactions that lead to the outcome of poverty. On the other hand, indirect channels often provoke vulnerability linkages and postulate the series of connection between exposure of

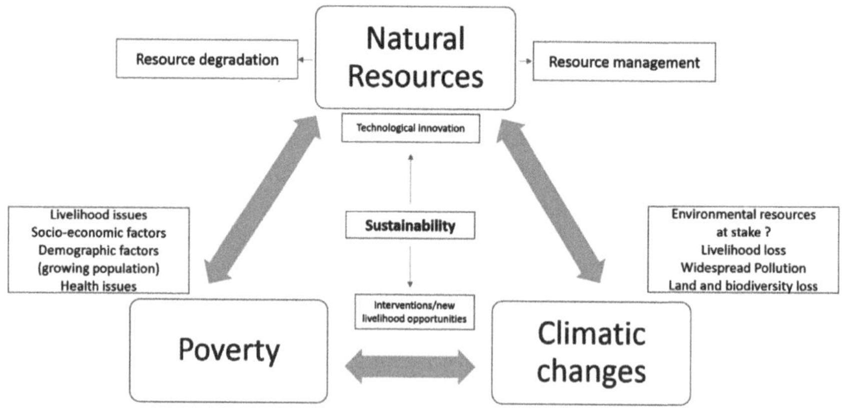


**FIGURE 8.1**   Interaction and vicious effects of poverty, climatic changes and resources.

climate and poverty. Indirect channels are influenced by a number of factors such as decision-making processes and economic as well as social factors. Indirect channels also participate in poverty such as poor health or political encounter, highlighting that climate change intersects with the socioeconomic impacts and environmental stressors.

The escalation of food prices and agricultural productivity can be seen as direct avenues which influence poverty. The estimation of these channels through climate projection models, evaluation of crop production and economic models to study how the change in climate, elevation in temperatures and alteration in the average pattern of rainfall may affect the yield of agriculture and food prices. The people living in urban areas are more exposed to increasing food prices, as they spend a large amount of their income on food. On the other hand, dispossessed communities living in rural areas are also susceptible to food insecurity as the price of food increases. For people living in rural areas, agricultural producers residing in areas influenced by climate-related shock, including prolonged drought periods, decreased productivity and loss of income, can openly participate in food insecurity and poverty. The producers not residing in affected areas, increasing the food prices possibly mean high market prices which may support increased production- and consumption-related costs. Climate-related shock waves also have the possibility to participate in food insecurity within many developed countries mostly subdued by social safety nets. The impacts of climate change on poverty are highly diversified, having the biggest poverty-increasing impacts among the urban and non-agricultural house holdings, mostly seen in low-income countries of Africa and South Asia. The weather volatility also influences the number of people that drop below the poverty line; hence, the poverty is directly linked with weather volatility.

Climate change is expected to have some favorable consequences. Increased water availability, for example, may assist water-scarce regions such as Southeast Asia. However, developing countries are likely to bear the brunt of climate change's negative consequences. This is due to the economic importance of climate-sensitive sectors (such as agriculture and fisheries) in these nations, as well as their low human, institutional and financial capacity to foresee and respond to climate change's direct and indirect effects. In general, least developed nations (LDCs) are most vulnerable in tropical and subtropical locations. As a result, the countries with the fewest resources are the poorest.

## 8.2.2 Indirect Impacts

Indirect channels of climate change are more noticeable and are more exposed to vulnerabilities. Much research-oriented studies have been carried out to determine the chains of causality which are not only about biophysical impacts but are also interceded by uncountable social, cultural and organizational factors at different spatial scales. Resource-reliant societies have a wide-ranging experience of information about managing climate volatility. Climate change combined with changing economic landscapes can aggravate poverty by discouraging traditional coping approaches such as crop rotation and expanding livelihood (Chen et al. 2009). Ecosystem services such as soil, water regulation as well as biodiversity are also thought to be influenced by climate change in which poor populations are truly dependent. Poor individuals in comparison with wealthier individuals are more subjected to vulnerability, as the richer people are capable to substitute nature capital and can depend on fossil energy sources. Ecosystem services are important to the subsistence of crop yield, livestock foraging, fishing and hunting.

Impacts on physical and mental well-being signify another way through which climate change can indirectly participate to destitution. Besides physical toll, psychological health may be able to also be the victim of climate change. The impacts on mental well-being may contain personal knowledge of injury arisen from severe events and disclosure of media events which leads to stress. Climate change can destabilize culture as well as identity by reducing practicability of traditional livelihoods. Climate change can act as a risk multiplier enhancing the possibility of political volatility and violent conflicts that significantly proliferate poverty, comprising damage to properties, income and ecosystem services. Many rural communities shift to urban areas, where they are more contacted to novel types of climatic threats. No doubt, migration is seen as a constructive adaption to climate changes, but the unprivileged people are not able to migrate and are more stuck in environmentally degraded areas. Other climate-migration-poverty connections are also possible such as increasing the food prices aggravates rural poverty by decreasing the payments by urban workers.

Natural resources and linked ecosystem services emphasize human life existence. Natural resources (as minerals, air, land, water, fisheries, forests, flora and fauna) which are managed sustainably have established improved trust in modern years particularly based on equity focus and reduction of poverty-hunger within the SDGs (Sustainable Development Goals) (Gereta & Roskaft 2010).

Consequently, it has been contended that natural resource management ought to endow to poverty alleviation, while natural resources must be consumed in an equitable manner to boost human welfare. Almost three to four people who live in rural areas are mostly poor, where they depend on natural resources mainly for their livelihoods. Poverty is indeed now a global threat, which creates a vast impact on citizens around the globe. In 2002, according to World Bank, 90% of the world's 1 billion people are making less than 1$ per day and are depending on the forests while facing daily hardships that determine their endurance. The link between poverty alleviation and NRM (natural resource management) narrates how the poorest citizens of the world are dependent upon the natural resources for their existence which examines the economic, social and governmental factors that describe this significant relation and also shows how the sensible use of them can provide us the origin for efficient poverty-reduction strategies (World Bank 2000).

There's a report by international development agencies in 2002 which describes that about 90% of the 15 million people that work on the world's waters were largely poor, small-scale fishers. Poor people from rural areas depend mostly on natural capital. More than 7 in 10 poor people live in the rural regions of Africa with most engaged in resource-based activities like fishing, small-scale farming, production of livestock, hunting, logging and artisanal mining. For primary source of income, poor people depend on harvests while recede on natural resources when all other sources fail for income (USAID 2006).

### 8.2.3  SOCIAL PREFERENCE AND NATURAL RESOURCE DEPLETION

Various ecologists have raised the argument that the natural resource extinction is a consequence of our own selfish behavior and biased view of ecosystem. To gratify human desires, various people behave like they are self-sufficient that gives the essential resources to sustain life. According to this context, it is a materialistic greed, or overconsumption, that is the main cause of resource depletion problem. Moreover, economists also have identified in turn about the situations under which resource depletion can be the result of lucid behavior. Poor countries mostly have biomass-reliant survival economies. Various rural populations in poor countries rely on natural resources; hence, it might be a thought that the residents of these populations would have sustainability as an economic and social goal. On the other hand, if sociopsychological work and pressure of social standard of work are well-built, the employment can also be a vital goal. Attaining sustainability along with full employment is not an issue if IRB (initial resource base) of community is sufficient and RGC (resource generative capacity) is adequately high; if not, there might be a clash between full employment and sustainability (Dasgupta & Maler 1995).

### 8.2.4  ENVIRONMENTAL STANDARDS AND INCOME GENERATION

Natural ecosystems possess numerous features that make them more accessible and attractive as an income source to the poor rural people. Environmental resources are widely distributed and renewable which are often found in common areas where the access is easy for poor people without owning the property. Natural resources promote cohesion and reinforce the safety net for the whole community. Poor people generally have restricted access to financial and physical capital. Additionally, poor and rich people use natural resources in various ways. Rich people derive environmental income from the natural resources rather than poor people. Numerous studies conducted have shown that small-scale activities in the livestock, forest, agriculture, mining and fishing sectors can certainly contribute 15%–17% of rural household income and additional bigger values for their subsistence. The rural poor, lacking cash, have no choice other than to rely on common natural resources for their food, medicines and firewood. With low liability to threat, the rich can concentrate selectively on one or two activities like agriculture and grazing to improve their savings (USAID 2006).

### 8.2.5  TECHNOLOGICAL INEQUALITY AND RESOURCE DEPLETION

Rapid increase in inequality is frequently seen to spare with resources overconsumption—for instance, degradation of land in Bangladesh, overextraction of water in Pakistan, fishing in Lake Victoria and forest harvest activities in sub-Saharan Africa. However, access to ecosystem services given by the common natural resources alleviates poverty, but extensive access to technology by rich people might stimulate resource depletion, therefore broadening the assets gap. Increase in local inequality might lead to degradation of resources and significant transitions like unpredicted rise in poverty and ecological resource degradation.

Dilemma with the poverty persistence is that it is not only overwhelming for people but can also lead to consistent inequality over time within a growing society. Various theories try to explain the long-term actions of inequality. Several case studies demonstrate how resource usage with technological access can worsen both poverty and inequality. For instance, in Pakistan, traditional inequality in water access across the regions and between the social classes has cited exceptional stress on water resources (Mustafa et al. 2013).

Era of modern technology insisted wealth accumulation in the form of commercial agriculture, where tube wells aided water overconsumption and other dynamic but water-rigorous production options such as cement, leather, textiles, sugarcane and fertilizers (Shah 2008). Therefore, this results in the migration of small tenant farmers from the resources (Rahman 2014).

The accumulation of wealth by the elites is attained by controlling and prying natural resources on which poor communities also depend typically, such as forest depletion for timber or another land use, commercial agriculture through exploitation of groundwater sources and overfishing for the exports (Godoy et al. 2010).

Consequently, rapid wealth growth in unequal societies by the elites can put moderate pressure on the natural resources, therefore disturbing the reliable poor community. Extensive resource consumption impacts the dependent sustenance on ecological resources (Daily 1997).

## 8.3  CLIMATE CRISIS IN THE FORM OF COSTS

The observed repercussions of a warming planet are increasing the pressure to act on the diminishing glaciers, decreasing crop productivity and irreversible sea level rise due to ice sheet collapse. The understanding that postponing mitigation increases mitigation costs adds to the urgency for immediate action. Delaying emission reduction measures means that $CO_2$ concentrations will continue to rise. The elevated levels of $CO_2$ concentrations and temperatures are linked to disproportionately greater losses. Climate change becomes a self-reinforcing loop after the global temperature hits a tipping point (Furman et al. 2015).

Climate costs can be controlled by taking steps to protect coasts and lowland urban areas from elevating sea levels and flooding, as well as preventing farm yields from dropping, owing to changing climatic trends. Cities are at the heart of economic activity; thus, more efforts are needed to protect them against storms and flooding. Extreme weather events and climate change have regularly rated among the top five global risks in terms of probability and impact since 2011, yet a very slow progress has been seen in terms of climate action and catastrophe mitigation.

Despite having the proof of the economic costs of climate change, a very small number of politicians have been elected to national government after promising to address the issue. This poses a perplexing problem. Climate mitigation will lag without a political mandate, and without action, uncontrolled climate change would harm lives and livelihoods while stifling economic progress. As a result, it's critical to seize every opportunity to change people's minds and leverage windows of opportunity. As India and Indonesia have already proved, the drop in oil prices that began in the second half of 2014 gives a once-in-a-lifetime chance to eliminate these subsidies without involving the economic and political consequences of high oil prices (WEF 2015).

Another chance exists in the ongoing global debates on climate goals and commitments. In contrast to their history of failure, these conferences have shown signs of life, with the US and the People's Republic of China committing to substantial aims, even if not in a binding form, as previously stated. The new deal reached at the COP21 climate conference in Paris in December 2015, which will go into effect in 2020, represents a new beginning. The key question arising here is whether the looming climate disaster will prompt timely national and international environmental action. This will happen if we recognize that the hazards are both at the regional as well as global level, that they influence the present and the future and that economic growth will be hampered by climate inactivity rather than action.

## 8.4 CLIMATE CHANGE AND NATURAL RESOURCE MANAGEMENT

### 8.4.1 CONCEPTUALIZING POVERTY REDUCTION THROUGH NRM INTERACTION

Natural resource management is indeed crucially important. The means by which societies deal with the supply of natural resources, upon which they depend for their subsistence and development, are assumed as NRM. Humans are primarily dependent on natural resources for endurance while assuring the sustainable consumption and preservation of nature for the better civilization.

There are numerous views about the interaction between resource management and poverty. Some view increasing populations as negatively affecting limited natural resources with the technology alleviating the type and extent of impact. Accordingly, poverty is seen as a "driver" of environmental degradation and biodiversity loss. Inability of a poor individual to accumulate wealth from resources can lead to environmental overexploitation. Inconsistency in the poverty-environment interactions brought the development of "asset-based approach" to reduce poverty (USAID 2006).

Poverty is a critical phenomenon which affects the citizens globally. Though during the three past decades, the world has made vital progress with reference to the reduction of the rate of people living under the poverty line and abolished all form of poverty which continues to create a dreadful challenge. About 37% of 2 billion global populations in 1990s lived under poverty line of $1 per day (World Bank 2016).

Conversely, the Millennium Development Goal of halving the proportion of intense poverty by 2015 were met in 2010 effectively. In 2013, according to the world development indicator 2017, almost 10% of the world's population (about 0.8 million) lived in severe poverty (World Bank 2017).

Poor individuals face daily hardships that decide their existence. Various development populations comprising of banks, governmental agencies and nongovernmental organizations (NGOs) hunt for improvement of livelihoods of impecunious citizens through different poverty-reduction strategies. Poverty-reduction strategies of every country try to address root causes of poverty, its different extents and its detrimental effect on the people bound in fierce poverty cycle.

From the origin of mankind, every human society is linked with the ecological processes and their healthy systems for their daily life subsistence. The reliant quality of human beings on nature is becoming intense and diverse, gradually adapting various forms in different ages. Furthermore, in the context of rural poor people, the reliability is more important than in any other case. There are many countries in the entire African continent where over 7 in 10 poor individuals live in the rural regions. Problems in natural resource management and poverty reduction define their linkage between them. This linkage between poverty and natural resource management is dynamic, complex and diverse. To integrate improved natural resources and poverty reduction in the developmental strategies and programs, some basic remittance is inadequately required (Raggl 2017).

Some of the general considerations for development of programs and strategies to reduce povertre discussed next:

a) Accounting the immense significance of common pool resources for the poor individuals:

Natural resources all around the world, including biodiversity, water, soil, ecosystems and all other living things, involve natural capital. It is stated that poor people have the highest share of natural capital compared to the rich people. The service or industrial sector in poor countries is comparatively small; therefore, poor people have limited access to them. Thus, it is really important to consider natural resources as the main assets for the poor.

b) Highlight the challenges, capacities and choices of poor rural livelihoods in the management of natural resources:

Natural and social capital typically includes the portfolio of poor people. Pro-poor approaches mainly focus on the natural resources. The poor-friendly development strategies are planned on such a manner which increases the poor people's income and return from common pool resources. Poor livelihoods are at the core of the asset-based approach.

c) Reinforce the usage of natural resources for pro-poor economic growth:

Pro-poor economic growth is the rate and growth pattern that enhances the ability of poor people to participate in and contribute to the economic growth and get entertained by the benefits later (OECD 2008). Natural resource use can enhance the poor income and also reduce poverty, thereby sustaining growth. It is mandatory to improve the NRM practices for lasting pro-poor economic growth. Both renewable and non-renewable resources are essential to economical activities in numerous ways. Renewable energy is indeed a fundamental element of economy that plays a vital role in giving poor people energy access. Natural resources are specifically significant in booming the economic growth in developing countries. In Indonesia and Cameroon, wealth is accumulated from forestry, while in the Pacific islands and Mauritania, fishery is a basic source of income. Although in less economically stabled countries—Asia, Africa and Latin America—environmental tourism is getting recognition gradually. However, in Botswana and Kuwait, extraction of mineral adds enough contribution to the overall economic growth. It is assessed that natural capital rationalizes about 26% of the total capital in low-revenue countries, 13% of total capital in middle-revenue countries and just 2% of capital in OECD (Organization for Economic Cooperation and Development) or industrialized countries. Thus, natural-resource-based enterprises and industries give a better employment source and good revenue opportunities (OECD 2008).

d) Creating a broad range of constructive by-products through natural resource usage at the local, national and even global levels:

Additionally, natural resources also generate various services, such as water purification services by the wetlands or water cycle regulation by watersheds. One of the most important and timely services from these assets is carbon sequestration by soil or forests which assists and mitigates the climate change.

e) Poor-people-friendly natural resource policies, institutions and economic structures:

It is obvious that the productivity of the poor people can be adversely affected by the policies, institutions and different economic structures. Unsuitable polices may reduce and evade the productivity of poor people from natural resource consumption which results in the difficulty to access natural resources and control their mechanisms to convert them into wealth.

f) Improving the efficacy of different policies for beneficial and effective natural resource management:

Environmental and economic policies are vital for the overall society development. Efficient and effective policies can cause increase in the production of these assets. Consequently, various policies and programs require analyzing the particular market conditions, asset insurance, property and land rights, system of approaches to credit, etc. Advancing the efficacy of such programs and policies in the management of natural resources can increase the return from natural assets to manifold. Hence, it is pointless to say that all kinds of management of natural resources in a sustainable way can be a better direction for economic growth and poverty reduction (Ahmed Khan & Sultana 2020).

## 8.5 POVERTY REDUCTION THROUGH NRM APPROACHES

Problems associated with natural resource management are typically complex because of their dependence on different components like hydrological cycles, ecological cycles, geography, plants animals, climate, etc. All of these are dynamically interrelated. Alteration in any of them may have long-term effects which can be irreversible even. To conserve natural resources and mitigate poverty, some traditionally adopted approaches of natural resource management include the following:

### 8.5.1   Asset-Based Approach

Poverty includes more than income and wealth. It is a complex and versatile deficiency that affects people's different abilities and their well-being in general. Access to health, education, land, family, credit, justice, community support and other productive resources are all significant in developing a rational sustenance. Poverty has been narrated as the deprivation of various kinds of freedom—political, economic and social—and those choices that affect the livelihoods. For instance, political freedom can assist secure fair resource rights system, leading to equity and greater wealth. In this way, freedom is both the means and ends of development.

Asset-based approach states that poverty is a multidimensional process in space or time which offers reliable strategies to reduce the vulnerability and risks facing poor ménage and to improve their capability to participate in and take advantage from new economic opportunities by concentrating on their resources. This approach mainly focuses on generating stock of wealth which is available to poor and on their capabilities to manage the risks and vulnerability to attain the durable advancements in well-being. The significance of natural capital along with the complete stock of capital accessible to households is inclined to fluctuate inversely with income levels. The poorer the population of the country, the more important the part natural capital plays in arbitrating poverty outcome. Natural resources conduce to livelihoods by offering a buffer against the economic deficit or temporary dietary shortfalls while helping as basis of employment and wealth in crisis and serving as a voluntarily adaptable capital asset.

The poor, like others, deliberately deal with their multifaceted resource groups, balancing viability, related costs, trade and expected profits, allowing for each option. There are dominant accompaniments to the various assets: endowing in education when future wage earnings show significant gains, finding harmless savings methods and putting in social assets as protection against future risks making a living by leasing out houses or operating small businesses (USAID 2006).

### 8.5.2   Community-Based NRM Approach

CBNRM is a community-based natural resource management approach which basically combines generations of economic benefit with the conservation objectives for the rural communities (Thakadu 2005). This approach is based on the fundamental assumptions:

  i. Localities are better positioned to preserve natural assets
  ii. Localities will merely conserve an asset that best rewards the expenses of preservation
  iii. Localities will protect resources that are straightforwardly related to their personal lives

iv. It is clear that when a nearby local area's life quality is upgraded, their endeavors and obligations to guarantee the future prosperity of the assets are likewise improved (Ostrom et al. 1993)

### 8.5.3 Adaptive Management Approach

This is one of the most debated among other approaches. The concept behind this approach is to create a link between policy learning and implementation. It primarily shares the learning concept which involves more traditional accumulative purposeful trialing (Holling 1978). Adaptive management, nevertheless, adds a deliberate, explicit and official dimension to enclosing relevant problems and questions, undergoing different testing, assembling and crucially dealing with the results and revaluating the policy framework that previously triggered the inquiry considering the newly obtained knowledge (Ahmed Khan & Sultana 2020).

### 8.5.4 Integrated NRM Approach

Poverty reduction has various ways just like poverty has diverse appearances in different perspectives. Shortage of food and income relates with the absence of access to energy, water, flood protection, rights and acknowledgements. Amongst the paths by means of which agricultural research can boost the rural wealth, INRM deals with a complicated sequence of problems, with trades along with the issues that are in different stages of rejection, recognition, investigation, modernization, scenario amalgamation and generation of podium for the policy change (Lovell et al. 2002). INRM is integrated natural resource management which is being utilized comprehensively in various parts of globe. It's a phenomenon of managing natural assets in an organized way that comprises of multiple facets of natural resource utilization such as sociopolitical, economical and biophysical. It is initially used in the agriculture field to achieve the production targets. However, it has additional use in attaining goals of the broader community like poverty alleviation, sustainable conservation of environment and future generation welfare (van Noordwijka 2019).

World Resources Institute (WRI) proposed seven different steps to integrate the environmental resources more strongly into the efforts of poverty reduction:

- Orientation of an ecosystem and environmental returns: Highlight the value of ecosystem services as a driver of wealth for the poor.
- Income sustainability over time: Acquire a durable approach and consider the significance of developing forestry, fishery and agricultural sectors.
- Possession and accession of resources: Apprehend the central significance of land possession in rural poverty-reduction policy.
- CBNRM and decentralization: Decentralize power over the management of resources to experienced community groups and local authorities.
- Participation, bureaucratic rights and gender equality: Ground the policies in sophisticated contribution by the civil society. Accentuate free, prior and informed assent by the localities in different economic developmental projects.

- Standard monitoring of environment: Comprised of approaches to monitor every environmental situation to keep a track of effects of economic growth on the environmental returns.
- Targets, assessments and indicators: Signify environmental indicators and poverty to assess performance and give allotment for adaptive management (USAID 2006).

## 8.6   LINK BETWEEN ECOSYSTEM AND POVERTY REDUCTION

Few of the modern policies aspire to relate ecosystem and poverty reduction. Unluckily, current poverty evaluation and strategies for poverty reduction under approximate rural profit from the management of natural assets and underrate services of ecosystem as benefit for the poor. Formerly, development frequently highlights the high-input, export-based consumption of natural resources and government-financed industrialization. Unfortunately, these efforts were not specifically pro-poor. In context of forestry, extensive plantations and concessions followed a policy that destitute the access of the poor to basic resources and eventually not contributed to NDGs (national development goals). In the latest World Bank reviews, IMF (International Monetary Fund) and UN (United Nations) disclosed an unsure approval of the significance of a healthy ecosystem as a plus point for the poor. However, the UN Millennium Development Goals (MDGs) and Poverty Reduction Strategy Papers (PRSP) requisite by World Bank and IMF in return for credit assistance (USAID 2006).

### 8.6.1   Climate Change and Biodiversity

Climate change will have a wide range of effects on forests, depending on the species involved and other circumstances. Forest production will likely grow as $CO_2$ levels raise until other climate change effects, such as increased drought, forest fire, pests and exotic species, place extra stress on forests. The distribution and composition of tree species will continue to move northward as temperatures rise. In practice, this means that by 2100, the favored tree species will have shifted from maple and birch to oak and hickory. By the end of the century, certain species, such as red pine, may have totally disappeared from the Great Lakes region (Hughes 2000; Lawler et al. 2009; Parmesan 2006).

As climate change causes habitat zones to shift northward, most of the animal diversity may be forced to migrate north as well. Many species will be unable to migrate because of the Great Lakes, which will be a key barrier for those who are unable to adapt. Wetland-dependent wildlife may be the most vulnerable since rising evaporation rates may lower overall wetland coverage, further stressing those species. The Great Lakes region's fish populations may become less diversified. Warmer water temperatures are anticipated to cause a drop in cold-water fish populations with population enrichment in warm water. Lake stratification and added frequency of hypoxic environment may impair overall production in lakes and streams (IPCC 2012).

Lower intensities of water in the summer are more likely to hinder groundwater recharge, dry up tiny streams and reduce wetlands, resulting in worse water quality and fewer wildlife habitats. Climate change may compound existing land use consequences by initiating more recurrent and intense precipitation events. The natural flood-absorbing capacity of wetlands and floodplains are harmed by impervious surfaces, increasing the danger of flooding and erosion. Watersheds as well as other ecosystems near agribusinesses are being impacted as severe storms become more common and stronger.

Disturbances that cause modifications in wildlife habitat may also be a result of climate change effects on wildlife. Alterations in vegetation community composition as well as other ecosystem components, such as pest outbreaks and other instabilities like wildfire, are examples of climate-related changes that may have an indirect impact on wildlife. The kind of vegetation, which is frequently regarded as a key feature of land cover, is a key predictor of species occurrence. Plant species distribution is commonly thought to be driven by interactions between position and temperature, while other factors like as rainfall, disturbance and the contact between the communities may also play a role (Harsch et al. 2009).

Climate change is likely to impact the frequency, extent, duration and intensity of disruptions, although predicting exact results is difficult. Individual disturbances may alter landscape structure, affecting wildlife habitat in the process. Even though disturbance is a natural feature of the environment, climate-related variations in disturbance occurrence may have an impact on wildlife's ability to respond to and adapt to specific disturbances. Insect infestations in forest ecosystems are one sort of disturbance that climate change is thought to effect. Warmer winters and summers, as well as lower summer precipitation, are thought to have aided the mountain pine beetle. Climate change is likely to have an impact on the quantity and severity of forest fires. In the United States, forest fire intensity was expected to increase by 10% to 50%, but the details of the estimates varied amongst climate models (Inkley et al. 2004).

Exotic plants' propensity to infiltrate and actually out contest natural vegetation may be influenced by climate-related disturbance. Natural catastrophes such as hurricanes, floods and droughts, which occur more frequently and with greater intensity, may have a greater impact on changes in the vegetation community than climate-stimulated changes in individual species growth rates. Fast-growing exotics like the Chinese tallow may disrupt the formation of native forest ecosystems and limit the regeneration of plant species that wildlife depends on.

Concurrent human-caused factors of habitat loss may also amplify the effects of climate change. Many studies have looked at the direct consequences of land use disturbance on specific species. It has been discovered that land use intensification has an inverse relationship with various traits expressing plant function such as leaf volume, nutrient-uptake availability, reaction to disruption such as resprouting ability and time taken for reproduction, implying that land use intensification may reduce the overall affected ecosystem's resiliency (ability to successfully respond to stressors). If plant resiliency declines, it's probable that reliant wildlife will become more vulnerable to disturbances (Laliberté 2010).

For cover such as protection from predators, reproduction and food, wildlife species rely on vegetation qualities (pollen, seeds and fruits). Wildlife plays an immediate ecological and behavioral response to change in the patterns of the climate, such as species-range alterations. Many abiotic (physical obstacles, climate) and biotic (competition among predator and prey, population dynamics) variables influence the global distribution of species, both individually and in combination. Researchers have noticed changes in species distributions through time and geography that adapt in response to climate change. Climate change has caused temporal range shifts as well. Individual climate variables, such as change in spring temperature, may cause slow changes in wildlife behavior (Gaston 2003).

## 8.7 SDGS LINKED WITH POVERTY REDUCTION AND NATURAL RESOURCE MANAGEMENT

Poverty eradication and sustainable natural resource management have achieved considerable coverage and attention in all SDGs. All participant stakeholders and countries are promised to execute this framework in collaboration with in subsequent one-and-a-half decades. The main apparent link of poverty to SGD is marked in Goal 1 (that has five specific targets) which discuss about ending every kind of poverty by 2030. To achieve those mentioned targets in SDG 1, it is very significant to assure the essential enlistment of resources (including economical and efficient consumption of natural resources and secure policy guidelines at local, national and international levels). Additionally, Goal 15 endows with poverty reduction through the management of natural resources; SDG 15 is all about "life on land" which means to conserve, restore and endorse the sustainable utilization of terrestrial ecosystem, manage the forests sustainably, fight desertification, stop biodiversity loss and reverse degradation of land by 2030. There are nine particular targets to be achieved under Goal 15. Furthermore, countries' development and reduction of poverty are also emphasized by this development by combining ecosystems and biodiversity principles with local and national planning. Besides Goals 1 and 15, SDG 13 (combat the climate change) and SDG 14 (conserve life below water) are also obliquely associated with poverty reduction and play a significant role in the sustainable natural resource management activities (Ahmed Khan & Sultana 2020).

## 8.8 IMPACT OF CLIMATE CHANGE ON FUTURE HAZARDS

Natural disasters already wreak havoc on economies, societies and environments around the world. Natural hazards are predicted to worsen as a result of projected increases in the intensity and occurrence of severe events owing to climate change, as well as increased exposure and vulnerability of populations (UNISDR 2012). Natural catastrophes can have devastating effects on many communities, the environment and the economic prosperity of the countries affected. Extreme events are putting a strain on industries that are strongly linked to the climate, such as agricultural productivity, tourism as well as water scarcity (IPCC 2012). Over the last few decades, some types of climate events have become more common and severe with each passing year. Particular incidents can be traced back to human influences; the goal must be to reduce the terrible consequences of natural disasters. This can be accomplished in a variety of ways, including using early warning systems and minimizing underlying vulnerability and exposure of people and assets.

The number of cold days and nights has reduced, while the number of warm days and nights has increased. This is based solely on land data and on a worldwide scale. Heat waves have become more common in most parts of Europe, Asia and Australia. Furthermore, several places have seen a statistically significant rise in extreme events of rainfall, i.e., 95th percentile. This is in line with the rising temperature and reported water vapor levels in the atmosphere. Heavy precipitation events occurred more frequently and with greater severity in Europe and North America. However, depending on the region, there is still a lot of variety in precipitation trends. Droughts, floods and cyclone activity show fewer uniform trends. Some places, for example, have had more severe and prolonged drought periods such as those seen in southern Europe; however, others have seen a decline such as that in Central North America (IPCC 2013).

Extreme temperatures as well as severe precipitation events are expected to become more common as the climate warms. According to model forecasts, the number of exceptionally hot days and heat waves will return. At the end of the 21st century, the hottest day, which used to happen once every 20 years, is projected to happen once every other year in most parts of the world. The length, occurrence as well as intensity of heat waves are also expected to rise. It's crucial to remember that frigid, extreme events will continue to occur. Extremely hot and cold days, for example, can have an impact on city death rates. Severe precipitation events and elevation in the level of coastal high waters are expected to become more common in many parts of the world. During the winter, heavy precipitation events are expected to rise in higher attitudes, tropical regions and the Northern Hemispheric mid-latitudes (Peterson et al. 2012).

## 8.9  CONCLUSION

Resources are exposed to climate change in the 21st century, which has resulted in numerous stressors. Developing nations that are already vulnerable to several pressures are more affected by climatic disasters. Climate change is associated with profligacy in rich civilizations, poverty in poor nations and global environmental degradation. The increased weather anomalies bring increased extreme events like unpredicted rainfall, hurricanes and heat waves. Such incidents result in pushing the livelihood of poor population at stake. While poverty poses a greater threat to the management of natural resources and causes greater deterioration, climate change influences each place on every continent variously. Because the effects of climate change are already being felt, adaptation is necessary. However, this is a short-term solution that ignores related issues and falsely claims that the problem can be endured indefinitely. Since our issues span the globe, so must our overall governance.

## REFERENCES

Ahmed Khan, N., Sultana, N. (2020). *Natural resource management: an approach to poverty reduction.* SpringerLink, Cham. https://doi.org/10.1007/978-3-319-69625-6_111-1

Bai, Z.G., Dent, D.L., Olsson, L., Schaepman, M.E. (2008). Proxy global assessment of land degradation. *Soil Use Manag.* 24(3): 223–234.

Chen, I.-C., Shiu, H.-J., Benedick, S., Holloway, J.D., Chey, V.K., Barlow, H.S., Hill, J.K. and Thomas, C.D. (2009). Elevation increases in moth assemblages over 42 years on a tropical mountain. *Proc Natl Acad Sci*. 106: 1479–1483.

Daily, G.C. (1997). Nature's services: societal dependence on natural ecosystems. In *Nature's services: societal dependence on natural ecosystems*. Island Press, Washington, DC. https://doi.org/10.1017/S1367943098221123.

Dasgupta, P., Maler, K. (1995). Poverty, institutions, and the environmental resource-base. In J. Behrman, T. Srinivasan, ed., *Handbook of development economics* (Vol. IIIA). Elsevier Science, The Netherlands.

Furman, J., Shadbegian, R., Stock, J. (2015). *The cost of delaying action to stem climate change: a meta-analysis*. VOX CEPR's Policy Portal, 25 February. www.voxeu.org/article/cost-delaying-action-stemclimate-change-meta-analysis

Gaston, K.J. (2003). *The structure and dynamics of geographic ranges*. Oxford, United Kingdom: Oxford University Press.

Gereta, E., Roskaft, E. (2010). *Conservation of natural resources: some African and Asian examples*. Tapir Academic Press, Trondheim.

Godoy, R., et al. (2010). The effect of wealth and real income on wildlife consumption amongnative Amazonians in Bolivia: estimates of annual trends with longitudinal household data (2002–2006). *Anim Conserv*. 13, 265–274.

Harsch, M.A., Hulme, P.E., McGlone, M.S., Duncan, R.P. (2009). Are treelines advancing? A global meta-analysis of treeline response to climate warming. *Ecol Lett*. 12, 1040–1049.

Holling, C.S. (1978). *Adaptive environmental assessment and management*. Wiley, London.

Hughes, L. (2000). Biological consequences of global warming: is the signal already apparent? *Trends in Ecology and Evolution*. 15, 56–61.

Inkley, D.B., Anderson, M.G., Blaustein, A.R., Burkett, V.R., Felzer, B., Griffith, B., Price, J., Root, T.L. (2004). *Global climate change and wildlife in North America*. Wildlife Society Tech. Review 04-2. The Wildlife Society, Bethesda, MD, p. 26.

IPCC. (2007). Climate Change 2007: Synthesis Report. *Contribution of Working Groups I, II and III to the Fourth Assessment Report of the Intergovernmental Panel on Climate Change*. Core Writing Team, Pachauri, R.K. and Reisinger, A. (eds.). IPCC, Geneva, pp. 104.

IPCC. (2012). Managing the risks of extreme events and disasters to advance climate change adaptation. In Field, C.B., Barros, V., Stocker, T.F., Qin, D., Dokken, D.J., Ebi, K.L., Mastrandrea, M.D., Mach, K.J., Plattner, G.-K., Allen, S.K., Tignor, M., Midgley, P.M. eds., *A special report of working groups I and II of the intergovernmental panel on climate change (IPCC)*. Cambridge University Press, Cambridge, UK/New York, p. 582.

IPCC. (2013). *Approved summary for policy makers: twelfth session of working group I. Working group I contribution to the IPCC Fifth assessment report. Climate change 2013: the physical science basis*. Cambridge University Press, Cambridge, UK/New York.

IPCC. (2021). *The intergovernmental panel on climate change*. https://www.ipcc.ch/

Laliberté, E., Wells, J.A., DeClerck, F., Metcalfe, D.J., Catterall, C.P., Queiroz, C., Aubin, I., Bonser, S.P., Ding, Y., Fraterrigo, J.M., McNamara, S., Morgan, J.W., Sánchez Merlos, D., Vesk, P.A., Mayfield, M.M. (2010). Land-use intensification reduces functional redundancy and response diversity in plant communities. *Ecol Lett*. 13, 76–86.

Lam, H.Y., Yurchisin, J., Cook, S.C. (2016, November). Young adults' ethical reasoning concerning fast fashion retailers. In *International textile and apparel association annual conference proceedings* (vol. 73, no. 1). Ames, IA 50011, United States: Iowa State University Digital Press.

Lawler, J.J., Shafer, S.L., White, D., Kareiva, P., Maurer, E.P., Blaustein, A.R., Bartlein, P.J. (2009). Projected climate-induced faunal change in the western hemisphere. *Ecology*. 90, 588–597.

Lovell, C., Mandondo, A., Moriarty, P. (2002) The question of scale in integrated natural resource management. *Conserv Ecol.* 5(2), 25

Luce, C.P., Morgan, K., Dwire, D., Isaak, Z., Holden, B. (2012). *Rieman climate change, forests, fire, water, and fish: building resilient landscapes, streams, and managers USDA.* Fort Collins, CO: U.S. Department of Agriculture, Forest Service, Rocky Mountain Research Station, p. 207.

Millennium Ecosystem Assessment [MEA]. (2005). *Ecosystems and human wellbeing: biodiversity synthesis.* World Resources Institute, Washington, DC. p. 86.

Mustafa, D., Akhter, M., Nasrallah, N. (2013). *Understanding Pakistan's water-security nexus.* https://www.usip.org/publications/2013/05/understanding-pakistans-water-security-nexus

ODI. (2002). *Simon Maxwell, Famine and the failure of development?* Southern Africa. https://odi.org/en/events/famine-and-the-failure-of-development-southern-africa-2002/

OECD. (2008). *Natural resources and pro-poor growth: the economics and politics.* OECD, Paris.

Ostrom, E., Schroeder, L., Wynne, S. (1993). *Institutional incentives and sustainable development: infrastructure policies in perspective.* Westview Press, Oxford, UK.

Peterson, T.C., Stott, P.A., Herring, S. (2012). Explaining extreme events of 2011 from a climate perspective. *Bull Am Meteorol Soc* 93(7), 1041–1067.

Prüss, A., Kay, D., Fewtrell, L., Bartram, J. (2002). Estimating the burden of disease from water, sanitation, and hygiene at a global level. *Environ Health Perspect* 110(5), 537–542.

Raggl, A.K. (2017). *Natural resources, institutions, and economic growth: the case of Nigeria.* Paper No. WPS 8153. World Bank Group, Washington, DC.

Rahman, T. (2014). *The class structure of Pakistan* (M. Ravallion, Ed.). pp. 80, 344, 851–855.

Shah, T. (2008). Taming the anarchy: Groundwater governance in South Asia. *Taming Anar Groundwater* Govern South Asia. Oxford University Press, https://EconPapers.repec.org/RePEc:oxp:obooks:9780199065073.

Thakadu, O.T. (2005). Success factors in community based natural resources management in northern Botswana: lessons from practice. *Nat Res Forum.* 29(3), 199–212.

Umair Riaz, Z.A., Qamar uz Zaman, M., Mubashir, M., Jabeen, S.A., Zulqadar, Z., Javeed, S., Rehman, M.A., Qamar, J. (2018). Evaluation of groundwater quality for irrigation purposes and effect on crop yields: a GIS-based study of Bahawalpur. *Pakistan J Agric Res.* 31(1), 29–36.

UNISDR. (2012). *Number of climate-related disasters, 1980–2011—graphic.* www.preventionweb.net/english/professional/statistics. Accessed 20 March 2013.

United Nations Environment Progamme International Centre for Integrated Mountain Development Women, Energy and Water in the Himalayas Incorporating the Needs and Roles of Women in Water and Energy Management Project Learning. (2002). https://lib.icimod.org/record/7420/files/attachment_31.pdf

USAID. (2006). *Issues in poverty reduction and natural resource management.* United States Agency for International Development. Prepared by the Natural Resources Information Clearinghouse, an operation of Chemonics International Pennsylvania Avenue, NW Washington, DC.

van Noordwijka, M. (2019). Integrated natural resource management as pathway to poverty reduction: Innovating practices, institutions and policies. *Science Direct.* 172.

Warziniack, T.W., Elmer, M.J., Miller, C.J., Dante-Wood, S.K., Woodall, C.W., Nichols, M.C., Domke, G.M., Stockmann, K.D., Proctor, J.G., Borchers, A. (2018). *Borchers Effects of climate change on ecosystem services.* US Department of Agriculture, Forest Service, Fort Collins, CO.

WMO (World Meteorological Organization). (2013). *Powering our future with weather, climate and water.* WMO, Geneva.

World Bank. (2000). *Natural resource management.* World Bank, Washington, DC.

World Bank. (2016). *Poverty and shared prosperity report 2016: taking on inequality.* World Bank Group, Washington, DC.

World Bank. (2017). *World development indicators 2017.* World Bank, Washington, DC.

World Development Report. (2000/2001). The International Bank for Reconstruction and Development/The World Bank Attacking Poverty published by World Bank Oxford University Press, Washington, DC.

World Economic Forum. (2015). The Global Competitiveness Report 2015–2016, Global Competitiveness and Risks Team. World Economic Forum Geneva. https://www3.weforum.org/docs/gcr/2015-2016/Global_Competitiveness_Report_2015-2016.pdf

# 9 Soils' Role in Meeting Global Demand for Food, Water, and Bioenergy

*Chinmaya Kumar Swain*
*and Kiran Kumar Mohapat*

## 9.1 INTRODUCTION

Sustainable Development Goals (SDGs) of the United Nations (UN) can only be achieved with the help of soil science (United Nations 2020). However, soil knowledge cannot be applied to current difficulties without a qualified specialist with in-depth understanding of soil behavior and strong communication abilities. Some in the soil science community are worried that there isn't enough of an academic backbone to support the many people who work as scientists or perform research on soil science work that has direct relevance to achieving the SDGs (Brevik et al. 2020; Field et al. 2017; Diochon and Ghore 2016). Despite evidence that healthy soil is important for human health, its value is sometimes overlooked. Moreover, the food production must be increased globally by 70% to 100% by 2050 to fulfill the need of the growing populations of emerging economies (World Bank 2008). Productivity improvements must be realized in a changing and highly variable climate, with competing uses for limited natural resources (e.g., urbanization, industrialisation, biofuels, and recreation) and the need to protect ecosystem services and environmental quality. A multifaceted global strategy is needed to ensure food security, as well as a comprehensive strategy to protect soil resources, especially if we are to achieve the SDGs, including sustainable enhancement (e.g., Reduction of arable land). good planning is essential. Translating the potential of existing and new crop varieties into realistic yields depends on the environment, particularly the soil, on which crop growth depends. Integrated Soil Plant System Management (ISSM) maximises yields and minimises environmental impact by adapting cropping systems to local soil and environmental conditions and optimising nutrient application (Chen et al. 2014). Systematic experiments across China on three major cereal crops showed that ISSM improved yields. Current agricultural practices can be nearly doubled without increasing nitrogen fertilizer use, while significantly reducing reactive nitrogen losses and greenhouse gas emissions (Chen et al. 2014). In fact, UN Food and Agriculture Organization (FAO) estimates that the world will need to grow more food production by 60% between 2005 and 2050 to feed more than 9 billion people (Alexandratos and Bruinsma 2012). It is being explored how soil and land works in

DOI: 10.1201/9781003358169-9

providing the world's need for food, water, and energy. The two basic human requirements are a nutritious meal and access to clean water, so we will mostly concentrate on these factor.

### 9.1.1 Functions of Land, Soil Formation

About 30% of the surface of the earth is covered by land, one of the most valuable natural resources, and other areas with smaller percentages are inhospitable. In reality, the FAO of UN forecasts that in order to feed more than 9 billion people between 2005 to 2050, global food production will need to increase by 60% (Alexandratos and Bruinsma 2012). The top layer of the earth's crust's most biologically active, porous medium is the soil. One of the primary substrates of life on earth is soil, which serves as a source of water and nutrients, a filter for toxic wastes, a location for their breakdown, and a player in the cycle of carbon and other elements through the planet's ecosystem. Weathering processes driven by topographical, geologic, biological, and climatic factors have caused it to change. In reaction to various elements of the broader environment, such as vegetation and slope gradient, soil types vary at various scales.

Dokuchaev's discovery caused him to revaluate his prior beliefs about soil, which soon appeared to be out of date for his era, the late 19th century. Dokuchaev and a handful of his coworkers began to see soil as a natural organism deserving of in-depth study in and of itself. Soil forming factors that cause the formation of layers or horizons in the soil that are nearly parallel to the earth's surface. Different components work together to naturally grow the soil as a whole, from the soil's top to its deepest layers. The components are (1) the parent material, (2) relief/topography, (3) climate, and (4) time (Figure 9.1). A single parent material will develop into a different soil individual if it is exposed to several climes. A new soil will form if any

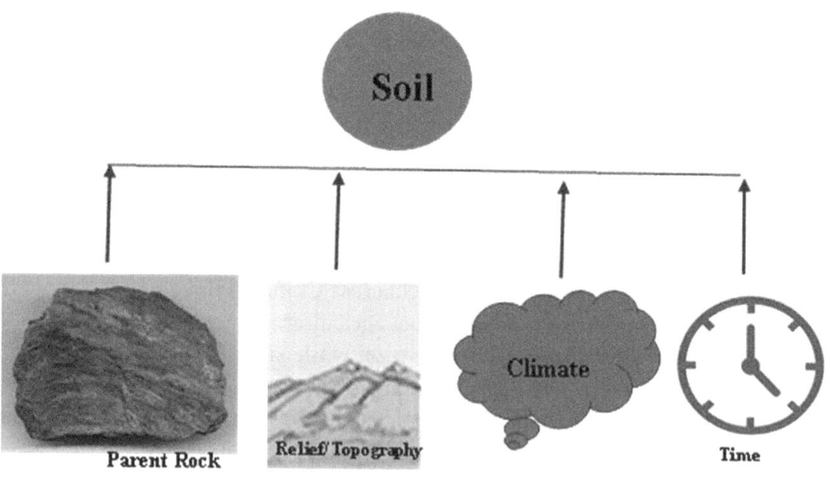


**FIGURE 9.1**   Different factors affecting soil formation.

one of the five components is altered but the other four stay unchanged. This process is known as "soil genesis". The term "functional land management" was put out by Schulte et al. (2014) as a more effective framework for maximising the performance of the five primary soil functions.

(1) Primary productivity, which provides fuel, food, fiber, and other necessities.
(2) The ability of soils to regulate and cleanse water for human consumption while maintaining the integrity of the ecosystem.
(3) The potential of the soil to retain carbon to control soil biological and physical processes as well as to partially offset greenhouse gas emissions.
(4) Nutrient cycling, notably the soil's capacity to sustainably house external nutrients received from sewage sludge and other organic waste as well as landless agricultural systems.

### 9.1.2　Land Management and SDGs

On the basis of SDGs, it is intended to stop the loss of biodiversity with managing forests sustainably, preventing desertification, and protecting, restoring, and promoting sustainable use of terrestrial ecosystems. To check the loss and halt soil biodiversity and land managemnet is utmost necessary. Soil biodiversity is critical for all of the soil processes mentioned earlier, including food production and groundwater resource protection. Soil biodiversity is also directly related to aboveground biodiversity, such as pests and diseases that threaten food production and storage. Chemical and biological soil components are likely to have an impact on human health, either directly or indirectly. It all depends on soil restoration (Brevik and Burgess 2013). However, the distribution and quality of the world's soil are crucial to the provision of commodities with services through sustainable agricultural land management om relation to food production (i.e., SDG no. 2). Through sustainable land management, especially agricultural land management, SDGs can be achieved. Figure 9.2 shows the goods and services that can be provided through sustainable farming, producing biomass, fiber, and bioenergy on topsoil with lots of clean groundwater below. Groundwater is often the only source of drinking water. Therefore, agricultural management and plant production (particularly pesticide use) in the topsoil must be coordinated quantitatively and qualitatively with groundwater production below. Additionally, surface runoff of soil particles into open water resources like lakes and ponds should be minimised or avoided by soil management. Agricultural management decreases the production of these gases, which is SDG no. 13, relating to the climate action which may be due to the direct gas exchange in the atmosphere with the help of soil and the release of trace gases that are related to climate change, such as carbon dioxide ($CO_2$), methane ($CH_4$), and nitrous oxide ($N_2O$). It should be controlled and minimised and climate moderated.

### 9.1.3　Role of Food Security and Its Complex Issues

Due to its emphasis on food security, sustainable food production is essentially dependent on the functioning of soil. Food security was described at the World

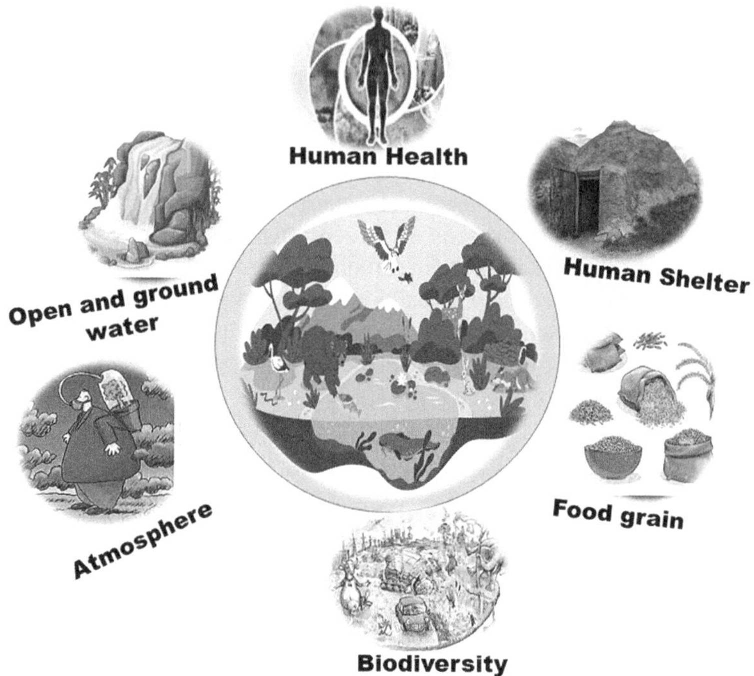


**FIGURE 9.2**  Sustainable farming producing biomass provided in terms of better goods and service.

Summit (1996) as a state where everyone always has physical, cultural, and socioeconomic access to an adequate supply of food that is safe and nutritious and meets their dietary reqirements for healthy life (Barrett 2010). Other available resources include land surface, terrain, soils, water, biota, and climate conditions. Important factors include agrochemicals (pesticides, fertilizers, etc.), human and financial resources, farming practices, land ownership, and food policy regimes, as well as the availability of agricultural technology. Due to diverse eating habits of human beings on worldviews, myths, religions, and general cultural perspectives, cultural accessibility and diversity are of the utmost importance on a worldwide scale. Last but not least, because the genetic composition of people around the world varies, different foods are physically available to different people in various ways, depending on their healt.

### 9.1.4  Relationship of Soils to Feed

The interaction of natural soil chemistry, physical characteristics, and climate—all factors that fall under the purview of soil categorization—determines the potential productivity of agricultural land to a great extent. When climate is taken into account, some soil types govern the world's food and feed production as a result of a number of physical and chemical characteristics that more intensely encourage plant growth. The soil type, i.e., Mollisols, are cultivated most intensively worldwide,

**TABLE 9.1**

**Different Soil Orders with its Taxonomy**

| Soil orders | Taxonomy |
| --- | --- |
| Aridisols | Soils with weak development in arid and semi-arid climates. |
| Entisols | Young soils devoid of recognisable diagnostic horizons |
| Gelisols | Permafrost-covered soils |
| Histosols | High-organic-matter-accumulating soils, such as fen and moor |
| Inceptisols | Soils that have only mildly weathered and are beginning to create diagnostic horizons |
| Mollisols | A horizons and organic matter content |
| Oxisols | Tropical and subtropical soils having an increase in Fe- and Al-oxides, a decrease in Si, and severely weathered clay minerals, such as kalinite |
| Spodosols | Accumulation of highly organic top layers, leaching of iron, aluminium, and organic acids and the development of topsoil horizons that are bleached |
| Ultisols | Acidic tropical and subtropical soils with a build-up of clay in the lower strata and very low base saturation |
| Vertisols | Soils strong in clay minerals that expand when wet and contract when dry, as well as tropical and subtropical locations with clearly defined wet and dry seasons |
| Alfisols | Strongly acidic, well-developed soils with strata of clay |
| Andisols | Formed on volcanic rocks, acid with dark-brown to black colors |

*Source:* soil survey staff (1999)

especially in the European and Asian countries (Table 9.1) (Hatfield et al. 2017). Surface soil organic matter tends to build up in Mollisols, which helps with nutrient supply as well as nutrient and water retention. Mollisols are less susceptible to erosion than some other typical soil types because they frequently occur in stable settings without significant topographic change (Hatfield et al. 2017; Eswaran et al. 2012). Numerous Inceptisols are intensively cultivated in Europe and South Asia, similar to Alfisols, because of the naturally favorable nutrient availability brought by high soil organic matter content or profuse weathering products (Hatfield et al. 2017; Eswaran et al. 2012). Together, Mollisols, Alfisols, and Inceptisols make around 17% of the world's land surface, whereas Mollisols alone make up about 15% (Eswaran et al. 2012). Entisols are often characterised by steep slopes which tend to sustain lower agricultural productivity with the likely exception of Entisols created from the latest alluvial deposits (Grossman 1983). Entisols dominate in areas of Australia and Africa and make up nearly 18% of the world's land surface (Eswaran et al. 2012). Dry land soils, or Aridisols (12% of the world's land surface), develop in typically non-arable environments, where crops cannot thrive. These soils frequently support grazing operations. High salt concentration limits the plant growth in some Aridisols which flooded irrigation can help (Nettleton and Peterson 1983). Livestock grazing management dominates Aridisols in areas of America, South Africa, Australia, and

Argentina (Eswaran et al. 2012; Asner 2004). The earth's land surface is made up of Andisols (≥ 1%) which are the result of recent volcanic activity and can support especially productive soils. Because Andisols are in tropical areas, deforestation for agricultural growth is a significant concern (Small and Naumann 2001).

### 9.1.5 Nutrients in Agricultural Soil

The availability of nutrients in soils varies greatly. A large portion of the world's highest native soil N content cropland shows the highest crop yields used primarily for the production of livestock feed or biofuel (Hatfield et al. 2017; Eswaran et al. 2012). The calories produced (i.e., > 10%) on cropland in the USA in 2019 were used for human consumption. The majority of these calories were lost due to the ineffective conversion of maize and soy's nutrients into energy when used as biofuel or livestock feed (Hatfield et al. 2017; Eswaran et al. 2012). Some of the highest rates of crop-embedded nutrient conversion may be seen in eastern and central sub-Saharan Africa, as well as in India, both of which have lower overall soil nutrient availability and output. In East Africa, sorghum, sugarcane, and tubers are planted. In India, rice, sugarcane, and legumes are farmed (Cassidy et al. 2013; FAO 2020; FAOSTAT 2021). Despite having some of the highest agricultural productivity rates in the world, oil palm planting predominates in Indonesia, where less than half of the calories produced are converted to food, highlighting the incredible impact of economics on croplands (Condro et al. 2020).

## 9.2 LAND AND SOIL RESOURCE DISTRIBUTION

The capacity of a piece of land to carry out particular tasks is measured as its quality. It is typical for land to be classified based on its capacity to carry out the aforementioned duties. The enactment to biomass production with agricultural element of biomass is the main focus of the quality evaluation in the majority of systems (Mueller et al. 2010). According to their capability to perform the aforementioned tasks, lands, and soils are categorised, distinguishing between two key parameters: to create energy and their persistence that relates to its ability to recover a new balance following disturbance based on the two essential soil parameters, "performance" and "resilience". Beinroth et al. (2001) created an alternative land classification system with a focus on agricultural grain yield. Class I lands, which have optimum soils, are located in ideal conditions for agricultural production, have high productivity and resilience, have a high responsiveness to management, and have few restrictions, making up only 2.38% of the world's largest land surface. Despite this, more than 40% of the world's foodstuff is produced on Class I lands. There are a few minor limits (such Alfisols) on the 9.53% of the world's natural assets in classes II and III, but these may be quickly fixed and do not preclude long-term utilization of the lands. The world's population (over 50%) lives in land quality classes IV–VI (such as Oxisols, Vertisols, and Ultisols) which make around 34% of the planet's land area and are primarily in the tropics. Numerous restrictions affect these soils and the environment in which they grow at elevated temperatures that poorly affects the seed rates to poor nutrient availability that restricts the biomass output of annual crops. A little more than 9% of the land area is made up of Class VII lands, which include shallow soils like Entisols and Inceptisols, soils with a lot of salt in them like Aridisols, and soils with a lot of organic content (e.g., Histosols and Spodosols). Saline soils may be exploited with particular adapted cultivars and cropping strategies, but shallow soils are typically thought to be unsuited for agricultural cropping. Peatlands with high levels of organic matter are especially susceptible to deterioration and long-term loss from drainage. Almost 17% of the land's surface is made up of Class VIII lands. They are typically regarded as being unfavorable for agriculture since they have low temperatures (like Gelisols) and/or steep slopes. Over 28% of the land surface is covered by Class IX fields, which are made up of soils with insufficient moisture to sustain the production of annual crops. Almost 25% of the world's population lives on the 12% of land that has soil and land quality classifications I to III, where traded foods for the global market are produced.

## 9.3 SUSTAINABLE LAND MANAGEMENT: MEETING GLOBAL DEMANDS FOR COMMODITIES AND SERVICES

The protection of biodiversity and fresh water in the form of groundwater are other requirements for sustainable land management and for long-term availability of biomass such as food, forage, and renewable energy. Land management should regulate the flow of climate-related trace gases between soils and atmosphere in order to safeguard human health from tainted food supplies and clean open water resources like lakes and rivers from pollution. Since every raindrop that falls on the land must first pass through the soil before it can turn into groundwater, it is evident that managing the land sustainably is a difficult endeavour. For instance, there are many conflicts between agricultural output through the application of fertilizers and protection compounds in the plants on the top of the soil and the groundwater below. Additionally, the types and quantities of trace gases produced by soils, including $CO_2$, $CH_4$, and $N_2O$, which are linked to climate change, are influenced by the physical and chemical effects of agricultural machinery and fertilizer application on soil. Therefore, the SDG of managing agricultural soils affects the amount and quality of groundwater resources for human life. Land and soil management, to a lesser extent, is connected to the universal access of inexpensive, dependable, sustainable, and modern energy since it entails the production of biomass as a basis for the development of bioenergy, like biofuel from carbohydrates like grains, biodiesel from vegetable oils (canola and palm oil), and biogas (solid burning like straw and wood).

## 9.4 SOIL AND LAND THREATS COULD COMPROMISE MEETING THE NEED OF GLOBAL POPULATION

As soil responds to a variety of environmental factors, there is a serious concern that changes in the environment are endangering the normal functioning of soil (Scheffer et al. 2001). According to Foley et al. (2005), these changes are typically caused by or affected by humans. See also SDG number. Agriculture clearly affects the state of the environment globally (Tilman et al. 2001). Recent worldwide efforts to conserve soil have outlined a number of potential challenges to the soil's ability to

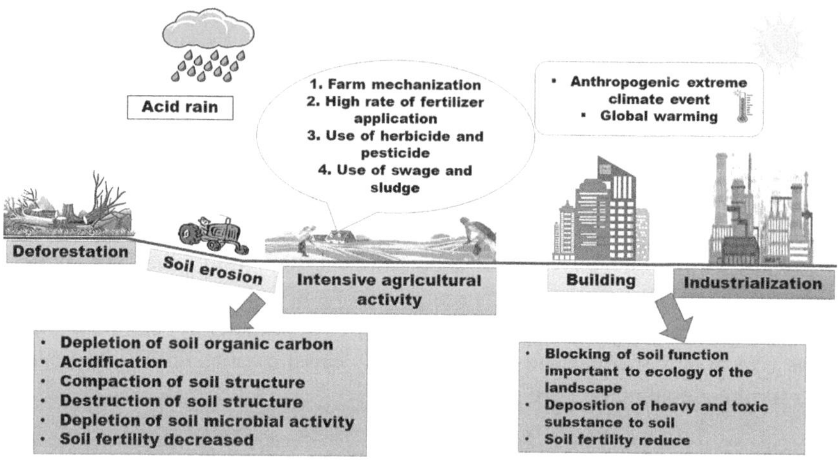


**FIGURE 9.3**   Threats to sustainable soil by human intervent

carry out its tasks. Examples of threats to the sustainable use of soil include (a) sealing, (b) erosion, (c) compaction, (d) organic matter, (e) loss of biodiversity, (f) pollution, (g) salinization and (h) landslides (noted by the Commission of the European Communities 2002). These human-created risks are present around the world, and while the European Communities Commission focused on Western Europe when highlighting them, their relative relevance depends on the region and kind of land use (Figure 3). These risks exist in part because we expect the soil to carry out a variety of tasks, often several tasks at once. Consequently, the soil's capacity to carry out its regular tasks is frequently drastically reduced (Lal 2009; Pretty 2008). These dangers are increasingly considered to be especially pertinent to soils' capacity to produce biomass, including the delivery of biofuels.

### 9.4.1   TRANSPORTATION-RELATED URBANISATION, INDUSTRIALISATION, AND SOIL CLOSURE

Soil sealing or closure is the process of enclosing soil to prepare it for modern infrastructure, such as roads, homes, and other land projects. When a piece of land is sealed off, the soil loses the ability to do the majority of its jobs, including screening polluting substances and absorbing rainwater for infiltration. Additionally, sealed regions alter the surface pattern of water flow, resulting in flooding and inundations and worsening biodiversity fragmentation. The nighttime view of the earth from space, which clearly demonstrates that the majority of high-quality land are expressed as a global distribution of croplands (Foley et al. 2005) and are connected to the areas of high levels of urbanization, can be used to examine the extent to which soil is being sealed off throughout the world. According to The Earth Institute (2005), regions with a significant association to high-quality land are those in North America, Western Europe, and Japan, where there is a noticeable increase in urbanization. The growing urbanization in China and the Indian subcontinent also makes this element apparent. Since our predecessors picked the best soils for settlement to ensure the fulfilment of basic necessities, such as food, fiber, and water, the sealing of the best soil attributes has its roots in the history of the founding of settlements. These early villages have developed into significant urban agglomerations over. Therefore, any further urbanization also results in the sealing of more prime land. The transition from rural to urban lifestyles, which reduces biomass output and jeopardises food security, is a second effect of sea.

### 9.4.2 Soil Losses through Land-Dwelling

In order to produce bricks, tiles, and other building materials, agriculture soils with a greater texture (i.e., clay and silt) and lacking carbonates are dug up to 5 m deep, sometimes even further. The majority of the time, these are created in mobile furnace buildings close to the excavation sites. The high groundwater levels in these places are filling the excavation holes, which are usually several hectares in size, and are creating small lakes or water ponds. The agricultural soils that have been excavated are permanently gone for the production of fiber and food. Due to the lack of additional soil resources for agricultural production in the area, this process cannot be stopped. These regions have some of the greatest population densities and population growth rates in the world; hence, food shortage issues are likely too.

### 9.4.3 Improper Land and Soil Management Leads to Soil Degradation

Along with sealing and excavation, there were other forms of soil degradation that also occurred, including compaction, erosion, organic matter reduction, biodiversity loss, pollution, land degradation, soil salinity, nutrient extraction, and flooding (Blum 2008). Steep slopes, temperature (severe weather conditions), improper use of land, low vegetation coverage, and occasionally ecological calamities like tectonic activities or forest fires all contribute to erosion, a natural geological process (Bai et al. 2008). Additionally, soil's inherent characteristics including their silty texture and small organic matter concentration make them vulnerable to erosion (Boardman and Poesen 2006). Although erosion being a natural phenomenon, some human activities, particularly the use of unsustainable agricultural land, may cause erosion rates to dramatically increase (Lal 2001).

Loss of soil functions is a consequence of soil erosion, and at its worst, the soil itself is lost. Even a little quantity of topsoil loss can have a major detrimental effect on the soil's capacity to withstand elevated amounts of biomass production (Den Biggelar et al. 2003). It was establishing a relationship between both the erosive damage of naturally improved topsoil and the potential consequences on global carbon budget (Lal 2005). Extra effects result from the soil being transferred to watercourses, which introduces contaminants, increases water turbidity, and causes additional environmental harm by its deposition downstream. Heavy machinery or excessive grazing on agricultural land can mechanically compact the soil, especially when the soil is wet (Horn and Peth 2011). In addition, the need to reduce compaction by reducing traffic when soils are moist and vulnerable to damage can be noticed.

These was a potential contributing element. The role of soils in meeting food, water, and bioenergy global demands to the increasing vulnerability to compaction skiing and walking tourism (hiking) in sensitive locations also add to the issue, and compaction on construction sites is a frequent occurrence. The amount of pore space between soil particles is reduced during compaction, and as a result, some or all of the soil's ability to absorb water is lost. It is challenging to undo soil compaction in deeper soil layers (Horn et al. 2000). Soil structure is harmed by compaction, water storage capacity, bioactivity and fertility reduction, and biomass production (Clarke et al. 2008). Additionally, the soil cannot be easily penetrated by water during periods of severe rain. Surface water runoff is the outcome, and this can cause degradation or even flooding and inundation. More critically, though, surface-level soil and water retention facilities won't be replaced by runoff water. The majority of agricultural systems rely on near-surface stored water to keep plants growing during dry seasons.

The components of soil organic matter are humus, live creatures (fungi, bacteria, earthworms, and other soil fauna), and organic matters (plant roots, leaves, other plant material, dead animals, and excretions). Organic matter is continuously added to the soil in natural systems and breaks down. Plant nutrients are released during this decomposition and frequently absorbed by plants, producing an almost restricted cycle. In addition, carbon is expelled into the atmosphere as $CO_2$, where it is subsequently captured during photosynthesis. Key soil functions are maintained in large part by organic matter, which also significantly influences soil fertility and erosion resistance (Lal 2002). This reduces the quantity of pollution that diffuses from the soil to the water by helping to agglomerate soil particles and boost the physical and chemical buffer ability of the soil (Pulleman et al. 2000; Sonneveld et al. 2008).

### 9.4.4 REDUCTION IN BIODIVERSITY OF SOILS

The world's largest gene bank is found in soil, which also serves as a home for a wide range of other living things. Due to the interaction of the factors that affect how soil is formed, such as geoscience, weather, natural vegetation, and topography, there is a great variation in the characteristics of soils and of soil organisms (Bardgett et al. 2005; Gardi and Jefferey 2009). We don't fully understand the organisms that live in soil. Our understanding of the relative scale of the soil biota, particularly the role played by microbes, is quite restricted (Pimentel et al. 2006). The physical, chemical, and particularly biochemical characteristics required for soil fertility are maintained in large part by the bacteria, fungus, protozoa, and other microorganisms living in the soil (Barrios 2007; Brussaard et al. 2007). Larger species like snails, crabs, and small arthropods diminish the quantity of organic matter wastes, which are subsequently transported to deeper soil strata, where they become more resilient and further deteriorated by microorganisms.

Soil contaminations have the potential to damage or eliminate one or more of the functions of soils as well as to contaminate groundwater and surface water resources. When pollutants are present in soils over a specific threshold, the food chain, human health, various ecosystems, and other natural resources are all negatively impacted. It is crucial to take into account not only the concentration of soil pollutants but also their environmental attributes and exposure pathways for both human and ecological health in order to assess the potential effects of soil pollutants. Differentiating between soil contamination caused by dispersed sources and from clearly contained sources is a typical practice (local or point-source contamination). The Technical Working Group on Soil Contamination's study, which was created in compliance with the European Commission's Contextual Strategy for Soil Protection, provides an extensive summary of the origins and spread of contaminated soils throughout Europe (Van Camp et al. 2004). These practices put land and water at risk. The processing or removal of wastewater and the application of explosives are related to the risk in mining. The processing or removal of wastes and the usage of specific chemical substances are related to the danger in mining. Diffuse pollution is typically linked to insufficient waste and sewage recycling and treatment, specific farming techniques, and air deposition. Many of the diffuse atmospheric pollution sources are caused by contaminants that are emitted as by-products of industry, transportation, and agriculture. The build-up of airborne pollutants causes the release of metals (such as cadmium, lead, arsenic, and mercury), acidifying contaminants (such as $SO_2$ and NOx), and a variety of organic substances (such as polycyclic aromatic hydrocarbons and dioxins and PCBs) into soils.

Along with the linkages between the soil and atmosphere, there are connections between both soil and hydrosphere. In a research from northern France emphasised the necessity of taking into account the impact of diffuse contaminants, which may be caused by the distribution of pesticides on water resources. Due to the frequently composite relationships between the soil and hydrological systems, the impacts of this pollution may take some time to manifest in the composition of the groundwater. Therefore, it is crucial to take into account the nature of these connections when making efforts to manage contamination and when remediating polluted soils and water resources. There are two ways that contaminated soil might affect food production. Because the pollutants prevent growth, the first immediate effect is a decrease in the soil's ability to produce. A second influence on food production becomes contaminated during growth and becomes unfit for human consumption. If allowed to continue without proper planning and supervision, the previously mentioned rising urbanization and industrialization have the potential to contaminate the soil and have an even greater influence on the world's food output (Blum 1998).

### 9.4.5 Mining of Soil Nutrients

Despite not being recognized a threat in the European context, soil protection legislation and soil nutrient mining became major threats to food production in a lot of the tropics (De Jager et al. 2001). Much of the agricultural output on the continent is grown in conditions where ecosystems are stressed, soil fertility is low, and modern inputs like mineral fertilizers, plant protection agents, and improved crop varieties are rarely used. This practice of letting soil rest and "repair" by not cultivating it has a long history in Africa. Due to the growing strain on land caused by both the increasing population and the exclusion of indigenous communities from certain areas of the landscape, the length of fallow seasons has been decreased and few cases have been eliminated entirely. This trend often results in the "mining" or depletion of the soil's nutrient reserve, especially when no or very little fertilizer is employed.

Communities strive to use marginal lands, which often have lower intrinsic fertility and other controls for crop development to maintain overall food production levels, but this practice destroys land with important ecological value.

## 9.5 GLOBAL POPULATION GROWTH, MODIFICATIONS, AND MIGRATION IN FOOD HABITS AS WELL AS LIFESTYLE

Soil losses (including sealing and excavation), soil compaction, organic matter reduction, and the availability of food and water include pollution, salinity, desertification, mineral extraction, floods, and landslides, jeopardize global processes. These increase the need for both space and food while simultaneously posing a threat to the basis of food production, having a knock-on effect on the commodities and services that land and soil provide. Changes in nutrition and lifestyle show how society can adapt to an increase in population or the urbanization of rural areas. Increased energy demand and consumption, as well as the loss of natural resources for the expansion of food and fiber production only add to the difficulties already caused by the usage of fossil fuels. Along with the current economic situation, newly emerging economic practices on food production and trade trend towards increasing biofuel production pose an additional threat to food security.

Over 70% of the world's population growth, which is increasing by about 80 million people annually, is concentrated in South and Southeast Asia, Central America, and South America (Lutz et al. 2008; IIASA 2007). In the perspective of sustainable development, Lutz et al. (2014) offered new options for the increase of the global population. The number of undernourished persons globally grew by almost 9% since 1990, despite a more than 12% rise in food production (Barrett 2010). Lal (2010) observed that the number of people experiencing food insecurity increased from 854 million in 2007 to over 1020 million in 2009 due to changes in market conditions and food pricing (IFPRI 2008; FAO 2008). According to FAOSTAT (2011), there were around 926 million hungry individuals. All of these statistics only account for those who are protein/calorie malnourished; they do not account for the approximately 2 billion people who are iron or iodine malnourished or the approximately 750 million people who suffer from other particular nutritional deficiencies (WHO 2000).

With the population growing from roughly 2.5 billion to 6 billion from 1950 to 2000, the 20th century saw gains in global food production of nearly 250%. The "green revolution success" and the addition of vast quantities of fossil energy into the form of fertilizers and plant protection compounds (Borlaug 2007) made this possible (Blum and Eswaran 2004; Godfray et al. 2010; FAOSTAT 2011). However, population growth started to outpace productivity development in the last decade of the 20th century. As a result, the amount of food produced per person decreased slightly over time. Additionally, the green revolution had little effect in some regions due to nutrient-poor soils, restricted availability to fertilizers, plant protection products, and inadequate agricultural tools (Sanchez and Swaminathan 2005; Scherr 1999). According to St. Clair and Lynch (2010), between 1990 and 2030, the world's need for food is anticipated to rise two- to fivefold. In addition, due to population expansion, growing soil losses from urbanization, and land and soil degradation, the per capita amount of land used for food production, which is currently 0.27 ha, continues to decline (Satterthwaite et al. 2010).

## 9.6 FOOD SECURITY A BIG THREAT TO THE CHANGING CLIMATE

The biological functions in terrestrial and aquatic ecosystems are being impacted by climate change, which pose various threats to food supply. Due to water and air erosion, flooding, landslides, and salinization, the losses and degradation of soil increase as extreme weather events occur more frequently (Alley et al. 2003; Haron and Dragovich 2010). Plant growth is hampered by abrupt transitions between wet and dry conditions, particularly during persistent droughts. Fresh water for irrigation is becoming more scarce, and households, businesses, and industries are competing for the limited amount of water available for biofuel production (Rosegrant and Cai 2002). In addition, the production of food and biofuels contributes to climate change by speeding up urbanization, industrialization, transportation, and deforestation for agricultural land usage, which reduces soil carbon sequestration. Intensive agriculture for feed and biofuel produces rising amounts of methane and $N_2O$, which influences climate change more than $CO_2$ through soils and other organisms (Blum et al. 2010). The increase in areas with limited water supplies such as near East Mediterranean and other areas of the world is already having a substantial impact on the production of food through irrigation (IPCC 2014; FAOSTAT 2011; IWMI 2007). Contrarily, the hike of temperature—particularly in the northern regions of America, Asia, and Europe, with plentiful freshwater resources—may somewhat counterbalance the drop in food production in regions with inadequate water availability as well as those with serious soil degradation and loss.

## 9.7 GLOBAL ECONOMIC CHANGES AND NEW MONETARY PATTERNS IN THE FOOD MARKETING

When the global economic crisis began in the first decade of the 21st century, it sparked the development or realization of new business models and strategies in food production and marketing (Piesse and Thirtle 2009). Food became the first commodity to be widely speculated on derivative transactions at worldwide and local food chains (Williams 2001; Garcia and Leuthold 2004), causing major price volatility and concerns with food availability (IFPRI 2008). Land grabbing, the practice of acquiring or leasing vast areas of land for the purpose of cultivating food or biofuels, has also increased dramatically in recent years, especially in underdeveloped nations. Between 2006 and 2009, over 3 billion acres of land were seized for food production in Africa, Asia, and Latin America. Von Braun and Meinzen-Dick (2009) and Adésínà (2010) claim that derivative transactions—which use agricultural items, notably food—are a type of economic speculation in Africa. However, land grabbing has the potential to significantly change the agrarian and rural geographies of states. Direct issues with fertility are also anticipated, particularly when food and biofuel production is maximized to generate as much revenue as possible from leased land (Davis et al. 2014). Many developed nations and multinational corporations have acknowledged the impossibility of creating new land, as seen by their practice of acquiring property. Due to the depletion of fertile land inside their own ecosystems and the increasing local and global demands for food and energy, secured land in other countries and locations has become a coveted resource.

Production of biofuel such as gaseous biogas, ethanol, wood, and straw has become commercially attractive due to the consistently rising costs for oil and gas in the form of fossil energy. Given the importance of reducing greenhouse gas emissions through the use of biofuels, this is especially true when monetary aid is provided (Goldemberg and Guardabassi 2009; Demirbas 2007; OECD 2007; Pimentel and Pimentel 2007). Natural vegetation cover, such as forests, is routinely cleared in different parts of the globe, especially Asia and Latin America, to produce new surfaces for the production of biofuels, notably sugarcane and plants producing oil (Fargione et al. 2008). Popp (2010) emphasized the "trilemma" between environmental, energy, and food security. Due to losses and degradation through erosion and other forms of deterioration, as well as the consistently diminishing surface area for food production, biofuel production competes with production and security at least for medium to long term (Pimentel et al. 2009; Tilman et al. 2009). The full removal of organic matter during agricultural biofuel production and the return of inadequate organic matter to the soil to support biodiversity are harmful to soil quality (Cruse et al. 2010; Lal and Stewart 2010). The struggle for scarce water resources is exacerbated. There are numerous parts of the world where there isn't enough water to sustain agricultural output (Bernades 2008; Blum et al. 2010).

## 9.8 SUMMARY AND FUTURE OUTLOOK

The "green revolution" and other significant advancements, particularly the use of mineral fertilizers and plant protection substances, allowed for a significant rise in food production in the 20th century, but it was unable to halt the situation in which an average of 1 billion people today suffer from hunger, primarily due to a lack to food. A total of 4 billion people are considered to be the prey of malnutrition because of additional specific dietary deficiencies. Impacts from human activity, particularly shifting patterns of land use at local and global dimensions, are posing an increasing danger to the eco-friendly production and supply of food. The greatest threat to food security comes from soil losses brought on by urbanization, industrialization, transportation, desertification, and flooding, which also pose a serious threat to food security. Food security is also threatened by factors other than the shrinkage of arable land that is productive and the degradation of the remaining productive land. Around 80 million people are introduced to the globe's population every year, and 80 million–100 million people move from rural to urban areas every year, leaving them dependent on food markets because they can no longer grow their own food. Direct threats to food security from climate change include rising soil losses and degradation, primarily as a result of harsh weather. Agriculture that relies on irrigation and rainfall is being threatened in many areas by a decline in freshwater resources. In the meantime, the production of biofuels is posing a significant threat to resources like space, energy, and water. Demand for food and fiber is rising both locally and globally.

This demonstrates how far we still have to go in order to accomplish SDG no. 2's goals of ensuring universal access to water and electricity, as well as eradicating hunger, food security, improved nutrition, and sustainable agriculture. If new methods for producing biomass aren't put into practice, it will endanger millions of citizens and worsen hunger, particularly in developing countries. These methods should focus on producing food both locally and globally, protecting against groundwater pollution, and producing enough biomass to meet local and international energy needs.

## REFERENCES

Adésínà, J.O. 2010. Re-appropriating matrifocality: Endogeneity and African gender scholarship. *African Sociological Review/Revue Africaine de Sociologie* 14(1), 2–19.

Alexandratos, N., and Bruinsma, J. 2012. *World agriculture towards 2030/2050: The 2012 revision*. http://doi.org/10.22004/ag.econ.288998.

Alley, R.B., Marotzke, J., Nordhaus, W.D., Overpeck, J.T., Peteet, D.M., Pielke, R.A. Jr., Pierrehumbert, R.T., et al. 2003. Abrupt climate change. *Science* 299, 2005–2010.

Asner, G.P., Elmore, A.J., Olander, L.P., Martin, R.E. and Harris, A.T. 2004. Grazing systems, ecosystem responses, and global change. *Annu Rev Environ Resour* 29, 261–299.

Bai, Z.G., Dent, D.L., Olsson, L. and Shaepman, M.E. 2008. Proxy global assessment of land degradation. *Soil Use Manag Soil Use Manage* 24, 223–234.

Bardgett, R.D., Usher, M.B. and Hopkins, D. 2005. *Biological diversity and function in soils*. Cambridge: Cambridge University Press.

Barrett, C.B. 2010. Measuring food security. *Science* 327, 825–828.

Barrios, E. 2007. Soil biota, ecosystem services and land productivity. *Ecol Econ* 64, 269–285.

Beinroth, F.H., Eswaran, H. and Reich, P.F. 2001. Global assessment of land quality. In *Sustaining the global farm*, ed. D.E. Stott, R.H. Mohtar, and G.C. Steinhard, 569–574. Selected papers from the 10th international soil conservation organisation meeting, 24–29 May 1999, West Lafayette.

Bernades, G. 2008. Future biomass energy supply: The consumptive water use perspective. *Int J Water Resour Dev* 24, 235–245.

Blum, W.E.H. 1998. Soil degradation caused by industrialisation and urbanisation. In *Towards sustainable land use, Advances in geoecology 31*, vol. I, ed. H.-P. Blume, H. Eger, E. Fleischhauer, A. Hebel, C. Reij and K.G. Steiner, 755–766. Reiskirchen: Catena Verlag.

Blum, W.E.H. 2008. Characterisation of soil degradation risk: An overview. In *Threats to soil quality in Europe*, ed. G. Tóth, L. Montanarella, and E. Rusco, 5–10. JRC scientific and technical reports EUR 23438 EN, Ispra.

Blum, W.E.H. and Eswaran, H. 2004. Soils for sustaining global food production. *J Food Sci* 69(2), 37–42.

Blum, W.E.H., Gerzabek, M.H., Hackländer, K., Horn, R., Rampazzo-Todorovic, G., Reimoser, F., Winiwarter, W., Zechmeister-Boltenstern, S. and Zehetner, F. 2010. Ecological consequences of biofuels. In *Soil quality and biofuel production, advances in soil science*, ed. R. Lal and B.A. Stewart, 63–92. Boca Raton: CRC Press.

Boardman, J. and Poesen, J. 2006. *Soil erosion in Europe*. Chichester: Wiley.

Borlaug, N.E. 2007. Feeding a hungry world. *Science* 318, 359.

Brevik, E.C. and Burgess, L.C. 2013. *Soils and human health*. Boca Raton/London/New York: CRC Press.

Brevik, E.C., Slaughter, L., Singh, B.R., Steffan, J.J., Collier, D., Barnhart, P. and Pereira, P. 2020. Soil and human health: Current status and future needs. *Air, Soil and Water Research*, 13, 1178622120934441.

Brussaard, L., de Ruiter, P.C. and Brown, G.G. 2007. Soil biodiversity for agricultural sustainability. *Agric Ecosyst Environ* 121, 233–244.

Cassidy, E.S., West, P.C., Gerber, J.S. and Foley, J.A., 2013. Redefining agricultural yields: From tonnes to people nourished per hectare. *Environmental Research Letters* 8(3), 034015.

Chen, S., Rotaru, A.E., Liu, F., Philips, J., Woodard, T.L., Nevin, K.P. and Lovley, D.R., 2014. Carbon cloth stimulates direct interspecies electron transfer in syntrophic co-cultures. *Biores Tech* 173, 82–86.

Clarke, M., Creamer, R., Deeks, L., Gowing, D., Holman, I., Jones, R., Palmer, R.C., Potts, S.G., Rickson, R.J., Ritz, K. and Smith, J., 2008. Scoping study to assess soil compaction affecting upland and lowland grassland in England and Wales. *Final project (BD2304) report to Defra.*

Condro, A.A., Setiawan, Y., Prasetyo, L.B., Pramulya, R. and Siahaan, L. 2020. Retrieving the national main commodity maps in Indonesia based on high-resolution remotely sensed data using cloud computing platform. *Land* 9(10), 377.

Cruse, R.M., Cruse, M.J. and Reicosky, D.C. 2010. Soil quality impacts of residue removal for biofuel feedstock. In *Soil quality and biofuel production, advances in soil science*, ed. R. Lal and B.A. Stewart, 45–62. Boca Raton: CRC Press.

Davis, J., Mengersen, K., Bennett, S. and Mazerolle, L. 2014. Viewing systematic reviews and meta-analysis in social research through different lenses. *Springer Plus* 3, 1–9.

De Jager, A., Onduru, D., Van Wijk, M.S., Vlaming, J. and Gachini, G.N. 2001. Assessing sustainability of low-external-input farm management systems with the nutrient monitoring approach: A case study in Kenya. *Agricult Sys* 69, 99–118.

Demirbas, A. 2007. Progress and recent trends in biofuels. *Progress in Energy and Combustion Science* 33, 1–18. http://dx.doi.org/10.1016/j.pecs.2006.06.001

Den Biggelar, C., Lal, R., Wiebe, K., Eswaran, H. and Breneman, V. 2003. The global impact of soil erosion on productivity II: Effects on crop yields and production over time. *Advances in Agronomy* 81, 49–95.

Diochon, M. and Ghore, Y. 2016. Contextualizing a social enterprise opportunity process in an emerging market. *Social Enterprise Journal* 12(2), 107–130.

The Earth Institute. 2005. Inside The Earth Institute a monthly e-newsletter https://www.earth. columbia.edu/sitefiles/file/about/newsletter/2005/october05.pdf

Eswaran, H., Reich, P. and Beinroth, F. 2001. Global desertification tension zones. In *Sustaining the global farm*, ed. D.E. Stott, R.H. Mohtar, and G.C. Steinhardt, 24–28. Selected papers from the 10th international soil conservation organisation meeting, 24–29 May 1999, West Lafayette.

Eswaran, H., Reich, P.F. and Padmanabhan, E. 2012. World soil resources opportunities and challenges. In *World soil resources and food security*, eds. B. Lal, R. Stewart, 29–52, Boca Raton, Floride, USA: CRC Press.

FAO. 2008. *The state of food insecurity in the world*. Rome: Food and Agriculture Organization of the United Nations.

FAO. 2020. *The state of food and agriculture*. Overcoming water challenges in agriculture, Rome. https://doi.org/10.4060/cb1447en

FAOSTAT. 2011. *FAOSTAT database on agriculture*. http://faostat.fao.org/. Accessed 3 January 2012.

FAOSTAT, 2021. *FAOSTAT statistics database*. Rome: Food and Agriculture Organization of the United Nations (FAO). www.fao.org/faostat. Accessed 2 January 2021.

Fargione, J., Hill, J., Tilman, D., Polasky, S. and Hawthorne, P. 2008. Land clearing and the biofuel carbon debt. *Science* 319, 1235–1238.

Field, E.H., Jordan, T.H., Page, M.T., Milner, K.R., Shaw, B.E., Dawson, T.E., Biasi, G.P., Parsons, T., Hardebeck, J.L., Michael, A.J. and Weldon, R.J., 2017. A synoptic view of the third Uniform California Earthquake Rupture Forecast (UCERF3). *Seismolog Res Lett* 88(5), 1259–1267.

Foley, J.A., Defries, R., Asner, G.P., Barford, C., Bonan, G., Carpenter, S.R., Chapin, F.S. et al. 2005. Global consequences of land use. *Science* 309, 570–574.

Garcia, P. and Leuthold, R.M. 2004. A selected review of agricultural commodity futures and options markets. *Eur Rev Agric Econ* 31, 235–272.

Gardi, C. and Jefferey, S. 2009. *Soil biodiversity*. JRC scientific and technical research series. http://eusoils.jrc.ec.europa.eu/ESDB_Archive/eusoils_docs/other/EUR23759.pdf. Accessed 3 January 2012.

Godfray, H.C.J., Beddington, J.R., Crute, I.R., Haddad, L., Lawrence, D., Muir, J.F., Prett, J., Robinson, S., Thomas, S.M. and Toulmin, C. 2010. Food security: The challenge of feeding 9 billion people. *Science* 327, 812–817.

Goldemberg, J. and Guardabassi, P. 2009. Are biofuels a feasible option? *Energy Policy* 37(1), 10–14.

Grossman, G.M. 1983. *International trade, foreign investment, and the formation of the entrepreneurial class* (No. w1174). Washington: National Bureau of Economic Research.

Gutierrez, A. and Baran, M. 2009. Long-term transfer of diffuse pollution at catchment scale: Respective roles of soil and the saturated and unsaturated zones (Brévilles, France). *J Hydrol* 369, 381–391.

Haron, M. and Dragovich, D. 2010. Climatic influences on dryland salinity in central west New South Wales, Australia. *J Arid Environ* 74(10), 1216–1224.

Hatfield, R.D., Rancour, D.M. and Marita, J.M. 2017. Grass cell walls: A story of cross-linking. *Front Plant Sci* 7, 2056.

Horn, R. and Peth, S. 2011. Mechanics of unsaturated soils for agricultural applications. In *Handbook of soil science*, 2nd ed., ed. P.M. Huang, Y. Li, and M.E. Sumner, 3-1–3-30. Boca Raton: CRC Press.

Horn, R., van den Akker, J.J.H. and Arvidsson, J. 2000. Subsoil compaction—distribution, processes and consequences. *Adv Geoecol* 32.

IFPRI. 2008. *International food prices: The what, who, and how of proposed policy action*. Washington, DC: International Food Policy Research Institute.

IIASA. 2007. A world of simultaneous population growth and shrinking unified by accelerating aging. *Popul Netw Newslett* 39, 1–3.

IPCC. 2014. *Climate change 2014: "Mitigation of climate change"*. Cambridge: Cambridge University Press.

IWMI. 2007. *Water for food, water for life: A comprehensive assessment of water management in agriculture*. London: International Water Management Institute.

Lal, R. 2001. Soil degradation by erosion. *Land Degrad Devel* 12, 519–539.

Lal, R. 2002. Soil carbon dynamics in cropland and rangeland. *Environ Poll* 116(3), 353–362.

Lal, R. 2005. Soil erosion and carbon dynamics. *Soil Till Res* 81, 137–142.

Lal, R. 2009. Soils and food sufficiency: A review. *Agron Sustain Devel* 29, 113–133.

Lal, R. 2010. Managing soils and ecosystems for mitigating anthropogenic carbon and advancing global food security. *Bio Science* 60(9), 708–721.

Lal, R. and Stewart, B.A. (eds.). 2010. *Soil quality and biofuel production, advances in soil science*. Boca Raton: CRC Press.

Lutz, A.F., Immerzeel, W.W., Shrestha, A.B. and Bierkens, M.F.P. 2014. Consistent increase in High Asia's runoff due to increasing glacier melt and precipitation. *Nature Climate Change* 4(7), 587–592.

Lutz, W., Sandersen, W. and Scherbov, S. 2008. The coming acceleration of global population ageing. *Nature* 451, 716–719.

Mueller, L., Schindler, U., Mirschel, W., Shepherd, T.G., Ball, B.C., Helming, K., Rogasik, J., Eulenstein, F. and Wiggering, H. 2010. Assessing the productivity function of soils: A review. *Agron Sustain Devel* 30, 601–614.

Nettleton, W.D. and Peterson, F.F. 1983. Chapter 5 Aridisols. In *Pedogenesis and soil taxonomy*, eds. L.P. Wilding, N.E. Smeck, G.F. Hall, 165–215. Academic Press, Cambridge, Massachusetts, USA: Elsevier.

OECD. 2007. *Biofuels: Is the cure worse than the disease?* Paris: Organization for Economic Cooperation and Development.

Piesse, J. and Thirtle, C. 2009. Three bubbles and a panic: An explanatory review of recent food commodity price events. *Food Policy* 34, 119–129.

Pimentel, D., Marklein, A., Toth, M.A., Karpoff, M.N., Paul, G.S., McCormack, R., Kyriazis, J. and Krueger, T. 2009. Food versus biofuels: Environmental and economic costs. *Human Ecology* 37, 1–12.

Pimentel, D., Petrova, T., Riley, M., Jacquet, J., Ng, V., Honigman, J. and Valero, E. 2006. Conservation of biological diversity in agricultural, forestry and marine systems. In *Focus on ecology research*, ed. A.R. Burk. Hauppauge: Nova Science.

Pimentel, D. and Pimentel, M. 2007. *Food, energy and society*. Boca Raton: Taylor & Francis.

Pretty, J. 2008. Agricultural sustainability: Concepts, principles and evidence. *Phil Trans Royal Soc B* 363, 447–465.

Pulleman, M.M., Bouma, J., van Essen, E.A. and Meijles, E.W. 2000. Soil organic matter content as a function of different land use history. *Soil Sci Soc Am J* 64, 689–694.

Rosegrant, M.W. and Cai, X. 2002. Global water demand and supply projections. Part 2: Results and prospects to 2025. *Water International* 27, 170–182.

Sanchez, P.A. and Swaminathan, M.S. 2005. Hunger in Africa: The link between unhealthy people and unhealthy soils. *Lancet* 365, 442–444.

Satterthwaite, D., McGranahan, G. and Tacoli, C. 2010. Urbanisation and its implications for food and farming. *Phil Trans Royal Soc B* 365, 2809–2820.

Scheffer, M., Carpenter, S., Foley, J.A., Falke, C. and Walker, B. 2001. Catastrophic shifts in ecosystems. *Nature* 413, 591–596.

Scherr, S.J. 1999. *Soil degradation: A threat to developing-country food security by 2010?* Food, agriculture and the environment discussion paper 27. Washington, DC: International Food Policy Research Institute.

Schulte, R., Jordan, L.C., Morad, A., Naftel, R.P., Wellons III, J.C. and Sidonio, R. 2014. Rise in late onset vitamin K deficiency bleeding in young infants because of omission or refusal of prophylaxis at birth. *Pediatric Neurology* 50(6), 564–568.

Small, C. and Naumann, T. 2001. The global distribution of human population and recent volcanism. *Global Environ Change Part B Environ Hazard* 3(3), 93–109.

Soil Survey Staff. 1999. *Soil taxonomy: A basic system of soil classification for making and interpreting soil surveys, U.S. Department of Agriculture handbook 436*, 2nd ed. Washington, DC: U.S. Government Printing Office.

Sonneveld, M. P. W., Schröder, J.J., de Vos, J.A., Monteny, G.J., Mosquera, J., Hol, J.M.G., Lantinga, E.A., Verhoeven, F.P.M. and Bouma, J. 2008. A whole-farm strategy to reduce environmental impacts of nitrogen. *J Environ Qual* 37(1), 186–195.

St. Clair, S.B. and Lynch, J.P. 2010. The opening of Pandora's Box: Climate change impacts on soil fertility and crop nutrition in developing countries. *Plant Soil* 335, 110–115.

Tilman, D., Fargione, J., Wolff, B., D'Antonio, C., Dobson, A., Howarth, R., Schindler, D., Schlesinger, W., Simberloff, D. and Swackhamer, D. 2001. Forecasting agriculturally driven global environmental change. *Science* 292, 281–284.

Tilman, D., Socolow, R., Foley, J.A., Hill, J., Larson, E., Lynd, L., Pacala, S., Reilly, J., Searchinger, T., Somerville, C. and Williams, R. 2009. Beneficial biofuels—the food, energy and environment trilemma. *Science* 325, 270–271.

United Nations. 2020. *UN policy brief: The impact of COVID-19 on women*, 9 April. https://reliefweb.int/sites/reliefweb.int/files/resources/policy-brief-the-impact-of-covid-19-on-women-en.pdf

Van Camp, L., Bujarrabal, B., Gentile, A.R., Jones, R.J.A., Montanarella, L., Olazabal, C. and Selvaradjou, A.-K. 2004. *Reports of the technical working groups established under the thematic strategy for soil protection, vol. IV: Contamination and land management (EUR 21319 EN/4)*. Brussels, Belgium: Joint Research Centre, European Commission. Also available at http://ec.europa.eu/environment/soil/pdf/vol4.pdf

Von Braun, J. and Meinzen-Dick, R. 2009. *"Land grabbing" by foreign investors in developing countries: Risks and opportunities.* IFPRI Policy Brief13, www.ifpri.org/sites/default/publications/bp013all.pdf

Williams, J.C. 2001. Commodity futures and options. In *Handbook of agricultural economics, vol. 1B: Marketing, distribution and consumers*, ed. B.L. Gardner and G.C. Rausser, 745–816. Amsterdam: Elsevier Science B.V.

World Health Organization (WHO). 2000. *Nutrition for health and development—a global agenda for combating malnutrition.* Geneva: World Health Organization.

# 10 The Nexus between Environmental Damage, Poverty, and Climate Change in Hard-to-Reach Areas

## A Somber Tale of the 21st Century

*Amitava Panja, Punam Bhattacharjee, Subhradip Bhattacharjee, Dinesh Kumar, Govind Makarana, Manish Kushwaha, M.R. Yadav, Rakesh Kumar, and Vishnu D. Rajput*

## 10.1 INTRODUCTION

> There's one issue that will define the contours of this century more dramatically than any other, and that is the urgent threat of a climate change.
>
> **Barack Obama**

Human civilization has witnessed two major world wars, i.e., World War I and World War II, within a century. Climate change's impacts are the only things capable of igniting a new global conflict. Such a phenomena is unavoidable and has long-term repercussions; hence, it is aptly classified as an urgent global challenge. Climate change according to IPCC refers to "a change in the state of the climate that can be identified (e.g., by using statistical tests) by changes in the mean and/or the variability of its properties and that persists for an extended period, typically decades or longer" (IPCC, 2018). It has a significant influence in global warming and environmental harm due to the effects of susceptibility and biodiversity loss caused by climate change. Climate change's negative effects on the environment have intensified to such a degree that combating them is a central component of world policy today. The climatic conditions of this planet are dynamic. However, the dramatic climate change over the past century is cause for grave alarm. It is one of the most serious and consequential concerns ever confronted by humanity and is continually



DOI: 10.1201/9781003358169-10

altering the probability and scale of natural disasters (Gardoni et al., 2016). United Nations Framework Convention on Climate Change has described climate change as a process that is ascribed directly or indirectly to human actions, altering the global atmosphere over comparable time periods, and observed alongside natural climatic variability. Climate change's negative consequences on human civilization have already been initiated by overexploitation and poor management of natural resources. Thus, a mindset that views it as an issue for the future can never be successful. This is not only a threat to sustainability; it is also currently occurring. Various impacts of climate change include rise of global temperature, warming of ocean water, shrinkage of ice sheets, retreating of glaciers, decline in snow cover, rise in sea level, increase in extreme events such as cyclone, droughts, landslides, acidification of ocean water, and many more (NASA, 2022).

Climate change is impacting each and every one living on the planet in some way or another. This is irrespective of developed or developing country and may be with a difference in type and magnitude of losses. It is anticipated that within a few decades, billions of people, particularly those living in underdeveloped nations, will suffer the lethal effects of climate change, including food and water shortages, health difficulties, and many others. Impacts of climate change can be both in direct and indirect ways. Direct impacts of climate change include various climate led disasters such as droughts, cyclones, floods, avalanches, landslides, heatwave, etc. The indirect effects of climate change can be more severe than their direct counterparts. It includes a decline in air quality, food insecurity, availability and quality of fresh drinking water, and productivity of rural industries. It is also a threat to public health and a hindrance to crop production, livestock husbandry, and fisheries, among other things (Figure 10.1). Nevertheless, effects faced by marginal people, living in climatically vulnerable regions, is far more compared to other sections of the society. Communities residing in climate-vulnerable locations and reliant on

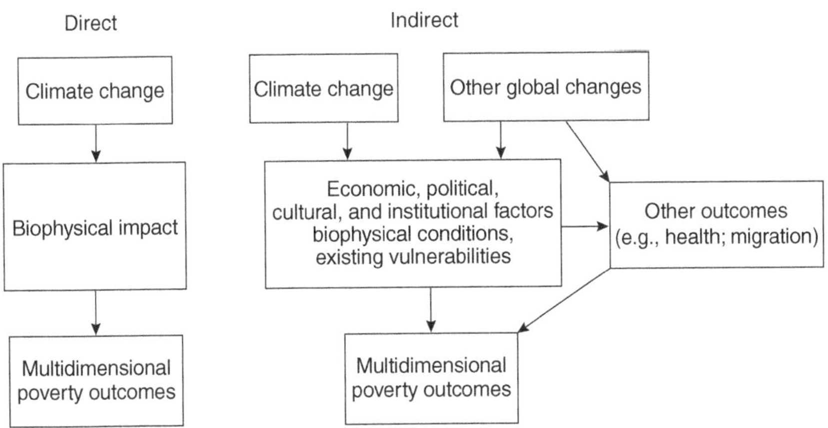


**FIGURE 10.1**   Direct and indirect impacts of climate change.

**Source: Leichenko and Silva (2014)**

climate-sensitive livelihoods are not only more susceptible to the effects of climate change but also less able to implement adaption strategies (Asfaw et al., 2018). The repercussions of climate change are not limited to natural, physical, and biological losses; rather, they extend to the triple bottom line, i.e., people, the planet, and profits. In addition to massive damage of natural and physical assets, the effects of climate change and climate-driven disasters on humans (particularly those living in climatically susceptible locations) are great and devastating. Marginal people, low in resources and reliant on industries severely impacted by climate change for their livelihood, suffer tremendous social and economic losses. Researchers have acknowledged that climate change's vulnerabilities, impacts, and responses are intricately intertwined with the political, social, and economic processes that perpetuate poverty (Eriksen & O'Brien, 2007; Kelly & Adger, 2000; Rayner & Malone, 2001).

This chapter has tried to focus on the interaction effect of climate change, environmental damage, and poverty in three major climatic vulnerable scenarios across the world viz. tropical rainforest and other forest lands, high mountain areas, and coastal mangrove areas. All these three zones mentioned earlier are highly vulnerable to climate change and at the same time are home to millions of marginal people who lack in resources and information and always stand at the threat of the dreadful consequences of climate change.

## 10.2  SCENARIO 1: TROPICAL RAINFOREST AND OTHER FOREST LANDS

### 10.2.1  RAINFOREST

A rainforest, as its name indicates, is a type of forest that receives a heavy amount of rainfall around the year. This particular type of forest is generally located near the equator and is the habitat of the enormous amount of flora and fauna. The atmosphere and weather in such areas largely remain the same throughout the year and can be best described by the tree canopy's presence of high temperature, high humidity, and low sunlight penetration. The most prominent feather of a rainforest is the high amount of rainfall which could reach up to 2,000–10,000 mm per year, making it the wettest ecosystem on earth. The rainforest is also believed to be the oldest and continuously evolving ecosystem on our planet.

The appearance of the word "rainforest" in literature is quite old, most probably first mentioned by Schimper in 1903, although sometimes the word does not properly justify the weather and ecological condition of the area. To be more precise, the rainforest in the majority of cases indicates the relationship of a forest with an abundance of rainfall; however, many rainforests around the world do not fall in this group. In Australia, there are long strips of rainforest along the side of rivers that although lie within the humid climate belt, do have a pronounced yearly dry season. On the other hand, the colder rainforests of North America have little similarity with their equatorial counterpart which remain warm throughout the year. Despite these complexities, as there is no other alternative word available, the rainforest is widely used in literature.

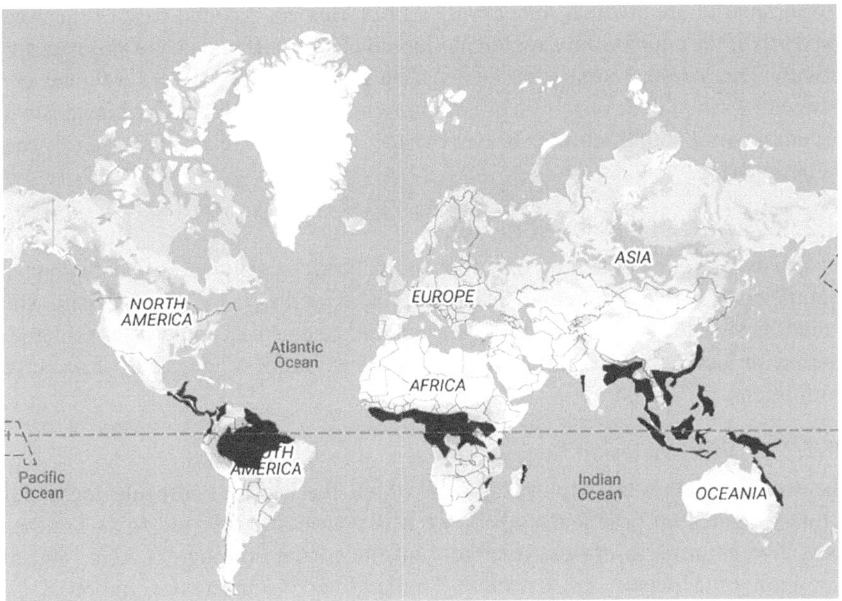


**FIGURE 10.2** Global distribution of tropical rainforests.

The rainforest by large is divided into three groups: tropical rainforest, subtropical rainforest, and temperate rainforest. However, once again, these terms are more of an environmental term rather than a description of a forest. This often leads to unnecessary complexities when authors and researchers sometime report a tropical rainforest in subtropical areas. As a large majority of the rainforest lies within the 10° of the equators, it should be marked as a tropical rainforest. The forest vegetation in all these three groups is occupied by tall and woody evergreen trees often taller than 30 meters or more. The trees form such a thick canopy that it is impossible to see the ground from above the trees. The prominent species of rainforest are *Euphorbiaceae, Leguminosae, Myrtaceae, Burseraceae, Lauraceae, Myristicaceae, Anacardiaceae,* and *Annonaceae*; however, considerable local variation is also present. The diameter of the tree trunks is generally slender to the proportion of the height of the tree, while only the dominant crown spreads substantially and other crowns remain small. The area which lies in close vicinity of the swamp land is occupied by the trees which have prop roots.

## 10.2.2 Geographical Distribution of Tropical Rainforest

The tropical rainforest of the world is divided into three major regions: American, African, and Indo Malesian.

### 10.2.2.1 American Rainforest Region or Neotropics

Three distinct rainforest subregions can be found on the continent of South America. The largest by far is Amazonia, which must be regarded as largely intact due to its size. A smaller area, which sometimes reaches Mexico, is found west of the Andes and north of the equator. This region has already badly suffered due to anthropogenic activity. The smallest area is a slender strip of land along Brazil's Atlantic coast between 14° and 21° S; very little of it remains now. On the bigger Caribbean islands, it is unknown if actual rainforests ever existed; those that are still there hardly merit the designation.

### 10.2.2.2 African Rainforest Region

It consists of four zones, all of which have been partially destroyed, along the Atlantic coast between roughly 10° N and 5° S, as well as in the Congo basin, which extends east to the mountains. East African outlier enclaves could be considered to be more or less authentic rainforests but not the woods of Madagascar or its current reminiscent.

### 10.2.2.3 Indo Malesian Region

The former Malaysian Archipelago, to which the Malay Peninsula technically belongs, makes up practically all of what is commonly referred to as Southeast Asia. The genuine rainforest is largely missing from continental Asia, with the exception of Malaya, southernmost Thailand, and southwest Cambodia. Two identical nuclei can be identified in this area. Begin with Borneo-Malaya, with exceptions in the Andaman Islands, Ceylon, and the less developed forests of southwest India, Sumatra, and the Philippines among the regions. The second is New Guinea, which also includes the less diverse forests on the islands to its west and east. However, there is one exception: a small area of rainforest that is still present along tropical Australia's eastern coast. In the Malesian region, larger islands tend to have a higher species diversity. True rainforest, however, can only occupy a small portion of most islands in Malesia as a whole due to its steep terrain. New Guinea's forests are still mostly intact, in contrast to the recent severe depletion of the forests in Malaya, Sumatra, Borneo, and the Philippines (Myers, 1989).

### 10.2.3 Ecological Character of Tropical Rainforest

The study of ecological characteristics of rainforests has attracted global attention since the great naturalist of the 19th century Wallace divided the zone into three to five biogeographical regions based on the organism of interest. As discussed earlier, although geographically, the zone is divided into three groups, on an ecological basis; the same area can be divided into five groups. These zones are America, Africa, Southeast Asia, Madagascar, and New Guinea with smaller stretches of Australia, South Asia, and many smaller tropical islands. The difference in distinctiveness in geographical and ecological zones is much more essential when it comes to conserving natural resources and endangered species. The ecological characteristics of all the rainforests around the world are reminiscent of part of the ancient supercontinent of the Mesozoic Era known as "Gondwanaland" which can be easily comprehended from the striking similarities of plant families and molecular characterization of dominant species of angiosperms (Corlett & Primack, 2006).

Many scientists and explorers have noted that the number of species seen in the tropics is far more than what they were used to seeing in their own countries for almost all types of life forms. The diversity is better observed and documented in the case of tree species, which is so diverse that on average, only 3.6 trees/ha was observed from the same species. Even the most dominant species, based on which the forest area is botanically identified, is hardly 15% of the total population. It has been estimated that tropical rainforest around the world have around 175,000 species of plants which is roughly two-thirds of the total plant species on earth. This species diversity comes with some amazing benefits such as protection from pest and diseases which otherwise are very common in hot and humid climate and complementing different nutrient cycling and plant nutritional demand.

The abundance of fallen tree leaves in tropical rainforest also helps in the development and higher multiplication of herbivore insects, as leaves are more nutritious than any other part of the plant. However, this is not always true, as soils from many of the tropical rainforests are relatively nutrient poor; hence, less amount of energy is present in plant leaves. This might act as a defense mechanism against various herbivore pests. Another mechanism which is also often deployed by the plants of tropical rainforest against the pests is the presence of secondary metabolites such as phenolics in plant parts (Kursar & Coley, 2003; Mali & Borges, 2003). The soils of these areas are highly weathered and have lower nutrient retention capacity; however; the presence of organic matter due to the decomposition of plant leaves acts as a chelating agent and also as a source of nutrients which is transferred to the plant via symbiotic association of soilborne fungi.

Not only in plants but the animal kingdom also has its own share of diversity in tropical rainforest. The highest level of diversity in the animal kingdom can be seen in primates. The Amazonian rainforest has around 70 primate species, African rainforest has 50 primate species, Southeast Asia has 25 primate species, while smaller Madagascar has 20 species. Apart from primates, the rainforests have higher diversity in carnivore species. These carnivorous animals play a crucial role in food web of rainforest. Their prey, which includes large terrestrial herbivores, monkeys, rodents, birds, reptiles, insects, and other smaller carnivores, remains within the population limit which the ecosystem can sustain. Prey animals also avoid locations and times of day where they might be more vulnerable to attacks which is known as "ecology of fear" (Ripple & Beschta, 2004). Carnivores indirectly affect the structure of plant communities by affecting the population sizes and behaviors of herbivorous animals. The impact of large carnivores on the population of smaller carnivores, which serve as the primary predators of birds and other small vertebrates, may be just as significant. Because the impacts of the elimination of large carnivores spread from top carnivores to plants in what is frequently referred to as a "trophic cascade," the removal of these animals might have a significant negative impact on the remainder of the rainforest community.

## 10.2.4 Anthropogenic Character of Tropical Rainforest

The distinctiveness of different rainforest area is not only limited to the plants and animals but indigenous people groups are also equally diverse and interested. The

**TABLE 10.1**

**Some of the Prominent Hunting-Gathering Communities from the Pygmy Tribe of the Congo Basin Rainforest**

| Name and alternative names | Country | Approximate population size |
|---|---|---|
| Baka/Bangombe/Bibayak/Babinga | Cameroon, Congo, Gabon | 30,000–40,000 |
| Bedzan/Tikar Pygmies | Cameroon | 400 |
| (Ba)Kola/(Ba)Gyeli | Cameroon | 4,000 |
| (Ba)Koya/(Ba)Kola | Gabon, Congo | 2,600 |
| (Ba)Bongo | Gabon | 3,000 |
| Mikaya/Bambenga | Congo | No data |
| (Ba)Aka | Central African Republic, Congo | 30,000–50,000 |
| (Ba)Twa/Konda Twa | Democratic Republic of Congo | No data |
| (Ba)Sua/(Ba)Mbuti | Democratic Republic of Congo | 26,000 |
| Asua | Democratic Republic of Congo | 10,000 |
| Efe/(Ba)Mbuti | Democratic Republic of Congo | 10,000 |

*Source:* Ichikawa (2014)

human existence on these rainforests is very old, sometimes older than 35,000 years which is as old as the cradle of civilization. Existence of such ethnic groups are quite common in the Amazonian rainforest. Over the course of time, these tribes have developed and shaped their lifestyle in such a manner which resembles nothing to the city dwellers of modern time. For the larger period of history, their lifestyle evolved from the hunting-gathering style which on later stages partially converted to shifting cultivation. The seeping of modern civilization and colonization of non-indigenous people is further influencing the lifestyle of the indigenous population and pushing them towards adoption of conventional farming and rearing practices. As the peculiar cultures of the indigenous people are linked with the lifestyle of the population, the change of the occupation has the direct effect on extinction of local cultures.

Anthropologists have three of the oldest societies flourishing in tropical rainforest and along the equator; these are in the Amazon basin, Congo basin, and tropical islands of Southeast Asia (Sumatra, Borneo, and New Guinea). These are among the last tribes which have come into contact with the modern world (van Solinge, 2010). One of the world's last remaining unspoiled natural areas, New Guinea's deep and largely unexplored forest, is home to hundreds of indigenous tribes. In the same manner, the Pygmy tribe of the Congo basin rainforest is the largest and most diverse group of the last remaining hunting-gathering community on earth (Ichikawa, 2014). In this larger group, experts have identified 15 groups which are ethnolinguistically different with a total population of 250,000 to 350,000 individuals. These communities are now facing existential crisis due to multifaceted issues ranging from habitat change due to climate change issues to threat from loggers who are lurking into the rainforest for the woods.

### 10.2.5 Climate Change in Rainforest: A Complex Scenario

Before bringing the angle of climate change in rainforest, we must have to understand how the rainforest is one of the essential cogwheels of climate engine of earth. From 145 million to 66 million years ago, the Cretaceous geological period lasted. Since then, the rainforest is shaping the climatic scenario of our planet. Amazon, the largest rainforest on earth, is responsible for the 15% of global terrestrial photosynthesis (Malhi et al., 2008). The evaporation and condensation of water from the Amazon rainforest is one of the primary driving forces of water cycle in the Northern Hemisphere. However, the whole climatological phenomenon based on the Amazon rainforest is jeopardized by human activities which are accelerating the climate change phenomenon. Each year, the Amazon rainforest is shrinking by a size of more than 25,000 km$^2$, 80% of which is within the boundary of Brazil. From January to June 2021, almost 17% of the region's rainforest was lost, resulting in a deforestation area that is roughly 3,610 square kilometers, or four times the size of New York City (Zaman, 2022). In the Southern Brazilian Amazon, deforestation rates are so high that if current trends continue, they could reach up to 56% by 2050 (Leite-Filho et al., 2021).

The major threat currently posed by the Amazon rainforest is the growing interest in soybean cultivation by local population followed by cattle ranching. Soybean cultivation is in one end playing a strong role in the economic growth of Brazil, while on the other, it is meeting the global soy demand but at a cost of depleting Amazon's forest land. After the deforestation, it has been seen that approximately 6% of the land is used for crop production, 62% from the ranching, and 32% for regrowing of the vegetation (Malhi et al., 2008). However, the anthropogenic activity in Amazon is not limited to these two activities; rather, several other activities such as logging, hunting, mining, and occurrence of fire incidents are also responsible for deforestation.

### 10.2.6 Signs of Climate Change Phenomenon in the Amazon Rainforest

Amazon rainforest is currently facing an unprecedented level of global warming threat which is not seen anywhere in the tropical zone. During the years 1975–2004, the temperature in the Amazon, especially in the southern and eastern frontiers, has risen at a rate of 1°C/decade. Interestingly, the same area is also known as the "arc of deforestation" (Costa et al., 2022). The temperature incremental rate is much higher than the global average incremental rate; the severity of the scenario can be easily felt during the dry season of the area. This area has floodplain forests, which have just been termed as the "Achilles heel of Amazonian forest resilience" due to their extreme sensitivity to drought and warming. In observational and reanalysis data, it has been discovered that warming trends, severe droughts, and negative trends in dry season rainfall occur simultaneously in this area. It is of no surprise that the regular occurrence of heat waves and surface dryness is suspected due to the anthropogenic actions (Costa et al., 2022).

Numerous past studies supported the negative impact of Amazonian deforestation and its risk-raising effect on climate change. The Amazon is regarded as one of the most important carbon sinks in the world. However, due to widespread deforestation, which causes climate change, this sink is no longer safe from environmental risks. Large perennial trees are significant sources of long-term carbon trapping; however, climate change is posing a great threat to it. Simulated projections indicate that germination of tree species will significantly decline, which is not so prominent in the case of shrubs and herbs. This situation leads to the succession of tropical forest by savanna-like vegetation in the Amazon region, which might also cause the partial or complete habitat loss in that area (Silva et al., 2021).

### 10.2.7　Effect of Anthropogenic Activity on Climate Change Effect of Amazon

Human activity has a significant role in the climate change–mediated devastation of biodiversity of the Amazon rainforest. A recent study conducted by Zaveri et al. (2022) pointed out that human activity in the Amazon rainforest is causing the release of aerosol nanoparticles in the atmosphere at a higher rate. Previously, these aerosol nanoparticles were considered to be too small to form cloud condensation nuclei (CCN) which could play a role in climate change. However, a recent model analysis by the authors described earlier suggests that anthropogenic aerosols have a noticeable impact on the life cycle of clouds which could lead to climate change. Another prominent threat is the anthropogenic origin of forest fire in the Amazon. Deforestation with an aim to convert the land to cropland or ranch is often associated with removing bushes and trees with fire. Given that Amazonian species have mostly developed in the absence of fire, a complete loss of habitat is virtually guaranteed in such a situation (Feng et al., 2021). Due to increased edge density, elevated regional temperatures, and decreased precipitation, deforestation makes up a significant source of ignition and makes the surviving forest more flammable (Barlow et al., 2020). A simpler relation among deforestation, fire, drought, and climate change can be stated as a higher population load leads to the expansion of crop and ranchland to the forest which is the primary cause of deforestation. As climate change is making the Amazon drier, it is becoming more prone to fire, while on the other hand, fire incidents are increasing the occurrence of the climate change phenomenon. The occurrence of prolonged dry season and its relation with deforestation can be hypothesized in a manner shown in Figure 10.3.

The agricultural practice in the Amazon is largely a slash-and-burn type in which dry seasons have a multitude of advantages; hence, the farming community is not much concerned regarding the frequent occurrence of dry spells. The livestock growers also use fire to remove the woody plants, converting the land to grazing land. Moreover, during the dry season, fire often escapes the intended area and spreads to the larger area. The incidents of fire alter the water cycle of the area, as evapotranspiration is much higher in forest land than in any other area; decreasing forest land translates to the reduction of evapotranspiration, hence, altering the whole water cycle.

### 10.2.8　Effect of Climate Change on the Amazon Rainforest

#### 10.2.8.1　Effect on Agriculture/Rearing/Farming

The Amazon rainforest has been converted into pastureland, transforming it from a forest that consumes methane to one that produces it. The resulting greenhouse gas

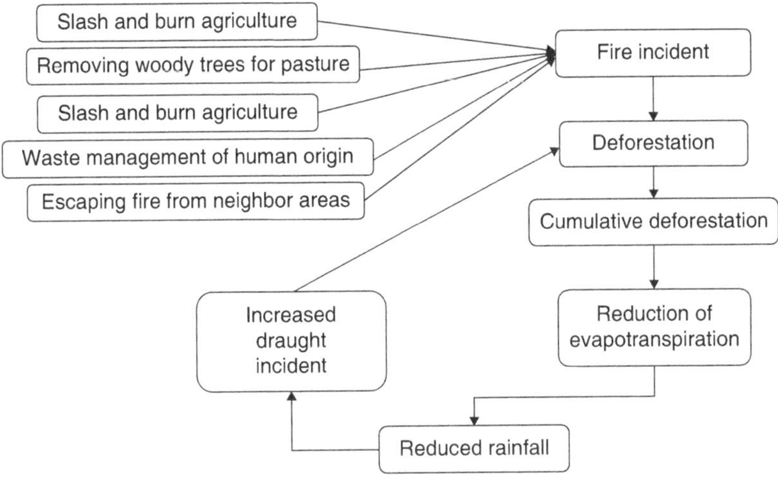


**FIGURE 10.3** Cyclic pattern of deforestation-draught relation in the Amazon rainforest.

**Source: Staal et al. (2020)**

emissions represent 50% of Brazil's greenhouse gas production and have already surpassed national fossil fuel emissions by more than 20%, at the same time making the country the world's fourth largest carbon emitter (Fonseca de Souza et al., 2022).

### 10.2.8.2 Microbial Community

The human activity in rainforests not only affects the plant and tress but it also alters even the microbial community too. As all the biological cycles pass through the microbial world, even a minute change in its community structure leads to a more drastic change in ecology. Understanding the microbial community structure in soil is a very complex task and only became possible after the rapid development in the field of metagenomics, especially shotgun metagenomics which enabled better interpretation of microbial actions up to the molecular level. The molecular-level investigations have shed light to the fact that changing in land use planning has a direct effect on fungal and bacterial community structure (De Castro et al., 2008). Fungi play a variety of intricate and crucial roles in soil that are essential to the biome's continued functionality. Along with interacting with other microbes, the fungi can participate in the cycling of nutrients and form harmful or symbiotic relationships with plants and animals. It has been observed that the fungal community structure is more closely linked with the plant community of the same area than with environmental and geographical factors. Deforestation leads to changes in the fungal community structure because of the changes in the plant community. Experiment results indicate that the fungal community of the secondary forest is more similar to the fungal community of the primary forest as compared to the converted pasture land (Mueller et al., 2014).

The higher rhizospheric microbial species richness in forest areas than in agriculture, pasture, and agroforestry links between plant and fungal communities across a deforestation chrono sequence in the Amazon rainforest stem in the Amazon is due to a higher diversity of plant species and its interaction with the microbial world. On a flip side, low diversity in cultivated land is due to lesser plant diversity (Fracetto et al., 2013). In other occasions, the changing of land use planning did not alter the community diversity; rather, it changed the functionally of microbial communities due to the change in relative abundance. The reason could be due to the necessity for the adaptation and functioning under transformed geochemistry (Fracetto et al., 2013).

As a result, the nutrient cycling, especially phosphorus acquisition, is highly dependent on fungal action and is bound to change along with the decomposition of lignin-rich organic matter which also largely depends on the fungal community. Among different soil biochemical factors, the primary factors which might be responsible for the difference in the microbial community structure between pasture soil and forest land of the Amazon are increased pH, copper, iron, and zinc concentration in soil (Jesus et al., 2009). The organic carbon concentration in converted pasture land was found to be higher than the forest soil, which might be giving an edge to copiotrophic microorganisms with diverse mechanisms of carbohydrate metabolism. Lignin content of soil organic matter is usually high in forest areas, as it is found in higher content in perennial trees as compared to the grassland which has a higher amount of hemicellulose. Metagenomic analysis of the soil of these two areas also indicated a higher abundance of the lignin degradation gene in forest areas and cellulose and hemicellulose degrading genes in the grassland (Kroeger et al., 2018). The change of lignin content in soil is drastic, as the large trees of the Amazon rainforest have a lignin content of about 40%; however, when converted to pasture land, the lignin content has been decreased to 5%–10% in forage grasses (Mauri et al., 2015). In such scenarios (grassland/pasture), the abundance of Firmicutes increases, as they are responsive to increased organic matter in soil. Kroeger et al. (2018) have also found out that the conversion of a forest area to grassland leads to the complete loss of archaea Thaumarchaeota due to the increased activity of biological nitrification inhibitors (BNIs), especially brachialactone, which is a cyclic diterpene. Brachialactone inhibits ammonia monooxygenase (amo) and hydroxylamine oxidoreductase (hao), further resulting the inhibition of ammonia-oxidizing bacteria and archaea (Moreta et al., 2014).

### 10.2.9 MIGRATION

Climate change is becoming one of the key driving forces behind the decision of migration in recent time. A global trend can be observed where migration is taking place because of the declining crop productivity in challenged areas due to climate change, reduced forest area, and fluctuation of crop yield (Gray, 2009; Massey et al., 2010). In addition to these factors, frequent heat waves due to deforestation and forest fire and more frequent occurrences of diseases are also behind the migration in the Amazon rainforest (Confalonieri, 2000). Conversion of forest land to agricultural land especially to soybean cultivation and ranching resulted in the improvement of the economic condition of the local community; it resulted in higher awareness regarding education and improvement of livelihood. Increased interest in education among teenagers and the youth population resulted in the migration towards urban areas (rural-urban shift). However, once they reach to urban areas, they have very little resource to stay in a modest environment; hence, poverty is very common among first-generation migrants (VanWey et al., 2011).

## 10.3  SCENARIO 2: HIGH MOUNTAIN AREA: HIMALAYAN FOOTHILLS

The natural evolution of a piece of land that leads to a high-altitude land having a certain special ecological characteristic is called a mountain. Mountains may differ in appearance and altitude. Some have green rounded peaks that climb up to 2,000 feet (600 m), whereas others have sharp, snow-covered rocks that soar into tens of thousands of feet. Since a certain height has not been uniformly established, there is no precise mountain definition in terms of elevation; some regions of the globe consider 1,000 feet (300 m) to be a mountain, while others place the amount at 2,000 feet (600 m). Despite this ambiguity, a mountain is a formation or landform that rises substantially above the surrounding terrain. Typically, mountains have steeper slopes and are considered taller than hills. In addition, mountains are often found in mountain ranges or mountain belts, which consist of a sequence of peaks. These qualities characterize mountains in general. A landform is a world in many ways that are a component of the terrain or land.

The Himalayan region is a special place with diverse landscapes, many ethnic groups, and an extensive range of flora and animals. It includes portions of China's Yunnan Province, Sagaing Region, Bhutan, sections of Myanmar's Chin and Rakhine states, and portions of India's east and northeast. Due to its undeveloped industrialization and weak infrastructure connections, it has a low GDP and per capita income. In addition, it has seen high maternal and infant mortality rates, climatic change,

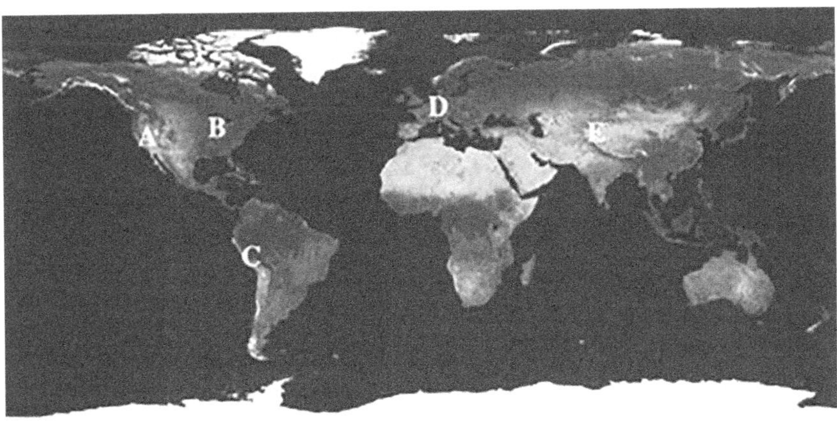


**FIGURE 10.4**  Topography of the earth's land surface. Some major mountain belts are marked: (A) Cascades, (B) Appalachians, (C) Andes, (D) Alps, and (E) Himalayas.

natural catastrophes, political unrest, and armed wars, all of which have affected the population's livelihoods and general well-being (Hazarika & Banerjee, 2017).

These culturally diverse Himalayan villages rely on subsistence agriculture for their survival, and livestock, particularly cows, bulls, and buffaloes, play a crucial role in the agricultural system by ploughing and fertilizing the fields, providing income through the production of milk, and serving as a social safety net. Selling of livestock for post-flood recovery is common, and livestock meat is also an important source of food during the shortage of staple crops (Hazarika & Banerjee, 2017). The Himalayan region, also called the biodiversity hotspot, has numerous ethnic communities with different cultures and habits, and to unite all these groups, democratization is the finest option. However, due to the vast cultural diversity of the Himalayan region, democratization is a difficult challenge in many areas. Ethnic rivalry, identity assertion, and the need for ethnic homelands to defend, preserve, and advance diverse communities' political and socioeconomic rights have been amplified by the democratization process (Arora & Jayaram, 2017).

### 10.3.1 EVALUATION AND DISTRIBUTION OF HIMALAYAN POPULATION

East Asia, which extends physically to the Pamir Plateau in the west, the Himalayan Mountains in the southwest, Lake Baikal in the north, and the South China Sea in the south, is home to many peoples, civilizations, and languages. Multiple disciplines, including genetics, archaeology, linguistics, and ethnology, are tasked with reconstructing the natural history of East Asians. Geneticists corroborate East Asians' recent African ancestry. Anatomically modern humans originated in Africa roughly 50,000 years ago and moved to East Asia through the south. Following the end of the Last Glacial Maximum roughly 12,000 years ago, rice and millet were domesticated in the south and north of East Asia, respectively, allowing for the expansion of human populations and the formation of linguistic and ethnic communities. These Neolithic people closely connected the current genetic architecture and language families. (Yu & Li, 2021). Plant usage, collection, and transmission were closely linked to cultural legacy, but the ecogeographical environment impacts how plants are used and gathered. The diverse knowledge of plant collection, usage, and transmission among the inhabitants of the Himalayan region was largely ascribed to the region's legacy. The lack of agreement on plant usage may be caused by cultural variety, varying accessibility, physiographic variability, and the uniqueness of the biodiversity. More evaluation of these factors might improve the ecology and cultural well-being of the area since different perspectives on plant use may contribute to diversifying plant use techniques in line with the livelihood, culture, and environment (Kunwar et al., 2018).

### 10.3.2 CHALLENGES IN HIMALAYAN UPHILL DUE TO CLIMATE CHANGE

Mountains have delicate ecosystems, especially when it comes to farming and climate change. However, there is little empirical knowledge on the possible effects of climate change for many such geographies. In the Himalayan area, the vast majority of farmers are affected by climate change, which has negative effects on family income, poverty rates, and wheat yields while also increasing dependency on wood and non-timber forest products (Ali et al., 2017).

The climate of the Himalayan uphill area is impacted by global climate change, which significantly influences the region's water supplies, agriculture, and economy. The western Himalayas demonstrate a rise in seasonal low temperatures of 2°C to 18°C. Three mountain ranges in the western Himalayas have a rising tendency for seasonal highest and lowest temperatures, whereas one range has a decreasing trend (Karakoram). Seasonal precipitation patterns are declining throughout these ranges (Saxena et al., 2022).

Due to incongruent policy and practice goals, the west Himalayan forests, which have existed for over nine decades, are losing their lustre. With the right measures, the situation may be improved to make these forests more important to the local population and the rising global restoration and Sustainable Development Goals (Pathak et al., 2021). The Hindu Kush Himalayan (HKH) region includes places where flood hazards are quite high. Numerous non-climate stressors, including poverty, environmental and climatic shocks, and poor infrastructure, pose problems for the area (Dasgupta et al., 2020). Various circumstances may have influenced the decision to stop engaging in agro-livestock operations. Although the effects of climate change are complicated due to their cascading effects on already low-income and marginalized households, they are still having an impact on agro-ecology due to changes in weather patterns, an increase in invasive species and crop-livestock pests, and labour migration abroad as a result of decreased farm output, harming the general mountainous region's agricultural environment (Pandey, 2019).

### 10.3.3 Migration and Climate Change

Globally, migration for labour continues to be a means of sustenance, particularly in light of the unparalleled pace of environmental change. During the winter, farmers in the high Himalayas relocate when agricultural operations are diminished. Migration is mostly driven by structural poverty rather than the effects of climate change since males from low-caste groups who live in impoverished homes with tiny landholdings and significant food insecurity travel for employment in the winter. Hill people have a distinct understanding of how climate change and fluctuation affect food production negatively. In this situation, the poorest families find it dangerous to cultivate their own land.

Additionally, the old method of sharecropping, which assisted them in reducing food shortages, has lost its profitability. Therefore, in the setting of climate change, more families are expected to engage in seasonal migration, and those who are currently travelling are likely to do so for longer lengths of time. Currently, many migrants labour in low-paying, unskilled jobs that barely allow them to save enough money to pay off the debt their family accrued because of food shortages. Because they lack the social connections, educational background, and financial resources necessary to meet the administrative and financial requirements for more economically advantageous long-term overseas migration, these farmers are likely to be restricted to the same pattern even in the future. Therefore, it seems unclear that migration would significantly help rural agricultural communities in the Himalayas create livelihood resilience to climate change (Gautam, 2017; Bisht et al., 2016). Women, the poor, and socially excluded groups are significantly worse off regarding climate change's effects. However, current adaptation strategies and viewpoints give little thought to the unique requirements of these communities. These groups examine the variations in multidimensional poverty levels between various socioeconomic groupings. Rural women are comparatively more exposed to the negative consequences of climate change than males. Improve outcomes or positive effects of adaptation planning, organizational and policy procedures, and the ability of women and disadvantaged groups must be improved (Khadka et al., 2015). Fragmented methods and a mismatch between international agreements and agreements are results of the complexity of climate governance at several levels, which causes delays in progress. In order to effectively manage climate change, there must be extensive discussion, action, and funding. It is also advisable to build institutional frameworks for climate governance (Koshy et al., 2014).

The Hindu Kush Himalayan (HKH) area is particularly susceptible to the effects of climate change. Building community capacity is essential to implementing an adaptive plan to combat the negative consequences of climate change. Enhancing local community capacity in financial literacy, flood preparation, and disaster risk reduction has a big impact and is closely related to more effective resilient planning techniques and practices (Ali et al., 2022). Mountains provide 40% of world commodities and services. The Hindu Kush Himalayan region provides water, biodiversity, niche products, hydroelectricity, wood, mineral resources, and leisure. HKH is rich in cultures, ethnicities, and traditional knowledge systems. Poverty is rife despite abundant natural resources and civilizations. The HKH is a hotspot for climate change consequences, although the area lacked data in the 2007 IPCC assessment report. Since then, data gaps have been filled in biodiversity, cryosphere, and climate change. This chapter explains the HKH area and discusses models for climate change (Sharma, 2012).

Nepal is one of the landlocked nations with unique physiographic features and rough terrain. In the future, Nepal's natural resources may provide substantial benefits. Climate change may significantly impact Nepal's water resources and other sectors. Water is mostly derived from summer monsoon rains and the melting of the Himalayan highlands' vast snow and glacier reserves. There are definite indications of substantial warming in the current environment. The nation has an annual average trend of 0.06°C. Gradually increasing warming rates are seen at higher altitudes. The bulk of Nepal's glaciers has shrunk rapidly as a consequence of climate change. In addition, the probable effects of climate change on other sectors, including agriculture, biodiversity, human health, and livelihoods, are substantial (Shrestha & Aryal, 2011).

### 10.3.4 The Vulnerable Group in the Himalayan Region

The inhabitants' lives in the central Himalayan countryside mostly depend on the agricultural, animal husbandry, and forestry industries, all of which are intertwined. The contribution of forest products and services to maintaining agricultural and zoological output is significant. Numerous features of the middle Himalayan area make it more susceptible to the effects of climate change. The area is particularly susceptible to climate change due to its high population density, extreme poverty, and poor climate risk resistance (Sharma, 2012). The bulk of impoverished, marginalized, and traditional communities rely significantly on natural resources, especially in the hilly areas, for their livelihoods. It is anticipated that changes in resource availability, which are exacerbated by climate concerns, would have significant effects. These dangers might undo the improvements in livelihoods and poverty reduction and obstruct efforts to achieve the targeted national development objectives (Shrestha & Aryal, 2011). The International Panel on Climate Change (IPCC) has determined that the climate system will unavoidably change. Climate change has been predicted to affect the average temperature, the frequency and severity of severe weather events, the pattern of precipitation, sea level rise, and glacier retreat. Although it is difficult to forecast and quantify how climate change may affect socioeconomic systems, it is generally established that one important aspect of human well-being—food security—is at risk. In the middle Himalayas, these climate change–related concerns pose a danger to over 70% of the rural residents who rely heavily on rainfed agriculture. Human security and food security go hand in hand with the rural poor. Extreme droughts or unexpectedly large rains have often resulted in fatalities as well as the deterioration of the natural resource base, which has increased poverty overall and in low-income groups in particular. As a result, those with low incomes or impoverished are most exposed to the effects of climate change (Maikhuri et al., 2013).

### 10.3.5 Changes in Livelihood Pattern

For aboriginal populations, forest resources are their main source of income. These are regarded as the green jewel of the Himalayan area because they are rich in biodiversity. This system sustains life is under significant biotic stress and has grown susceptible due to mounting anthropogenic demand (Datta & Behera, 2022). Due to landslides and soil erosion among forest and non-forest land use and land-cover groups, a sizable area is under stress and the forest ecosystem has become fragile. The Indian Himalayan region was also shown to have significant levels of climate fluctuation. Large trees have been recklessly felled as part of a hydropower project development, resulting in frequent landslides and soil erosion at the dam and powerhouse sites. In addition, between 1964 and 2009, the region saw many earthquakes with magnitudes of around 4.0. Therefore, these occurrences affect the fragility of forests and the decline in biodiversity, whether directly or indirectly. In the last several years, the maximum average temperature grew by 0.05°, the lowest by 0.07°C, and precipitation totaled 1.48 mm (Kanwar & Kuniyal, 2022).

For isolated populations in the Himalayan, springs provide an important source of reliable, high-quality freshwater supplies. They also support the Himalayas' diverse ecosystems and abundant wildlife. The demands of 60%–70% of the Himalayan inhabitants are directly met by springs in terms of their homes and means of subsistence. Despite this, it has been noted that in the last several decades, around 60% of low-discharge springs have declined. In addition, reports have connected nitrate and faecal coliform pollution to fertilizer use, open defecation, and septic tanks. The survival of these essential resources has been further endangered by a high level of urbanization, which includes 500 expanding townships and 8–10 significant cities, leading to a serious water crisis in the Himalayas. Several sustainable development objectives may be attained, and improved water availability and livelihoods can result from spring revitalization. The effectiveness of such attempts is, however, hampered by several difficulties. The inadequate Knowledge of intricate groundwater (spring) systems and their connections with human cultures is a basic constraint (Verma & Jamwal, 2022). India's agriculture, which relies heavily on the monsoon, is very susceptible to the effects of climate change, and poor policies exacerbate socioeconomic problems in rural areas. While the little study is done to understand the socioeconomic components, especially the views of farmers, many current policies and efforts to mitigate the consequences of climate change are concentrated on the technical aspects of adaptations. Most farmers know climate change and its effects on agriculture and the economy. Most farmers in each of the eight NER states of India have dealt with unexpected weather conditions during the last 10 to 15 years. There is a pressing need for action since many people think that man is mostly to blame for climate change (Bhalerao et al., 2022).

Most families in the sub-basin area of the western Himalayas have seen an increase in the frequency of floods, landslides, droughts, livestock illnesses, and crop pests in recent years and have linked this to climate change. These changes have resulted in poor agricultural output and revenue, notably in Eastern Brahmaputra, where a sizeable percentage reported a fall in the production of almost all cash and staple crops, leading to extremely low farm income. As a result, families' reliance on outside food products delivered from plain regions has grown, especially in the Upper Indus. Following risks, families experience temporary food insecurity as a result of disruptions to their local food systems and sources of income, as well as limited food supplies from other regions. People in this area adjust their agricultural methods and animal management strategies to deal with this. The need for the policy tools to achieve sustainable food security is based on the agro-ecological potential and chances to increase agricultural resilience and variety of livelihoods in that region (Hussain et al., 2016).

Some ethnic groups in the Himalayan area were particularly susceptible to the ongoing industrialization efforts—climate change, industrialization, and their expanded participation in it. Additionally, the establishment of industry in the chosen area has had several negative effects, including worsening air quality due to industrial emissions like greenhouse gas (GHG) emissions and the generation of largely untreated waste, which has further contributed to serious environmental and health issues. The Himalayan regions had a greater rise in health-related problems, particularly respiratory illnesses (Devi & Singh, 2021).

At the local and regional levels, numerous effects of climate change on farmers' livelihoods have been noted. The Himalayan area is fully aware of the impact of climate change, as shown by the fundamental understanding of local farmers. Increased temperatures, agricultural illnesses brought on by the climate, a rise in the frequency of pests, drought and floods, and a reduction in rainfall are all clear signs of climate change for the basin. These variables were thought to negatively affect agricultural output (89.4%), human health (82.5%), livestock (68.7%), and vegetation (52.1%), among other things. A rising temperature and a declining trend in rainfall were the two main climatic trends detected for all of the physiographic areas (Paudel et al., 2021). Moreover, the security of the water resources system in the Himalayan area is now challenged by changes in socioeconomic growth and climatic circumstances. The worry included the effects of socioeconomic and climatic changes on the livelihood of future water security in India's Himalayan basin (Dau & Kuntiyawichai, 2021).

### 10.3.6 Livelihood Threats to Ethnic Groups

Climate changes in high altitudes restrict the amount of agriculture possible in the western Himalayas. Using the Gaddi tribe as an example, many households in the Gaddi tribal society seek economic options that mix small-scale agriculture with nomadic pastoralism since agriculture alone is inadequate to ensure family reproduction. However, there are limitations to flexibility, and many Gaddi families are now very susceptible to climate change due to their reliance on grazing resources, rainfed irrigation, and increased susceptibility to severe weather events. Gaddis are being forced into new types of work regularly since they can no longer depend on their traditional livelihood of pastoralism and agriculture. Building roads or working on hydro projects is a risky, unpredictable job that necessitates leaving rural homes and travelling to isolated mountain passes in the Himalayas. The effects of climate change vary by class, caste, and ethnicity in rural Himachal Pradesh; those who are wealthy may choose to leave agriculture. Gaddi households are being driven down new and perilous paths into an increasingly uncertain future due to their long-standing agro-pastoralism occupying an increasingly unstable niche (Axelby & Bulgheroni, 2021).

Mountains are delicate and dangerous terrains with large cryosphere regions. Hindu Kush Himalayan communities depend on the cryosphere. Communities in the area have undergone many changes, including those triggered by climate change, with immediate implications on their lives and livelihoods. Langtang Valley communities in Nepal are suffering socioeconomic changes affecting young ambitions. The increasing distance between civilization and the surrounding cryosphere affects local knowledge transmission and development. Changes in the cryosphere and socioeconomic realm have homogenized livelihood sources, with tourism emerging as the primary source. The local populace now depends on imported food. Long-term dangers for local lives and livelihoods include a rising reliance on tourists, imports for food and other basic requirements, and a lack of a risk mitigation plan (Tuladhar et al., 2021).

### 10.3.7 Human-Animal Conflict Due to Climate Change

A felid known as the snow leopard inhabits the rocky alpine portions of several mountain ranges in South and Central Asia. Due to climate change and the significantly quicker rate of global warming in South Asian mountain ranges, this lonely species requires expansive expenses for its ranges, but as the hydrology of the region changes and tree line moves, its habitat will become smaller and more fragmented. The population of its prey will diminish due to the change of the alpine flora by vegetation and competition with domestic goats, which increases the possibility of confrontation and competition with the common leopard. A smaller lone snow leopard may experience severe stress due to the common leopard's bigger size, greater adaptability, and social organization. The habitat would become smaller, and snow leopards would travel upward or northward to regions in Central Asia; however, their anticipated migratory tendencies are uncertain (Kazmi et al., 2022). In Nepal's Himalayan area, blue sheep are the main prey of the endangered snow leopard. The Sagarmatha (Mt. Everest) National Park, where blue sheep are now missing, has lately seen a return of the snow leopard population, and there is evidence of snow leopard livestock predation there. The reintroduction of blue sheep to this region might be a solution to the potential conflict between people and nature; however, climatic adaptability will play an important role. 7,343 square kilometers (or 49%) of Nepal's 14,603 square kilometers were discovered to have blue sheep. The alpine meadow, pasture, and grassland favored by blue sheep are those with mean annual precipitation between 200 and 1,000 mm, heights between 3,300 m and 5,100 m, and soil combinations of haplumbrepts, dystrochrepts, and cryumbrepts (Aryal et al., 2013).

The Kangchenjunga Landscape, a significant store of biodiversity, confronts many difficulties due to many change-causing factors. One such problem that cuts over social, economic, environmental, national, and international boundaries among the three nations of Bhutan, India, and Nepal is human-wildlife conflict (HWC) brought on by aberrant climatic conditions. It makes it a complicated transboundary issue. A human-made element, namely, the distance to roadways, is the strongest predictor of human-wildlife conflict in around 19% of the landscape's area. Certain protected regions are more vulnerable than others. High HWC zones include the Terai-Duars savanna and grasslands ecoregion (43%) and the Himalayan subtropical pine forest ecoregion (63%), respectively. Due to greater forest fragmentation in low- and mid-elevation zones, conflicts are more likely to arise. Patchy protected areas are also insufficiently large to accommodate large mammals like elephants and tigers. The conflict between humans and animals is shown to differ depending on the elevation and temperature of the terrain, and it is strongly connected with forest fragmentation in the mid hills. So to address the conflict between humans and animals, a comprehensive strategy at the landscape level is required. An efficient way to lessen human-wildlife conflict is to reconnect excellent habitats by repairing fragmented inter- and intranational regions (Sharma et al., 2020).

## 10.4  SCENARIO 3: COASTAL AND MANGROVE AREA

Although there isn't a single, precise definition of a coastal area, it is generally accepted to refer to the region where land and water converge. Around 600 million people, accounting for almost 10% of the global population dwell in coastal areas, which is less than 10 meters above mean sea level (MSL). Also, 2.4 billion people (approx.), accounting for almost 40% of the global population, live within 100 km (60 miles) of the coast (United Nations, 2017), which is also predicted to be home for 50% of the population by 2030 (Small & Nicholls, 2003). The population in the coastal area across the world is increasing (Neumann et al., 2015). Coastal systems consist of both natural and human systems. Beaches, barriers, rocks, dunes, coral reefs, deltas, lagoons, mangrove forests, etc., are examples of natural systems. Human systems include human settlements, infrastructures, agriculture, and transport system. Together, they create a closely connected socio-ecological system (Hopkins et al., 2012). A growing number of natural and human systems are in danger as a result of continuous climate change and its effects.

### 10.4.1 Signs of Climate Change

The world's coastal regions are suffering greatly at the hands of climate change. Coastal systems and low-lying areas are among the many locations that are being impacted by climate change, and they are becoming more and more vulnerable throughout the 21st century (Nicholls et al., 2007). Coastal regions face a number of definite climate disasters such as tropical cyclones, tsunamis, coastal flooding, etc. Coastal areas are particularly vulnerable to changes in rainfall patterns, incursion of saline, variations in sea surface temperature, and rising water levels that cause flooding and inundation. To be precise, climate change is having an impact on this region both directly and indirectly. Direct effects include damage to emergency infrastructure, loss of life owing to frequent tropical cyclones, economic and ecological losses, coastal erosion, destruction of agricultural fields, livestock deaths, etc. Indirect impacts, which include effects like food insecurity, infrastructure destruction, large debt loads on the poor, etc., are just as destructive as direct effects. Impacts of climate change and resultant disasters due to it are serious threats to sustainable livelihood and development. It hampers all the assets included in sustainable livelihood framework viz. human, social, natural, physical, and economical. Loss of lives is the most tragic incident that happens due to climate change. Besides, there is also spread of many communicable diseases due to the floods, water pollution, etc., making the situation more complex. Malnutrition is also one of the major indirect effects affecting the human asset due to the effects of climate change. From economical point of view, people who are lacking in terms of resources, including people under poverty line, small and marginal farmers, fisheries etc., suffer the most. Loss of livelihoods, particularly from the primary sector, creates huge unemployment, and they need to become fully dependent on government aids and grants. In addition, this type of situation is also a huge blow to the psychological state of mind of the sufferers, which often remains unsaid. It reduces their risk-bearing abilities, adaptation capacities, rate of adoption, innovativeness, etc. Thus, growth in population, urbanization, and developmental activities following the path shall make the coastal regions across the world more vulnerable (Hallegatte et al., 2013).

Various direct and indirect impacts due to the impact of climate change as faced by coastal regions is mentioned next (Table 10.2).

### 10.4.2 Actions Responsible for Damage Accelerating Climate Change

Both climate-related drivers and human-related drivers are responsible for the acceleration of climate change in the coastal region, hence increasing the vulnerability of the coastal community.

**TABLE 10.2**

**Direct and Indirect Impacts of Climate Change in Coastal Regions**

| Climatic events | Direct impact | Indirect impact |
|---|---|---|
| Increase in sea level | • Flooding of coastal and low-lying areas<br>• Coastal erosion | • Loss of assets<br>• Salinity intrusion<br>• Loss of human settlement<br>• Degradation of agricultural fields<br>• Huge economic loss |
| Cyclone and storm surge | • Uprooting of trees<br>• Loss of mangrove ecosystem<br>• Flood<br>• Damage to infrastructures such as dykes and embankments, roads, and bridges<br>• Damage to agricultural products (standing/homestead crop)<br>• Death of livestock animals | • Destruction of houses particularly *kaccha* houses<br>• Damage to power supply<br>• Food insecurity<br>• Trap of debt<br>• Unavailability of fresh surface and groundwater<br>• High economic cost |
| Floods | • Loss of assets<br>• Unavailability of basic amenities<br>• Damage to infrastructures<br>• Sharp decline in agricultural activities | • Environmental damage<br>• Spread of infectious disease<br>• Food insecurity |
| Coastal erosion | • Loss of dry land<br>• Loss of soil fertility<br>• Damage to infrastructure | • Decline in agricultural productivity<br>• Food insecurity<br>• Loss of ecosystem particularly mangrove |

### 10.4.2.1 Climate-Related Drivers

There are several climate-related drivers which are responsible for the acceleration of climate change in the coastal region. Some of the major climate-related drivers are tropical cyclones, rise in sea level, sea surface temperature, coastal floods, erosion, etc. Climate-related drivers are having a strong inter-linkage, and one incidence is often either the cause or effect of another. For example, increase in sea surface temperature is often considered as one of the prominent reasons for the generation and growth of a cyclone. Cyclone, again, is a cause for heavy torrential rainfall, flood, coastal erosion, etc.

#### 10.4.2.1.1 Tropical Cyclone

Tropical cyclones are large cyclonically rotating wind systems that are generally forming over tropical or subtropical oceans and mainly concentrate during summer or early autumn months in either or both hemispheres (Korty, 2013). They are the deadliest coastal hazard-causing colossal loss of lives and huge economic tolls due to its devastating winds, floods, huge waves, torrential rainfall, etc. This is such a coastal disaster which brings with itself all other climate-related drivers of climate change. There are almost 90 tropical cyclones formed every year across the world (Murakami et al., 2013). According to the International Workshop on Tropical Cyclones, the intensity and frequency of tropical cyclones have amplified in the previous 35 years as a result of the ongoing warming of sea surface water, which has been beyond the threshold temperature of 28°C as a result of climate change (Dasgupta et al., 2010; Bielli et al., 2021). Over the last 50 years, 1942 disasters have occurred due to tropical cyclones, killing approx. 8 lakhs of people and causing $1,407.6 billion in economic losses—an average of 43 deaths and $78 million in damages every day (World Meteorological Organization, 2020).

Tropical cyclone is a deadly coastal hazard in both developed as well as developing countries. However, it is more vulnerable in developing countries. Developing countries like India and Bangladesh, where agriculture still plays the role of being the major employment sector, have the majority of small and marginal farmers. Coastal regions in such countries are also at par with the natural context with more than 90% of the farmers in coastal regions belonging to smallholder farming communities. They generally are dependent on mixed farming, including three major enterprises viz. crop farming, livestock farming, and aquaculture or fishery. Tropical cyclones bring with it not only immediate disaster, especially for the agricultural sector in coastal regions, but also a long-lasting consequence that trail over. With increase in frequency as well as intensity of the tropical cyclones across the world, with a major reason being the increase in sea-surface temperature, there is a catastrophe in all three categories popularly known as the "triple bottom line," i.e., economic, ecological, and social. Furthermore, the time period to get over the consequences by adopting various adaptation measures and building climate resilient pathways is decreasing, thus making the situation more complex. One example that can be mentioned here is coastal West Bengal is comprising the largest mangrove forest in the world, i.e., Sundarban Biosphere Reserves. It is also known as the "cyclone capital of India." It faced Cyclone "Amphan," a super cyclone formed over the Bay of Bengal, in May 2020. It was a huge catastrophe, causing widespread damage in the region, vandalizing natural resources, built up infrastructures, etc. Again, in May 2021, Cyclone "Yaas" attacked the same region, and again, a turmoil of similar nature was observed.

### 10.4.2.1.2  Coastal Erosion

Another important driver for climate change in coastal region is coastal erosion. It is having direct impacts on the coastal community, hampering regular source of income and income-generating activities. Coastal erosion leads to the destruction of houses, agricultural fields, grazing land, etc., putting the stake of the poor people in coastal areas at high risk. It also causes salinity intrusion, decreasing soil fertility and soil productivity, which in the long run has a high toll on the food security of the region as well as snatches away the livelihood of many, causing high unemployment (Islam et al., 2012). Erosion of the riverbanks, especially during the monsoon season, has a long-standing effect that is difficult to manage for the coastal community (Alam, 2016). Loss of houses and basic amenities due to the coastal erosion and riverbank erosion enhances the vulnerability of the people.

### 10.4.2.1.3  Sea Level Rise

Sea level rise is both a cause and effect of climate change. It happens due to combined factors, including severe storms, tides, glacier melting, and wind waves. It is also a salient cause for coastal erosion and inundation. This is a threat to the coastal community as well as for the marine lives due to the loss of balance in their ecosystem and their breeding ground. Mangrove forests are also highly hampered with rise in sea level. Increase in sea level is gradually destroying the natural resources, agriculture, water resources and ecosystem, and aquaculture, increasing the salinity of the region by flooding in the coastal region and thus, worsening the effect of climate change on smallholder farming community.

### 10.4.2.1.4  Sea Surface Temperature

There has been a significant rise in the temperature of the sea surface over the last 30 years along more than 70% of the world's coastlines (Lima & Wethey, 2012). Like other climate-related drivers, it is also one of the potential causes of tropical cyclones and has both direct and indirect effects on the natural as well as human resources and natural resources of the coastal community. It melts the sea ice at higher latitudes and bleaches the coral reefs, leading to their mortality. Besides, it also increases algal bloom in the sea water, a threat to the marine ecosystem as a direct effect. And as an indirect effect, it multiplies the loss of livelihood of the fisheries. Increase in sea surface temperature also induces the migration of poleward species. Increase in sea surface temperature also plays a destructive role in reducing fresh water and groundwater quality on which coastal communities are directly dependent for daily use as well as agricultural activities. They also use it for the cultivation of freshwater fishes, a potential target of increase in sea surface temperature.

### 10.4.2.2  Human-Related Drivers

Beside climate-related drivers, climate change in the coastal region is also highly impacted as a result of various human activities. The coastal region is subjected to a diverse range of human-related and anthropogenic activities (Crain et al., 2009), which in interaction with climate-related drivers have a confounding effort to increase the impacts of climate change in the region.

Socioeconomic developmental activities carried out by human beings plays a catalytic role in climatic hazards in the coastal region. Various socioeconomic development activities include activities for occupational needs, activities for educational purposes, cultural purposes, building of infrastructures, etc., having impacts in disturbing the balance of the coastal ecosystem. Faster growth in coastal population automatically leads to booming of industrialization, urbanization, and coastal migration. There is a gradual change in the land use pattern of the region and are driven by an amalgamation of all such factors, including social, economic, and institutional. Continuous socioeconomic development in the region also has a role to play in influencing the capacity to adapt of the people living in coastal regions. The top five countries in the world, classified based on the populations living in coastal and low-lying areas, are Bangladesh, China, Vietnam, India, and Indonesia (Jongman et al., 2012), majority of which are developing countries and are newly industrialized countries.

The destruction of coastal wetlands for carrying out activities such as agriculture, human settlement, etc., makes coastal communities more vulnerable towards climate change and its associated consequences. The ecosystem in wetlands acts as a buffer to severe impacts of storm, tidal waves, coastal erosion, etc. For example, the mangrove forest located in coastal West Bengal in India and in coastal Bangladesh plays a crucial role in minimizing the losses and damages due to the impacts of tropical cyclones. Agricultural residues include runoff water, which contains various chemical nutrients after getting dumped in sea water or rivers in coastal regions. This causes eutrophication, and as a result, there is a decomposition of organic matter. This is a primary cause of decrease in oxygen concentration in the marine ecosystem, also known as hypoxia. The gradual decline of oxygen concentration and the simultaneous spike in sea water temperature, decreasing oxygen solubility in water (Shaffer et al., 2009), lead to an increasing threat for marine ecosystems. Although it is a secondary driver for climate change, it is locally very important. Waste generated from industrial activities as well as from households and other human activities gets dumped into rivers and oceans in coastal regions, creating pollution and loss of marine lives. Human activities such as creating dams and mining of sand and gravels in river channels contribute to the significant decline in sediment delivery and speed up coastal erosion.

## 10.4.3 EFFECT OF CLIMATE CHANGE IN THE COASTAL REGION

The coastal community is impacted by climate change in a variety of ways. Its effects are shown on the triple bottom line, which includes the social, economic, and environmental aspects. People with poor socioeconomic background are more likely to be impacted by the effects of climate change in coastal communities. Communities belonging to climatically vulnerable regions are not only more vulnerable to climate related hazards but are also able to pursue less adaptation measures since adaptation interventions adopted by these community depend upon various factors, such as demographic factor, economic factor, technological factor, social factor, etc.

### 10.4.3.1 Effect on Environmental Aspects

#### 10.4.3.1.1 Impacts on Beaches and Coastlines

Vulnerable coastal and marine areas as well as the structure and operation of their ecosystems could suffer greatly from the effects of climate change. Sea level rise (1.7 mm/year) alters the form of coastlines, exacerbates coastal erosion, causes flooding, and increases subsurface saltwater intrusion. Vulnerable coastal and marine areas as well as the structure and operation of their ecosystems could suffer greatly from the effects of climate change. The coastal ecology will suffer greatly as a result, with the destruction of mangrove forests, coastal species, and natural food habits and a decline in the supply of fresh surface water and groundwater. Increase in the sea level shall also lead to salinity intrusion and loss of natural fertility of the soil. Infrastructures like dykes and embankments are once again under danger due to the increase in the frequency and intensity of tropical cyclones that strike the coastal region, as well as their aftereffects. Floods have a higher likelihood of occurring when dykes and embankments are damaged. Under increasing sea levels, certain coastal systems may be able to move inland, but others will face coastal squeeze, which happens when an eroding shoreline comes close to hard, immovable features like seawalls or resilient natural cliffs. Due to the consequent sediment deficiency, in these situations, the beaches will narrow, which will have negative effects, including destroying habitats and lowering the capacity of various organisms to survive (Jackson & McIlvenny, 2011).

### 10.4.3.1.2  Impacts on Wetlands

The key ecosystem of the coastal areas consists of vegetated coastal habitats and coastal wetlands. Seagrass beds, marshy areas, and mangrove forests are examples of coastal wetlands. According to numerous studies, the coastal habitat is gradually deteriorating as a result of rising sea levels and more intense wave activity (Alongi, 2008). Wetlands, which have historically served as a buffer zone to protect coastal communities from various climatic risks, including storms, floods, and cyclones, become more susceptible to the effects of climate change and lose their ability to do so. It also leads to the loss of carbon stored in the sediments. Organic deposits emit 0.04 to 0.28 PgC per year as a result of the loss of coastal wetlands and seagrass meadows (Pendleton et al., 2012).

### 10.4.3.1.3  Impact on Coral Reef

Coral reefs are shallow-water habitats consisting of calcium carbonate released by algae and corals that form reefs. Climate change poses serious threat to the one of the most diverse ecosystems. A serious effect of climate change on this environment is the mass bleaching of coral reefs, which is compounded by an increase in sea surface temperature (Kleypas et al., 2008). Maintaining a balance between calcium carbonate production and erosion is essential for coral reef survival. Ocean acidification poses a threat to the entire system since it promotes calcium carbonate–erosive processes and prevents the synthesis of the mineral (Andersson & Mackenzie, 2011; Kroeker et al., 2013; Wisshak et al., 2012), leading to a poorly cemented coral reef (Manzello et al., 2008). Coral reefs are, therefore, vanishing more quickly as a result of climate change.

### 10.4.3.1.4  Impact on Coastal Aquifers

Groundwater networks that cross land-ocean boundaries are known as coastal aquifers. This plays a strategic role in supplying of fresh water to the coastal community, particularly to small islands. Disasters brought on by climate change, such as cyclones, floods, severe rain, rising sea levels, etc., are important triggers for salinity intrusion in the coastal region. This lowers the quality of coastal aquifers, and the lack of fresh surface and groundwater has emerged as one of the primary issues in coastal communities. In addition to climate change, human-caused factors including excessive groundwater extraction, mining, urbanization, pollution, etc., have had a substantial impact on the water quality in coastal aquifers.

### 10.4.3.2  Effect on Economic Aspects

#### 10.4.3.2.1  Impact on Livelihood of the Coastal Community

Community members who live in coastal areas rely heavily on the primary sector for their source of income. Agriculture and related industries like crop, livestock, and fisheries are included in the primary sector. For residents in coastal communities, the tourism industry is also a key source of income. All of the livelihood alternatives present in the primary sector are seriously threatened by climate change. There are two basic categories of economic effects of climate change on coastal communities: direct and indirect effects (Bosello et al., 2012). Direct effects include the whole or partial devastation of homestead crops, the mass death of livestock, and a scarcity of food and fodder (Goswami et al., 2021). Aftermath consequences of cyclone including flood, erosion, etc., causes salinity intrusion and loss of soil fertility. It also leads to pest invasion of the crop and reduces productivity of agricultural as well as horticultural crops (Singh et al., 2019), as well as loss of human lives. Indirect impacts of climate disasters on coastal community are more deadly than the former. Food insecurity prevails over the region for a significant period of time. Nevertheless, scarcity of essential commodities in the market hampers the daily lifestyle of the people living in the coastal community. Smallholder farmers also fall into the trap of debt. They are unable to repay their loan which they had borrowed for their agricultural and allied activities. The situation is further worse in developing countries such as India, Bangladesh, etc., where majority of the loan amount borrowed is from unorganized financial sources with a high amount of loan interest levied on the amount. Other indirect impacts of climate change on the agricultural sector of the coastal region includes pest invasion and loss of soil fertility. This increases the application of chemical fertilizers and pesticides which again triggers eutrophication and hampers the environment.

### 10.4.3.2.2 Impact on Coastal Tourism

Coastal tourism is one of the largest contributors of tourism industry across the globe. More than 100 nations profit from the recreational value that their coral reefs provide, generating $11.5 billion in international tourism (Burke et al., 2011). However, climate change and its consequences pose a serious threat to this dynamic industry. It gets directly impacted due to the loss of infrastructures, roads, bridges, beaches, etc. Indirect impact of climate change on coastal tourism includes gradual destruction of coral reefs, coastal erosion, loss of mangrove biodiversity, loss of coastal wildlife, etc. Also, there is a short-term change in the perception of the tourists soon after the occurrence of any climate disaster, such as a cyclone, flood, storm surge, etc., and this brings an overall effect of the financial reserve of the people belonging to the coastal community, where tourism is a significant livelihood option (Phillips & Jones, 2006).

### 10.4.3.2.3 Impact on Human Settlement

Human settlement is highly affected due to the consequences of climate change. Gradual loss of dry land to soil erosion, submergence, and damage on built environments is due to extreme climate events. It proves to be more vulnerable for the resource-poor person belonging to the coastal community, particularly smallholder farmers, fishermen, etc. Climate change also deteriorates water quality and disrupts energy supply, transportation, and communication systems. They lack financial reserves to rebuild their lost property and often fall in the trap of indebtedness and left with no other option other than migration.

### *10.4.3.2.4 Impact on Industrialization*

Industrialization status of a region depends on various factors such as availability of land, skilled and unskilled labourers, smooth transportation system, uninterrupted power and water supply, sewage system, etc. Consequences of climate change and related disasters are threats to all the requisite factors for industrialization. Damage on available infrastructures is having a serious devastation effect on the operations of all the sectors related to industrialization due to extreme weather events such as high precipitation, storm surges, and floods (Handmer et al., 2012). Besides, labourers associated with industries in the coastal region, particularly in developing countries, opt for migration to big cities due to severe impacts on human settlement.

### 10.4.3.3 Effect on Social Aspects

### *10.4.3.3.1 Impact on Health and Well-Being*

Coastal areas are generally overpopulated and thus is often a potential target of climate change and consequences with respect to the health and well-being of the people belonging to coastal communities. Climate change and disasters often hamper the availability and quality of the basic necessities of the people, including nutritious food, fresh drinking water, emergency care, etc. An increase in sea water levels, salinity intrusion, deteriorating quality of fresh groundwater, and many other related hazards enhance the risk of health vulnerability due to the spread of germs and microbes and cause various infectious diseases such as cholera, diarrhea, malaria, etc. Climate change is also an important reason for the decline in soil fertility, and the situation facilitates the use of more and more chemical fertilizers and pesticides, causing significant reduction in the nutrition value of the agricultural products. It is also a burning reason for an increase in eutrophication. Climate disasters, such as cyclones, floods, high storm surges, tsunamis, etc., bring the situation to the extreme worst, and the marginal and resource-poor people often have their houses washed away, with no availability of food, drinking water, and health care amenities. This situation is highly responsible for the spread of various deadly vector-borne diseases.

### *10.4.3.3.2 Impact on Psychology*

Psychology is something which is always very underrated, and particularly in the aspect of climate disaster, it is under the least consideration for advocating policies. There have been numerous disaster risk reduction approaches to meet the negative impacts of climate change and climate disasters in the coastal community. However, majority of them focused mainly on physical components and mitigation and adaptation strategies and have critically missed out human factors, social factors, and cultural factors (Mercer et al., 2007). Vital disaster preparedness measures of a community depend on major three themes, i.e., personal protection, practical preparedness, and social preparedness. Communities living in climatically vulnerable regions and depending on climatic sensitive livelihoods are not only more vulnerable to climate change impacts but are also less able to pursue adaptation measurers (Asfaw et al., 2019). Adaptation interventions generally advocated under policy for intervention should be identified and implemented based on participatory approach method and not by top-bottom approach since adaptation interventions get influenced by various factors such as technological factors, economic factors, demographic factors, environmental factors, and social factors. What often remains unsaid is that although belonging from a same coastal community, people are heterogenous in nature. This is particularly true for resource poor persons depending on climate sensitive livelihood such as smallholder farmers, fishermen, livestock farmers, and daily wage labourers. They differ from each other with respect to knowledge, traditional belief, risk-bearing abilities, adaptation behavior, etc. Henceforth, their outlook on the disaster and its adaptation strategies also varies. If suitable intervention measure for overcoming the climate change, disaster, and consequences are not advocated, it may hamper the psychology of the affected and constraints shall prevail in overcoming the psychological barrier of the climate change.

## 10.5  CONCLUSION

The cause and impact of climate change goes far beyond what is generally perceived by the common people or even by the scientific community. The threats of climate change issues perceived by different places are often linked with the inherent vulnerability of such places, and often, the most fragile ecosystems and its members bear the highest cost. Hence, three of the most challenging places—i.e., deep rainforests, high mountains, and mangrove areas—are getting severely affected by climate change consequences. The phenomenon often becomes very complex, as causes and impacts are intervened together. While the climate change brings drastic change to the ecosystem and habitat of those areas, the people tend to migrate in other places, leading to even more pressure on the environment. The insecurity of livelihood, migration, and conflicts with wildlife are some of the indirect yet grave consequences of climate change.

## ACKNOWLEDGEMENT

VDR acknowledge the support of Strategic Academic Leadership Program of the Southern Federal University ("Priority2030")

## REFERENCES

Alam, G. M. M. (2016). *An assessment of the livelihood vulnerability of the riverbank erosion hazard and its impact on food security for rural households in Bangladesh*. PhD Thesis, February, 1–216. https://eprints.usq.edu.au/34160/1/Alam_2016_whole.pdf

Ali, A., Khan, M. Z., Khan, B., & Ali, G. (2022). Migration, remittances and climate resilience: Do financial literacy and disaster risk reduction orientation help to improve adaptive capacity in Pakistan? *Geo Journal*, 88, 595–61. https://doi.org/10.1007/s10708-022-10631-6

Ali, A., Rahut, D. B., Mottaleb, K. A., & Erenstein, O. (2017). Impacts of changing weather patterns on smallholder well-being: Evidence from the Himalayan region of northern Pakistan. *International Journal of Climate Change Strategies and Management*, 9(2), 225–240. https://doi.org/10.1108/IJCCSM-05-2016-0057

Alongi, D. M. (2008). Mangrove forests: Resilience, protection from tsunamis, and responses to global climate change. *Estuarine, Coastal and Shelf Science*, 76(1), 1–13. https://doi.org/10.1016/J.ECSS.2007.08.024

Andersson, A. J., & Mackenzie, F. T. (2011). Effects of ocean acidification on benthic processes, organisms, and ecosystems. In *Ocean acidification*. Oxford Academic. https://doi.org/10.1093/oso/9780199591091.003.0012.

Arora, V., & Jayaram, N. (2017). Democratisation in the Himalayas: Interests, conflicts, and negotiations. In *Democratisation in the Himalayas: Interests, conflicts, and negotiations*. Taylor and Francis. https://doi.org/10.4324/9781315276984

Aryal, A., Brunton, D., & Raubenheimer, D. (2013). Habitat assessment for the translocation of blue sheep to maintain a viable snow leopard population in the Mt Everest Region, Nepal. *Zoology and Ecology, 23*(1), 66–82. https://doi.org/10.1080/21658005.2013.765634

Asfaw, A., Simane, B., Bantider, A., & Hassen, A. (2019). Determinants in the adoption of climate change adaptation strategies: evidence from rainfed-dependent smallholder farmers in north-central Ethiopia (Woleka sub-basin). *Environment, Development and Sustainability, 21*(5), 2535–2565. https://doi.org/10.1007/S10668-018-0150-Y/TABLES/8

Asfaw, A., Simane, B., Hassen, A., & Bantider, A. (2018). Variability and time series trend analysis of rainfall and temperature in northcentral Ethiopia: A case study in Woleka sub-basin. *Weather and Climate Extremes, 19*, 29–41. https://doi.org/10.1016/J.WACE.2017.12.002

Axelby, R., & Bulgheroni, M. (2021). Old ways and new routes: Climate threats and adaptive possibilities in the Indian Himalayas. In *Climate change in the global workplace: Labour, adaptation and resistance*. Taylor and Francis. https://doi.org/10.4324/9780367822903-7

Barlow, J., Berenguer, E., Carmenta, R., & França, F. (2020). Clarifying Amazonia's burning crisis. *Global Change Biology, 26*(2), 319–321. https://doi.org/10.1111/GCB.14872

Bhalerao, A. K., Rasche, L., Scheffran, J., & Schneider, U. A. (2022). Sustainable agriculture in Northeastern India: How do tribal farmers perceive and respond to climate change? *International Journal of Sustainable Development and World Ecology, 29*(4), 291–302. https://doi.org/10.1080/13504509.2021.1986750

Bielli, S., Barthe, C., Bousquet, O., Tulet, P., & Pianezze, J. (2021). The effect of atmosphere-ocean coupling on the structure and intensity of tropical cyclone Bejisa in the Southwest Indian Ocean. *Atmosphere, 12*(6), 688. https://doi.org/10.3390/ATMOS12060688

Bisht, J. K., Meena, V. S., Mishra, P. K., & Pattanayak, A. (2016). *Conservation agriculture: An approach to combat climate change in Indian Himalaya*. Springer. https://doi.org/10.1007/978-981-10-2558-7

Bosello, F., Nicholls, R. J., Richards, J., Roson, R., & Tol, R. S. J. (2012). Economic impacts of climate change in Europe: Sea-level rise. *Climatic Change, 112*(1), 63–81. https://doi.org/10.1007/S10584-011-0340-1/TABLES/8

Burke, L., Reytar, K., Spalding, M., & Perry, A. (2011). Reefs at Risk Revisited. In *Defenders* (Vol. 74, Issue 3). www.pubmedcentral.nih.gov/articlerender.fcgi?artid=3150666&tool=pmcentrez&rendertype=abstract

Confalonieri, U. (2000). Environmental change and human health in the Brazilian Amazon. *Global Change and Human Health, 1*(2), 174–183. https://doi.org/10.1023/A:1010081206165

Corlett, R. T., & Primack, R. B. (2006). Tropical rainforests and the need for cross-continental comparisons. *Trends in Ecology and Evolution, 21*(2), 104–110. https://doi.org/10.1016/j.tree.2005.12.002

Costa, D. F., Gomes, H. B., Silva, M. C. L., & Zhou, L. (2022). The most extreme heat waves in Amazonia happened under extreme dryness. *Climate Dynamics, 59*(1), 281–295. https://doi.org/10.1007/S00382-021-06134-8

Crain, C. M., Halpern, B. S., Beck, M. W., & Kappel, C. V. (2009). Understanding and managing human threats to the coastal marine environment. *Annals of the New York Academy of Sciences, 1162*, 39–62. https://doi.org/10.1111/J.1749-6632.2009.04496.X

Dasgupta, P., Sahay, S., Prakash, A., & Lutz, A. (2020). Cost effective adaptation to flood: sanitation interventions in the Gandak river basin, India. *Climate and Development*, *12*(8), 717–729. https://doi.org/10.1080/17565529.2019.1682490

Dasgupta, S., Huq, M., Khan, Z. H., Ahmed, M. M. Z., Mukherjee, N., Khan, M. F., & Pandey, K. (2010). Vulnerability of Bangladesh to cyclones in a changing climate potential damages and adaptation cost. *World*, *16*(April), 54. https://doi.org/10.1596/1813-9450-5280

Datta, P., & Behera, B. (2022). Assessment of adaptive capacity and adaptation to climate change in the farming households of Eastern Himalayan foothills of West Bengal, India. *Environmental Challenges*, *7*. https://doi.org/10.1016/j.envc.2022.100462

Dau, Q. V, & Kuntiyawichai, K. (2021). Assessment of potential impacts of climate and socio-economic changes on future water security in the Himalayas, India. *Geographia Technica*, *16*(2), 1–18. https://doi.org/10.21163/GT_2021.162.01

De Castro, A. P., Quirino, B. F., Pappas, G., Kurokawa, A. S., Neto, E. L., & Krüger, R. H. (2008). Diversity of soil fungal communities of Cerrado and its closely surrounding agriculture fields. *Archives of Microbiology*, *190*(2), 129–139. https://doi.org/10.1007/S00203-008-0374-6

Devi, N., & Singh, H. (2021). Socio-environmental and industrial vulnerability analysis of the communities in the Shiwalik region of Indian Himalayas foothills. *Trees, Forests and People*, *6*. https://doi.org/10.1016/j.tfp.2021.100155

Eriksen, S. H., & O'Brien, K. (2007). Vulnerability, poverty and the need for sustainable adaptation measures. *Climate Policy*, *7*(4), 337–352. https://doi.org/10.1080/14693062.2007.9685660

Feng, X., Merow, C., Liu, Z., Park, D. S., Roehrdanz, P. R., Maitner, B., Newman, E. A., Boyle, B. L., Lien, A., Burger, J. R., Pires, M. M., Brando, P. M., Bush, M. B., McMichael, C. N. H., Neves, D. M., Nikolopoulos, E. I., Saleska, S. R., Hannah, L., Breshears, D. D., . . . Enquist, B. J. (2021). How deregulation, drought and increasing fire impact Amazonian biodiversity. *Nature*, *597*(7877), 516–521. https://doi.org/10.1038/s41586-021-03876-7

Fonseca de Souza, L., Alvarez, D. O., Domeignoz-Horta, L. A., Gomes, F. V., de Souza Almeida, C., Merloti, L. F., Mendes, L. W., Andreote, F. D., Bohannan, B. J. M., Mazza Rodrigues, J. L., Nüsslein, K., & Tsai, S. M. (2022). Maintaining grass coverage increases methane uptake in Amazonian pastures, with a reduction of methanogenic archaea in the rhizosphere. *Science of The Total Environment*, *838*, 156225. https://doi.org/10.1016/J.SCITOTENV.2022.156225

Fracetto, G. G. M., Azevedo, L. C. B., Fracetto, F. J. C., Andreote, F. D., Lambais, M. R., & Pfenning, L. H. (2013). Impact of Amazon land use on the community of soil fungi. *Scientia Agricola*, *70*(2), 59–67. https://doi.org/10.1590/S0103-90162013000200001

Gardoni, P., Murphy, C., & Rowell, A. (2016). Risk analysis of natural hazards: Interdisciplinary challenges and integrated solutions. In *Risk analysis of natural hazards: Interdisciplinary challenges and integrated solutions*. Springer International Publishing. https://doi.org/10.1007/978-3-319-22126-7_1/COVER

Gautam, Y. (2017). Seasonal migration and livelihood resilience in the face of climate change in Nepal. *Mountain Research and Development*, *37*(4), 436–445. https://doi.org/10.1659/MRD-JOURNAL-D-17-00035.1

Goswami, R., Roy, K., Dutta, S., Ray, K., Sarkar, S., Brahmachari, K., Nanda, M. K., Mainuddin, M., Banerjee, H., Timsina, J., & Majumdar, K. (2021). Multi-faceted impact and outcome of COVID-19 on smallholder agricultural systems: Integrating qualitative research and fuzzy cognitive mapping to explore resilient strategies. *Agricultural Systems*, *189*, 103051. https://doi.org/10.1016/J.AGSY.2021.103051

Gray, C. L. (2009). Environment, land, and rural out-migration in the Southern Ecuadorian Andes. *World Development*, *37*(2), 457–468. https://doi.org/10.1016/J.WORLDDEV.2008.05.004

Hallegatte, S., Green, C., Nicholls, R. J., & Corfee-Morlot, J. (2013). Future flood losses in major coastal cities. *Nature Climate Change, 3*(9), 802–806. https://doi.org/10.1038/nclimate1979

Handmer, J., Honda, Y., Kundzewicz, Z. W., Arnell, N., Benito, G., Hatfield, J., Mohamed, I. F., Peduzzi, P., Wu, S., Sherstyukov, B., Takahashi, K., Yan, Z., Vicuna, S., Suarez, A., Abdulla, A., Bouwer, L. M., Campbell, J., Hashizume, M., Hattermann, F., . . . Yamano, H. (2012). Changes in impacts of climate extremes: Human systems and ecosystems. In *Managing the risks of extreme events and disasters to advance climate change adaptation: Special report of the intergovernmental panel on climate change* (Vol. 9781107025, pp. 231–290). IPCC. https://doi.org/10.1017/CBO9781139177245.007

Hazarika, S., & Banerjee, R. (2017). Gender, poverty and livelihood in the eastern Himalayas. In *Gender, poverty and livelihood in the Eastern Himalayas*. Taylor and Francis. https://doi.org/10.4324/9781315107943

Hopkins, T. S., Bailly, D., Elmgren, R., Glegg, G., Sandberg, A., Støttrup, J. G., Hopkins, T. S., Bailly, D., Elmgren, R., Glegg, G., Sandberg, A., & Støttrup, J. G. (2012). A systems approach framework for the transition to sustainable development: Potential value based on coastal experiments. *Ecology and Society, Published Online* (September 29, 2012), *17*(3). https://doi.org/10.5751/ES-05266-170339

Hussain, A., Rasul, G., Mahapatra, B., & Tuladhar, S. (2016). Household food security in the face of climate change in the Hindu-Kush Himalayan region. *Food Security, 8*(5), 921–937. https://doi.org/10.1007/s12571-016-0607-5

Ichikawa, M. (2014). Forest conservation and indigenous peoples in the Congo Basin: New trends toward reconciliation between global issues and local interest. In *Hunter-Gathers of the Congo Basin* (pp. 321–342). Routledge. https://doi.org/10.4324/9780203789438-12

IPCC. (2018). Annex I: Glossary. In V. Masson-Delmotte, P. Zhai, H.-O. Pörtner, D. Roberts, J. S. Roberts, P. R. Shukla, A. Pirani, W. Moufouma-Okia, C. Pean, R. Pidcock, S. Connors, J. B. R. Matthews, Y. Chen, X. Zhou, M., Gomis, E. Lonnoy, T. Maycock, M. Tignor, & T. Waterfield (Eds.), *Global warming of 1.5°C. An IPCC special report on the impacts of global warming of 1.5°C above pre-industrial levels and related global greenhouse gas emission pathways, in the context of strengthening the global response to the threat of climate change* . . . (pp. 541–562). Cambridge University Press. https://doi.org/10.1017/9781009157940.008

Islam, M. S., Hasan, T., IR Chowdhury, M. S., Rahaman, M. H., & Tusher, T. R. (2012). Coping techniques of local people to flood and river erosion in Char Areas of Bangladesh. *Journal of Environmental Science and Natural Resources, 5*(2), 251–261. https://doi.org/10.3329/JESNR.V5I2.14827

Jackson, A. C., & McIlvenny, J. (2011). Coastal squeeze on rocky shores in northern Scotland and some possible ecological impacts. *Journal of Experimental Marine Biology and Ecology, 400*(1–2), 314–321. https://doi.org/10.1016/J.JEMBE.2011.02.012

Jesus, E. D. C., Marsh, T. L., Tiedje, J. M., & Moreira, F. M. D. S. (2009). Changes in land use alter the structure of bacterial communities in Western Amazon soils. *The ISME Journal, 3*(9), 1004–1011. https://doi.org/10.1038/ismej.2009.47

Jongman, B., Ward, P. J., & Aerts, J. C. J. H. (2012). Global exposure to river and coastal flooding—long term trends and changes. *Global Environmental Change, 22*(4), 823–835. https://doi.org/10.1016/J.GLOENVCHA.2012.07.004

Kanwar, N., & Kuniyal, J. C. (2022). Vulnerability assessment of forest ecosystems focusing on climate change, hazards and anthropogenic pressures in the cold desert of Kinnaur district, Northwestern Indian Himalaya. *Journal of Earth System Science, 131*(1). https://doi.org/10.1007/s12040-021-01775-z

Kazmi, F. A., Shafique, F., Hassan, M. U., Khalid, S., Ali, N., Akbar, N., Batool, K., Khalid, M., & Khawaja, S. (2022). Ecological impacts of climate change on the snow leopard (Panthera unica) in south asia [Impactos ecológicos da mudança climática no leopardo-da-neve (Panthera uncia) no sul da ásia]. *Brazilian Journal of Biology*, 82. https://doi.org/10.1590/1519-6984.240219

Kelly, P. M., & Adger, W. N. (2000). Theory and practice in assessing vulnerability to climatechange andfacilitating adaptation. *Climatic Change*, *47*(4), 325–352. https://doi.org/10.1023/A:1005627828199

Khadka, M., Rasul, G., Bennett, L., Wahid, S. M., & Gerlitz, J.-Y. (2015). Gender and social equity in climate change adaptation in the Koshi basin: An analysis for action. In *Handbook of climate change adaptation*. Springer. https://doi.org/10.1007/978-3-642-38670-1_78

Kleypas, J. A., Danabasoglu, G., & Lough, J. M. (2008). Potential role of the ocean thermostat in determining regional differences in coral reef bleaching events. *Geophysical Research Letters*, *35*(3), 3613. https://doi.org/10.1029/2007GL032257

Korty, R. (2013). Hurricane (Typhoon, Cyclone). In P. T. Bobrowsky (Ed.), *Encyclopedia of natural hazards: Encyclopedia of earth sciences series* (pp. 481–493). Springer. https://doi.org/10.1007/978-1-4020-4399-4_175

Koshy, K., Stevenson, L. A., Boonjawat, J., Campbell, J. R., Ebi, K. L., Lotia, H., & Zondervan, R. (2014). Climate and society. *Advances in Global Change Research*, *56*, 199–252. https://doi.org/10.1007/978-94-007-7338-7_5

Kroeger, M. E., Delmont, T. O., Eren, A. M., Meyer, K. M., Guo, J., Khan, K., Rodrigues, J. L. M., Bohannan, B. J. M., Tringe, S. G., Borges, C. D., Tiedje, J. M., Tsai, S. M., & Nüsslein, K. (2018). New biological insights into how deforestation in amazonia affects soil microbial communities using metagenomics and metagenome-assembled genomes. *Frontiers in Microbiology*, *9*(July), 1635. https://doi.org/10.3389/FMICB.2018.01635/BIBTEX

Kroeker, K. J., Kordas, R. L., Crim, R., Hendriks, I. E., Ramajo, L., Singh, G. S., Duarte, C. M., & Gattuso, J. P. (2013). Impacts of ocean acidification on marine organisms: Quantifying sensitivities and interaction with warming. *Global Change Biology*, *19*(6), 1884–1896. https://doi.org/10.1111/GCB.12179

Kunwar, R. M., Fadiman, M., Cameron, M., Bussmann, R. W., Thapa-Magar, K. B., Rimal, B., & Sapkota, P. (2018). Cross-cultural comparison of plant use knowledge in Baitadi and Darchula districts, Nepal Himalaya. *Journal of Ethnobiology and Ethnomedicine*, *14*(1). https://doi.org/10.1186/s13002-018-0242-7

Kursar, T. A., & Coley, P. D. (2003). Convergence in defense syndromes of young leaves in tropical rainforests. *Biochemical Systematics and Ecology*, *31*(8), 929–949. https://doi.org/10.1016/S0305-1978(03)00087-5

Leichenko, R., & Silva, J. A. (2014). Climate change and poverty: Vulnerability, impacts, and alleviation strategies. *Wiley Interdisciplinary Reviews: Climate Change*, *5*(4), 539–556. https://doi.org/10.1002/WCC.287

Leite-Filho, A. T., Soares-Filho, B. S., Davis, J. L., Abrahão, G. M., & Börner, J. (2021). Deforestation reduces rainfall and agricultural revenues in the Brazilian Amazon. *Nature Communications*, *12*(1), 1–7. https://doi.org/10.1038/s41467-021-22840-7

Lima, F. P., & Wethey, D. S. (2012). Three decades of high-resolution coastal sea surface temperatures reveal more than warming. *Nature Communications*, *3*(1), 1–13. https://doi.org/10.1038/ncomms1713

Maikhuri, R. K., Rawat, L. S., Nautiyal, S., Negi, V. S., Pharswan, D. S., & Phondani, P. (2013). Promoting and enhancing sustainable livelihood options as an adaptive strategy to reduce vulnerability and increase resilience to climate change impact in the Central Himalaya. *Environmental Science and Engineering*, 555–574. https://doi.org/10.1007/978-3-642-36143-2_32

Malhi, Y., Roberts, J. T., Betts, R. A., Killeen, T. J., Li, W., & Nobre, C. A. (2008). Climate change, deforestation, and the fate of the Amazon. *Science*, *319*(5860), 169–172. https://doi.org/10.1126/SCIENCE.1146961/SUPPL_FILE/MALHI_SOM.PDF

Mali, S., & Borges, R. M. (2003). Phenolics, fibre, alkaloids, saponins, and cyanogenic glycosides in a seasonal cloud forest in India. *Biochemical Systematics and Ecology, 31*(11), 1221–1246. https://doi.org/10.1016/S0305-1978(03)00079-6

Manzello, D. P., Kleypas, J. A., Budd, D. A., Eakin, C. M., Glynn, P. W., & Langdon, C. (2008). Poorly cemented coral reefs of the eastern tropical Pacific: Possible insights into reef development in a high-$CO_2$ world. *Proceedings of the National Academy of Sciences of the United States of America, 105*(30), 10450–10455. https://doi.org/10.1073/PNAS.0712167105/ASSET/839C0BB9-3995-4ECD-9B40-EB2C590F9ABA/ASSETS/GRAPHIC/ZPQ9990839920007.JPEG

Massey, D. S., Axinn, W. G., & Ghimire, D. J. (2010). Environmental change and out-migration: Evidence from Nepal. *Population and Environment, 32*(2), 109–136. https://doi.org/10.1007/S11111-010-0119-8

Mauri, J., Helena Techio, V., Chamma Davide, L., Laís Pereira, D., Souza Sobrinho, F., & José Pereira, F. (2015). Forage quality in cultivars of Brachiaria spp.: Association of lignin and fibers with anatomical characteristics. *AJCS, 9*(12), 1148–1153.

Mercer, J., Dominey-Howes, D., Kelman, I., & Lloyd, K. (2007). The potential for combining indigenous and western knowledge in reducing vulnerability to environmental hazards in small island developing states. *Environmental Hazards, 7*(4), 245–256. https://doi.org/10.1016/J.ENVHAZ.2006.11.001

Moreta, D. E., Arango, J., Sotelo, M., Vergara, D., Rincón, A., Ishitani, M., Castro, A., Miles, J., Peters, M., Tohme, J., Subbarao, G. V., & Rao, I. M. (2014). Biological nitrification inhibition (BNI) in Brachiaria pastures: A novel strategy to improve eco-efficiency of crop-livestock systems and to mitigate climate change. *Tropical Grasslands-Forrajes Tropicales, 2*(1), 88–91. https://doi.org/10.17138/TGFT(2)88-91

Mueller, R. C., Paula, F. S., Mirza, B. S., Rodrigues, J. L. M., Nüsslein, K., & Bohannan, B. J. M. (2014). Links between plant and fungal communities across a deforestation chronosequence in the Amazon rainforest. *The ISME Journal, 8*(7), 1548–1550. https://doi.org/10.1038/ismej.2013.253

Murakami, H., Wang, B., Li, T., & Kitoh, A. (2013). Projected increase in tropical cyclones near Hawaii. *Nature Climate Change, 3*(8), 749–754. https://doi.org/10.1038/NCLIMATE1890

Myers, N. (1989). The tropical rain forest: A first encounter: by M. Jacobs. 1988, Springer, Berlin, 295 pp. illustrated. DM64 -/£21.95 softback. ISBN 3-450-17996-8. *Forest Ecology and Management, 28*(1), 71–72. https://doi.org/10.1016/0378-1127(89)90075-3

NASA. (2022). Evidence: How do we know climate change is real? *Global Climate Change: Vital Signs of the Planet.* https://climate.nasa.gov/evidence/

Neumann, B., Vafeidis, A. T., Zimmermann, J., & Nicholls, R. J. (2015). Future coastal population growth and exposure to sea-level rise and coastal flooding—a global assessment. *PLOS One, 10*(3). https://doi.org/10.1371/journal.pone.0118571

Nicholls, R. J., Poh Wong, P., Burkett, V., Codignotto, J., Hay, J., McLean, R., Ragoonaden, S., Woodroffe, C. D., Abuodha, P., Dronkers, J., Wong, P., Burkett, V., Codignotto, J., Hay, J., McLean, R., Ragoonaden, S., Woodroffe, C., Parry, M., Canziani, O., . . . Hanson, C. (2007). Coastal systems and low-lying areas. In *Contribution of working group II to the fourth assessment report of the intergovernmental panel on climate change.* IPCC.

Pandey, R. (2019). Farmers' perception on agro-ecological implications of climate change in the Middle-Mountains of Nepal: A case of Lumle Village, Kaski. *Environment, Development and Sustainability, 21*(1), 221–247. https://doi.org/10.1007/s10668-017-0031-9

Pathak, R., Thakur, S., Negi, V. S., Rawal, R. S., Bahukhandi, A., Durgapal, K., Barola, A., Tewari, D., & Bhatt, I. D. (2021). Ecological condition and management status of Community Forests in Indian western Himalaya. *Land Use Policy, 109.* https://doi.org/10.1016/j.landusepol.2021.105636

Paudel, B., Wang, Z., Zhang, Y., Rai, M. K., & Paul, P. K. (2021). Climate change and its impacts on farmer's livelihood in different physiographic regions of the trans-boundary koshi river basin, central himalayas. *International Journal of Environmental Research and Public Health*, *18*(13). https://doi.org/10.3390/ijerph18137142

Pendleton, L., Donato, D. C., Murray, B. C., Crooks, S., Jenkins, W. A., Sifleet, S., Craft, C., Fourqurean, J. W., Kauffman, J. B., Marbà, N., Megonigal, P., Pidgeon, E., Herr, D., Gordon, D., & Baldera, A. (2012). Estimating Global "Blue carbon" emissions from conversion and degradation of vegetated coastal ecosystems. *PLOS One*, *7*(9), e43542. https://doi.org/10.1371/JOURNAL.PONE.0043542

Phillips, M. R., & Jones, A. L. (2006). Erosion and tourism infrastructure in the coastal zone: Problems, consequences and management. *Tourism Management*, *27*(3), 517–524. https://doi.org/10.1016/J.TOURMAN.2005.10.019

Rayner, S., & Malone, E. L. (2001). Climate change, poverty, and intragenerational equity: The national level. *International Journal of Global Environmental Issues*, *1*(2), 175–202. https://doi.org/10.1504/IJGENVI.2001.000977

Ripple, W. J., & Beschta, R. L. (2004). Wolves and the ecology of fear: Can predation risk structure ecosystems? *Bio Science*, *54*(8), 755–766. https://doi.org/10.1641/0006-3568(2004)054[0755:WATEOF]2.0.CO;2

Saxena, A., Buettner, W. C., Kestler, L., & Kim, Y.-S. (2022). Opportunities and barriers for wood-based infrastructure in urban Himalayas: A review of selected national policies of Nepal. *Trees, Forests and People*, *8*. https://doi.org/10.1016/j.tfp.2022.100244

Shaffer, G., Olsen, S. M., & Pedersen, J. O. P. (2009). Long-term ocean oxygen depletion in response to carbon dioxide emissions from fossil fuels. *Nature Geoscience*, *2*(2), 105–109. https://doi.org/10.1038/ngeo420

Sharma, E. (2012). Climate change and its impacts in the Hindu Kush-Himalayas: An introduction. *Community, Environment and Disaster Risk Management*, *11*, 17–32. https://doi.org/10.1108/S2040-7262(2012)0000011008

Sharma, P., Chettri, N., Uddin, K., Wangchuk, K., Joshi, R., Tandin, T., Pandey, A., Gaira, K. S., Basnet, K., Wangdi, S., Dorji, T., Wangchuk, N., Chitale, V. S., Uprety, Y., & Sharma, E. (2020). Mapping human–wildlife conflict hotspots in a transboundary landscape, Eastern Himalaya. *Global Ecology and Conservation*, *24*. https://doi.org/10.1016/j.gecco.2020.e01284

Shrestha, A. B., & Aryal, R. (2011). Climate change in Nepal and its impact on Himalayan glaciers. *Regional Environmental Change*, *11*(Suppl. 1), 65–77. https://doi.org/10.1007/s10113-010-0174-9

Silva, I. M. S., Calvi, G. P., Baskin, C. C., dos Santos, G. R., Leal-Filho, N., & Ferraz, I. D. K. (2021). Response of central Amazon rainforest soil seed banks to climate change—simulation of global warming. *Forest Ecology and Management*, *493*, 119224. https://doi.org/10.1016/J.FORECO.2021.119224

Singh, P. K., Papageorgiou, K., Chudasama, H., & Papageorgiou, E. I. (2019). Evaluating the effectiveness of climate change adaptations in the world's largest mangrove ecosystem. *Sustainability*, *11*(23), 6655. https://doi.org/10.3390/SU11236655

Small, C., & Nicholls, R. J. (2003). A global analysis of human settlement in coastal zones. *Journal of Coastal Research*, *19*(3), 584–599. www.jstor.org/stable/4299200

Staal, A., Flores, B. M., Aguiar, A. P. D., Bosmans, J. H. C., Fetzer, I., & Tuinenburg, O. A. (2020). Feedback between drought and deforestation in the Amazon. *Environmental Research Letters*, *15*(4), 044024. https://doi.org/10.1088/1748-9326/AB738E

Tuladhar, S., Pasakhala, B., Maharjan, A., & Mishra, A. (2021). Unravelling the linkages of cryosphere and mountain livelihood systems: A case study of Langtang, Nepal. *Advances in Climate Change Research*, *12*(1), 119–131 https://doi.org/10.1016/j.accre.2020.12.004

United Nations. (2017). Factsheet : Marine pollution. *The Ocean Conference.* www.un.org/sustainabledevelopment/wp-content/uploads/2017/05/Ocean-fact-sheet-package.pdf

van Solinge, T. B. (2010). Deforestation crimes and conflicts in the Amazon. *Critical Criminology, 18*(4), 263–277. https://doi.org/10.1007/S10612-010-9120-X

VanWey, L. K., Guedes, G. R., & D'Antona, Á. O. (2011). Out-migration and land-use change in agricultural frontiers: Insights from Altamira settlement project. *Population and Environment, 34*(1), 44–68. https://doi.org/10.1007/S11111-011-0161-1

Verma, R., & Jamwal, P. (2022). Sustenance of Himalayan springs in an emerging water crisis. *Environmental Monitoring and Assessment, 194*(2). https://doi.org/10.1007/s10661-021-09731-6

Wisshak, M., Schönberg, C. H. L., Form, A., & Freiwald, A. (2012). Ocean acidification accelerates reef bioerosion. *PLOS One, 7*(9), e45124. https://doi.org/10.1371/JOURNAL.PONE.0045124

World Meteorological Organization. (2020). *Tropical cyclones.* World Meteorological Organization (WMO). https://public.wmo.int/en/our-mandate/focus-areas/natural-hazards-and-disaster-risk-reduction/tropical-cyclones

Yu, X., & Li, H. (2021). Origin of ethnic groups, linguistic families, and civilizations in China viewed from the Y chromosome. *Molecular Genetics and Genomics, 296*(4), 783–797. https://doi.org/10.1007/s00438-021-01794-x

Zaman, K. (2022). Environmental cost of deforestation in Brazil's Amazon Rainforest: Controlling biocapacity deficit and renewable wastes for conserving forest resources. *Forest Ecology and Management, 504,* 119854. https://doi.org/10.1016/J.FORECO.2021.119854

Zaveri, R. A., Wang, J., Fan, J., Zhang, Y., Shilling, J. E., Zelenyuk, A., Mei, F., Newsom, R., Pekour, M., Tomlinson, J., Comstock, J. M., Shrivastava, M., Fortner, E., Machado, L. A. T., Artaxo, P., & Martin, S. T. (2022). Rapid growth of anthropogenic organic nanoparticles greatly alters cloud life cycle in the Amazon rainforest. *Science Advances, 8*(2). https://doi.org/10.1126/SCIADV.ABJ0329/SUPPL_FILE/SCIADV.ABJ0329_SM.PDF

# 11 Urban Soil Management for Improved Resource-Efficiency of Developing Cities

*Mahima Dixit, Immanuel Chongboi Haokip,
Bhabani Prasad Mondal, Muskan Porwal,
Debabrata Ghoshal, Abhishek Mandal,
and B.P. Dhyani*

## 11.1 INTRODUCTION

More than 50% of the global population already dwells in urban habitats, which can be defined as locations possessing a population of 10,000 or greater residents (DEFRA 2005), while this proportion is expected to rise about to over 70% by 2050 (United Nations 2019). As the urban population expands, the capacity of the urban environments to create desirable habitats and support resilient ecosystem is becoming increasingly dubious (Biggs et al. 2012). Urbanization is the driving force behind several problems. As a result, cities have a key influence in determining environmental trends as a consequence of the rising percentage of the world's population living in urban areas and the high intensity of urban dwellers' activities (Marcotullio et al. 2008). People have lost contact with the soil and its life-sustaining services as the world continues to urbanize. To meet the growing population, urbanization continues to transform (sometimes informally and uncoordinatedly) fertile agricultural lands and green spaces into urbanized regions (Arezki et al. 2015). Consistent urbanization, growing demands, system shortfalls caused by inadequate and poorly developed infrastructure, severe dysfunctional land and housing markets leading to informal settlements, transportation and mobility challenges, socioeconomic issues, and environmental degradation are among the greatest obstacles to resource efficiency and the sustainable management of natural resources that must be addressed. Specifically, this increases the strain on water supply and sanitation, energy supply and efficiency, waste recycling and resource recovery, land usage, and food security (Bay and Lehmann 2017). In addition to the hazards to human health caused by air pollution and climate change, it is now more imperative than ever to assess the capacity of urban settings to preserve ecosystem services (eSs) (Jacob and Winner 2009, Heaviside et al. 2017, O'Donnell and Thorne 2020), i.e., the advantages that people get from ecosystems. Soil plays a critical role in the provision of various essentials (eSs) (Dominati et al. 2010; Adhikari and Hartemink 2016; Jónsson and Davíðsdóttir 2016; Greiner et al. 2017), and within the soil science community, the relevance of soil in delivering eSs in urban environments is being recognized more. (Lehmann and Stahr 2007, Pavao-Zuckerman 2008; Lal and Stewart 2017; Ziter and Turner 2018; Bray and Wickings 2019). Urban soils are basically soils which are located within city or urban places. Urban soils basically include the arena named as SUITMA (Soils of Urban, Industrial, Traffic, Mining, and Military Areas), which are characterized as soils drastically altered by human activity in their compositions and functions. They might comprise of both strongly transformed soils and soils of pseudo-natural nature (Morel et al. 2015). In metropolitan regions, urban soil is the foundation for several eSs that contribute to human urban resilience and human well-being (Gómez-Baggethun et al. 2013; Haase et al. 2014; McPhearson et al. 2015). Urban soil when managed properly can serve as an efficient resource to cater the needs of human beings and also maintain environmental integrity. Both locally and globally, these services are the ones counting to nutrient cycling and enhanced carbon (C) storage. Locally, they encompass mitigating floods, buffering of urban heat island effect, aiding in air pollution capturing, providing physical strength to infrastructures, growing food in urban areas, and providing accessible green spaces for emotional well-being and physical health. Many of the eSs that may be found in non-urban soils can also be found in urban soils (Pavao-Zuckerman 2012; Morel et al. 2015; Pouyat et al. 2020). While, compared to eSs in non-urban soils, our understanding of how to quantify them is now somewhat restricted.

## 11.2  DEFINITION AND NATURE OF URBAN SOIL

The knowledge of biochemical properties, features, and other critical qualities of urban soils is crucial for comprehending nutrient cycling in urban environments and establishing a database for the management of urban soils (Lorenz and Kandeler 2005). Zemlyanitskiy (1963) was the first one to use the term "urban soil" to characterize the features of heavily disturbed soils in metropolitan settings or urban regions. Later, Craul (1992) defined urban soil as "a soil material with a non-agricultural, man-made surface layer more than 50 centimeters thick that was generated by blending, filling, or by the contamination of land surface in urban and suburban areas." The earlier concept is comparable to prior definitions by Bockheim (1974) and Craul and Klein (1980). Evans et al. (2000) and, subsequently, Capra et al. (2015) adopted the term "anthropogenic soil" and put urban soils in the larger framework of human-altered soils as opposed to confining the concept to highly populated suburban and urban regions alone. To acknowledge a broader set of observations, Effland and Pouyat (1997), Lehmann and Stahr (2007), and Morel et al. (2017) have expanded the definition of urban soils to include relatively undisturbed soils that have been modified by urban changes in the environment, such as the accumulation of atmospheric pollutants.

## 11.3  MAJOR CHARACTERISTICS AND PROPERTIES OF URBAN SOILS

For the application of best management practices, the basic characteristics of urban soils should be known well. Urban soils show a varied nature when compared to non-urban soils. Soil conditions in urban areas correspond to a spectrum of anthropogenic effects, ranging from relatively reduced impact (native forest or grassland soil) or oblique urban environmental impacts (remnant forest stands) to those deduced from human-created materials (landfills), sealed by impermeable surfaces (asphalt), or modified by physical disturbances and management (residential yards) (Pouyat et al. 2020). Despite the high levels of disturbance normally encountered by most urban soils, they have the capacity to support host plant, animal, and microbial life and to mediate hydrological and biogeochemical cycles, much like their rural counterparts (Pouyat et al. 2010). The resulting communities of soil organisms are a unique combination of native species surviving or flourishing in urban areas and introduced species from other regions. Therefore, urban soil management for ecosystem services demands an interdisciplinary approach. According to Setälä et al. (2014), urban soils have numerous and often contradictory functions within various types of urban ecosystems. All of these functions must be examined separately for urban environments. The functions of urban soils often differ greatly from those of their rural equivalents, owing not just to their changed properties but also to their position within the metropolitan environment. Consider a compacted soil that has been deteriorated by building development or destruction as an example. Compaction (a change in function) impairs the permeability of this soil, and its placement in a mostly impermeable environment leads in the soil absorbing much more water through runoff (due to landscape position). Other essential soil functions are particular to urban environments. For instance, they offer a strong foundation for constructions like houses and roadways. In addition, urban soils offer physical support and an accessible place for subsurface infrastructure. They may play a part in waste processing, whether from septic systems or recycling programs for food and yard waste. All soils are capable of accumulating nutrients such as phosphorus (P) and nitrogen (N), which, if transferred to surface waterways, may result in ecologically detrimental algal blooms. In addition to storing a variety of macro and microartifacts, soils also retain considerable amounts of urban-associated toxicants (e.g., lead and arsenic) (Rossiter 2007). These soils may constitute a threat to public health if people are exposed to them when they are utilized for other reasons (e.g., urban agriculture or recreations).

## 11.4  PHYSICAL CHARACTERISTICS OF URBAN SOILS

Soil physical characteristics comprise of texture (particle size), structure (ped form), density, plasticity, porosity, permeability, temperatures, and moisture content. Urbanization modifies the physical characteristics of soils. For numerous years, material inflows into cities have surpassed production. This has resulted in the overall elevation of the land in older districts and the filling of rivulets, canals, and valleys and various other areas of cities. Given the ongoing changes and additions to the urban ecology, it is not surprising that urban soils are distinct. Craul (1985) was maybe the first to comprehensively examine the changes in urban soils.

### 11.4.1  SOIL SEALING AND COMPACTION

Compaction is one of the most prevalent soil physical characteristics in urban settings. It is the most severe kind of physical degradation, characterized by a decrease in soil volume followed by a rise in the soil's bulk density, tighter packed of solid particles, and decreased porosity (Glinski 1990; Jim 1993; Yang et al. 2004).

Compaction increases soil strength, resulting in greater root resilience for urban plants (Jim 1998). Furthermore, soil compaction reduces the interaction between soil and water, slowing the diffusion of both oxygen ($O_2$) as well as carbon dioxide ($CO_2$). Soil compaction has several harmful consequences on the urban ecology as a result of the aforementioned processes, including the following:

1. reduced water infiltration and replenishment of the groundwater system;
2. increased surface runoff and the emergence of urban floods;
3. increased pollutant loading in basins of water bodies;
4. increased heat island effect in cities;
5. changed soil temperature, decreased microorganism activity, and transformation of nutrients; and
6. varied plant growth.

Globally, urban soil compaction may have a significant impact on the urban ecosystem and often lead to a reduced life quality of the urban inhabitants. Soil sealing happening through the construction of urban infrastructure, which comprises all types of pavements and structures—i.e., impermeable surfaces consisting of asphalt and materials like concrete—is widespread and quite common in urban settings (Breuste and Qureshi 2011). Sealing reduces the potential sustainability of the environment.

## 11.4.2 Soil Structure

Vegetation removal, topsoil relocation, stocking up, cementation, development, or soil replenishment are all urban land development activities that can have a substantial influence on soil aggregation (Wick et al. 2009). Supposedly, removal of topsoil and replacement reduced the fraction of macroaggregates in surface soils by around 29% on average as discovered by Chen et al. (2014). In urban areas, soil structure is weak, with little aggregation. It is also seen that fine particles deficiency impedes the establishment of a robust soil structure due to the absence of aggregating agents and integrating components between coarse grains. The urbanization-induced soil compaction may affect the arrangement of soil aggregates and pore space. Whereas areas having low organic material content and silty to sandy structure contribute to low aggregation and poor soil structure (Six et al. 2004).

## 11.4.3 Infiltration Rate

Compaction decreases soil porosity, increasing soil bulk density and influencing soil water transport. Compaction and sealing diminish the water-retaining ability of soil and regulation of the water balance. In urban locations, the overall soil water storage capacity, the usable water storage capacity, and the short-term water storage capacity decreased steadily as soil compaction rose, whereas the ineffective water storage capacity grew gradually (Yang and Zhang 2008).

### 11.4.4 SURFACE CRUSTING AND WATER REPELLENCE

In urban areas, a thin crust forms over the surface of the earth which is produced by a number of causes, the most significant of which are foot and transportation activities and the resulting absence of groundcover vegetation (Marcotullio et al. 2008). The lack of groundcover diminishes the stability impact of roots and shoots, the lightening and binding action of the rhizosphere, and the enhancement organic matter to soil. After, the raindrop ceases to fall and the splashing power decreases it, disintegrates soil aggregates, and makes extremely fine sand, silt, and clay particles to fill the nearby pores leading to soil compaction (Hillel 2012). Compaction and addition of tiny particles restrict water penetration and the exchange of $CO_2$ and $O_2$ between the soil and the atmosphere. Atmospheric accumulation of petroleum-based aerosols and particles on the surface of the soil may also generate water-repellent waxy and greasy compounds.

## 11.5 CHEMICAL CHARACTERISTICS OF URBAN SOILS

Understanding of chemical parameters of urban soils showed their varying chemical properties. These disparities are a caused by the setup in which the soils lie and the materials enter them.

### 11.5.1 REACTION OF URBAN SOILS

The pH values of urban soils are often distinct from those of their natural equivalents. In most circumstances, pH levels in urban settings are higher, or more alkaline (Craul and Klein 1980). There are several reasons for urban soils' increased pH levels. For example, the use of chlorides of sodium and calcium salt as sidewalks and street de-icing salts (Craul 1992). The second factor is the watering of the urban soils with water enriched with calcium salts. Atmospheric pollution accounts for the third explanation. Typically, ash residue contains calcium. This becomes further evident by the fact that pH levels fall with increasing depth. The fourth explanation is the release of calcium from weathering construction wastes, such as bricks, cement, plaster, and so on, or the washing of calcium off building facades by contaminated acid precipitation. Alkaline soils of cities not only cause nutritional imbalances but also alter metal speciation and activity.

### 11.5.2 HEAVY METALS

Heavy metals accumulation in urban soil is a result of both endogenous and external deposition. Multiple studies have shown that heavy metals are more frequently prominent in urban soils than those in nearby agricultural and forest soils as a consequence to the increased uptake of heavy metals due to urbanization (Tiller 1992; Chen et al. 1997; Stroganova et al. 1998; Manta et al. 2002; Lu et al. 2003). Heavy metals found in the environment are called exogenous, and their origins may be traced back to places like homes, rubbish dumps, cars, mines, factories, and the combustion of fossil fuels (Tiller 1992; Stroganova et al. 1998; Wong et al. 2006). Just as the majority of lead (Pb) deposition in urban soils may be attributed to the burning of leaded gasoline. Road traffic was shown to be a major contributor to metal pollution in an experiment using Pb isotope techniques by Cicchella et al. (2008). Coal combustion was the primary source of mercury (Hg) and arsenic (As), whereas industrial processes were the primary source of cadmium (Cd) (Li et al. 2001; Yang et al. 2011). According to many studies, the major contaminating elements are Hg, Zn, Pb, and Cu, which are typically known as "urban heavy metals," whereas the accumulation of other heavy metals is minimal compared to what was said previously.

### 11.5.3 Organic Pollutants

Organic pollutants such as persistent organic pollutants (POPs) pose a major threat to the quality of air and water, in addition to ecosystem and health of human beings. Since these comprise of semi-volatile and volatile organic compounds, they may enter into water or air sources by means of leaching, diffusion, or volatilization, owing to a given concentration gradient (Xue et al. 2002; Tang, 2003). POPs may be found in urban soils from a variety of different sources, including petrogenic materials, combusted coal residues, burnt biomass, sources related to Coke tar, creosote, and vehicle emissions (Wang et al. 2013). According to researches done in Delhi, India, the concentration of polycyclic aromatic hydrocarbons (PAHs) was much higher (4–15 times) in urban soils than the national norms (Khillare et al. 2014). Human influence is the primary source of PAHs in the environment. Tang et al. (2005) showed that PAHs in urban soil mostly arise from pyrogenic sources. According to molecular constitutions, however, other researches revealed that traffic emissions might be the primary source of PAHs in soils (Wilcke et al. 1999; Wang et al. 2013; Yu et al. 2014).

## 11.6 BIOLOGICAL CHARACTERS OF URBAN SOILS

### 11.6.1 Biodiversity

Urban sprawl's impact on wildlife is an area of little knowledge. Savard et al. (2000) proposed ideas for fostering biodiversity in man-made environments. In light of urbanization's expected expansion, McDonald et al. (2008) showed changes to ecoregions, endangered species, and conservation areas. They also demonstrated that urbanization is a major threat to biodiversity, with 8% of terrestrial vertebrate species on the International Union for Conservation of Nature Red List considered in danger due to human activity in 29 of the world's 825 ecoregions. As a result, urbanization is a major factor in the decline of animal populations (McKinney 2006).

### 11.6.2 Activity of Soil Biota

The various kind of contamination of urban soils endangers soil biota. There is evidence that microorganisms' diversity and abundance may change due to pollution in urban soil. Yang et al. (2001) observed that microbial respiration ratio in urban soil was higher, although the biomass count was much lower which demonstrate that there was a substantial elevation in organic carbon (OC) consumption by microbes in urban soil.

## 11.7   THE NEED FOR SOIL MANAGEMENT IN URBAN AREAS

Soil is critical to the evolution of human civilization. The public's attention is centered on soil as a foundation for homes, industrial and commercial uses, infrastructure, leisure spaces, and food production (Figure 11.1). However, soils in urban areas provide considerably more than these obvious benefits. The following serve as a main reason for managing soils, for example, sustaining biological activity, diversity, and productivity; controlling and reformatting solute and water flow; sieving, buffering, denigrating, immobilizing, and detoxifying toxic substances arising from municipal and industrial by-products as well as atmospheric depositions; storing and cycling nutrients and other elements within the earth's biosphere; producing renewable primary products; microclimate control; and socioeconomic support. As a result, there is now a lack of public understanding for crucial soil processes that ensure the quality of human existence and the environment. By applying proper soil management practices, the urban soils can be harboured to the best of their capabilities and could be used for increasing resource utilization.

## 11.8   MANAGEMENT OF URBAN SOILS FOR EFFICIENT RESOURCE UTILIZATION

The global community is paying close attention to protecting and restoring soil since it is continually deteriorated by human activities. Conventions like the Rio Summit,

**FIGURE 11.1**   Recourse recovery of urban soils showing multiple areas where urban soils hold potential to enhance resource reserves, including areas like power generation, urban farming, and regions where the condition is beyond reclamation. They prove to be an excellent building material or source of power generation. Modified and redrawn from open access.

**Source: Kumar and Hundal (2016)**

the United Nations Framework, and the Kyoto Protocol show that the international community is concerned about this problem (Hannah 2011). Sustainable agriculture, population growth, and human well-being all depend on well-managed soil (White et al. 2015). The availability of adequate quantities of all necessary mineral elements in soil is ensured through careful management. Plants may be poisoned by too much of them, while not having enough of them causes them to be unavailable. Soil microbes provide mineral nutrients and are supported by good soil management, which aids in carbon sequestration. The incorporation of biomass into soil management practices decreases the inclination of soil erosion, boosts the performance of soil organisms, increases the variety of plant and animal life, and reduces water use (Lal 2004). And soil management practices like mulching, composting, intercropping, nitrogen cycling, cover crop usage, and zero tillage all contribute to higher organic matter content of soil and escalate soil health (White et al. 2015). These days, sustainable methods are crucial for keeping up and enhancing soil quality, which is essential for meeting the food demands of future generations (Figure 11.2). As a result of factors including erosion and nutrient imbalance, topsoil is being lost in many areas of the globe (Ranjan et al. 2021). Moreover, urban soils all around the world are becoming increasingly polluted with heavy metals. Soil erosion is caused by several factors, including climate change, storms, and water runoff. Several sustainable soil management options attempt to contribute to long-term food security by fixing degraded soils via coordination with several integrated sustainable and eco-friendly agronomic practices. Therefore, certain management practices have been employed on the urban soils for their upliftment and enhancement of the utilization of resources of the urban region.

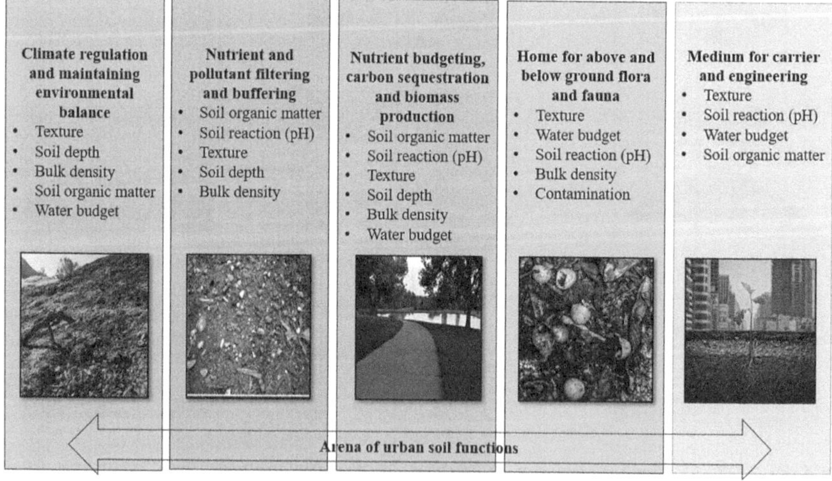


**FIGURE 11.2** The functions and dynamics of urban soils in various fields such as household, environmental, and ecological arenas. Redrawn with the permission.

**Source: Foldal et al. (2022)**

The following stages are advised for the actual execution of urban soil management:

- Collecting information on land use pattern, soil health, and quality, including the extent of contamination.
- Evaluating soil quality and current land use patterns and executing urban development concepts and plans.
- Defining protocols for soil protection.
- Underlining thresholds for acceptable land consumption and the resulting soil management needs.
- Choosing and implementing the most viable urban soil management techniques and instruments.
- Goal achievement evaluation, in order to assess the consequences of its implementation, an urban soil management concept requires continuous monitoring of development. The monitoring findings allow for process adjustments and goal-directed soil management.

### 11.8.1 Urban Soil Management Projects

Protecting long-term sustainability of soil resources, both in relation to quality and quantity, is essential for soil management in rapidly developing metropolitan regions. The only way to realize this objective is to include it into the urban planning process. Regional government and municipal authorities are actively involved in the planning of urban areas, and thus, they are the intended audiences for urban soil management system. Consultants and government organizations working in urban planning and environmental protection are also targeted. A larger audience, including municipal council members and the general public, is the intended objective of awareness-raising efforts. Effective soil management may be defined, designed, and developed with the use of proper methodologies and best practice examples. Artificial intelligence–based software tools should be created and deployed in light of this. Soil awareness should be promoted using well-elaborated information, arguments, and strategies. Testing the efficacy of the urban soil management strategies and methodologies and then putting them into practice after analysis is the best course of action (Blümlein et al. 2012).

### 11.8.2 Use of Artificial Intelligence for Soil Management

Incorporating soil challenges into urban design requires a solid foundation of soil knowledge. Despite its availability, non-soil specialists, including planners and decision-makers, often struggle to make sense of and make good use of the data. A user-end oriented system that provides risk assessment methods and detailed evaluation should provide data on soil quality, contamination extent, texture, and penetration rate accessible, along with targeted interpretations. Over the last several decades, GIS-based digital technologies have been created to precisely monitor a variety of physical and chemical soil parameters, including soil moisture, soil roughness, and soil texture (Bousbih et al. 2019). Together, GIS and remote sensing provide an accurate digital soil map that may be stored for years to come (Trubina et al. 2019). The digitized environmental covariates are utilized to make predictions about soil characteristics throughout the sampling process. These methods provide details on the soil microbial population, plant density, mineralogy, and parental stock (Bernardi et al. 2017). Soil erosion, water distribution, and the effects of sunlight are only few of the topics covered under the digital evaluation model (DEM) (Florinsky 2016). Since the geophysical sensors employed by soil scientists in landscape and field studies are quick, simple to use, and inexpensive, these methods are more efficient and have lowered the burden. Soil management is made easier due to GIS, which pinpoints optimal application times for fertilizers and water, as well as the moisture and nutritional content of soils (Lakhiar et al. 2018). Therefore, these resources have the potential to be invaluable in the efficient administration of urban green spaces (UGSs) and soil.

### 11.8.3 Soil Remediation and Heavy Metal Contamination Treatment

Remediation of contaminated soil may enhance the well-being and health of urban soils, as it reduces the biotic inhabitant's exposure to toxins through several pathways. For example, urban soils usually show their alkaline nature due to the accumulation of various salts and industrial residues. In this case, liming the soil and other remediation techniques diminish the bioavailability of hazardous metals like cadmium and lead in polluted soils and, subsequently, in the hazardous urban dust. Specific soil-grown plants possess the capacity in deactivating particles that are hazardous to human health. *In-situ* remediation is an option that has been explored using several techniques. When the affected region is vast and the pollution load is low, *ex-situ* cleanup might be prohibitively expensive, but *in-situ* remediation can significantly reduce these costs (Hettiarachchi and Pierzynski 2004). Human exposure hazards to polluted soil may be reduced by remediation treatment, which also has the added benefit of restoring the soil's natural ecological function. Nonetheless, it is feasible that remediated areas won't be as fruitful as those that were never poisoned (Hinojosa 2004; de Mora et al. 2010). Most *in-situ* remediation strategies include replanting and amending the soil at problem areas to restore stability (chemical, physical, and phyto-stabilization). The goal of stabilization is to lessen the spread of contaminated soils and the accessibility and mobility of contaminants. Reducing bioavailability and mobility makes toxins far less of a concern to people and ecosystems and stops pollution from spreading to other places via erosion, leaching, or the food chain (Vangronsveld et al. 2009).

### 11.8.4 Urban Green Spaces for Urban Soil Management for Mitigating Climate Changes

Proper management of urban soils may help in mitigating the excessive climate changes occurring in urban areas. The wastelands can be utilized for plantation purposes or the setting up of urban greens, which has diverse benefits. A crucial component of every city's landscape is urban green spaces (UGSs) or urban greeneries (UGs) which help in keeping city dwellers safe. To relax and improve one's health, urban green spaces (UGSs) are a warm amenity, especially for city people. Urban

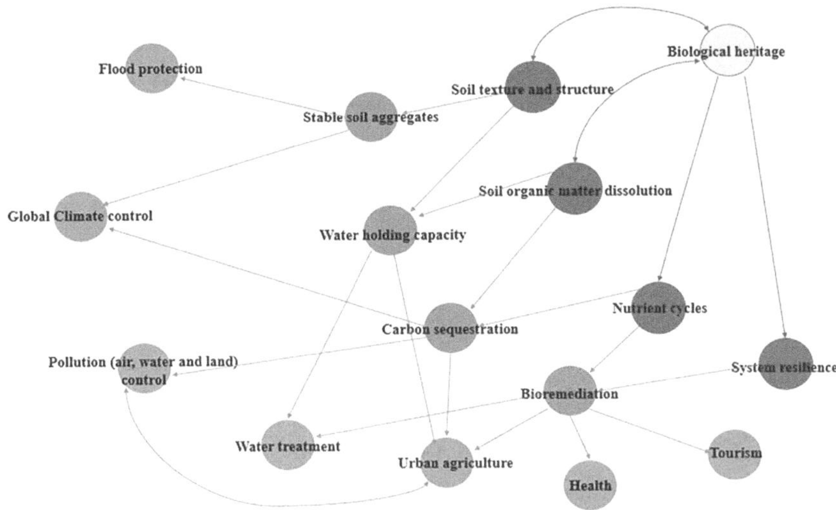



**FIGURE 11.3** Multiple domains of urban greens; urban greens hold a wide potential in many areas as shown in the figure which are interconnected (solid lines) in their own way. Redrawn and modified with permission.

**Source: Guilland et al. (2018)**

greens act as home to a diverse species of flora and fauna, which contribute to their aesthetic appeal (Tribot et al. 2018). UGSs also provide a venue for city dwellers to engage in sport and soil fitness. Urban green spaces, if managed in a wise way, can be home to numerous activities. These activities are interconnected and thus work as a whole for the enhanced resource availability (Figure 11.3). There is 83% more reported social engagement in green urban environments than in grey urban areas. Also, green areas in urban housing developments are a fantastic way to bring people together. As plants filter the air by eliminating dust particles and hazardous pollutants, UGs want to do more than just provide oxygen; they also aim to manage urban air and create a microclimate of cities (Czaja et al. 2020). The same may be said about UGs, which aid in suppressing unwanted noise. Several plant species are very vulnerable to harmful gases created by automobiles and environmental pollutants, which inhibit plant growth and disrupt the metabolism and photosynthesis process. One of the key causes for the decline of UGs is the rise in urban pollution. Therefore, it is important to prioritize sustainable agroecosystems and the careful design and management of green spaces in urban areas. This may be achieved via the use of cutting-edge GIS tools in the urban environment (Rakhshandehroo et al. 2016).

## 11.8.5 Enhancing Soil Fertility of Urban Soils for Food Production

Subsoils in urban areas may have degraded biological, chemical, and physical properties, making them less fertile than soils in rural areas. To sum up, the characteristics of urban soils can make it tough to cultivate a crop (Bartens et al. 2012). The soil organic matter (SOM) content, pollutant levels, and moisture profile of urban soils are all factors that contribute to their potential poor physical attributes (Lehmann and Stahr 2007). One of the benefits of urban ecosystems is the opportunity to improve soil quality for agricultural use. The use of cover crops, careful digging and tillage, and introduction of organic supplements, such as ample agricultural leftovers or post-consumer wastes, are all effective ways to increase soil fertility (Cogger and Brown 2016). Recycled water, yard wastes, food scraps, and municipal biosolids may be processed before being utilized as additions to enhance urban areas (UA) soils (Brown and Goldstein 2016). Recycling urban waste streams may improve soil fertility, and underutilized land can be put to good use (Smit and Pilifosova 2001). This helps in maintaining the resource loop of urban areas.

### 11.8.6 Improving Biological Properties of Urban Soils for Improving Crop Cover

Soil fertility and UA production can be improved by cultivating a physical and chemical environment conducive to biotic activity in the soil. Best management practices (BMPs) include building raised beds in areas where chemical and physical constraints cannot be removed and adding OM as a food resource for soil organisms to improve habitat conditions. In addition to this ecosystem engineers, involving a wide range of soil biota (like earthworms, termites, and ants) deserve special attention because they play crucial roles in establishing habitats for other organisms (microfauna, mesofauna, and microorganisms), and manipulating their activities will direct soil processes along predetermined pathways (Lavelle et al. 2016). According to a study, long-term UA management in Stuttgart, Germany, produced garden soils that were rich in SOC and N, were home to abundant microbial biomass, and were brimmed with metabolic activity (Lorenz and Kandeler 2005). However, very little is known about how UA and other urban soil uses influence soil microbial communities (Silva et al. 2010).

### 11.8.7 Treatment of Wastes Before Release into the Urban Ecosystem

Urban soils get contaminated due to a variety of factors, including manufacturing, transportation systems that are both dense and filthy, the use of fossil fuels, and the discharge of several chemicals into municipal waste streams (Meharg 2016). In addition to toxins from suspended soil dust, metropolitan air already includes high amounts of nitrogen oxides, sulphur oxides, hydrocarbons, and particles from automobile exhausts. It is well established that air pollution lowers urban crop yields (Thomaier et al. 2015), but the effects of consuming polluted foods are poorly understood (Meharg 2016). Only lately has urban science garnered greater attention notably with reference to urban health and individual well-being (Editorial Nature 2016).

### 11.8.8 Management or Mitigation of Soil Loss from Urban Areas

Management and mitigation practices can be applied to various soil types depending upon their classification. Practically, irreversible loss of all soil functions is caused by soil sealing or deterioration. Due to the impossibility of entirely mitigating or compensating soil function loss, it is crucial to minimize soil consumption to the greatest extent feasible. While soil recovery can be impactful in mitigating soil loss—for example, removal of sealing, decontamination of soil from potent toxins, cultivating disturbed soil with remediating plant species, avoiding toxic landfills, enhancing soil biological activities, replacement of topsoil of the affected areas, irrigation with good-quality water (restoration of natural biotope), extensification in the use of arable land, loosening the soil of deep layers, decompaction, preventing erosion, management of soil reaction, etc.—in general, majority of soil-recovery methods used to rehabilitate profiles of soil are favorable to soil biodiversity and the whole environment because they enhance conditions for plant and soil organism development. It should be noted, though, that often, soils with severe characteristics exist (such as wet, arid, or nutrient-deficient) which are very hard to restore.

### 11.8.9 CHEMICAL TREATMENT FOR THE MANAGEMENT OF URBAN SOILS

Urban soils are characterized by varied chemical properties as compared to natural region soils. These soils serve as a sink for various industrial by-products as well as the households and city waste. By the application of certain chemical treatments, the deteriorated or the futile land can be used for other purposes, for example, increasing the green areas and other engagements. Preserving and restoring degraded soil should be prioritized. Techniques such as the application of chemical treatment, vapor extraction, washing the soil, phytoremediation, and biosorption serve as viable options. Reactions can be determined by severity of the problems, the number of sources involved, and the efficacy and suitability of the current remediation techniques for dealing with industrial and more mobile source levels. Even after application of the following techniques, rapidly industrializing regions of the emerging world act as an area of concern.

### 11.8.10 MANAGEMENT OF SOC STOCKS TO IMPROVE URBAN SOIL QUALITY

The properties of urban soils can be maintained or improved for UA by managing SOC stocks (Lorenz and Lal 2015). In addition to water filtration, erosion control, soil strength and stability, nutrient conservation, pollutant denaturing and immobilization, habitat and energy for soil organisms, and pest and disease regulation, the SOC stock also supports critically important ecosystems derived from the soil, which directly affect food production and soil fertility (Adhikari and Hartemink 2016; Franzluebbers 2010; Lal et al. 2012). By decreasing tillage and other soil-disturbing operations, for instance, the SOC stock on UA sites may be safeguarded against losses due to erosion. In urban regions, composted municipal solid waste, biosolids, and collected nutrients from wastewater might be readily accessible to boost SOC stocks and, by extension, fertility and ecosystem services. Diverting urban garbage from landfills and recycling it into compost is one way to increase agricultural yields and other eSs. This may be done on a small scale in individual homes or on a larger one in municipalities (Kumar and Hundal 2016).

### 11.8.11 RECYCLING OF SOLID AND LIQUID URBAN WASTE

Composts, fertilizers, and soil supplements made from urban residues have several possible applications. There are many different ways to do this, such as with a home garden, turf fertilization, urban agriculture, green stormwater infrastructure, tree planting, or even with artificial topsoil. Despite the lack of a standardized national policy for integrating residuals into urban planning, there are a few notable cases in which the significance of these materials has been acknowledged. Use of composts or other residual-based soil additives is prevalent in urban agriculture due to concerns about soil quality and soil safety (Brown and Goldstein 2016). Compost may be applied to gardens via a municipality's typical waste management infrastructure; the municipality can offer training and some finances to develop local collecting and composting operations, or collection and composting can be done independently of any municipal help.

### 11.8.12 SOIL CONSERVATION PRACTICES

Deforestation, urbanization, infrastructure, mining, and the development of agriculture are only few of the human activities that have contributed greatly to soil degradation during the past several decades. Massive soil deterioration has also been brought on by natural calamities, including floods, water logging, soil erosion, and soil crusting (Chalise et al. 2019). Maintaining soil fertility and quality has been a worldwide focus in recent years. The preservation of the microbial and other organismal variety that contributes to soil fertility is a primary goal of the many oil conservation approaches now in use across the globe. Soil conservation and sustainability can be practiced for the efficient management of urban area soils. These include practices involving minimum tillage, contour farming, intercropping, strip cropping, agro-astrological practices, windbreaks, and grassed waterways (Khan et al. 2022). By conserving urban soils, a cascade of resource opens up in the domain of urban soils which enhance the resource utilization and availability (Figure 11.4).

**FIGURE 11.4**  A futuristic approach of urban soils, including the various fields which have not been explored or should be explored for better understanding of urban soils.

## 11.9 CONCLUSION AND PATH AHEAD

Rapid urbanization, anthropogenic activities, and population boom have been the major factors acting behind soil deterioration. Maintaining soil resources for satisfying the global necessities call for the widespread use of eco-friendly practices that boost yields. Sustainable agricultural techniques like strip cropping, contour cropping, windbreaks, shelterbelts, buffer strips, and grassed waterways help in preventing both soil and water erosion, while techniques like intercropping, farm waste composting, mulching, residue recycling, crop rotation, and other sustainable agroecosystem techniques improve soil structure, minimize environmental pollution, improve soil nutrient status, and increase soil biodiversity. Moreover, applying the soil engineering approaches enhanced soil physical properties, rate of nutrient uptake, organic matter content, soil water-holding capacity, and rate of water penetration. Sustainable practices employed in soil amendment not only boost agricultural productivity but also conserve soil resources for the upcoming generations. Therefore, in order to enhance soil-based UA (best management practices) BMPs are required, and this information must be communicated among soil managers, policymakers, and local communities. One important part of this information sharing is educating people about the natural capital that urban soils supply (Robinson et al. 2013) and the eSs they give to urban populations, all of which contribute to better human health and well-being.

## REFERENCES

Adhikari, K., and Hartemink, A. E. (2016). Linking soils to ecosystem services—A globa review. *Geoderma*, 262, 101–111.

Arezki, R., Deininger, K., and Selod, H. (2015). What drives the global "land rush"? *The World Bank Economic Review*, 29(2), 207–233.

Bartens, J., Basta, N., Brown, S., Cogger, C., Dvorak, B., Faucette, B., and Toor, G. (2012). Soils in the city—A look at soils in urban areas. *CSA News*, 57, 4–13.

Bay, J. H. P., and Lehmann, S. (Eds.). (2017). *Growing compact: Urban form, density and sustainability* (1st ed.). Boca Raton, Floride, USA: Routledge.

Bernardi, A. C. D. C., Grego, C. R., Andrade, R. G., Rabello, L. M., and Inamasu, R. Y. (2017). Spatial variability of vegetation index and soil properties in an integrated crop-livestock system. *Revista Brasileira de Engenharia Agrícola e Ambiental*, 21, 513–518.

Biggs, R., Schlüter, M., Biggs, D., Bohensky, E. L., Burn Silver, S., Cundill, G., and West, P. C. (2012). Toward principles for enhancing the resilience of ecosystem services. *Annual Review of Environment and Resources*, 37, 421–448.

Blümlein, P., Medved, P., Vernik, T., and Vrščaj, B. (2012). *Soil in the city: URBAN soil management strategy*. City of Stuttgart-Department for Environmental Protection and Central Europe Programme co-financed by the European Regional Development Funds (ERDF).

Bockheim, J. G. (1974). *Nature and properties of highly disturbed urban soils*. Philadelphia, Pennsylvania. Div. S-5, Soil Science Society of America, Chicago, IL.

Bousbih, S., Zribi, M., Pelletier, C., Gorrab, A., Lili-Chabaane, Z., Baghdadi, N., . . . and Mougenot, B. (2019). Soil texture estimation using radar and optical data from Sentinel-1 and Sentinel-2. *Remote Sensing*, 11(13), 1520.

Bray, N., and Wickings, K. (2019). The roles of invertebrates in the urban soil microbiome. *Frontiers in Ecology and Evolution*, 7, 359.

Breuste, J., and Qureshi, S. (2011). Urban sustainability, urban ecology and the society for urban ecology (SURE). *Urban Ecosystems*, 14, 313–317.

Brown, S., and Goldstein, N. (2016). The role of organic residuals in urban agriculture. In *Sowing seeds in the city* (pp. 93–106). Dordrecht: Springer.

Capra, G. F., Ganga, A., Grilli, E., Vacca, S., and Buondonno, A. (2015). A review on anthropogenic soils from a worldwide perspective. *Journal of Soils and Sediments*, 15(7), 1602–1618.

Chalise, D., Kumar, L., Spalevic, V., and Skataric, G., 2019. Estimation of sediment yield and maximum outflow using the IntErO model in the Sarada river basin of Nepal. *Water*, 11(5), 952.

Chen, T., Wong, M. H., and Wong, J. W. C. (1997). A study on heavy metal pollution in soils in Hong Kong. *Acta Geographica Sinica*-Chinese Edition, 52, 236–244.

Chen, Y., Day, S. D., Wick, A. F., and McGuire, K. J. (2014). Influence of urban land development and subsequent soil rehabilitation on soil aggregates, carbon, and hydraulic conductivity. *Science of the Total Environment*, 494, 329–336.

Cicchella, D., De Vivo, B., Lima, A., Albanese, S., McGill, R. A. R., and Parrish, R. R. (2008). Heavy metal pollution and Pb isotopes in urban soils of Napoli, Italy. *Geochemistry: Exploration, Environment, Analysis*, 8(1), 103–112.

Cogger, C., and S. Brown. (2016). Soil formation and nutrient cycling. In S. Brown, K. McIvor, and E. H. Snyder (eds.), *Sowing seeds in the city* (pp. 25–52). Dordrecht: Springer.

Craul, P. J. (1985). A description of urban soils and their desired characteristics. *Journal of Arboriculture*, 11(11), 330–339.

Craul, P. J. (1992). *Urban soil in landscape design.* New York: John Wiley & Sons.

Craul, P. J., and Klein, C. J. (1980). Characterization of streetside soils of Syracuse, New York. In *Metropolitan Tree Improvement Alliance (METRIA) proceedings* (Vol. 3, pp. 88–101).

Czaja, M., Kołton, A., and Muras, P. (2020). The complex issue of urban trees—Stress factor accumulation and ecological service possibilities. *Forests*, 11(9), 932.

de Mora, S., Tolosa, I., Fowler, S. W., Villeneuve, J. P., Cassi, R., and Cattini, C. (2010). Distribution of petroleum hydrocarbons and organochlorinated contaminants in marine biota and coastal sediments from the ROPME Sea Area during 2005. *Marine Pollution Bulletin*, 60(12), 2323–2349.

DEFRA (2005) Securing the Future—Delivering UK Sustainable Development Strategy [WWW Document]. https://www.gov.uk/government/uploads/system/uploads/attach ment_data/file/69412/pb10589-securing-the-future-050307.pdf

Dominati, E., Patterson, M., and Mackay, A. (2010). A framework for classifying and quantifying the natural capital and ecosystem services of soils. *Ecological Economics*, 69(9), 1858–1868.

Editorial. (2016). Metropolis now. *Nature*, 531, 275–276.

Effland, W. R., and Pouyat, R. V. (1997). The genesis, classification, and mapping of soils in urban areas. *Urban Ecosystems*, 1(4), 217–228.

Evans, C. V., Fanning, D. S., and Short, J. R. (2000) Human-influenced soils. *Agron Monogr*, 39, 33–67.

Florinsky, I. (2016). *Digital terrain analysis in soil science and geology.* Academic Press.

Foldal, C.B., Leitgeb, E., and Michel, K. (2022). Characteristics and functions of urban soils. In *Soils in urban ecosystem* (pp. 25–45). Singapore: Springer Singapore.

Franzluebbers, A. J. (2010). Will we allow soil carbon to feed our needs? *Carbon Management*, 1(2), 237–251.

Glinski, J. (1990). *Soil physical conditions and plant roots* (1st ed.). Boca Raton, Floride, USA: CRC Press. https://doi.org/10.1201/9781351076708

Gómez-Baggethun, E., Gren, Å., Barton, D. N., Langemeyer, J., McPhearson, T., O'farrell, P., . . . and Kremer, P. (2013). Urban ecosystem services. In *Urbanization, biodiversity and ecosystem services: Challenges and opportunities* (pp. 175–251). Springer, Dordrecht.

Greiner, L., Keller, A., Grêt-Regamey, A., and Papritz, A. (2017). Soil function assessment: Review of methods for quantifying the contributions of soils to ecosystem services. *Land Use Policy*, 69, 224–237.

Guilland, C., Maron, P.A., Damas, O., and Ranjard, L. (2018). Biodiversity of urban soils for sustainable cities. *Environmental Chemistry Letters*, 16, 1267–1282.

Haase, D., Frantzeskaki, N., and Elmqvist, T. (2014). Ecosystem services in urban landscapes: Practical applications and governance implications. *Ambio*, 43(4), 407–412.

Hannah, L. (2011). Mitigation: Reducing greenhouse gas emissions, sinks, and solutions. *Climate Change Biology*. Elsevier, 339–356.

Heaviside, C., Macintyre, H., and Vardoulakis, S. (2017). The urban heat island: Implications for health in a changing environment. *Current Environmental Health Reports*, 4(3), 296–305.

Hettiarachchi, G. M., and Pierzynski, G. M. (2004). Soil lead bioavailability and in situ remediation of lead-contaminated soils: A review. *Environmental Progress*, 23(1), 78–93.

Hillel, D., 2012. *Applications of soil physics*. Academic Press, Cambridge, Massachusetts, USA.

Hinojosa, M. B., García-Ruíz, R., Viñegla, B., and Carreira, J. A. (2004). Microbiological rates and enzyme activities as indicators of functionality in soils affected by the Aznalcóllar toxic spill. *Soil Biology and Biochemistry*, 36(10), 1637–1644.

Jacob, D. J., and Winner, D. A. (2009). Effect of climate change on air quality. *Atmospheric Environment*, 43(1), 51–63.

Jim, C. Y. (1993). Soil compaction as a constraint to tree growth in tropical & subtropical urban habitats. *Environmental Conservation*, 20(1), 35–49.

Jim, C. Y. (1998). Soil compaction at tree-planting sites in urban Hong Kong. In *The landscape below ground II: Proceedings of an International Workshop on Tree Root Development in Urban Soils* (pp. 166–178). Champaign: International Society of Arboriculture.

Jónsson, J. Ö. G., and Davíðsdóttir, B. (2016). Classification and valuation of soil ecosystem services. *Agricultural Systems*, 145, 24–38.

Khan, M. M., Akram, M. T., Khan, M. A., Al-Yahyai, R., Qadri, R. W. K., and Janke, R. (2022). Urban soils and their management: A multidisciplinary approach. In *Soils in urban ecosystem* (pp. 137–157). Singapore: Springer.

Khillare, P. S., Hasan, A., and Sarkar, S. (2014). Accumulation and risks of polycyclic aromatic hydrocarbons and trace metals in tropical urban soils. *Environmental Monitoring and Assessment*, 186(5), 2907–2923.

Kumar, K., and Hundal, L. S. (2016). Soil in the city: Sustainably improving urban soils. *Journal of Environmental Quality*, 45(1), 2–8.

Lakhiar, I. A., Jianmin, G., Syed, T. N., Chandio, F. A., Buttar, N. A., and Qureshi, W. A. (2018). Monitoring and control systems in agriculture using intelligent sensor techniques: A review of the aeroponic system. *Journal of Sensors*, 2018, 18.

Lal, R. (2004). Soil carbon sequestration impacts on global climate change and food security. *Science*, 304(5677), 1623–1627.

Lal, R., Lorenz, K., Hüttl, R. F., Schneider, B. U., and Braun, J. V. (2012). Terrestrial biosphere as a source and sink of atmospheric carbon dioxide. In *Recarbonization of the biosphere* (pp. 1–15). Dordrecht: Springer.

Lal, R., and Stewart, B.A. (2017). *Urban soils* (1st ed.). Boca Raton: CRC Press.

Lavelle, P., Spain, A., Blouin, M., Brown, G., Decaëns, T., Grimaldi, M., and Zangerlé, A. (2016). Ecosystem engineers in a self-organized soil: A review of concepts and future research questions. *Soil Science*, 181(3/4), 91–109.

Lehmann, A., and Stahr, K. (2007). Nature and significance of anthropogenic urban soils. *Journal of Soils and Sediments*, 7(4), 247–260.

Li, X., Poon, C. S., and Liu, P. (2001). Concentration and chemical partitioning of road dusts and urban soils in Hong Kong. *Applied Geochemistry*, 16, 1361–1368.

Lorenz, K., and Kandeler, E. (2005). Biochemical characterization of urban soil profiles from Stuttgart, Germany. *Soil Biology and Biochemistry*, 37(7), 1373–1385.

Lorenz, K., and Lal, R. (2015). Managing soil carbon stocks to enhance the resilience of urban ecosystems. *Carbon Management*, 6(1–2), 35–50.

Lu, Y., Gong, Z., Zhang, G., and Burghardt, W. (2003). Concentrations and chemical speciations of Cu, Zn, Pb and Cr of urban soils in Nanjing, China. *Geoderma*, 115(1–2), 101–111.

Manta Salviagio, D., Angelone, M., Bellanca, A., Neri, R., and Sprovieri, M. (2002). Heavy Metals in urban soils: A case study in the city of Palermo (Sicily), Italia. *Science of the Total Environment*, 300, 229–243.

Marcotullio, P. J., Braimoh, A. K., and Onishi, T. (2008). The impact of urbanization on soils. In *Land use and soil resources* (pp. 201–250). Dordrecht: Springer.

McDonald, R. I., Kareiva, P., and Forman, R. T. (2008). The implications of current and future urbanization for global protected areas and biodiversity conservation. *Biological Conservation*, 141(6), 1695–1703.

McKinney, M. L., 2006. Urbanization as a major cause of biotic homogenization. *Biological Conservation*, 127(3), 247–260.

McPhearson, T., Andersson, E., Elmqvist, T., and Frantzeskaki, N. (2015). Resilience of and through urban ecosystem services. *Ecosystem Services*, 12, 152–156.

Meharg, A. A. (2016). Perspective: City farming needs monitoring. *Nature*, 531(7594), S60–S60.

Morel, J. L., Burghardt, W., and Kim, K. H. L. (2017). The challenges for soils in the urban environment. Soils within Cities: Global approaches to their sustainable management-Composition, properties, and functions of soils of the urban environment. *Stuttgart: Catena Soil Science*, 1–6.

Morel, J. L., Chenu, C., and Lorenz, K. (2015). Ecosystem services provided by soils of urban, industrial, traffic, mining, and military areas (SUITMAs). *Journal of Soils and Sediments*, 15(8), 1659–1666.

O'Donnell, E. C., and Thorne, C. R. (2020). Drivers of future urban flood risk. *Philosophical Transactions of the Royal Society A*, 378(2168), 20190216.

Pavao-Zuckerman, M. A. (2008). The nature of urban soils and their role in ecological restoration in cities. *Restoration Ecology*, 16(4), 642–649.

Pavao-Zuckerman, M. A. (2012). Urbanization, soils and ecosystem services. In *Soil ecology and ecosystem services* (pp. 270–278). Oxford: Oxford University Press.

Pouyat, R. V., Page-Dumroese, D. S., Patel-Weynand, T., and Geiser, L. H. (2020). *Forest and rangeland soils of the United States under changing conditions: A comprehensive science synthesis* (p. 289). Springer Nature.

Pouyat, R. V., Szlavecz, K., Yesilonis, I. D., Groffman, P. M., and Schwarz, K. (2010). Chemical, physical, and biological characteristics of urban soils. *Urban Ecosystem Ecology*, 55, 119–152.

Rakhshandehroo, M., Yusof, M. J. M., Arabi, R., and Jahandarfard, R. (2016). Strategies to improve sustainability in urban landscape. *Journal of Landscape Ecology*, 9, 5–13.

Ranjan, R., Sharma, N. K., Kumar, A., Pramanik, M., Mehta, H., Ojasvi, P. R., and Yadav, R. S. (2021). Evaluation of maize-based intercropping on runoff, soil loss, and yield in foothills of the Indian sub-Himalayas. *Experimental Agriculture*, 57(2), 69–84.

Robinson, D. A., Hockley, N., Cooper, D. M., Emmett, B. A., Keith, A. M., Lebron, I., and Robinson, J. S. (2013). Natural capital and ecosystem services, developing an appropriate soils framework as a basis for valuation. *Soil Biology and Biochemistry*, 57, 1023–1033.

Rossiter, D. G. (2007). Classification of urban and industrial soils in the world reference base for soil resources (5 pp). *Journal of Soils and Sediments*, 7(2), 96–100.

Savard, J. P. L., Clergeau, P., and Mennechez, G. (2000). Biodiversity concepts and urban ecosystems. *Landscape and Urban Planning*, 48(3–4), 131–142.

Setälä, H., Bardgett, R. D., Birkhofer, K., Brady, M., Byrne, L., De Ruiter, P. C., . . . and Van der Putten, W. H. (2014). Urban and agricultural soils: Conflicts and trade-offs in the optimization of ecosystem services. *Urban Ecosystems*, 17(1), 239–253.

Silva, L., Cachada, A., Freitas, A. C., Pereira, R., Rocha-Santos, T., and Duarte, A. C. (2010). Assessment of fatty acid as a differentiator of usages of urban soils. *Chemosphere*, 81(7), 968–975.

Six, J., Bossuyt, H., Degryze, S., and Denef, K. (2004). Mycorrhizal symbiosis. *Soil and Tillage Research*, 79, 7–31.

Smit, B., and Pilifosova, O. (2001). Adaptation to climate change in the context of sustainable development and equity. In *Working group II: Impacts, adaptation and vulnerability*, Cambridge University Press, USA: IPCC Assessment Report, IPCC.

Stroganova, M., Miagkova, A., Prokofieva, T., and Skvortsova, I. (1998). *Soils of Moscow and urban environment*. Moscow, 1998, 150pp.

Tang, H. X. (2003). Environmental Nano pollutants and micro interface water quality process. *Acta Scientiae Circumstantiae*, 23(2), 146–155.

Tang, L., Tang, X. Y., Zhu, Y. G., Zheng, M. H., and Miao, Q. L. (2005). Contamination of polycyclic aromatic hydrocarbons (PAHs) in urban soils in Beijing, China. *Environment International*, 31(6), 822–828.

Thomaier, S., Specht, K., Henckel, D., Dierich, A., Siebert, R., Freisinger, U. B., and Sawicka, M. (2015). Farming in and on urban buildings: Present practice and specific novelties of Zero-Acreage Farming (ZFarming). *Renewable Agriculture and Food Systems*, 30(1), 43–54.

Tiller, K. G. (1992). Urban soil contamination in Australia. *Soil Research*, 30(6), 937–957.

Tribot, A. S., Deter, J., and Mouquet, N. (2018). Integrating the aesthetic value of landscapes and biological diversity. *Proceedings of the Royal Society B: Biological Sciences*, 285(1886), 20180971.

Trubina, L., Nikolaeva, O., and Mullayarova, P. (2019). The environmental monitoring and management of urban greenery using GIS technology. In *IOP conference series: Earth and environmental science* (Vol. 395, No. 1, p. 012065). Barnaul, Russian Federation: IOP Publishing, Russia.

United Nations. (2019). *World urbanization prospects 2018: Highlights (ST/ESA/SER. A/421)*. Retrieved October, 16, 2019.

Vangronsveld, J., Herzig, R., Weyens, N., Boulet, J., Adriaensen, K., Ruttens, A., and Mench, M. (2009). Phytoremediation of contaminated soils and groundwater: Lessons from the field. *Environmental Science and Pollution Research*, 16(7), 765–794.

Wang, X. T., Miao, Y., Zhang, Y., Li, Y. C., Wu, M. H., and Yu, G. (2013). Polycyclic aromatic hydrocarbons (PAHs) in urban soils of the megacity Shanghai: Occurrence, source apportionment and potential human health risk. *Science of the total Environment*, 447, 80–89.

White, L. J., Brözel, V. S., and Subramanian, S. (2015). Isolation of rhizosphere bacterial communities from soil. *Bio-protocol*, 5(16), e1569–e1569.

Wick, A. F., Stahl, P. D., Ingram, L. J., and Vicklund, L. (2009). Soil aggregation and organic carbon in short-term stockpiles. *Soil Use and Management*, 25(3), 311–319.

Wilcke, W., Lilienfein, J., do Carmo Lima, S., and Zech, W. (1999). Contamination of highly weathered urban soils in Uberlandia, Brazil. *Journal of Plant Nutrition and Soil Science*, 162(5), 539–548.

Wong, C. S. C., and Li, X. (2006). Thornton I. Urban environmental geochemistry of trace metals, review. *Environmental Pollution*, 142, 1–16.

Xue, Q., Liang, B., and Liu, X. L. (2002). Progress on organic contaminant transport and transform in soil. *Soil and Environmental Sciences*, 11(1), 90–93.

Yang, J. L., Wang, J. K., and Zhang, G. L. (2004). The compaction degradation in urban soil and its environmental impacts. *Chinese Journal of Soil Science*, 35(6), 688–694.

Yang, J. L., and Zhang, G. L. (2008). Loss of soil water capacity in urban areas and it's impacts on environment. *Soils*, 40(6), 992–996.

Yang, Y., Paterson, E., and Campbell, C. (2001). Accumulation of heavy metals in urban soils and impacts on microorganisms. *Huan jingkexue= Huanjingkexue*, 22(3), 44–48.

Yang, Z., Lu, W., Long, Y., Bao, X., and Yang, Q. (2011). Assessment of heavy metals contamination in urban topsoil from Changchun City, China. *Journal of Geochemical Exploration*, 108(1), 27–38.

Yu, G., Zhang, Z., Yang, G., Zheng, W., Xu, L., and Cai, Z. (2014). Polycyclic aromatic hydrocarbons in urban soils of Hangzhou: Status, distribution, sources, and potential risk. *Environmental Monitoring and Assessment*, 186(5), 2775–2784.

Zemlyanitskiy LT (1963). Characteristics of the soils in the cities. *Soviet Soil Science* 5:468–475

Ziter, C., and Turner, M. G. (2018). Current and historical land use influence soil-based ecosystem services in an urban landscape. *Ecological Applications*, 28(3), 643–654.

# 12 Potential Nexus Approach for Sustainable Soil Productivity Resource Management

*Prithwiraj Dey, B.S. Mahapatra, Biplab Mitra,
Arya Kumar Sarvadamana, Arnab Roy Chowdhury,
Abhishek Rana, and Diyan Mandal*

## 12.1 INTRODUCTION

It is not wrong to say that human civilization has reached an all-time high that has never been evidenced before because of the advancements in its food production system. The most primitive natural roadblock for a species to grow on earth is the limited food reserves. Human civilization has repeatedly broken this barrier by improving our food production systems. Today, the challenge is again ahead of the human race to manage its food systems sustainably. In order to fulfill the enormous needs of the expanding population for food-fiber-fodder, undue pressure on land resources has been placed for several decades (Godfray and Garnett, 2014). Numerous human activities, including the construction of extensive irrigation systems, deforestation, intense farming, and faulty land use have exacerbated the deterioration in the productivity of several agroecosystems. Today's agricultural systems are challenged by factors like increased salinization, desertification, loss of soil due to erosion, declining factor productivities, and development of complex multi-nutrient deficiencies in soil along with aberrant weather conditions and changing climatic scenarios (Rastgoo and Hasanfard, 2021). Declining productivity of agricultural systems may eventually result in more deforestation, excess grazing, enhanced soil erosion, deterioration in soil and environmental quality, and challenged nutritional security worldwide (Pimentel and Burgess, 2013).

Not only are stagnated production or soil health the current concerns but the environmental footprints of our systems have also become alarming these days. Agricultural systems, including forestry and land use worldwide, are responsible for nearly 24% of the total emission of greenhouse gases (GHGs), placing it in second place for emissions only after the electricity and heat generation sector (IPCC, 2014). Utilization of finite natural resources, such as water, is also consumed at a very high rate in our food systems. As far as the numbers are concerned, nearly 700 BCM of water per year is alone consumed by India's agriculture sector, the world's largest

DOI: 10.1201/9781003358169-12

243

agricultural water consumer. The water requirement of the farming sector in India almost doubled from 1975 to 2010 as the country witnessed the green revolution. China is the world's second-largest user, nearly extracting 385 BCM of water for agricultural activities in 2015 (Ritchie and Roser, 2017).

Apart from the use, these intensive systems with injudicious use of several inputs have also resulted in the river systems and groundwater being contaminated with several toxic chemicals, nitrates as well as arsenic, and so on (Mitra et al., 2021). The harm that has been caused or is being caused by modern agriculture has severe consequences for the environment and ecosystem quality. Urbanization, soil erosion, desertification, salinization, and industrialization have also been pushing the land resources available to agricultural systems to shrink with every passing day, whereas the pressure to feed the inflating population is going up by leaps and bounds. Such scenarios are making existing farming situations more intensive and resource-hungry. Such intensive systems without proper ecological balance may have detrimental consequences (Struik and Kuyper, 2017). In several parts of the world, green revolutions with modern input-intensive agricultural practices have resulted in severe land degradation, environmental pollution, reduced response to inputs, and decline or stagnation in productivity. In many places, groundwater has been depleted to catastrophic levels, whereas in other soil, quality has gone down, making the production systems practically less productive or even barren (FAO, 2018).

Under such backgrounds, it is high time to think again about the decisions we have taken to solve the problems of food production systems. Present-day issues need dynamic solutions that can boost the productivity of soil and food systems without deteriorating environmental quality or depleting the resource base. Resources, processes, and products must be seen within a single large picture to solve future problems. Three main determinants of such systems are energy, water, and food which may be looked at as a novel 'nexus' approach to address the interdependencies of these resource-process-product systems instead of managing each of these independently. Modern agro-techniques such as nanotechnology, drones, precision farming, machine learning, etc., can be amalgamated into rather classical management practices like integrated use of resources, sustainable diversification, and conservation agricultural practices. Such technology blends have proven their potential to address and manage present-day challenges in food production systems. With the aim of managing ecological balance, soil health, and the environment, such technologies can ensure a more sustainable future for the soil and food systems.

## 12.2  SOIL PRODUCTIVITY AND ITS DETERIORATION

The term 'soil productivity', or simply 'productivity', may be defined as soil's ability to generate specific crop yield under a specified set of management measures (Karlen, 2005). The decline in soil productivity can be attributed to several unfavorable changes in the soil's functional characteristics, including nutritional status, soil organic matter content, soil structure, texture, ionic balance, presence of physical or chemical contaminants, and so on. Globally, nearly 7.5 million ha of arable land becomes less productive or even unproductive each year as a result of several anthropogenic as well as natural causes (Aulakh and Sidhu, 2015). The issue of declining soil productivity is as ancient as agriculture itself. However, due to the bulging population of the world at a frightening rate over the past few decades, along with ever-decreasing land for agricultural use, its severity has escalated to a new high. Several countries are facing the challenges of crop yield stagnation as well as declining yield trends at the same time across the world.

One of the primary causes of the decline in the soil's ability to produce is soil erosion. The loss of topsoil should be addressed with utmost delegacy, as to date, the scientific community has not created any viable options to manufacture a single gram of topsoil. Forming a single millimeter of soil layer from bedrocks takes thousands of years on the anthropogenic scale. Several man-made activities may significantly accelerate soil erosion, such as faulty soil management, deforestation, etc. Methods not appropriate for a particular site, such as tillage along a slope or a lack of vegetation during a heavy downpour, excessive tillage, and absence of cover crops or crop residues/mulch, can significantly increase the amount of soil loss from the agricultural systems. In the Indian subcontinent only, more than 85 mha of the land area is thought to be eroding because of wind and water. Data on soil loss from the past few decades were examined, and it was found that soil erosion was occurring at a rate of nearly 16 t ha$^{-1}$ yr$^{-1}$, or 5334 M t yr$^{-1}$, on average from Indian agricultural lands (Narayana and Babu, 1983).

Faulty irrigation techniques and applying fertilizers with high salt index can be addressed as the primary anthropogenic causes of soil salinization. Although with the extension of irrigation facilities, agricultural productivity has gone up by many folds in several parts of the world, a faulty irrigation design and practice may turn that highly productive land into barren ones. Increasing problems of dam salinization are pretty common to notice near first-generation irrigation projects in several parts of the world. On the other hand, irrigating with groundwaters with high salt content may also turn the productive land to unfertile fallow with the build-up of salts from the salty irrigation water. Crop yields and system outputs are severely affected by the high amounts of soluble salts, exchangeable sodium, or both in the soils. The exchangeable sodium percentage (ESP) of alkali soils can reach up to 90%, and pH values can go up to 10.8, resulting in low levels of organic carbon, limited infiltration, destruction in the soil structure, and dispersion rendering low soil fertility conditions. The problem of salinization and development of sodicity in agricultural lands will inevitably get worse as long as saline water is continuously being used to irrigate crops without proper planning and good crop husbandry practices. According to reports, along with saline and sodic soils, acid soils impact production in about 7 mha of land in India, contributing to 9.4% of the country's total geographical area (Maji et al., 2008). Loss or depletion of soil organic matter is another problem that deteriorates soil productivity. The primary causes of the depletion of soil organic carbon are the removal or burning of crop residues, the absence of regular organic input, and intense agricultural practices (Aulakh, 2011).

Following the adoption of the green revolution, nutrient imbalance has become a significant issue. Application of nutrients in excess or unbalanced manner may put the environment, human health, and ecosystems at considerable risk. There are various ways of nutrient loss, including runoff, leaching, erosion, and volatilization to the air as $NH_3$, $N_2O$, $NO$, and $N_2$ or to water. The issue of nutrient loss and/or organic matter depletion may worsen in the future due to rising food needs to support the growing population. This issue of soil health exhaustion has previously been noted in several case studies, particularly in highly intensively farmed areas. Crop output and environmental protection may be significantly impacted by the nutritional imbalance in soils (Aulakh, 2011).

Due to human and geogenic influences, soil can also become contaminated with toxic chemicals and heavy metals. Their effect, however, varies depending on the soil depth, rainfall pattern, and several soil–agro-ecological considerations, including choice of crops, varietal sensitivity, management aspects, etc. Soils in the different corners of the globe include naturally occurring minerals that regulate the levels of geogenic contaminants in productive soils. For example, arsenic (As), fluoride (F), boron (B), uranium (U), and selenium (Se) may be naturally found in alluvial aquifers. In several cases, lifting these contaminants with irrigation water due to over depletion of natural aquifers or using industrial sewage water as an irrigation source may expose the crop and soil to these toxic contaminants, which may affect the soil's chemical and biological health parameters. Various researchers have recently reported arsenic contamination in rice-based systems in West Bengal, Punjab, Haryana states of India, and many parts of Bangladesh (Chakraborti et al., 2018; Chowdhury et al., 2000). Other studies have also highlighted the risks of contamination of agricultural systems with Cr, Cd, Pb, etc., heavy metals across the world (Alengebawy et al., 2021; C.R. et al., 2022).

## 12.3   UNSUSTAINABLE NATURAL RESOURCE UTILIZATION

Natural resources are materials or substances that are utilized by humans to satisfy their needs. Chapman and Roberts (1983) and Reijnders (2000) have described three types of natural resource reserves viz. continuous resources (solar and wind energy), renewable resources (forests and crops), and non-renewable resources (such as fossil fuels and mineral ores). The frequent use of non-renewable resources depletes the reserves since they are produced through prolonged geochemical processes. If sufficient time is not given for recovery, resources like water, fertile soil, and biodiversity might also be regarded as non-renewable. At the start of the 1970s, the Club of Rome first brought the commons' attention to the depletion of resources. If consumption continues at this rate, certain resource reserves will undoubtedly run out very soon. Even some fossil fuels and mineral ores may take longer to deplete. Still, other resources, including biodiversity and soil resources, are being deteriorated pretty quickly that there is a risk that their critical thresholds will be reached in near future.

Damage and depletion to resources like soil and biotic resources have particularly become concerning at present times. The 'health' of the global ecosystem has dropped by 30% over the past 25 years, as measured by estimates of the loss of forest area and freshwater and sea animal species (Loh et al., 1998). In the last five decades, 13% of the world's natural forest cover has vanished (Vancutsem et al., 2021). The most disturbing aspect is that there is no indication that this assault on biodiversity is worsening. When it comes to the depletion of scarce natural resources like water, the numbers are sufficient to show how severe the issue is. Today, water scarcity affects more than 700 million people in 43 nations. By 2025, it has been predicted that nearly 2 billion people will be residing in areas with total water shortage and that almost two-thirds of the world's population may experience moderate to severe water stress (Chu and Karr, 2017).

One factor that contributes to deforestation and makes it worse is poverty. Small and marginal farmers are responsible for around two-thirds of deforestation because they modify their farming practices and cut down trees for fuel and livelihood. However, most of the deforestation is due to the commercial harvesting of wood. The growing number of low-income individuals and the global demand for lumber add to the strain on the surviving forests. As the world's population grows, a rapid increase in demand for food, fiber, and agricultural land is quite expected. To meet the needs of a rapidly expanding population, almost all of the finest agricultural land has already been used. As a result, less appropriate lands are being brought under cultivation, which may cause increased soil erosion and loss of biodiversity due to faulty land use. The foundation of agricultural output is fertile soil, and 25% of all productive soils have been lost or deteriorated in the last 50 years. The least developed nations are the worst hit, and serious issues are predicted in several significant food-producing regions in third world countries. Because of how slowly soils naturally recover, restoration efforts in these nations are out of reach for most people at an individual level. The availability of adequate agricultural land is under more strain as the biofuel industry develops to meet the need for a sustainable fuel source. According to calculations made by the National Institute for Public Health and the Environment (RIVM) of the Netherlands, around 20%–25% of all the good agricultural land will be used for biofuel crops by 2050, if the need of energy is to be fulfilled by biofuel crops only. This large-scale conversion to biofuels will be prompted by policies aimed at reducing greenhouse gas emissions and the scarcity of fossil fuels (RIVM, 2000). Keeping the significant demand for land for food production, it seems improbable that this could occur. The need for both mineral and fossil resources is increasing on a global scale. There is no evidence that global initiatives to limit greenhouse emissions would prevent the usage of fossil fuels from increasing (RIVM, 2000). Closing the biogeochemical cycles can reduce the use of primary resources, but it is exceedingly challenging to meet the demand only by recycling.

The other vital reasons for resource depletion can be listed as follows:

- *Population explosion*—As the population grows, so do its needs, which further contribute to resource depletion.
- *Overconsumption and waste*—People tend to consume more and waste more as their standards of living rise.
- *Mining of oil and minerals*—Today's globe has a significant need for metals and minerals. The depletion of ores makes this a very serious issue.
- *Industrial and technological growth*—As technology grows, so does the need for resources.
- *Soil erosion*—This occurs as a result of deforestation and poor land use practices. As a result, soil loses valuable minerals and nutrients.
- *Pollution and contamination of resources*—Due to our careless attitude towards the environment, water and soil contamination are both rising alarmingly quick nowadays.

- Natural resources have already been contaminated due to pollution, and more and more resource bases are being increasingly contaminated.

When natural resources are short in supply, the cost of food, fuel, and energy increases. Even the cost of renewable resources goes up if they have to be transported from the point of generation to the locations of utilization. Water shortages may arise as a result of expanding infrastructure and population expansion. About 1 billion people do not have access to safe water as of right now. Due to the ongoing destruction of its natural resources, several countries are currently experiencing an ecological catastrophe. By 2025, India is anticipated to overtake China as the world's most populated country. Natural resources are in jeopardy due to the rapidly increasing population and ongoing overexploitation of land and water resources. Water, coal, non-fuel materials, energy, metals, as well as non-metallic materials are all becoming more and more scarce. Natural resources are distributed unevenly throughout the world. Natural resources must thus be used with extreme caution for maximum effect and must be conserved. In India only, water, soil, wind erosion, and other complicated environmental factors have harmed almost 14.7 mha of land (Shiferaw, 2002). Additionally, improper chemical fertilizer application has had a negative impact on soil fertility and minerals. Drought and flooding are affecting many areas of the world.

One must know that the depletion of natural resources is a grave concern. Some of the leading causes are the growing population and people's overwhelming demand for life's comforts. Deforestation, excessive consumption, and resource waste are other factors that contribute to depletion. The planet earth may soon run out of numerous natural resources if we do not address this issue immediately. This will significantly impact both the environment and human existence. Switching to alternative energy sources like solar and wind power and making responsible resource usage are some of the suggested answers. The idea of sustainable development thus must be embraced. We should use resources in a way that both meets our current requirements and leaves a sufficient quantity for usage by future generations. By guaranteeing sustainability in both developed and developing nations, economic and social development objectives may be met. But to address the issue of resource depletion, there must also be a shift in attitudes. The day when the very continuation of life on earth is threatened is not far off if our attitude towards natural resources does not change. It's time for us to accept that we are damaging the earth and its resources and that we must conserve them rather than use them irresponsibly. Only then will the planet be able to endure and retain its balance.

## 12.4   NEED FOR SUSTAINABLE MANAGEMENT OF PRODUCTIVITY

Soil functions can be expressed in terms of different biological, chemical, and physical characteristics of the soils (Balestrini et al., 2015). Ten significant risks to the realization of sustainable soil management (SSM) were highlighted in the status of the World's Soil Resources report. These risks include water and wind erosion, loss of soil organic carbon, nutrient imbalance, soil salinization, soil acidification, soil pollution, loss of soil biodiversity, compaction, sealing, and water logging (Erdogan et al., 2021). Depending on the geographical location, these risks vary in terms of presence and severity. SSM has the potential to directly as well as indirectly contribute towards the 2030 agenda for sustainable development to achieve the agreed goals and targets, such as the following:

- The challenge to eradicate hunger with malnutrition and ensure food security for an expanding population.
- The adaptation and mitigation of climate change, particularly per the Paris Agreement, which was agreed upon at the UNFCCC COP21, represents a significant assurance to address climate change.
- The dedication to halting desertification along with reducing the consequences of drought, particularly the drive to create a world free of land degradation while taking into account the possible advantages for everyone as per the most recent UNCCD COP12.
- Securing land following the guidelines dedicated to land, water, and forest management in the context of national food security.

A proper environment to promote SSM is fostered by the actions like the development of or improvements to policies that support inclusive SSM in agriculture. Environment-inclusive measures to adopt SSM should be linked to appropriate agricultural and environmental policies to maximize its adoption benefits. If these regulations currently exist, they can be updated as needed to incorporate SSM. Investing in SSM should be done more responsibly in accordance with the principles for responsible investments in agriculture and food systems. This includes increasing ethical investment and providing motivating incentives required to enhance sustainable soil management. It is conceivable to think about giving a good incentive to the stakeholders who use SSM principles and recognize the significance of ecosystem-services (Mehmet Tuğrul 2020).

## 12.5 SUSTAINABLE SOIL MANAGEMENT

### 12.5.1 Minimization of Soil Erosion

Several researchers worldwide have addressed that soil-erosion brought by wind and water is the biggest threat to worldwide production systems and the ecosystem services they supply (Steinhoff-Knopp et al., 2021). Soil erosion may potentially cause the loss of topsoil layers that contain organic and inorganic nutrient pools, the partial/complete loss of soil horizons, and the potential exposure of growth-restricting subsoil layers, in addition to off-site effects like harm to public infrastructure, poor water quality, and reservoir sedimentation. Human activities include decreasing plant and residue cover on soil, tillage, and other field operations, as well as lower soil stability, which may result in soil creep and landslides and accelerate soil erosion. Use of appropriate techniques, such as maintaining soil cover and minimal soil disturbances in terms of conservation tillage practices, should be continued to maintain a protective barrier of actively growing plants or other organic/inorganic wastes on the soil surface to reduce its erodibility (Lal, 2015). Land use changes remove the surface cover and cause soil carbon loss. Wetlands, water harvesting, buffer strips, riparian buffers, and cover crops should be employed where suitable to reduce water erosion on slopes and somewhat steep fields. These strategies include terrace building and upkeep, grassed waterways or vegetated buffer strips, contour planting, crop rotation, intercropping, agroforestry, strip cropping, etc. They also include cross-slope barriers such as contour bunds, grass strips, and stone lines. The erosion brought on by wind, especially by dust storms, should be mitigated and managed by utilizing natural (plants) or man-made (stone walls) wind barriers to slow the wind (Kumawat et al., 2021).

### 12.5.2 Sustaining Soil Organic Carbon

In order to preserve soil properties and stop soil deterioration, soil organic matter (SOM) is primarily central. Additionally, it is crucial for coping with and mitigating climate change that SOM should be stabilized or added to global stockpiles (Hatten and Liles, 2019). Inappropriate land use, poor crop management, or loss of soil organic carbon (SOC) can significantly reduce soil quality and health, whereas good SOC management improves soil health. There are many ways to increase the amount of organic matter in the soil. Crop residue management, grazing fodder rather than harvesting, organic and natural farming, integrating soil fertility and disease-pest management, applying manures or other carbon-rich wastes such as biochars, utilizing compost, spreading mulches, or permanently covering the soil are all examples of such practices. Residue burning and field fires should necessarily be avoided to prevent the loss of soil carbon and organic matter present on the soil surface (Bhuvaneshwari et al., 2019). Prolonged and consistent use of all sources of organic inputs to the fullest extent possible, including animal manures, appropriately treated sewage and sludge, green manures, composts, etc., which can act as a regular carbon input to the soil, may build up soil organic carbon. Decomposition and oxidation of soil organic matter can also be slowed down by using minimal or no tillage techniques, retention of residues, and cover cropping (Carter, 2005).

### 12.5.3 Promoting the Positive Nutrient Dynamics

Particularly relevant to the dynamics of nutrients in the continuum between soil, water, and plant systems, the sufficiency and utilization efficiency of plant nutrients can play a significant role in maintaining soil productivity. A proper nutrition schedule for crops should be developed based on crop requirements, regional soil qualities, circumstances, and weather patterns to tailor fit the soil–plant–agro-ecological conditions. By judicious recycling nutrients and integration of multiple sources of plant nutrients such as chemical fertilizers, organic manures, crop residues, other soil amendments, biofertilizers, etc., along with cutting-edge, efficiency-centric technologies like nanofertilizers, real-time nutrient management and plant nutrition schedules can be improved to supply a balanced nutrition (Mejias et al., 2021).

By preserving or improving soil organic matter, the inherent fertility of the soil and the natural nutrient cycles should be enhanced and preserved. Utilizing soil conservation techniques like regular inclusion of legumes, green and animal manures, and cover crops with minimal or no tillage to reduce the use of agrochemicals can improve soil fertility (Scavo et al., 2022). Through integrated systems like crop-livestock, crop-livestock-forest, and multicomponent-integrated systems, nutrient cycles may be effectively managed. Nutrient utilization efficiency should be increased by employing techniques like the application of balanced and appropriate soil organic and inorganic amendments, as well as innovative products (such as controlled release fertilizers, slow-release fertilizers, and nanofertilizers), as well as the recycling and reusing of nutrients (Sekaran et al., 2021). Field evaluations and soil and plant tissue tests must be employed to diagnose nutrient deficiencies properly and suggest proper correctional methods.

### 12.5.4 MANAGEMENT OF SALINIZATION, ALKALINIZATION, AND ACIDIFICATION

Salinization is the build-up of neutral salts that are water soluble in the soil. It results from high ET rates, inland sea water incursion, and human-induced processes (such as faulty irrigation practices). Evaporation losses can be improved by increasing the surface cover. Any instances of over-irrigation, specifically with salt water, must be avoided to reduce soil salinization (Vengosh, 2003). Under certain circumstances where the water source is saline, conjunctive use of poor-quality water with good-quality water can be resorted to. Other crop management strategies that may increase these soils' productivity include selecting tolerant crops and varieties, ensuring leaching requirements, using improved establishment methods such as furrow-irrigated, raised-bed method, and installing drip irrigation systems. Management of irrigation should offer enough water for plant growth and efficient drainage in order to avoid salinization problems. Water desalination can be a viable option when it is feasible, especially in water-scarce nations (He et al., 2021). Surface and subsurface drainage systems may be built up and maintained to control rising groundwater tables as a strategic option where a high groundwater level is a problem. The water balance in these areas must be considered in the design of these systems. Direct salt leaching, planting salt-tolerant plants, domesticating wild salt-tolerant plants for use in agri-pastoral systems, chemical amelioration, and the use of organic amendments are a few techniques for recovering saline soils. Alkali soils and saline-alkali soils with high exchangeable sodium need specific chemical amendments to balance their pH. Such amendments are gypsum, pyrite, elemental sulphur, etc. Also, cropping with tolerant crops like rice, Rhodes grass, etc., is possible even with alkali soil (Kumar and Sharma, 2020). Another problem associated with alkalinized soils is soil dispersion due to sodium. In such cases, if the soil is not calcareous, an external supply of Ca and organic matter to flocculate the soil should be ensured (Kumar and Sharma, 2020).

The leading causes of the acidity of agricultural and forest soils are generally natural, like excessive precipitation, very high temperature, and acidic parent materials; reasons that are anthropogenic in nature are the use of acidic fertilizers, continuous removal of basic cations, a reduction in the soil's ability to act as a buffer, or increased input of nitrogen and sulphur. Employing suitable amendments to reduce the acidity of the soil's surface and subsurface (such as lime, gypsum, and clean ash) can effectively reduce the soil acidity for crop production, per se. The use of organic amendments and balanced fertilization and careful selection of crops can maintain productivity even in challenging soil conditions (Urra et al., 2019).

### 12.5.5 Management of Soil Contamination

Agricultural inputs, by-products of atmospheric deposition, industries, unintentional spills, improper management of urban waste and wastewater, flood and irrigation water, and other factors can all introduce contaminants into soils. Plant toxicity and productivity losses and increased hazards to human and animal health due to build-up in the food chain are just a few examples of the adverse effects that it may result in. Establishing baseline levels, testing and monitoring, and analyzing the levels of contaminant to identify places that are likely to be polluted are necessary for managing local soil pollution (Astel et al., 2011). Remediation should be used to decrease hazards to people and natural systems, together with risk assessment and overall cost analysis. In order to ensure that recycled nutrients from treated wastewater or other waste materials used in agriculture have acceptable levels of pollutants and plant-available nutrients, they should be appropriately processed and analyzed before soil applications (Astel et al., 2011).

### 12.5.6 Management of Soil Compaction

The loss of soil structure and subsequent reduction of soil porosity caused by strains imposed by mechanical means is generally referred to as soil compaction. Due to the destruction of soil aggregates and compression, especially soil macro and mesoporous porosity reduces, thus affecting soil aeration, water holding, and soil hydraulic properties (Shaheb et al., 2021). Due to the decrease in the infiltration and percolation of water, the runoff potential of compact soils increases along with its erodibility. Compaction in the soil has a high mechanical impedance that restricts root development and can result in crusting on the surface, thus impacting seed emergence. It is important to prevent improper or excessive tillage from degrading the soil structure. Reduction in the quantity and frequency of soil mechanical manipulation activities, setting up controlled traffic systems, and only carrying out agricultural operations when the soil moisture content is within optimum limits can prevent the destruction of soil structure and soil compaction. Unhealthy tillage practices like puddling and rototilling should be wholly avoided to preserve the natural aggregation of soil particles (Zheng et al., 2018). Crops, agroforestry plants, and leguminous crops with strong tap roots systems should be rotated in the cropping system so that biological plowing actions can be utilized. To break the compaction in the plow layer, deep non-inverting tillage with subsoilers and chisel plows can be opted for once in 2–3 years in intensive and compaction-prone agricultural systems. Regular input of organic matter, crop residue, green manures, etc., also helps to maintain proper soil porosity and preserves soil structure (Sun et al., 2017).

## 12.6 SOIL BIOTA AS ECOLOGICAL DRIVERS OF PRODUCTIVITY

The important function of healthy soil concerning food production systems is its role in sustaining agricultural productivity. Healthy soil has the ability to promote and sustain crop productivity without resulting in soil degradation. Good soil health is of the utmost requirement for the sustainability of agricultural systems. Various soil ecological drivers are formed due to the interaction of optimum biotic (biological) and abiotic (physical and chemical) factors of soil. These soil health factors are inter-related, affecting the quality and quantity of agricultural produce and its long-term sustainability (Hornick, 1992).

Soil is a rich ecosystem containing diverse macro and microorganisms which make the soil a living system (Jacoby et al., 2017). The life within the soil remains unnoticed, as all the activities of soil organisms are hidden. Still, the importance of soil organisms in improving soil health and crop productivity is crucial (Neemisha, 2020). Bacteria, fungi, actinomycetes, and protozoa are some important microbes in the soil ecosystem that shape the soil health, quality, and material dynamics (Goel et al., 2020). Soil microbes with high metabolic activity decompose raw organic matter, releasing plant nutrients. They help in the recycling of all major nutrients viz. carbon, nitrogen, phosphorus, and other components, which improve the structural and functional properties of soil ecosystems and ultimately improve crop productivity (Dominati et al., 2010). Soil bacteria, the most numerically dominant soil biota, play a significant role in soil ecosystem balance. They are involved in organic matter decomposition, nutrient mineralization, and boosting plant growth through the secretion of several bioactive compounds, especially in the rhizospheric region. Nitrogen-fixing bacteria, such as *Rhizobium, Azotobacter, Azospirillum,* and *Rhodospirillum,* fix atmospheric nitrogen in the soil through symbiotic/non-symbiotic associations and thus reducing the need for nitrogen inputs in agricultural systems. Approximately $2.5 \times 10^{11}$ kg of nitrogen is annually added into the soil through biological nitrogen fixation (Cheng, 2008).

Ye et al. (2021) reported a significant role of the bacterial community in improving crop productivity in terms of maize yield from a four-year-long study due to several beneficial impacts of soil bacteria on agroecosystem quality parameters. Actinomycetes are another group of gram-positive bacteria that also have the potential to fix nitrogen in association with some non-leguminous plants (Bhatti et al., 2017). Actinomycetes live in close association with the rhizospheric zone, and through recycling organic waste and solubilizing phosphate, they help to improve soil fertility (Hozzein et al., 2019). It also has the potential to produce extracellular hydrolytic enzymes, which can catalyze organic matter decomposition in soil systems (Remya and Vijayakumar, 2008). Actinomycetes have also been identified for their growth-promoting activity on plants by stimulating the intensity of mycorrhizal formation in multiple field crops (Mohammadi and Lahdenpera, 1992). After soil bacteria, the next most crucial soil microorganism is soil fungi which dominate the soil by weight. Soil fungi can extend their filaments or hyphae out into the soil from plant roots and thus increase the effective absorbing surface of plant roots. Such structures may mobilize less mobile or immobile nutrients and water to plant roots. In return, the fungus gets nutrition from the host plants. Such symbiotic relationship between roots of higher plants and fungi is known as a mycorrhizal association. It plays a major role in nutrient and water absorption by plants and thus influences productivity (Hesami et al., 2014). Mycorrhizal fungi may act as a barrier for different pathogens and thus prevent crop plants from various diseases (Bagyaraj and Revanna, 2017). Another action performed by soil macro and micro fungi is the decomposition of organic matter. The activity of fungi becomes specifically essential for the decomposition of lignocellulosic materials, as fungi are equipped with ligninolytic enzymes that can break down complex lignin that is resistant to decomposition. Many fungal species absorb toxic metals (cadmium, copper, mercury, lead, and zinc) in their fruiting bodies and prevent toxicity of these soil pollutants in crop plants (Shah et al., 2021). Though the soil algae population is somewhat limited in aerobic soil ecosystems, it maintains soil fertility by adding organic matter to the soil, which is quite central to soil health and quality. Algae help in preventing soil erosion by joining up soil particles in a cohesive manner by secreting several cementing and aggregating agents and thus may help in the betterment of soil aggregation in terms of more water-stable aggregates and lesser soil erodibility by abrasive forces (Khan and Rao, 2019). Some algae, like blue-green algae (BGA), can fix atmospheric nitrogen in the soil. Such algae may be beneficial for biological nitrogen fixation, specifically in wet ecosystems such as rice. BGA has also been known to form a symbiosis with water fern *Azolla* for nitrogen fixation under harsh climatic conditions. Cultivating *Azolla* and making compost with it can add additional biomass and nitrogen to the compost. Otherwise, it can also be co-cultured under a lowland rice ecosystem for sustainable nitrogen management. The use of *Azolla* as a feed additive also benefits cattle by supplying proteins and growth-promoting substances.

Among the soil macrobiota, earthworms are also considered an integral part of the soil ecosystem due to their role in organic matter decomposition. Earthworm casting or excreta improves the physical properties of soil, as well as vermicast, and is known to contain several plant growth-promoting hormones and enzymes on the one hand and promote soil microbiological activities on the other (Gilbert, 2010). Such action of earthworms also aerates the soil and acts as a biological plow to the soil. Another important group among soil macrobiota is soil-dwelling termites which can play an essential role in organic matter decomposition. The termites perform partial digestion of plant matter in their gut and make it easier for other microorganisms to completely decompose this partially digested matter which ultimately releases nutrients in the soil for plant use. *Mastotermes* and *Nasutitermes* are some termites' genera that help decompose organic matter (Breznak, 2000). Marginal farmers often use termite mound soil to improve the soil's physical and chemical properties and enhance crop productivity, as it is a good source of nutrients, plant growth-promoting hormones, and organic matter (Lelisa Deke, 2016).

## 12.7   FOOD AND ENERGY NEXUS

'Nexus', a term that means 'to connect', has been the center of discussion in various scientific literature regarding sustainability in food production systems in recent times (Bazilian et al., 2011; Laurentiis et al., 2016; Lawford et al., 2013; Vogt et al., 2012). The term 'nexus' first came to light between food and energy in 1983, when several initiatives were introduced by the United Nations University (UNU) to depict the interrelations between food and energy (Tashtoush et al., 2019). The water-food-energy nexus has emerged as a new concept in recent times as a resource management option to achieve sustainable development goals. This integrates water, food, and energy, the key resources for sustaining lives in a singular approach. Because interdependencies exist among these, it is important to manage these together instead of each separately in isolation. With the world population growing beyond 7 billion and expecting to be 9 billion by 2050, there will be a 60% increase in global food demand, while energy consumption in the world is expected to get a boost of up to 50% by 2035 (FAO, 2014; IEA, 2010). This is accompanied by global challenges and issues like economic crisis, mismanagement of natural resources, climate change, increasing poverty, etc. These are again associated with economic and sociopolitical risks and unrest faced by generations one after another (Daher and Mohtar, 2015). In 2011, the World Economic Forum declared food, energy, and water as three upholders of the nexus approach (Daher and Mohtar, 2015). In the same year itself, the German Federal Government organized an international conference, 'The Water, Energy, and Food Security Nexus: Solutions for the Green Economy', in Bonn, which showed methods for acquiring sustainability, where the WEF nexus got focus as a global agenda for the first time (Leck et al., 2015). Since the Bonn conference, interconnections among three sectors as a whole have been called the 'water-energy-food (WEF) nexus'. The performance of a system can only be optimized when the functions of all its sub-systems are considered holistically rather than judging their individual performances in isolation (Newell et al., 2011). This indicates that the prosperity of one sector on its way to achieving sustainability will also impact the other sectors. The concept of the 'nexus' has managed to fetch attention, being an integrated approach. Its content applies to several linkages, i.e., food and energy (Liu et al., 2019; Mortada et al., 2018), water and energy (Dai et al., 2018), energy and food (Taghizadeh-Hesary et al., 2019), and water-energy-food (Hussien et al., 2017; Li et al., 2019).

### 12.7.1 Food-Energy-Water Interrelations

The energy sector is the second-largest, water-using sector after agriculture (Molajou et al., 2021). Water is required for energy generation, heating, cooling, resource extraction and refining, transportation, and bioenergy production (Daher and Mohtar, 2015). For instance, electricity generation in power plants needs water, or water is used as a coolant in thermal power plants (Kholod et al., 2021). Other uses of water include heating by generating steam, cleaning of equipment, etc. Thus, different equipment like chemical reactants, cooling towers, and heat exchangers depend on water for their use. After usage in all these activities, a large amount of wastewater is returned to water bodies, which can potentially impact the water quality. Another aspect where water is used indirectly in energy generation is the production of bio-fuel crops, a renewable energy source. Biofuel crops including cereals (corn, soybean), sugarcane, and vegetables need a huge amount of water for irrigation (Molajou et al., 2021). This bioenergy production is getting popular nowadays for combating climate change since it increases energy production and reduces GHG emissions (Chang et al., 2016). But it has also been observed that these biofuel crops are cultivated on the lands at the cost of removal of forests, emission of a considerable amount of carbon dioxide, and destruction of the ecosystems, along with the potential to cause biodiversity losses (Meehan et al., 2010).

On the other hand, energy is essential to supply water for agricultural, industrial, and domestic use. Extraction, treatment, transfer, and reuse—all these actions need energy. For lifting 100 m$^3$ of water per minute to 100 m height, more than 1.5 MW power is required. To get into numbers, 172, 312, 1,000, and 2,012 GWh of electricity are consumed by water transmission and supply, treatment, distribution, and wastewater treatment activities, respectively, in the USA alone. Different kinds of water sources, attributes of water obtained from those, and the system's efficiency determine energy consumption for using that water. For example, sea water treatment requires more amount of energy than that ground and surface water. In the early days of agriculture, irrigation was done in the fields using surface water only. Gradually use of groundwater increased, and this shift increased the energy expense alongside declining the groundwater level. And now, more and more energy is required to extract water from even lower levels of groundwater (Molajou et al., 2021).

Food production relies heavily upon energy and water inputs. FAO (2014) reported that agricultural food production and its supply chain on a global scale take around 30% of the total energy used globally. In current farming practice, most field operations like plowing, sowing, irrigation, and harvesting based on agricultural implements ranging from minuscule scale to large equipment depend on fuel or electricity (Molajou et al., 2021). A study showed that wheat planting in winter requires 3,400 million joules of energy. Also, 800 million and 650 million joules of energy are consumed for ploughing and harvesting most crops (Chang et al., 2016). Another vital aspect of energy use in food systems is the energy input in the transportation of food. To reduce the energy footprint of agricultural transport systems, localizing agricultural products has been proposed as a matter of focus in recent times to maintain the freshness and quality of the produce and reduce energy inputs. The use of chemical fertilizers for crop production also increases energy consumption. A considerable amount of energy is required to produce chemical fertilizers (Finley and Seiber, 2014). Researchers have shown that manufacturing fertilizer and pesticides alone take away 30% to 50% of the total energy used in agriculture. For the production of a kilogram of each nitrogen, phosphorus, and potassium fertilizer, energy spent is in the range of 34, 8, and 6 million joules, respectively (Asian Development Bank, 2013).

In the period of 2006–08, there was a significant hike in the average global price of different commodities. Consequently, a huge part of the global population could not meet their primary nutritional needs. Among several reasons for that price hike, one primary reason was the increase in world fuel prices. As a counter-response, several countries resorted to biofuel policies. This has received both positive and negative criticism. Biofuel policy advocates emphasize the fact that it has little impact on the environment, but again, this ignores the fact that the quantity of inputs used for biofuel production from 2006 to 2009 was a tiny amount of the globally consumed energy, but it was the year's supply of cereals for nearly 0.25 million people (Molajou et al., 2021). Therefore, food products should be looked at as mediums to provide food to people instead of treating those as a source of energy production, and the world biofuel sector needs to find a delicate balance between the production of food and fuel from the same scarce resource base (Dodds and Bartram, 2016). As already mentioned earlier, the sustainability of biofuels remains a serious question considering their water-intensive nature and soil and water degradation accompanied by excessive use of fertilizers. FAO predicts that the area for cultivating crops that supply fuel is projected to go up to 35 mha by 2030. Such a boom in biofuel production instead of food crops may create a serious concern for land resource use in the near future (Daher and Mohtar, 2015).

The links between water and food are expressed in terms of the need for water used in agriculture. A whopping 70% of the total global fresh water from different water sources is used in agriculture alone (FAO, 2014). With the increase in the population and technological development, water consumption has risen in recent times, but at the same time, available water resources have decreased (Molajou et al., 2021). Adopting different means, e.g., suitable crop varieties having more water productivity, resorting to micro-irrigation methods to irrigate the crops to optimize the production process, can decrease water consumption while increasing or maintaining the food production. Wasting of irrigation or rainwater should be prevented. Wasteful activities like the cultivation of summer rice with excessive irrigation water should be avoided, and in-situ water harvesting and groundwater recharge should be emphasized in agricultural systems. Another use of water includes raising livestock. Animal husbandry needs more water as compared to agricultural products (Chang et al., 2016). Sustainable livestock management is thus also required to reduce the agricultural water footprint as a whole.

Water is an essential input for food production and that is the reason why global water security gets affected by changes in patterns of food consumption and agricultural practices (Molajou et al. 2021). Reduction in the water quality is the most important ill effect of food production and processing systems on water. Chemical fertilizers and pesticides often contaminate irrigation water in agricultural lands. A certain amount of that again goes back to the water bodies through runoff or infiltration. In the food production process, wastewater is produced in large amounts, which deteriorates the water quality to a great extent. The magnitude of pollution has to be minimized by adopting different sustainable measures to decrease it. For example, through drip irrigation, water is applied only in the root zone in a discrete manner, which can be a helpful tool since there will hardly be any runoff issues. Groundwater aquifers should also be protected by assuring that polluted water doesn't enter the aquifers using suitable drains (Finley and Seiber, 2014; Slorach et al., 2020).

### 12.7.2 WEF Nexus: Potential Aide of Sustainable Goals

WEF nexus research and development can provide practical solutions and assistance to achieve Sustainable Development Goals. Out of 17 Sustainable Development Goals (SDGs), three are straightway connected to the elements of the WEF nexus, thereby making them hermetically associated (Daher and Mohtar, 2015; Mohtar and Lawford, 2016; Saladini et al., 2018). Such SDGs are as follows:

Goal 2: 'Zero Hunger', aiming for achieving food and nutritional security.
Goal 6: 'Clean Water and Sanitation', aiming to assuring clean water for livelihood and managing sanitation and water sustainably.
Goal 7: 'Affordable and Clean Energy', aiming for providing ensured access to affordable energy which is also sustainable.

Fusing the goals in such an integrated perspective can assist in acquiring other goals as well, such as zero poverty (Goal 1), sustainable economic growth (Goal 8), sustainable communities and cities (Goal 11), responsible consumption and production (Goal 12), and climate action (Goal 13). Therefore, the nexus approach is integrated into nature and is crucial for reaching the three goals. It may also act as means to analyze connections among other goals and achieve them altogether.

Since 2011, The WEF nexus approach has been largely promulgated, as security cannot be achieved for each sector separately; instead, all three sectors (water, energy, and food) have to be considered in connection with one another. The WEF nexus is a perspective in which its components rely upon each other while having inevitable reactions with other natural resources. The significance of this concept, by recognizing the interconnections among these three components, lies in showing ways to proficiently utilize and manage resources sustainably, reduce adverse effects on the environment, and combat climate change issues. Thus, the WEF nexus can be critical in achieving better coordination among these sectors.

## 12.8  EFFICIENCY-DRIVEN NOVEL TECHNOLOGIES

Higher input use efficiency decreases the cost of production, reduces spillage of inputs, especially chemical inputs, into the environment, and protects the environment from pollution. Ultimately, it enhances the profitability and sustainability of a production system. Several studies were conducted in different parts of the world to improve the efficiency of two primary crop production inputs, i.e., nutrients and water. Among the nutrients, nitrogen was mainly studied in the cereal-based production system, and several strategies were curved out through different technologies.

### 12.8.1  PRECISION FARMING

Precision farming (PF), or precision agriculture (PA), is a philosophy that avoids the blanket application of inputs, especially nutrients, water, agrochemicals, etc. Instead, temporal and spatial variability in soil and crops are taken into account, and inputs are applied as per the need of the crop and/or soil. Therefore, inputs are applied in precise amounts and in exact time to synchronize demand-and-supply balance. In PA, the entire field is bifurcated into small grids, also called 'management zones', depending on the physicochemical properties (pH, available N, P, K, secondary and micro nutrients, moisture regimes), yield limits, and other factors. Each zone adopts a different management strategy (rate of application, time of application, etc.) to fit the crop needs. Site-specific nutrient management (SSNM) is also used to address the indigenous heterogeneity of soil fertility in different sites and also crop-variety-specific nutrient demand with precise nutrient application considering the spatial variability. At a large scale, the adoption of PA requires several sophisticated technologies like remote sensing (RS), global positioning system (GPS), geographical information system (GIS), etc., but several hand-held sensors like chlorophyll meters (like SPAD meters), normalized difference vegetation index (NDVI) sensors like GreenSeeker, multispectral sensors, etc., and efficient tools like leaf color charts (LCC) are available for use in small-scale farm levels, showing encouraging results in bridging down the yield gaps and enhancing nutrient use efficiency, especially for nitrogen which is one of the most loss-prone nutrients in the system. The basic principle involved in all of these sensors or tools is that leaf chlorophyll concentration and leaf spectral behaviors are highly correlated with leaf nutrient content, indicating the crop's real-time nutrient demand. Though most of the works have been carried out in real-time nitrogen sensing largely and quite accurately, in recent times, it has also been demonstrated to accurately predict the demand of potassium and phosphorus along with nitrogen with the combined used of NDVI with normalized difference red-edge index (NDRE) sensors. Shukla et al. (2004) calibrated readings of LCC and SPAD in rice and wheat for optimizing yield and nitrogen recovery efficiency ($RE_N$) based on a two-year field study at Meerut, UP, India. They reported that the application of 20 kg N ha$^{-1}$ at LCC score $\leq$ 3 in tall traditional long-duration basmati rice and the application of 30 Kg N ha$^{-1}$ at LCC score $\leq$ 4 in short-duration inbreeds or medium-duration hybrid rice, skipping basal N application, significantly increased yield (13.1%, 10.9%, and 8.8% for three types of varieties, respectively) and $RE_N$ (26.0%, 30.2%, and 21.9% for three types of varieties, respectively) over fixed time N application at recommended doses of the respective varieties. Similarly, in the case of wheat, the application of 30 kg N ha$^{-1}$ at LCC score $\leq$ 4, skipping basal N application, significantly increased yield (14.2%, 16.3%, and 12.8% for the early, timely, and late sown wheat, respectively) and $RE_N$ (2.6%, 27.6%, and 30%, respectively, for three types of wheat varieties). Dobermann and Cassman (2002) reported that across the rice fields in Asia, a yield gain of 11% and an increase in $RE_N$ from 31% to 40% could be achieved through different SSNM practices.

Alongside nutrients, water has now become a precious input in crop production. Its productivity or use efficiency needs to be increased in crop production systems, particularly for high water-demanding crops (rice, sugarcane, vegetables, etc.). Rodrigues et al. (2022) conducted a two-year field study with different crop establishment methods and irrigation management in rice. They suggested the superiority of zero-tilled, direct-seeded rice (ZT-DSR) over conventionally tilled, direct-seeded rice (CT-DSR) for an advantage of 8%–10% higher water productivity (WP). They have also suggested that an application of irrigation water 72 hours after surface drying is better for enhancing WP (6%–8%); net energy use (16%–20%) over the application of irrigation at 20% depletion of available soil moisture for DSR. Several sophisticated water management technologies are being used worldwide, especially in high-value crops, to increase water use efficiency (WUE) and conserve water— for example, low-pressure sprinklers, low-energy precision application (LEPA), drip, micro-irrigation systems, variable rate irrigation (VRI) systems, etc. There are mainly two types of VRI systems studied adequately across the globe:

1. Variable rate lateral irrigation system consisting of the components like GPS receiver, specific software-based relays, and valves.
2. Center-pivot VRI system consists of components like a pivot control panel, control nodes, global navigation satellite system (GNSS), VRI control panel, variable frequency drive, solenoid valves, and a remote-control system (Neupane and Guo, 2019).

The performance of a variable rate drip irrigation system was assessed in the vineyard and found to increase yield by 17% and, at the same time, decrease water consumption by 20% (Neupane and Guo, 2019). Vories et al. (2017) assessed yield and WUE of rice cultivation with a VRI system using a center pivot having two management-allowed depletion (MAD) treatments (MAD 1: 10 mm application at 12 mm allowed soil water depletion [SWD]; MAD 2: 15 mm application at 19 mm SWD) and three VRI application (75%, 100%, and 125% of target water application [TWA] under each MAD). The treatment combination of MAD 1 (100% TWA) gave significantly higher grain yield than the mean yield performance, whereas the highest WUE was obtained from the treatment combinations (MAD 1 [75% TWA] and MAD 2 [75% TWA]) over the mean field performance. Such variable rate applicators of precision irrigation systems can be further equipped with ground-based or even airborne hyperspectral or multispectral and thermal imaging to accurately depict the real-time soil moisture and crop water stress in the field and formulate a decision support system based on the real-time inputs. Artificial intelligence and deep learning are also getting popular in increasing the modeling and predicting ability of precision farming technologies (Dey and Pandit, 2020; Pandit et al., 2021).

## 12.8.2 DRONE TECHNOLOGY

Drones are remotely operated unmanned aerial vehicles (UAVs), usually having sensors mounted on UAVs for the acquisition of real-time data from a relatively larger area within a very short time without deploying humans. In agriculture, a considerable gain in efficiency for field-level operations can be achieved through the use of this technology. For example, surveying and monitoring soil health include soil moisture content, terrain condition, the status of soil fertility, soil degradation, etc. It has profuse use in crop surveillance for monitoring crop nutrient status, an infestation of diseases and pests, etc. This technology ensures early detection of soil and crop health for its timely correction by the farmers and thus can potentially avoid huge damages to agricultural production. Drone-based technologies protect the environment by saving over the use of fertilizers, pesticides, and other agrochemicals. It also saves water, as drone-based sprayers generally operate on ultra-low volume (ULV) sprayers for uniform, precise spraying of materials on plants. Mogili and Deepak (2018) reviewed the use of UAVs in different farm operations like crop monitoring and agrochemical applications and pointed out that the use of UAVs can be of great help for implementing PA practices in real-time. For crop monitoring, thermal and multispectral cameras are generally mounted at the downside of the quadcopter. These cameras can take one snapshot per second, store them in memory, and transmit the images to the ground monitoring unit. These multispectral images are generally captured in five distinct bands under visible range having different wavelengths viz. blue (440–510 nm), green (520–590 nm), red edge (690–730 nm), and near infrared (NIR) (760–850 nm). The digital image captured by multispectral cameras is analyzed through NDVI, NDRE, SDVI, etc., indexes developed and tested by researchers all across the globe. With these indicators plotted in a single processed image, a farmer can easily identify the patch with sparse vegetation or poor plant health. Accordingly, corrective measures can only be taken in that patch, thus increasing the efficiency of field operation. The sprinkling system is usually attached to the lower side of the UAV, has a nozzle beneath the pesticide tank for spraying the chemical, and is controlled by a controller unit. It ensures a highly uniform spray of chemicals for hectares of the field in one go within a very short time and meager use of energy and human resources.

## 12.8.3 NANOTECHNOLOGY

While precision farming can bring down the spatiotemporal variability in systems to a significant extent, the use of novel nanotechnology in food production systems can cut down the use of the scarce resource as well as bring forward a way for sustainable farming solutions. Nanomaterials refer to any materials having one dimension within the 'nano' range, i.e., 1–100 nm, by the United States Environmental Protection Agency. But from the perspective of agriculture, nanomaterials may be of dimension 10–1000 nm possessing colloidal as well as particulate properties (Mukhopadhyay, 2014). Nanotechnology can be described as a set of specific techniques or procedures for using nanomaterials to perform precisely within a system through sophisticated devices. Presently, nanomaterials are used for the efficient nutrient delivery system to plants per plant demands. It is feasible to use nano-fabricated materials in aqueous suspension and hydrogel forms for their hazard-free storage and easy delivery system. Zerovalent iron nanoparticles, as well as nanoparticles from iron rust, are being used for the reclamation of soil polluted with pesticides, radioactive elements, heavy metals, etc., owing to the high adsorption affinity of these nanomaterials for organic molecules and heavy metals (Banik et al., 2020). There are further opportunities for nanomaterials for efficient drug delivery systems precisely at cellular levels within plant systems, developing sensors for early detection of pests and pathogens, and intelligent delivery of fertilizers and pesticides in agriculture (Mukhopadhyay, 2014). Shen et al. (2021) conducted a comparative study between nano-fabricated, controlled-release urea (CRU), designated as silica-modified, fluorinated, lauryl-methacrylate-containing polyacrylate (SFLPA) and a normal urea fertilizer in rice and reported that SFLPA significantly favored the yield (13.4%) and nitrogen use efficiency (24.3%) in rice over a three-split application of normal urea. Kumar et al. (2022) highlighted the result of a field trial with nanofertilizers (nano N, Cu, and Zn) in wheat and reported that combined application of 50% N along with the foliar application of nano N at 25 DAS, nano zinc at 38 DAS, and nano copper at 53 DAS significantly boosted grain yield of wheat over 100% N, P, and Zn. Even for reducing the environmental impacts of the manufacturing process as well as the cost of production of the nanomaterials, environmentally friendly green synthesis of several nanoparticles can be a potential option (Sahoo et al., 2021). The use of nanotechnology is not only limited to nano-nutrients with better absorption by plants. Several nano-formulations are being tested to achieve sophisticated control over release patterns. Nano-herbicides and insecticides are also under rigorous research to cut down the amount of agrichemicals used through achieving higher use efficiency of applied inputs.

Besides these, biotechnology may also take a lead role in genetic manipulation which is very important for bringing the sustainability of the agricultural systems.

## 12.9  AGRONOMIC MANAGEMENT FOR SUSTAINABLE PRODUCTION

Sustainable production systems aim at long-term stable food production without increasing the demand for various resources like nutrients, water, energy, light, etc. Hence, it focuses on conserving the future resource base and utilizes different inputs judiciously and efficiently without degrading the quality of crops, biodiversity, and the environment (Imadi et al., 2016). Sustainable production is driven by the principles of ecology and essentially integrates plant production practices with site-specific interventions having long-lasting benefits. It not only meets the demand for food and fiber for human beings but also enhances the quality of natural resources as well as the environment.

### 12.9.1  ORGANIC FARMING/ORGANIC-INORGANIC INTEGRATION

Managing farmlands with locally available organic inputs (manures, composts, biopesticides, bio-control methods, etc.), either solely or in combination with inorganic inputs (integrated system), is a sustainable approach to crop production which enhances the quality of the environment, promotes biodiversity of the ecosystem, and maintains a level of production for a more extended period. Organic farming promotes sustainability through crop synergism, nutrient recycling, soil and water conservation, enrichment of soil organic carbon (SOC), and so on. Das et al. (2014) conducted a study in a 30-year, long-term experiment in the eastern plain zone of Bihar, India, and found that a higher sustainable yield index (SYI) of rice (0.92–0.95) and wheat (0.71–0.74) under the rice-wheat cropping system can be attained after 27 cycles of rotation by substituting 50% N in rice with organic inputs (farmyard manure, wheat straw, or green leaf manure of *Sesbania aculeta*), and 100% recommended dose of fertilizers (RDF) with inorganic fertilizers had been applied in wheat, as compared to 100% RDF applied to both the crops. However, there is a concern about the yield stability of arable crops managed solely with organic inputs, though other aspects of sustainability, like biodiversity and environmental quality, are usually improved. Knapp and van der Heijden (2018) performed a meta-analysis of the temporal yield stability of three principal cropping systems *viz.* organic and conservation (no-tillage) agriculture vs conventional agriculture and concluded that organic agriculture has significantly inferior temporal yield stability (−15%) as compared to conventional agriculture. So future efforts should be attempted to reduce the temporal yield variability of organic agriculture systems considering the issue of food security in the developing world.

### 12.9.2  CONSERVATION AGRICULTURE

Conservation agriculture (CA) is a farming practice considered as most sustainable nowadays. It aims at conserving top fertile soil as well as regenerating degraded soil through its three main principles: minimum soil disturbance (with no or minimum tillage), permanent soil cover (retention of crop residue), and sustainable crop diversification. Sustainability in a CA-based production system is achieved in a number of ways like promotion of biodiversity and restoration of natural biological cycles both above and below the ground ecosystems, enhancement of soil organic carbon stock of arable soil profile, supporting good agronomic practices like timely sowing, better nutrient, water, and energy use efficiency, minimizing abiotic stresses like drought, heat, and salinity, better control of weeds on a long run and pest complex, reducing global warming potential and environmental footprints of cropping systems, nutrient enrichment of soil through legume inclusion, and above all, enhances marginal profitability of farmers through reduction of production cost. Recently, Kader et al. (2022) compared two levels of soil disturbances (conventional tillage [CT] and strip tillage [ST]) on unpuddled soil with two levels of residue load (low [15% of crop height] and high [30% of crop height]) and five levels of nitrogen (60%, 80%, 100%, 120%, and 140% recommended dose of nitrogen) in a nine-year rotation of rice-wheat-mungbean system of Bangladesh and found that the yield of rice and wheat were comparable between the two tillage systems for six and three years respectively, but the yield of mungbean crop was significantly increased in ST system form the first year itself, resulting in a 26% higher land equivalent ratio, which was significantly superior over the CT system. Similar improvement was noticed in the ST system for nitrogen use efficiency and partial factor productivity. The requirement of nitrogen fertilizer for maximum yield in the ST system was 10% higher in rice and 5% in mungbean because of the higher yield of the respective crops, but the demand for N was reduced by 5% in wheat. Interestingly, it was noticed that when the same yield level, as in the CT system, was compared under the ST system, the requirement of N was reduced by 50%–90%. It was also observed that the soil organic carbon stock in the top 15 cm soil depth was increased from 21.5 t ha$^{-1}$ to 30.5 t ha$^{-1}$ in the ST system due to less turnover of soil organic carbon (SOC) as well as carbon enrichment of soil due to residue retention. This proves CA-based ST system of production is high yielding in the long run (higher yield stability), low nutrient demanding for a comparable level of yield due to better nutrient use efficiency, and more eco-friendly due to improvement in SOC.

Partial nutrient balances for nitrogen and phosphorus were found positive with positive changes in soil organic C under rice-wheat and rice-maize managed under zero tillage systems in eastern Gangetic plains (Sinha et al., 2019). The study also indicated that specific remedial measures like liming and efficient fertilizer management could be advocated for the long-term sustainability of the systems. (Mitra et al., 2018) also reflected CA-based alternate crop establishment options like unpuddled transplanting of rice as a profitable, energy-efficient, and sustainable option for rice-based cropping systems. The impact of a CA-based system on GWP was assessed by Jahangir et al. (2022), who reported that a ST-based system with a higher residue load (30% of crop height) along with higher rates of N (180 kg ha$^{-1}$) resulted in 55% higher $N_2O$ emission than that of a CT-based system managed with lower residue (15% of crop height) along with lower rates of N (108 kg ha$^{-1}$). On the contrary, $CH_4$ emission was drastically reduced in CT-based ST system than CT system with low residue load. The ST plus lower residue along with medium rates of N (144 kg ha$^{-1}$) led to lower GWP by 39% over the same obtained under a CT-based system plus higher residue load along with a lower N rate (108 kg ha$^{-1}$) and 16% over a CT-based system. Such promising results show how adopting CA practices can effectively design environmentally sound, sustainable production systems.

### 12.9.3 CROP DIVERSIFICATION

Crop diversification is a critical cultural practice of sustainable agriculture and is essentially one of the main principles of CA. It refers to the inclusion of new crops or different varieties of the same crop and/or new enterprises in a pre-existing cropping pattern or farming system at the farm level or on a larger scale. Different agronomic practices, such as crop rotation or intercropping, may be followed in order to diversify a cropping pattern. Crop diversification ensures better resilience and provides more agronomic stability as well as enhanced spatial and temporal biodiversity in farms (Paroda, 2022). The resilience is increased due to reduced infestation of annual weeds, insects, and pathogens, less dependence on chemical fertilizers, especially nitrogenous fertilizers (cereal-legume rotation/mixing, intercropping), reduced erosion (introduction of cover crops), and enhanced soil fertility and soil biodiversity. In India, Andhra Pradesh, Maharashtra, and Gujarat are the states having the highest level of diversification, followed by West Bengal, Bihar, and Karnataka. In contrast, states like Odisha, Madhya Pradesh, and northeastern states still follow age-old cropping patterns and have a lower degree of diversification (Paroda, 2022).

Crop diversification strategy must be planned in view of all the major sustainability parameters like environmental aspects, economic viability, soil health, and resource use. The choice of high-value, low-input demanding crops may add more pennies to the farmers' pockets while simultaneously conserving the scarce resource bases.

**TABLE 12.1**

**Some Feasible Options for Crop Diversification in India.**

| States/Parts of India | Predominant crops/ Cropping system | Feasible options for diversification in specific areas |
|---|---|---|
| Punjab, Haryana, Uttar Pradesh | Rice-wheat | Rice (medium duration)-potato-green gram (Varanasi and Faizabad, Uttar Pradesh) |
| | | Rice-potato-sunflower (Jalandhar, Punjab) |
| | | Rice-vegetable pea-wheat-green gram (Karnal, Haryana) |
| Bihar | Rice-wheat | Rice (long duration)-chickpea + coriander-maize + cowpea (Bhagalpur, Bihar) |
| Southern India and Bihar | Chickpea (long duration) | Chickpea (short duration) in Andhra Pradesh (non-traditional areas), Bihar, Tamil Nadu, and Karnataka |
| Gujarat | Ground nut | Pigeon pea (short duration) |
| Central and peninsular region | Soybean-wheat, cotton-safflower, sorghum | Black gram + green gram (mixed cropping), pigeon-pea |
| Bihar (rainfed ecology), West Bengal, Odisha, Assam (low land, rainfed ecology) | Rice-fallow | Lentil, mustard, pea |

*Source:* Paroda (2022) and Singh et al. (2019)

Such crops include millets, pseudo-cereals, resource efficient, resilient crops, as well as high-value, alternative, nonconventional crops like dragon fruit, fiber flax, and industrial hemp (Dey et al., 2022, 2021b, 2021a). Inclusion of at least one legume crop in the system, especially rice-wheat system, can add benefits of atmospheric nitrogen and extra income. In such intensive systems, green manuring crops like dhaincha, sunhemp, etc., may also add biological nitrogen as well as biomass carbon to the soil and thus aid in ensuring the sustainability of the system (Mahapatra and Dey, 2022).

### 12.9.4 Altered Land Configuration

Alteration of land configuration can also be counted as a sustainable agronomic approach for diversification and the intensification of pre-existing cropping patterns for enhanced resource use efficiency and profitability. For example, the bun system of cultivation in Meghalaya, a northeastern state of India, refers to the cultivation of vegetable crops like carrot, cauliflower, tomato, cabbage, etc., in raised beds of 1 m wide, 8–10 m long, and 30–50 cm high, leaving the sunken space in between two beds unutilized. These raised beds are generally levelled off for the cultivation of rice after the vegetables grown on these beds are harvested. So construction and deconstruction of these beds incur recurring expenditures for the farmers, making the system unsustainable. A viable alternative, in this case, is the maintenance of a permanent raised bed (PRB) system for more profitability of the production system. Das et al. (2015) compared the outcome of the field demonstration with crop diversification on the PRB system with conventional farmers' practice which was mono-cropping of long-duration rice in flatland. In this PRB system, beds were prepared to a height of 50 cm with 1 m width (1:1 ratio) by the cutting and filling method. The beds' surfaces were leveled so that 50% of runoff from each of these beds could be disposed of to the interbeds' sunken areas (inter-plot water harvesting). In the sunken beds, rice-lentil/pea were grown, keeping the spacing of rice 20 cm × 20 cm and growing the succeeding crops in between the two rows of rice under zero tillage. In the raised beds, different vegetables were grown (potato, okra, tomato, carrot, brinjal, French bean, broccoli, etc.) as per the standard crop calendar of Meghalaya. Two to three rows of each vegetable were grown based on the spacing of the respective crop. In the raised beds, two brinjal-based systems (tomato-brinjal-cabbage and potato-brinjal cabbage) had higher system productivity; two broccoli-based systems *viz.* tomato-okra-broccoli and tomato-brinjal-broccoli gave the higher rice-equivalent yield over farmers' practice. Such practice may also bring a sustainable increase in farm income, thus ensuring a better standard of living and livelihood security to the most vulnerable geopolitical areas.

## 12.10  CONCLUSION

Improved agro-technologies and innovations have the potential to make the food system sustainable for the world. However, those technologies have to be adequately adopted by the farmers. If the developer of any agro-technology does not incorporate the indigenous know-how from the farmers or rectifies and retrofits the technology for the end clientele adaptation, every effort may go in vain. The development of technology is essential but so is its transfer. For efficient transfer of the technologies, tailor fitting each broad technology to the village level is of utmost importance. In the world, most farm households are small farms; thus, it is high time to tailor fit technologies to small farm situations. For example, the dryland area predominates world agriculture; however, generally, most of the newer technologies are only being tested for irrigated conditions. So targeted technologies are the need of the day, which cater to the site and clientele's specific requirements for a fine fit. Developing newer resource-conserving technologies is more critical in resource-poor and challenged ecosystems. Resource-efficient agricultural systems can be developed by integrating different approaches in various dimensions to solve present-day problems. Moreover, agricultural systems should be designed in such a way that they can maintain harmony with mother nature and don't destroy it. Sustainability is only possible when food systems care about concerning ecology and ecosystems. By amalgamating newer technologies with traditional know-how under the nexus approach, goals of conserving resources and sustaining production are quite possible and achievable in the very near future.

## REFERENCES


Alengebawy, A., Abdelkhalek, S.T., Qureshi, S.R., Wang, M.Q., 2021. Heavy metals and pesticides toxicity in agricultural soil and plants: Ecological risks and human health implications. *Toxics* 9, 1–34. https://doi.org/10.3390/TOXICS9030042

Asian Development Bank, 2013. *Thinking about water differently: Managing the water-food-energy nexus*. Metro Manila, Philippines: Mandaluyongy, 1–47.

Astel, A.M., Chepanova, L., Simeonov, V., 2011. Soil contamination interpretation by the use of monitoring data analysis. *Water, Air, and Soil Pollution* 216, 375–390. https://doi.org/10.1007/s11270-010-0539-1

Aulakh, M., 2011. Integrated soil tillage and nutrient management-A way to sustain crop production, soil-plant-animal-human health and environment. *Journal of the Indian Society of Soil Science* 59, 23–26.

Aulakh, M.S., Sidhu, G.S., 2015. Soil degradation in India: Causes, major threats, and management options. In: *MACRO Symposium 2015-Next Challenges of Agro-Environmental Research in Monsoon Asia*. National Institute for Agro-Environmental Sciences (NIAES), Tsukuba, Japan, pp. 151–156.

Bagyaraj, D.J., Revanna, A., 2017. Soil biodiversity: Role in sustainable horticulture. In: Peter, K. v (Ed.), *Biodiversity in Horticultural Crops*. New Delhi, India: Daya Publishing House, pp. 1–18.

Balestrini, R., Lumini, E., Borriello, R., Bianciotto, V., 2015. Plant-soil biota interactions. In: *Soil Microbiology, Ecology and Biochemistry*. Academic Press, Cambridge, Massachusetts, USA, pp. 311–338. https://doi.org/10.1016/B978-0-12-415955-6.00011-6

Banik, S., Dey, P., Pandit, P., 2020. Effect of different insecticides on yield, yield attributes and pollinator behaviour of different pigeon pea varieties. *Legume Research*, 46(2), 211–214. https://doi.org/10.18805/LR-4387

Bazilian, M., Rogner, H., Howells, M., Hermann, S., Arent, D., Gielen, D., Steduto, P., Mueller, A., Komor, P., Tol, R.S.J., Yumkella, K.K., 2011. Considering the energy, water and food nexus: Towards an integrated modelling approach. *Energy Policy* 39, 7896–7906. https://doi.org/10.1016/j.enpol.2011.09.039

Bhatti, A.A., Haq, S., Bhat, R.A., 2017. Actinomycetes benefaction role in soil and plant health. *Microbial Pathogenesis* 111, 458–467. https://doi.org/10.1016/j.micpath.2017.09.036

Bhuvaneshwari, S., Hettiarachchi, H., Meegoda, J., 2019. Crop residue burning in India: Policy challenges and potential solutions. *International Journal of Environmental Research and Public Health* 16, 832. https://doi.org/10.3390/ijerph16050832

Breznak, J.A., 2000. Ecology of prokaryotic microbes in the guts of wood- and litter-feeding termites. In: *Termites: Evolution, Sociality, Symbioses, Ecology*. Netherlands, Dordrecht: Springer, pp. 209–231. https://doi.org/10.1007/978-94-017-3223-9_10

Carter, M.R., 2005. Conservation tillage. In: *Encyclopedia of Soils in the Environment*. Academic Press, Cambridge, Massachusetts, USA, pp. 306–311. https://doi.org/10.1016/B0-12-348530-4/00270-8

Chakraborti, D., Singh, S.K., Rahman, M.M., Dutta, R.N., Mukherjee, S.C., Pati, S., Kar, P.B., 2018. Groundwater arsenic contamination in the ganga river basin: A future health danger. *International Journal of Environmental Research and Public Health* 15. https://doi.org/10.3390/IJERPH15020180

Chang, Y., Li, G., Yao, Y., Zhang, L., Yu, C., 2016. Quantifying the water-energy-food nexus: Current status and trends. *Energies (Basel)* 9, 65. https://doi.org/10.3390/en9020065

Chapman, P.F., Roberts, F., 1983. *Metal Resources and Energy*. Printed in England by The Thetford Press Ltd, Thetford, Norfolk: Butterworths.

Cheng, Q., 2008. Perspectives in biological nitrogen fixation research. *Journal of Integrative Plant Biology* 50, 786–798. https://doi.org/10.1111/j.1744-7909.2008.00700.x

Chowdhury, U.K., Biswas, B.K., Chowdhury, T.R., Samanta, G., Mandal, B.K., Basu, G.C., Chanda, C.R., Lodh, D., Saha, K.C., Mukherjee, S.K., Roy, S., Kabir, S., Quamruzzaman, Q., Chakraborti, D., 2000. Groundwater arsenic contamination in Bangladesh and West Bengal, India. *Environmental Health Perspectives* 108, 393–397. https://doi.org/10.1289/EHP.00108393

Chu, E.W., Karr, J.R., 2017. Environmental impact: Concept, consequences, measurement ☆. In: *Reference Module in Life Sciences*. Academic Press, Cambridge, Massachusetts, USA. https://doi.org/10.1016/B978-0-12-809633-8.02380-3

C.R., V., Sankhla, M.S., Singh, P., Jadhav, E.B., Verma, R.K., Awasthi, K.K., Awasthi, G., Nagar, V., 2022. *Heavy Metal Contamination of Food Crops: Transportation via Food Chain, Human Consumption, Toxicity and Management Strategies*. Environmental Impact and Remediation of Heavy Metals. London: IntechOpen. https://doi.org/10.5772/INTECHOPEN.101938

Daher, B.T., Mohtar, R.H., 2015. Water–energy–food (WEF) Nexus Tool 2.0: Guiding integrative resource planning and decision-making. *Water International* 40, 748–771. https://doi.org/10.1080/02508060.2015.1074148

Dai, J., Wu, S., Han, G., Weinberg, J., Xie, X., Wu, X., Song, X., Jia, B., Xue, W., Yang, Q., 2018. Water-energy nexus: A review of methods and tools for macro-assessment. *Applied Energy* 210, 393–408. https://doi.org/10.1016/j.apenergy.2017.08.243

Das, A., Layek, J., Ramkrushna, G.I., Patel, D.P., Choudhury, B.U., Chowdhury, S., Ngachan, S. v., 2015. Raised and sunken bed land configuration for crop diversification and crop and water productivity enhancement in rice paddies of the north eastern region of India. *Paddy and Water Environment* 13, 571–580. https://doi.org/10.1007/s10333-014-0472-9

Das, A., Sharma, R.P., Chattopadhyaya, N., Rakshit, R., 2014. Yield trends and nutrient budgeting under a long-term (28 years) nutrient management in rice-wheat cropping system under subtropical climatic condition. *Plant, Soil and Environment* 60, 351–357. https://doi.org/10.17221/46/2014-PSE

Dey, P., Mahapatra, B.S., Juyal, V.K., Pramanick, B., Negi, M.S., Paul, J., Singh, S.P., 2021a. Flax processing waste—A low-cost, potential biosorbent for treatment of heavy metal, dye and organic matter contaminated industrial wastewater. *Industrial Crops and Products* 174. https://doi.org/10.1016/J.INDCROP.2021.114195

Dey, P., Mahapatra, B.S., Pramanick, B., Kumar, A., Negi, M.S., Paul, J., Shukla, D.K., Singh, S.P., 2021b. Quality optimization of flax fibre through durational management of water retting technology under sub-tropical climate. *Industrial Crops and Products* 162. https://doi.org/10.1016/J.INDCROP.2021.113277

Dey, P., Mahapatra, B.S., Pramanick, B., Pyne, S., Pandit, P., 2022. Optimization of seed rate and nutrient management levels can reduce lodging damage and improve yield, quality and energetics of subtropical flax. *Biomass Bioenergy* 157. https://doi.org/10.1016/J.BIOMBIOE.2022.106355

Dey, P., Pandit, P., 2020. Relevance of data transformation techniques in weed science. *Journal of Research in Weed Science* 3, 81–89.

Dobermann, A., Cassman, K.G., 2002. Plant nutrient management for enhanced productivity in intensive grain production systems of the United States and Asia. *Plant Soil* 247, 153–175. https://doi.org/10.2307/24123904

Dodds, F., Bartram, J., 2016. *The Water, Food, Energy and Climate Nexus: Challenges and an Agenda for Action*. Boca Raton, Floride, USA: Routledge.

Dominati, E., Patterson, M., Mackay, A., 2010. A framework for classifying and quantifying the natural capital and ecosystem services of soils. *Ecological Economics* 69, 1858–1868. https://doi.org/10.1016/j.ecolecon.2010.05.002

Erdogan, H.E., Havlicek, E., Dazzi, C., Montanarella, L., Van Liedekerke, M., Vrščaj, B., Krasilnikov, P., Khasankhanova, G., Vargas, R., 2021. Soil conservation and sustainable development goals (SDGs) achievement in Europe and central Asia: Which role for the European soil partnership? *International Soil and Water Conservation Research* 9, 360–369. https://doi.org/10.1016/j.iswcr.2021.02.003

FAO, 2014. *The Water-Energy-Food Nexus: A New Approach in Support of Food Security and Sustainable Agriculture*. Rome, Italy: Food and Agriculture Organization of United Nations.

FAO, 2018. *Polluting Our Soils Is Polluting Our Future | FAO Stories | Food and Agriculture Organization of the United Nations* [WWW Document]. Food and Agriculture Organization of the United Nations. URL www.fao.org/fao-stories/article/en/c/1126974/ (accessed 9.23.22).

Finley, J.W., Seiber, J.N., 2014. The nexus of food, energy, and water. *Journal of Agricultural and Food Chemistry* 62, 6255–6262. https://doi.org/10.1021/jf501496r

Gilbert, H.J., 2010. The biochemistry and structural biology of plant cell wall deconstruction. *Plant Physiology* 153, 444–455. https://doi.org/10.1104/pp.110.156646

Godfray, H.C.J., Garnett, T., 2014. Food security and sustainable intensification. *Philosophical Transactions of the Royal Society B: Biological Sciences* 369, 20120273. https://doi.org/10.1098/rstb.2012.0273

Goel, R., Soni, R., Suyal, D.C., 2020. *Microbiological Advancements for Higher Altitude Agro-Ecosystems & Sustainability*. Springer Singapore, Singapore. https://doi.org/10.1007/978-981-15-1902-4

Hatten, J., Liles, G., 2019. A 'healthy' balance—The role of physical and chemical properties in maintaining forest soil function in a changing world. In *Developments in Soil Science* (vol. 36). Academic Press, Cambridge, Massachusetts, USA, pp. 373–396. https://doi.org/10.1016/B978-0-444-63998-1.00015-X

He, C., Liu, Z., Wu, J., Pan, X., Fang, Z., Li, J., Bryan, B.A., 2021. Future global urban water scarcity and potential solutions. *Nature Communications* 12, 4667. https://doi.org/10.1038/s41467-021-25026-3

Hesami, E., Farshidi, A., Sadaterbrahimi, F., Ali, T., 2014. The role of soil organisms on soil stability: A review. *International Journal of Current Life Sciences* 4, 10328.

Hornick, S.B., 1992. Factors affecting the nutritional quality of crops. *American Journal of Alternative Agriculture* 7, 63–68. https://doi.org/10.1017/S0889189300004471

Hozzein, W.N., Abuelsoud, W., Wadaan, M.A.M., Shuikan, A.M., Selim, S., al Jaouni, S., AbdElgawad, H., 2019. Exploring the potential of actinomycetes in improving soil fertility and grain quality of economically important cereals. *Science of The Total Environment* 651, 2787–2798. https://doi.org/10.1016/j.scitotenv.2018.10.048

Hussien, W.A., Memon, F.A., Savic, D.A., 2017. An integrated model to evaluate water-energy-food nexus at a household scale. *Environmental Modelling & Software* 93, 366–380. https://doi.org/10.1016/j.envsoft.2017.03.034

IEA, 2010. *World Energy Outlook 2010—Analysis—IEA*. Paris: IEA.

Imadi, S.R., Shazadi, K., Gul, A., Hakeem, K.R., 2016. Sustainable crop production system. In: *Plant, Soil and Microbes*. Cham: Springer, pp. 103–116.

IPCC, 2014. Climate change 2014: Impacts, adaptation, and vulnerability. In: *Contribution of Working Group II to the 5th Assessment Report of the Intergovernmental Panel on Climate Change*. Cambridge: Cambridge University Press.

Jacoby, R., Peukert, M., Succurro, A., Koprivova, A., Kopriva, S., 2017. The role of soil microorganisms in plant mineral nutrition—current knowledge and future directions. *Frontiers in Plant Science* 8. https://doi.org/10.3389/fpls.2017.01617

Jahangir, M.M.R., Bell, R.W., Uddin, S., Ferdous, J., Nasreen, S.S., Haque, M.E., Satter, M.A., Zaman, M., Ding, W., Jahiruddin, M., Müller, C., 2022. Conservation agriculture with optimum fertilizer nitrogen rate reduces GWP for rice cultivation in floodplain soils. *Frontiers in Environmental Science* 0, 291. https://doi.org/10.3389/FENVS.2022.853655

Kader, M.A., Jahangir, M.M.R., Islam, M. R., Begum, R., Nasreen, S.S., Islam, Md R., Mahmud, A. al, Haque, M.E., Bell, R.W., Jahiruddin, M., 2022. Long-term conservation agriculture increases nitrogen use efficiency by crops, land equivalent ratio and soil carbon stock in a subtropical rice-based cropping system. *Field Crops Research* 287, 108636. https://doi.org/10.1016/J.FCR.2022.108636

Karlen, D.L., 2005. Productivity. In: *Encyclopedia of Soils in the Environment*. Academic Press, Cambridge, Massachusetts, USA, pp. 330–336. https://doi.org/10.1016/B0-12-348530-4/00241-1

Khan, A., Rao, T.S., 2019. Molecular evolution of xenobiotic degrading genes and mobile DNA elements in soil bacteria. In: *Microbial Diversity in the Genomic Era*. Academic Press, Cambridge, Massachusetts, USA, pp. 657–678. https://doi.org/10.1016/B978-0-12-814849-5.00036-8

Kholod, N., Evans, M., Khan, Z., Hejazi, M., Chaturvedi, V., 2021. Water-energy-food nexus in India: A critical review. *Energy and Climate Change* 2, 100060. https://doi.org/10.1016/j.egycc.2021.100060

Knapp, S., van der Heijden, M.G.A., 2018. A global meta-analysis of yield stability in organic and conservation agriculture. *Nature Communications* 9, 1–9. https://doi.org/10.1038/s41467-018-05956-1

Kumar, A., Singh, K., Verma, P., Singh, O., Panwar, A., Singh, T., Kumar, Y., Raliya, R., 2022. Effect of nitrogen and zinc nanofertilizer with the organic farming practices on cereal and oil seed crops. *Scientific Reports* 12, 6938–6938. https://doi.org/10.1038/S41598-022-10843-3

Kumar, P., Sharma, P.K., 2020. Soil salinity and food security in India. *Frontiers in Sustainable Food Systems* 4. https://doi.org/10.3389/fsufs.2020.533781

Kumawat, A., Yadav, D., Samadharmam, K., Rashmi, I., 2021. Soil and water conservation measures for agricultural sustainability. In: *Soil Moisture Importance*. London: Intech Open. https://doi.org/10.5772/intechopen.92895

Lal, R., 2015. Restoring soil quality to mitigate soil degradation. *Sustainability* 7, 5875–5895. https://doi.org/10.3390/su7055875

Laurentiis, V. De, Hunt, D., Rogers, C., 2016. Overcoming food security challenges within an energy/water/food nexus (EWFN) approach. *Sustainability* 8, 95. https://doi.org/10.3390/su8010095

Lawford, R., Bogardi, J., Marx, S., Jain, S., Wostl, C.P., Knüppe, K., Ringler, C., Lansigan, F., Meza, F., 2013. Basin perspectives on the Water–Energy–Food Security Nexus. *Current Opinion in Environmental Sustainability* 5, 607–616. https://doi.org/10.1016/j.cosust.2013.11.005

Leck, H., Conway, D., Bradshaw, M., Rees, J., 2015. Tracing the water–energy–food nexus: Description, theory and practice. *Geography Compass* 9, 445–460. https://doi.org/10.1111/gec3.12222

Lelisa Deke, A., 2016. Soil physic-chemical properties in termite mounds and adjacent control soil in Miyo and Yabello districts of Borana Zone, Southern Ethiopia. *American Journal of Agriculture and Forestry* 4, 69. https://doi.org/10.11648/j.ajaf.20160404.11

Li, M., Fu, Q., Singh, V.P., Ji, Y., Liu, D., Zhang, C., Li, T., 2019. An optimal modelling approach for managing agricultural water-energy-food nexus under uncertainty. *Science of the Total Environment* 651, 1416–1434. https://doi.org/10.1016/j.scitotenv.2018.09.291

Liu, D., Zhang, J., Zeng, Y., Shen, Y., 2019. Modeling the physical nexus across water supply, wastewater management and hydropower generation sectors in river–reservoir systems. *Water (Basel)* 11, 822. https://doi.org/10.3390/w11040822

Loh, J., Randers, J., MacGillivray, A., Kapos, V., Groombridge, B., Jenkins, M., UK, J.E., CB, C., 1998. *Living Planet Report 1998*. WWF International, Gland, Switzerland.

Mahapatra, B.S., Dey, P., 2022. Integrated management practices for incremental wheat productivity. *New Horizons in Wheat and Barley Research* 367–392. https://doi.org/10.1007/978-981-16-4134-3_13

Maji, A.K., Reddy, O.G.P., Meshram, S., 2008. Acid soil map of India. In: *Annual Report 2008*. ICAR-National Bureau of Soil Survey and Land Use Planning, Nagpur, India. New Delhi: National Printers.

Meehan, T.D., Hurlbert, A.H., Gratton, C., 2010. Bird communities in future bioenergy landscapes of the Upper Midwest. *Proceedings of the National Academy of Sciences* 107, 18533–18538. https://doi.org/10.1073/pnas.1008475107

Mehmet Tuğrul, K., 2020. Soil management in sustainable agriculture. In: *Sustainable Crop Production*. London: Intech Open. https://doi.org/10.5772/intechopen.88319

Mejias, J.H., Salazar, F., Pérez Amaro, L., Hube, S., Rodriguez, M., Alfaro, M., 2021. Nano-fertilizers: A cutting-edge approach to increase nitrogen use efficiency in grasslands. *Frontiers in Environmental Science* 9. https://doi.org/10.3389/fenvs.2021.635114

Mitra, B., Chowdhury, A.R., Dey, P., Hazra, K.K., Sinha, A.K., Hossain, A., Meena, R.S., 2021. Use of agrochemicals in agriculture: Alarming issues and solutions. *Input Use Efficiency for Food and Environmental Security* 85–122. https://doi.org/10.1007/978-981-16-5199-1_4

Mitra, B., Patra, K., Bhattacharya, P.M., Chowdhury, A.K., 2018. Unpuddled transplanting: A productive, profitable and energy-efficient establishment technique in rice under Eastern sub-Himalayan plains. *Oryza (Cuttack)* 55, 459–466.

Mogili, U.R., Deepak, B.B.V.L., 2018. Review on application of drone systems in precision agriculture. *Procedia Comput Sci* 133, 502–509. https://doi.org/10.1016/J.PROCS.2018.07.063

Mohammadi, O., Lahdenpera, M.L., 1992. Mycostop Biofungicide in practice. In: *10th International Symposium on Modern Fungicides and Antifugal Compounds*. Thuringia, Germany, pp. 1–7.

Mohtar, R.H., Lawford, R., 2016. Present and future of the water-energy-food nexus and the role of the community of practice. *Journal of Environmental Studies and Sciences* 6, 192–199. https://doi.org/10.1007/s13412-016-0378-5

Molajou, A., Afshar, A., Khosravi, M., Soleimanian, E., Vahabzadeh, M., Variani, H.A., 2021. A new paradigm of water, food, and energy nexus. *Environmental Science and Pollution Research*. https://doi.org/10.1007/s11356-021-13034-1

Mortada, S., Najm, M.A., Yassine, A., Fadel, M. El, Alamiddine, I., 2018. Towards sustainable water-food nexus: An optimization approach. *Journal of Cleaner Production* 178, 408–418. https://doi.org/10.1016/j.jclepro.2018.01.020

Mukhopadhyay, S.S., 2014. Nanotechnology in agriculture: Prospects and constraints. *Nanotechnology, Science and Applications* 7, 63. https://doi.org/10.2147/NSA.S39409

Narayana, D.V., Babu, R., 1983. Estimation of soil erosion in India. *Journal of Irrigation and Drainage Engineering* 109, 419–434. http://dx.doi.org/10.1061/(ASCE)0733-9437(1983)109:4(419)

Neemisha. 2020. Role of soil organisms in maintaining soil health, ecosystem functioning, and sustaining agricultural production. In: Giri, B., Varma, A. (Eds.), *Soil Health. Soil Biology* (vol. 59). Cham: Springer. https://doi.org/10.1007/978-3-030-44364-1_17

Neupane, J., Guo, W., 2019. Agronomic basis and strategies for precision water management: A review. *Agronomy* 9, 87. https://doi.org/10.3390/AGRONOMY9020087

Newell, B., Marsh, D.M., Sharma, D., 2011. Enhancing the resilience of the Australian national electricity market taking a systems approach in policy development on JSTOR. *Ecology and Society* 16, 26268903.

Pandit, P., Dey, P., Krishnamurthy, K.N., 2021. Comparative assessment of multiple linear regression and fuzzy linear regression models. *SN Computer Science* 2. https://doi.org/10.1007/S42979-021-00473-3

Paroda, R., 2022. Crop diversification for sustainable agriculture. *Ecology, Economy and Society–the INSEE Journal* 5, 15–21. https://doi.org/10.37773/EES.V5I1.611

Pimentel, D., Burgess, M., 2013. Soil erosion threatens food production. *Agriculture* 3, 443–463. https://doi.org/10.3390/AGRICULTURE3030443

Rastgoo, M., Hasanfard, A., 2021. Desertification in agricultural lands: Approaches to mitigation. In: Yajuan Zhu, Qinghong Luo and Yuguo Liu, (Eds.), *Deserts and Desertification*. London. https://doi.org/10.5772/INTECHOPEN.98795

Reijnders, L., 2000. A normative strategy for sustainable resource choice and recycling. *Resources, Conservation and Recycling* 28, 121–133. https://doi.org/10.1016/S0921-3449(99)00037-3

Remya, M., Vijayakumar, R., 2008. Isolation and characterization of marine antagonistic actinomycetes from west coast of India. *Medicine and Biology* 15466, 13–19.

Ritchie, H., Roser, M., 2018. *Water Use and Stress*. Published online at Our WorldInData.org. https://ourworldindata.org/water-use-stress

RIVM, 2000. *Nationale Milieuverkenning 5, 2000–2030* | RIVM [WWW Document]. URL www.rivm.nl/publicaties/nationale-milieuverkenning-5-2000-2030 (accessed 9.23.22).

Rodrigues, C., Pratap, V., Dass, A., Dhar, S., Babu, S., Singh, V.K., Singh, Raj, Krishnan, P., Sudhishri, S., Bhatia, A., Kumar, S., Choudhary, A.K., Singh, Renu, Kumar, P., Kumar Sarkar, S., Kumar Verma, S., Kumari, K., San, A.A., 2022. Co-implementation of tillage, precision nitrogen, and water management enhances water productivity, economic returns, and energy-use efficiency of direct-seeded rice. *Sustainability* 14, 11234. https://doi.org/10.3390/SU141811234

Sahoo, S.K., Dwivedi, G.K., Dey, P., Praharaj, S., 2021. Green synthesized ZnO nanoparticles for sustainable production and nutritional biofortification of green gram. *Environmental Technology & Innovation* 24. https://doi.org/10.1016/J.ETI.2021.101957

Saladini, F., Betti, G., Ferragina, E., Bouraoui, F., Cupertino, S., Canitano, G., Gigliotti, M., Autino, A., Pulselli, F.M., Riccaboni, A., Bidoglio, G., Bastianoni, S., 2018. Linking the water-energy-food nexus and sustainable development indicators for the Mediterranean region. *Ecological Indicators* 91, 689–697. https://doi.org/10.1016/j.ecolind.2018.04.035

Scavo, A., Fontanazza, S., Restuccia, A., Pesce, G.R., Abbate, C., Mauromicale, G., 2022. The role of cover crops in improving soil fertility and plant nutritional status in temperate climates. A review. *Agronomy for Sustainable Development* 42, 93. https://doi.org/10.1007/s13593-022-00825-0

Sekaran, U., Lai, L., Ussiri, D.A.N., Kumar, S., Clay, S., 2021. Role of integrated crop-livestock systems in improving agriculture production and addressing food security—A review. *Journal of Agriculture and Food Research* 5, 100190. https://doi.org/10.1016/j.jafr.2021.100190

Shah, K.K., Tripathi, S., Tiwari, I., Shrestha, J., Modi, B., Paudel, N., Das, B.D., 2021. Role of soil microbes in sustainable crop production and soil health: A review. *Agricultural Science and Technology* 13, 109–118. https://doi.org/10.15547/ast.2021.02.019

Shaheb, M.R., Venkatesh, R., Shearer, S.A., 2021. A review on the effect of soil compaction and its management for sustainable crop production. *Journal of Biosystems Engineering* 46, 417–439. https://doi.org/10.1007/s42853-021-00117-7

Shen, Y., Zhou, J., Du, C., Zhou, Z., 2021. Hydrophobic modification of waterborne polymer slows urea release and improves nitrogen use efficiency in rice. *Science of the Total Environment* 794, 148612–148612. https://doi.org/10.1016/J.SCITOTENV.2021.148612

Shiferaw, B., 2002. *Poverty and natural resource management in the semi-arid tropics: Revisiting challenges and conceptual issues.* Working Paper Series no. 14. undefined.

Shukla, A.K., Ladha, J.K., Singh, V.K., Dwivedi, B.S., Balasubramanian, V., Gupta, R.K., Sharma, S.K., Singh, Y., Pathak, H., Pandey, P.S., Padre, A.T., Yadav, R.L., 2004. Calibrating the leaf color chart for nitrogen management in different genotypes of rice and wheat in a systems perspective. *Agronomy Journal* 96, 1606–1621. https://doi.org/10.2134/AGRONJ2004.1606

Singh, D.N., Bohra, J.S., Banjara, T.R., 2019. Diversification of rice-wheat cropping system for sustainability and livelihood security. In: Rathore, S.S., Shekhawat, K., Rajanna, G.A., Upadhyay, P.K., Singh, V.K. (Eds.), *Crop Diversification for Resilience in Agriculture and Doubling Farmer's Income.* New Delhi, India: ICAR-Indian Agricultural Research Institute, p. 210.

Sinha, A.K., Ghosh, A., Dhar, T., Bhattacharya, P.M., Mitra, B., Rakesh, S., Paneru, P., Shrestha, S.R., Manandhar, S., Beura, K., Dutta, S., Pradhan, A.K., Rao, K.K., Hossain, A., Siddquie, N., Molla, M.S.H., Chaki, A.K., Gathala, M.K., Islam, M.S., Dalal, R.C., Gaydon, D.S., Laing, A.M., Menzies, N.W., Sinha, A.K., Ghosh, A., Dhar, T., Bhattacharya, P.M., Mitra, B., Rakesh, S., Paneru, P., Shrestha, S.R., Manandhar, S., Beura, K., Dutta, S., Pradhan, A.K., Rao, K.K., Hossain, A., Siddquie, N., Molla, M.S.H., Chaki, A.K., Gathala, M.K., Islam, M.S., Dalal, R.C., Gaydon, D.S., Laing, A.M., Menzies, N.W., 2019. Trends in key soil parameters under conservation agriculture-based sustainable intensification farming practices in the Eastern Ganga Alluvial Plains. *Soil Research* 57, 883–893. https://doi.org/10.1071/SR19162

Slorach, P.C., Jeswani, H.K., Cuéllar-Franca, R., Azapagic, A., 2020. Environmental sustainability in the food-energy-water-health nexus: A new methodology and an application to food waste in a circular economy. *Waste Management* 113, 359–368. https://doi.org/10.1016/j.wasman.2020.06.012

Steinhoff-Knopp, B., Kuhn, T.K., Burkhard, B., 2021. The impact of soil erosion on soil-related ecosystem services: Development and testing a scenario-based assessment approach. *Environmental Monitoring and Assessment* 193, 274. https://doi.org/10.1007/s10661-020-08814-0

Struik, P.C., Kuyper, T.W., 2017. Sustainable intensification in agriculture: The richer shade of green. A review. *Agronomy for Sustainable Development* 37, 1–15. https://doi.org/10.1007/S13593-017-0445-7/FIGURES/3

Sun, X., Ding, Z., Wang, X., Hou, H., Zhou, B., Yue, Y., Ma, W., Ge, J., Wang, Z., Zhao, M., 2017. Subsoiling practices change root distribution and increase post-anthesis dry matter accumulation and yield in summer maize. *PLoS One* 12, e0174952. https://doi. org/10.1371/journal.pone.0174952

Taghizadeh-Hesary, F., Rasoulinezhad, E., Yoshino, N., 2019. Energy and food security: Linkages through price volatility. *Energy Policy* 128, 796–806. https://doi.org/10.1016/j. enpol.2018.12.043

Tashtoush, F.M., Al-Zubari, W.K., Shah, A., 2019. A review of the water–energy–food nexus measurement and management approach. *International Journal of Energy and Water Resources* 3, 361–374. https://doi.org/10.1007/s42108-019-00042-8

Urra, Alkorta, Garbisu, 2019. Potential benefits and risks for soil health derived from the use of organic amendments in agriculture. *Agronomy* 9, 542. https://doi.org/10.3390/ agronomy9090542

Vancutsem, C., Achard, F., Pekel, J.F., Vieilledent, G., Carboni, S., Simonetti, D., Gallego, J., Aragão, L.E.O.C., Nasi, R., 2021. Long-term (1990–2019) monitoring of forest cover changes in the humid tropics. *Science Advances* 7. https://doi.org/10.1126/SCIADV. ABE1603

Vengosh, A., 2003. Salinization and saline environments. In: *Treatise on Geochemistry.* Academic Press, Cambridge, Massachusetts, USA, pp. 1–35. https://doi.org/10.1016/ B0-08-043751-6/09051-4

Vogt, K., Patel-Weynand, T., Shelton, M., Vogt, D.J., Gordon, J., Mukumoto, C., Suntana, Asep. S., Roads, P.A., 2012. *Sustainability Unpacked.* Boca Raton, Floride, USA: CRC Routledge. https://doi.org/10.4324/9781849776653

Vories, E., Stevens, W.G., Rhine, M., Straatmann, Z., 2017. Investigating irrigation scheduling for rice using variable rate irrigation. *Agricultural Water Management* 179, 314–323. https://doi.org/10.1016/J.AGWAT.2016.05.032

Ye, Z., Li, J., Wang, J., Zhang, C., Liu, G., Dong, Q., 2021. Diversity and co-occurrence network modularization of bacterial communities determine soil fertility and crop yields in arid fertigation agroecosystems. *Biology and Fertility of Soils* 57, 809–824. https://doi. org/10.1007/s00374-021-01571-3

Zheng, H., Liu, W., Zheng, J., Luo, Y., Li, R., Wang, H., Qi, H., 2018. Effect of long-term tillage on soil aggregates and aggregate-associated carbon in black soil of Northeast China. *PLoS One* 13, e0199523. https://doi.org/10.1371/journal.pone.0199523

# 13 The Ecosystem Approach and Environmental Justice Nexus in Natural Resource Management

*Vikas and Rajiv Ranjan*

## 13.1 INTRODUCTION

The ecosystem approach is an integrated and sustainable administration of natural resources, including soil, water, fossil fuels, minerals, sunlight, air, flora, fauna and microorganisms, that promotes fair and equitable sharing, conservation and imperishable use of natural assets. Overexploitation and use of natural assets by developed countries causes severe damage to the environment and a heavy toll on low-income countries and their people. Environmental justice is the struggle of social activists, environmentalists and altruistic humankind to give an unblemished and thriving environment to the people regardless of race, color, nationality, income status and active participation of all people to develop, implement and enforce environmental law, rules and policies. There is a complementarity between the ecosystem approach and the conception of environmental justice in natural resource management).

## 13.2 ECOSYSTEM APPROACH

Industrialisation, urbanisation and intensive agriculture causes severe damage to the entire environment and disturbed ecosystems at different levels (Weiskopf et al. 2020, Chase et al. 2020, Massey 2004, Bulte et al. 2005, Singh et al. 2017). Global pollution, climate change, global warming, and acid rain are consequences of insensitive and arrogant behavior of human beings towards nature (Grennfelt et al. 2020, Cohen-Shacham et al. 2019, Khafaie et al. 2019, Cronin et al. 2009). As billions of people are living their lives ignoring global warming, pollution and climate change, this earth will be close to uninhabitable at end of this century (Wallace-Wells 2019). And considering the horrific truth that there is no planet B (Barners 2021, Liang et al. 2020), we need a scientific approach of resolving these environmental issues. And here, the ecosystem approach emerges in an important role. The ecosytem approach is centered on the applications of scientific approaches emphasising on the magnitude of the organisation of living organisms, functioning of ecosystems, their crucial processes,



DOI: 10.1201/9781003358169-13

interaction with other organisms and their environment and how they can be utilised and sustained for future generations (Queffelec et al. 2009, Waylen et al. 2014).

## 13.3  ECOSYSTEM APPROACH AND NATURAL RESOURCE MANAGEMENT

Biological diversity is the most significant natural resource. A healthy land and water is essential to flourish biological diversity, and these are also natural resources (Cardinale et al. 2012). Ecosystem approach considers these natural resources interdependent and shares synergy between them (Forget et al. 2001). The variety of life on earth has been threatened, productive land is shrinking and deserting and water resources are receding and polluting because of human activities (Peters et al. 2000; Taylor-Brown et al. 2019, Wang et al. 2015, Alam et al. 2019) These natural resource concerns have been awakening the social activists, environmentalists and intellectuals from across the globe. And as a result, during the Earth Summit held in Rio de Janeiro, Brazil, on June 5, 1992 (Oza 1992), the Convention on Biological Diversity (CBD) welcomed all the countries to participate in a holistic mission (Tinker 1995). And at the end of December 1993, 167 countries had penned the treaty (Cropper 1993). As of now, the Convention on Biological Diversity (CBD) has almost global participation with 193 countries (Chandra et al. 2011). Only four members of the United Nations (United States, South Sudan, Andorra and Holy See) are not involved in the CBD.

The Convention on Biological Diversity (CBD) has three main objectives: to conserve variety of life on earth, sustainable use of resources of biological diversity and equal benefits sharing coming out from the usage of genetic resources (Halewood et al. 2017). These objectives are also sharing beliefs of environmental fairness (Martin et al. 2013). The objectives of CBD are linking the ecosystem and its components, if any action taken in an ecosystem or at a location with rich diversity, consequence may be unpredicted and unexpected in this circumstances ecosystem approach provides a powerful strategy to unify the management of living resource, land and water that encourages conservation and equal sharing in a sustainable way. The structure of the ecosystem approach and its application helps in attaining the three intentions of the convention and partially fulfill the need of environmental justice, as shown in Figure 13.1.

## 13.4  ECOSYSTEM APPROACH: PRINCIPLES OF CONVENTION ON BIOLOGICAL DIVERSITY

**Principle 1: Land use, water use and living resource management are societal decisions**

Indigenous and local people and their cultural diversity are the main constituents of the ecosystem approach, so the management of land, water and living resources are the choice of its people. The iconic example of saving natural resources by its indigenous people is the Chipko Movement in India (Shiva & Bandyopadhyay 1986). And it also promotes environmental justice.

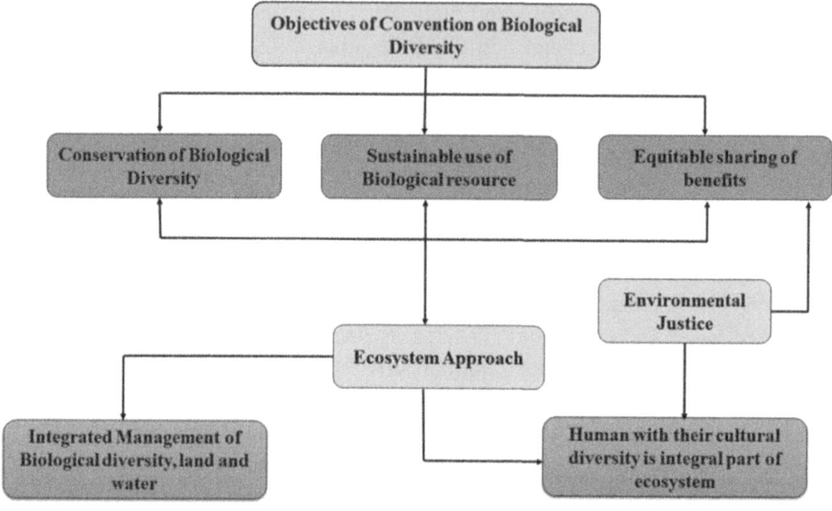


**FIGURE 13.1**   Sharing features and complementarities of natural resource, ecosystem approach and environmental justice.

## Principle 2. Decentralizing management to the suitable lower level

Ecosystem approach and environmental justice encourages the participation of all people of its community. When management and decision-making are decentralised to the lowest level, it leads to greater efficiency and decision concerns all the aspects of environmental issues and justifies the need of every part of the ecosystem.

## Principle 3. The effects of the ecosystem managers' activities (actual or potential) on adjoining parts ecosystems should be considered

The effects of management interventions in ecosystems often are unknown or unpredictable, so potential effects should be considered and carefully analyzed. For institutions involved in decision-making, this may require new arrangements that can estimate actual potential.

## Principle 4. Understanding and managing an ecosystem in an economic framework is usually needed to recognize potential gains from management

Any such ecosystem-management program should do the following:

(i)   Eliminating or reducing market distortions that negatively impact biodiversity;
(ii)  Conserving biodiversity and promoting sustainable resource management; and
(iii) Analyze the ecosystem's costs and benefits in so far as feasible.

Beneficiaries of conservation rarely pay back, and those who cause environmental problems usually escape responsibility.

### Principle 5. A priority target of the ecosystem approach should be to conserve ecosystem structure and function so that ecosystem services can be maintained

The health and stability of an ecosystem are determined by interactions within and among species, as well as interactions between species and the environment. In order to maintain biological diversity over the long run, it is more important to conserve and, where appropriate, restore these connections and processes than simply to protect living species.

### Principle 6. Ecosystems need to be maintained within their limit of ability to function

It is important to evaluate environmental factors that limit natural productivity, ecosystem framework, function and diversity as part of evaluating the likelihood of achieving management objectives. Various factors may have a negative impact on ecosystem functioning. Accordingly, administration should be judiciously cautious.

### Principle 7. Ecosystem research should be conducted at the proper spatial and temporal level

In order to attain the objectives, the greet should be constrained by appropriate spatial and temporal levels. Managers, scientists, users and indigenous and local people will establish the operational limits of management. Wherever possible, it is crucial that these areas are connected. Genes, species and ecosystems interact and integrate within an ecosystem based on the hierarchy of biodiversity.

### Principle 8. Ecosystem admins' objectives should be based on the continuing sequencing of processes and lag effects of varying temporal scales in ecosystems

The dynamics of ecosystems are identified by different lag effects and temporal scales. Inherently, this is constrained with human nature, which naturally prefers short-term gains over long-term benefits.

### Principle 9. The fact that change is inevitable must be recognized by management

The composition and abundance of species in ecosystems change. Therefore, management must adapt. Ecosystems are not only subject to their inherent dynamics of change but also to a complicated set of risks, potentials and uncertainties, "surprises" in the domains of the human, biological and environmental.

### Principle 10. Ecosystem approaches must integrate conservation and utilise biological diversity in an appropriate manner and seek an appropriate balance

It is vital not only for its intrinsic value but also because it provides essential ecosystem services for which we all rely. Traditionally, biological diversity has been treated as either protected or unprotected. In the future, it is necessary to move towards more pliable situations. The conservation and management of ecosystems are considered in context along with the implementation of a continuum of measures, such as protected ecosystems and man-made ecosystems.

> **Principle 11. It is imperative to take into account all aspects of the ecosystem approach, including scientific and native knowledge, innovation and practice**

Developing effective ecosystem management strategies requires input from all sources. Affected areas should share all relevant information with all stakeholders and any decisions required under Article 8(j) of the CBD.

> **Principle 12. Scientific disciplines and all applicable sectors of society should entail in the ecosystem approach**

Biological diversity management involves many interactions, side effects and consequences, so experts and stakeholders at all levels should be involved, including local, national, regional and international.

### 13.4.1 ENVIRONMENTAL JUSTICE

The majority of industrial facilities, garage dumping and processing of dangerous chemicals are typically located in poor and minority communities (Bullard 1993, Cutter et al. 1996, Boone et al. 1999). Residents are relatively unprotected from pollutants around them. So here, environmental justice comes into existence (Coglianese 2001, Adeol 2011). It is a very simple concept which says people shouldn't be subjected to more pollution because of their race, religion, national origin or income level (Checker 2005). Environmental justice is the practice of preserving, fulfilling and respecting the individual and group identities, needs and dignity in a way that promotes self-actualisation and community empowerment (Carruthers et al. 2008, Beltrán et al. 2016). Environmental justice is the chasing of equal justice and equal shielding under the law for all environmental enactment and regulations without discriminating on the basis of race, ethnicity or socioeconomic status (Brisman 2007, Bullard et al. 2000; Cutter 1995). Today, it is evident that pollution and climate change is a global issue since many countries emit more carbon and pollutants, and countries with low incomes are the ones who pay the price. In this scenario, environmental justice becomes more important.

### 13.4.2 AN OVERVIEW OF ENVIRONMENTAL JUSTICE

According to decades of research, Black and Brown people are most at risk for environmental harm. Primarily, the environmental justice movement led by Latinos, Asians and Pacific Islanders, African-Americans and Native Americans focuses on statistical truth: Black or Brown people and the poor dwell, work and play in the most polluted places in America. Yet the timeline of the initiation of environmental equity is not clear, and there are only a few instances that indicate where environmental justice finds its place in society. Few incidences are described in Table 13.1.

**TABLE 13.1**

**Incidence and Reasons for the Rise of Environmental Justice**

| S. No. | Incidence | Year | Major Activities |
|---|---|---|---|
| 1 | Latino farmworkers for workplace right and protection from pesticide | Early 1960s | It was Chavez who fought for worker's rights, such as the right to be protected from harmful pesticides in California's San Joaquin Valley (Shaw 2010). |
| 2 | African-American student protest in Houston | 1967 | A trash dump in Houston was responsible for the death of a child. Students of African-American origin took to the streets to protest the dump (Bullanrd 2018). |
| 3 | West Harlem protest | 1968 | In West Harlem, New York City, residents fought a sewage treatment plant without success (Checker 2011). |
| 4 | Memphis Sanitation Strike | 1968 | As a result of unfair treatment and concerns about environmental justice, the Memphis Sanitation Strike took place in Memphis, Tennessee (Estes 2000). |
| 5 | Bean vs Southwestern Waste Management Corporation and NECAG Formation | 1979 | The fight against landfill began in Houston, Texas. Northeast Community Action Group (NECAG) was formed by residents. Linda McKeever Bullard, an attorney for NECAG, filed a class action lawsuit to stop the landfill's construction (Bullard 2001). |
| 6 | A sit-in against the PCB landfill in Warren County, North Carolina | 1982 | It was the second time African-Americans protested against a polychlorinated biphenyl landfill in Warren County, North Carolina. In many ways, the Environmental Justice Movement was sparked by this event (Bullard 1990). |
| 7 | General Accounting Office Conducts Study | 1983 | This study is said to have "galvanized the environmental justice movement and provided empirical support for the claims for environmental racism." According to the General Accounting Office, at least 26% of the population of hazardous waste landfills examined were African-American, and their families earned less than the poverty line (Cutter 1995). |
| 8 | Toxic Waste in the United States | 1987 | Toxic Waste in the United States was released by the United Church of Christ Commission on Racial Justice. UCC investigated the racial/economic composition of host communities and hazardous waste site locations (Bullard et al. 2004). |
| 9 | West Harlem Environmental Action | 1988 | WE ACT (West Harlem Environmental Action) was founded. New York–based environmental justice group WE ACT was established to promote health, safety and the environment for communities of color (Vallentyne et al. 1988) |

*(Continued)*

**TABLE 13.1** *(Continued)*

**Incidence and Reasons for the Rise of Environmental Justice**

| S. No. | Incidence | Year | Major Activities |
| --- | --- | --- | --- |
| 10 | Indigenous Environmental Network | 1990 | By addressing issues of environmental and economic justice through the development of economically sustainable communities, indigenous peoples formed the Indigenous Environmental Network (IEN) (Network 2001) |
| 11 | Race and Environmental Hazards Conference at the University of Michigan | 1990 | As a result of this conference, environmental justice became a legitimate academic field of study, leading to a team of academics and activists led by Bryant and Mohai to advise the US EPA on environmental justice policy (Mohai et al. 1990). |
| 12 | EPA Administrator Creates the NEJAC (National Environmental Justice Advisory Council) | 1993 | EPA Administrator Carol M. Browner created NEJAC. There are public meetings held by NEJAC on environmental justice issues throughout the country. |
| 13 | Clark Atlanta University's Environmental Justice Resource Center was established | 1994 | The Environmental Justice Resource Center (EJRC) was formed at Clark Atlanta University (an HBCU) by Dr. Robert Bullard, who served as the center's director (Bullard et al. 2000). |
| 14 | EPA and CEQ Collaborate on NEPA to Advance Environmental Justice | 1997 | Guidelines under the NEP (National Environmental Policy) Act were developed by the CEQ (Council on Environmental Quality) in consultation with the environmental protection agency and other affected agencies (Outka 2006). |
| 15 | Warren County PCB Landfill is Remediated | 2001 | In North Carolina, the PCB landfill community in Warren County secured to remediate the site and develop an economic development plan (Zavestoski 2007). |
| 16 | EPA Offices Begin Implementing EJ Action Plans | 2002 | The EPA Environmental Justice Executive Steering Committee began developing the Environmental Justice Action Plans in fiscal year 2003 (London et al. 2008) |
| 17 | Toxic Waste and Race at Twenty | 2007 | Toxic Wastes and Race at Twenty found a higher concentration of people of color around hazardous waste facilities than the United Church of Christ 1987 study (Bullard 2007). |
| 18 | Plan Environmental Justice 2014 | 2011 | EPA is integrating environmental justice into all its programs, policies and activities through this road map, not a rule or regulation (Boone et al. 2014). |
| 19 | Final draft of EJ 2020 is introduced | 2016 | Environmental Justice 2020 was released for public comment both as a draft framework and a final draft, and thousands of comments were received from communities and stakeholders (Salcido 2016) |

## 13.5  COMPLEMENTARITY BETWEEN ENVIRONMENTAL JUSTICE AND NATURAL RESOURCE MANAGEMENT

Interaction of human population and environment is very complicated (Young et al. 2006). It is necessary to investigate how people interact with the environment and policies, economics and politics affecting resource usage (Taylor et al. 1992). A thriving ecosystem is dependent on its ability to generate functions and services. Rapid consumption and production leads to an intensified overuse of natural resources, and it causes pollution (Chen 2007) that is likely to affect certain groups of people due to inequality in society, such as racial, religious and national origin groups (Pellow 2005). To prevent the overuse and abuse of people and environment, rules and regulations are being developed. Depending on local conditions and specific problems, the best rules and appropriate mechanisms may differ. Regulations and laws may exercise management control directly with legislation setting and enforcing environmental standards (Heyes 2000). Taxes or permits may also be used to exercise control. Paying for ecosystem services (PES) is one of those methods; in exchange for managing their lands to provide ecological services, farmers are offered incentives (Kerr et al. 2014).

## 13.6  NATURAL RESOURCE MANAGEMENT (NRM)

Population growth, intensive agricultural expansion, economic development and the destruction of natural habitat are adversely affecting the environment and natural resources due to uncontrolled urbanisation and industrialisation (Ray et al. 2011). The concept of utilising major natural resources, including land, fisheries, minerals, water, air, forests and wild flora and fauna, in a sustainable manner is called natural resource management (Singh et al. 2017).

During the late 19th century, accelerated industrialisation gave importance to management of natural resources. According to a World Bank report, 30% of the world's irrigated land, 40% of its rainfed agricultural land and 70% of its rangeland are affected by erosion, salinization and compaction. More than 900 million people in 100 countries are affected by the deforestation of 11% of the earth's vegetation over the past 45 years (Marques 2020). Overconsumption of natural resources should be managed in order to keep the ecosystem balanced and to avoid further destruction of the environment.

### 13.6.1  NATURAL RESOURCE MANAGEMENT, ENVIRONMENTAL JUSTICE AND ECOSYSTEM APPROACH: SYNERGIES

The major component of environmental justice is sharing links with the principle of ecosystem approach. The first principle of ecosystem approach—relevant stakeholders' involvement in determining the design, protection and management of ecosystems—is important for acknowledging justice as recognition of environmental justice (Paloniemi et al. 2015, Martin et al. 2014, Alam et al. 2019). Distributive justice is a component of environmental justice which shares complementarities of

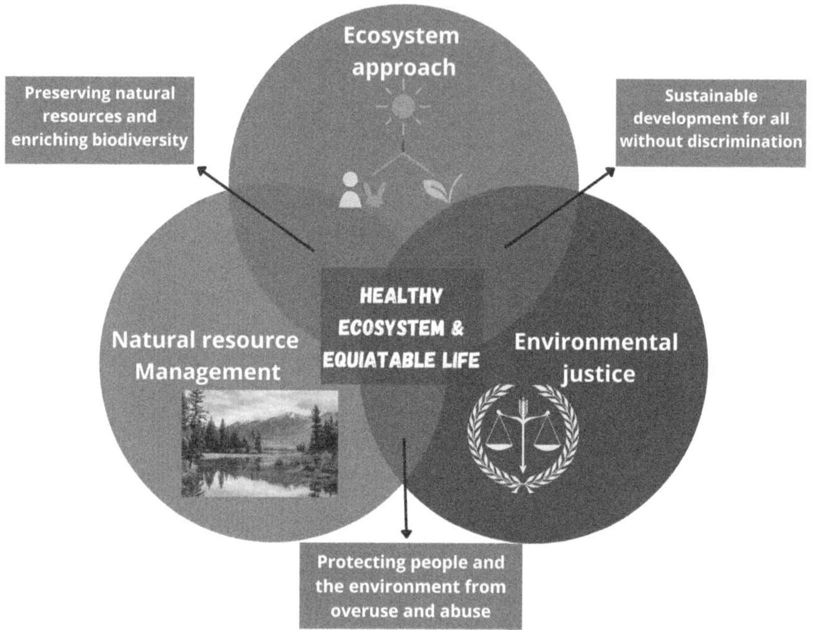


**FIGURE 13.2** Synergies between natural resource management, ecosystem approach and environmental justice.

the fourth principle of ecosystem approach which focuses on equity sharing, partnerships and value chains (Fleurbaey et al. 2014, Alam et al. 2019).

There is a synergistic relationship between procedural justice and the second principle of ecosystem approach which talks about decentralisation of management and participation of stakeholders as well as sharing the idea of Principle 12, which says the creation of ecosystem-based conservation initiatives involves multidisciplinary professionals and scientists. Intergenerational equity and precautionary principle are also components of environmental justice, sharing links with principles 3, 6, 7, 8, 9 and 10 of the ecosystem approach, respectively (Maltby 2000, Shepherd et al. 2004). Figure 13.2 shows a synergistic relationship between environmental justice, ecosystem approach and natural resource management. As an example, we can clearly see the links between ecosystem approach and the aspect of environmental justice from the standpoint of achieving equity and justice for people and nature in the international arena.

## 13.7 CONCLUSION

Peace is impossible as long as the poor, the socially oppressed, the disparate and the environment are being abused by the powerful and mighty. Environmental justice is about ensuring that no group suffers disproportionate environmental burdens or does not receive a fair share of the environmental benefits. Communities and governments can use ecosystem approaches as a means of tackling the distributional effects of pollution on the poor and inequality between the rich and poor as well as social factors that have led to disparate development outcomes among countries and among the poor. Almost all other problems of our national life are rooted in the overuse and misuse of natural resources. It is possible to enhance the quality of life and solve the problems by improving and implementing ecosystem and environmental justice approaches.

## ACKNOWLEDGEMENT

Our sincere thanks to the director of Dayalbagh Educational Institute, Agra, for providing us with infrastructure.

## REFERENCES

Adeola, F. O. (2011). *Hazardous wastes, industrial disasters, and environmental health risks: Local and global environmental struggles*. Springer.

Alam, S., & Mohammad, S. N. (2019). The ecosystem approach and environmental justice nexus in natural resource management. *Chinese Journal of Environmental Law*, *3*(1), 47–84.

Beltrán, R., Hacker, A., & Begun, S. (2016). Environmental justice is a social justice issue: Incorporating environmental justice into social work practice curricula. *Journal of Social Work Education*, *52*(4), 493–502.

Boone, C. G., & Modarres, A. (1999). Creating a toxic neighborhood in Los Angeles County: A historical examination of environmental inequity. *Urban Affairs Review*, *35*(2), 163–187.

Boone, C. G., Fragkias, M., Buckley, G. L., & Grove, J. M. (2014). A long view of polluting industry and environmental justice in Baltimore. *Cities*, *36*, 41–49.

Brisman, A. (2007). Crime-environment relationships and environmental justice. *Seattle Journal for Social Justice*, *6*, 727.

Bullard, R. D. (2001). Environmental justice in the 21st century: Race still matters. *Phylon (1960–)*, *49*(3/4), 151–171.

Bullard, R. D. (2018). *Dumping in Dixie: Race, class, and environmental quality*. Routledge.

Bullard, R. D. (Ed.). (1993). *Confronting environmental racism: Voices from the grassroots*. South End Press.

Bullard, R. D., & Johnson, G. S. (2000). Environmentalism and public policy: Environmental justice: Grassroots activism and its impact on public policy decision making. *Journal of Social Issues*, *56*(3), 555–578.

Bullard, R. D., & Wright, B. H. (1990). The quest for environmental equity: Mobilizing the African-American community for social change. *Society & Natural Resources*, *3*(4), 301–311.

Bullard, R. D., Johnson, G. S., & Torres, A. O. (Eds.). (2004). *Highway robbery: Transportation racism & new routes to equity*. South End Press.

Bullard, R., Mohai, P., Saha, R., & Wright, B. (2007). Toxic waste and race at twenty. *A report prepared for the United Church of Christ and Witness Ministries*, Cleveland, OH.

Bulte, E., Gerking, S., List, J. A., & De Zeeuw, A. (2005). The effect of varying the causes of environmental problems on stated WTP values: Evidence from a field study. *Journal of Environmental Economics and Management*, *49*(2), 330–342.

Cardinale, B. J., Duffy, J. E., Gonzalez, A., Hooper, D. U., Perrings, C., Venail, P., . . . Naeem, S. (2012). Biodiversity loss and its impact on humanity. *Nature*, *486*(7401), 59–67.

Carruthers, D. V., & Carruthers, D. V. (Eds.). (2008). *Environmental justice in Latin America: Problems, promise, and practice*. MIT Press.

Chandra, A., & Idrisova, A. (2011). Convention on biological diversity: A review of national challenges and opportunities for implementation. *Biodiversity and Conservation*, *20*(14), 3295–3316.

Chase, J. M., Blowes, S. A., Knight, T. M., Gerstner, K., & May, F. (2020). Ecosystem decay exacerbates biodiversity loss with habitat loss. *Nature*, *584*(7820), 238–243.

Checker, M. (2005). Polluted promises. In *Polluted promises*. New York University Press.

Checker, M. (2011). Wiped out by the "greenwave": Environmental gentrification and the paradoxical politics of urban sustainability. *City & Society*, *23*(2), 210–229.

Chen, J. (2007). Rapid urbanization in China: A real challenge to soil protection and food security. *Catena*, *69*(1), 1–15.

Coglianese, C. (2001). Social movements, law, and society: The institutionalization of the environmental movement. *University of Pennsylvania Law Review*, *150*(1), 85–118.

Cohen-Shacham, E., Andrade, A., Dalton, J., Dudley, N., Jones, M., Kumar, C., . . . Walters, G. (2019). Core principles for successfully implementing and upscaling Nature-based Solutions. *Environmental Science & Policy*, *98*, 20–29.

Cronin, R., & Pandya, A. (2009). Natural resources and the development-environment dilemma. *Exploiting Natural Resources. The Henry L. Stimson Centre*, 63.

Cropper, A. (1993). Convention on biological diversity. *Environmental Conservation*, *20*(4), 364–364.

Cutter, S. L. (1995). Race, class and environmental justice. *Progress in Human Geography*, *19*(1), 111–122.

Cutter, S. L., & Solecki, W. D. (1996). Setting environmental justice in space and place: Acute and chronic airborne toxic releases in the southeastern United States. *Urban Geography*, *17*(5), 380–399.

Estes, S. (2000). "I am a man!": Race, masculinity, and the 1968 Memphis sanitation strike. *Labor History*, *41*(2), 153–170.

Fleurbaey M., Kartha, S., Bolwig, S., Chee, Y. L., Chen, Y., Corbera, E., Lecocq, F., Lutz, W., Muylaert, M. S., Norgaard, R. B., Okereke, C., & Sagar, A. D. (2014). Sustainable development and equity. In O. Edenhofer, R. Pichs-Madruga, Y. Sokona, E. Farahani, S. Kadner, K. Seyboth, A. Adler, I. Baum, S. Brunner, P. Eickemeier, B. Kriemann, J. Savolainen, S. Schlömer, C. von Stechow, T. Zwickel, & J. C. Minx (Eds.), *Climate change 2014: Mitigation of climate change. Contribution of working group iii to the fifth assessment report of the intergovernmental panel on climate change*, Cambridge University Press. https://www.ipcc.ch/site/assets/uploads/2018/02/ipcc_wg3_ar5_chapter4.pdf

Forget, G., & Lebel, J. (2001). An ecosystem approach to human health. *International Journal of Occupational and Environmental Health*, *7*(2), 1–40.

Grennfelt, P., Engleryd, A., Forsius, M., Hov, Ø., Rodhe, H., & Cowling, E. (2020). Acid rain and air pollution: 50 years of progress in environmental science and policy. *AMBIO*, *49*(4), 849–864.

Halewood, M., López Noriega, I., Ellis, D., Roa, C., Rouard, M., & Sackville-Hamilton, N. R. (2017). Potential implications of the use of digital sequence information on genetic resources for the three objectives of the Convention on Biological Diversity. *A submission from CGIAR to the Secretary of the Convention on Biological Diversity (CBD)*.

Heyes, A. (2000). Implementing environmental regulation: Enforcement and compliance. *Journal of Regulatory Economics*, *17*(2), 107–129.

Kerr, J., Vardhan, M., & Jindal, R. (2014). Incentives, conditionality and collective action in payment for environmental services. *International Journal of the Commons*, *8*(2).

Khafaie, M. A., Sayyah, M., & Rahim, F. (2019). Extreme pollution, climate change, and depression. Environmental Science and Pollution Research, 26(22), 22103–22105.

Liang, J., Park, S., & Zhao, T. (2020). Representative bureaucracy, distributional equity, and environmental justice. *Public Administration Review, 80*(3), 402–414.

London, J. K., Sze, J., & Liévanos, R. S. (2008). Problems, promise, progress, and perils: Critical reflections on environmental justice policy implementation in California. *UCLA Journal of Environmental Law and Policy, 26,* 255.

Maltby, E. (2000, August). Ecosystem approach: From principle to practice. In *Ecosystem service and sustainable watershed management in North China International Conference, Beijing, PR China* (Vol. 205, pp. 205–224).

Marques, L. (2020). Water and soil. In *Capitalism and environmental collapse* (pp. 65–96). Springer.

Martin, A., Gross-Camp, N., Kebede, B., McGuire, S., & Munyarukaza, J. (2014). Whose environmental justice? Exploring local and global perspectives in a payments for ecosystem services scheme in Rwanda. *Geoforum, 54,* 167–177.

Martin, A., McGuire, S., & Sullivan, S. (2013). Global environmental justice and biodiversity conservation. *The Geographical Journal, 179*(2), 122–131.

Massey, R. (2004). Environmental justice: Income, race, and health. *Global Development and Environment Institute,* 1–26.

Mohai, M., Bertóti, I., & Révész, M., 1990. XPS study of the state of oxygen on a chemically treated glass surface. *Surface and Interface Analysis, 15*(6), 364–368.

Network, I. E. (2001). *Indigenous environmental network.* Indigenous Environmental Network.

Outka, U. (2006). NEPA and environmental justice: Integration, implementation, and judicial review. *Boston College Environmental Affairs Law Review, 33,* 601.

Oza, G. M. (1992). The Earth Summit 1992. *Indian Forester, 118*(5), 338–343.

Paloniemi, R., Apostolopoulou, E., Cent, J., Bormpoudakis, D., Scott, A., Grodzińska-Jurczak, M., . . . Pantis, J. D. (2015). Public participation and environmental justice in biodiversity governance in Finland, Greece, Poland and the UK. *Environmental Policy and Governance, 25*(5), 330–342.

Pellow, D. (2005). Environmental racism: inequality in a toxic world. In M. Romero, & E. Margolis (Eds.), *The Blackwell companion to social inequalities* (pp. 147–164). Blackwell Publishing Ltd. https://doi.org/10.1002/9780470996973.ch8.

Peters, N. E., & Meybeck, M. (2000). Water quality degradation effects on freshwater availability: Impacts of human activities. *Water International, 25*(2), 185–193.

Queffelec, B., Cummins, V., & Bailly, D. (2009). Integrated management of marine biodiversity in Europe: Perspectives from ICZM and the evolving EU Maritime Policy framework. *Marine Policy, 33*(6), 871–877.

Ray, S., & Ray, I. A. (2011). Impact of population growth on environmental degradation: Case of India. *Journal of Economics and Sustainable Development, 2*(8), 72–77.

Salcido, R. E. (2016). Reviving the Environmental Justice Agenda. *Chicago-Kent Law Review, 91,* 115.

Shaw, R. (2010). *Beyond the fields: Cesar Chavez, the UFW, and the struggle for justice in the 21st century.* University of California Press.

Shepherd, G., & Union mondiale pour la nature. Commission on ecosystem management. (2004). *The ecosystem approach: Five steps to implementation.* IUCN.

Shiva, V., & Bandyopadhyay, J. (1986). The evolution, structure, and impact of the Chipko movement. *Mountain Research and Development,* 133–142.

Singh, R. L., & Singh, P. K. (2017). Global environmental problems. In *Principles and applications of environmental biotechnology for a sustainable future* (pp. 13–41). Springer.

Singh, R. L., & Singh, P. K. (2017). Global environmental problems. In *Principles and applications of environmental biotechnology for a sustainable future* (pp. 13–41). Springer.

Taylor, P. J., & Buttel, F. H. (1992). How do we know we have global environmental problems? Science and the globalization of environmental discourse. *Geoforum, 23*(3), 405–416.

Taylor-Brown, A., Booth, R., Gillett, A., Mealy, E., Ogbourne, S. M., Polkinghorne, A., & Conroy, G. C. (2019). The impact of human activities on Australian wildlife. *PloS One*, *14*(1), e0206958.

Tinker, C. (1995). A new breed of treaty: The Untied Nations convention on biological diversity. *Pace Environmental Law Review*, *13*, 191.

Vallentyne, J. R., & Beeton, A. M. (1988). The 'ecosystem' approach to managing human uses and abuses of natural resources in the Great Lakes basin. *Environmental Conservation*, *15*(1), 58–62.

Wallace-Wells, D. (2019). *The uninhabitable earth* (pp. 271–294). Columbia University Press.

Wang, J., Wang, K., Zhang, M., & Zhang, C. (2015). Impacts of climate change and human activities on vegetation cover in hilly southern China. *Ecological Engineering*, *81*, 451–461.

Waylen, K. A., Hastings, E. J., Banks, E. A., Holstead, K. L., Irvine, R. J., & Blackstock, K. L. (2014). The need to disentangle key concepts from ecosystem-approach jargon. *Conservation Biology*, *28*(5), 1215–1224.

Weiskopf, S. R., Rubenstein, M. A., Crozier, L. G., Gaichas, S., Griffis, R., Halofsky, J. E., . . . Whyte, K. P. (2020). Climate change effects on biodiversity, ecosystems, ecosystem services, and natural resource management in the United States. *Science of the Total Environment*, *733*, 137782.

www.cbd.int/ecosystem/principles.shtml

Young, O. R., Lambin, E. F., Alcock, F., Haberl, H., Karlsson, S. I., McConnell, W. J., . . . Verburg, P. H. (2006). A portfolio approach to analyzing complex human-environment interactions: Institutions and land change. *Ecology and society*, *11*(2).

Zavestoski, S. (2007). *Transforming environmentalism: Warren County, PCBs, and the origins of environmental justice*. Rutgers University Press.

# 14 Soil
## *A Potential Source for Mitigating Food, Water and Bioenergy Crisis*

*Kumar Chiranjeeb, Sachin Kumar, and Ranbir Singh Rana*

## 14.1 INTRODUCTION

There are unprecedented challenges and opportunities facing humanity in the 21st century. Among scientists who focus on natural resources, the need and opportunities for innovative descriptions of those resources, their condition with respect to humanity's use of those resources and how they move through time are obvious. The "nexus approach" to food, energy and water is a set of newer approaches that have gained global attention (Hamiche et al. 2016; Cairns and Krzywoszynska 2016; Wichelns 2017; Al-Saidi and Elagib 2017). The soil water usage with its nexus system has increased six times more in the past 100 years, and it is likely to increase more in the 21st century (Wada et al. 2016). A host of newer approaches to addressing global resource security are based on the "water-energy-food nexus" first introduced at the 2011 World Economic Forum (Lal 2013; Lal 2016; Mohtar et al. 2016; Wichelns 2017). It is an approach aimed at examining natural resources and how they interact and addressing challenges related to food, water and energy security (Chang et al. 2016). In ancient civilizations, the interrelationship between food, energy and water was understood implicitly as well as explicitly (Lal 2013, 2016). Water and soil are essential components of an agroecosystem's food production strategy. As part of the overall nexus concept, soil processes and soil organic carbon (SOC) management have not been integrated (Howard 1947). Food is an important output, while water and energy are major inputs in the food, energy and water paradigm (Lal et al. 2017). Water consumption pattern accounts for about 60%–80% for irrigated agriculture (Gerbens-Leenes et al. 2009), and the pattern accounts for 90% in arid and semi-arid regions (Bazilian et al. 2011). Even though 40% of world's food production is from 20% of irrigated land (FAO 2016), the water use efficiency is very much low. Moreover, biofuel production consumes a significant amount of water. For production of ethanol from corn grains, there is a requirement of 4 gallons per gallon of water, whereas biodiesel is estimated at about 3 gallons per gallon (Pate et al. 2007). The amount of water needed for the production of biofuels varies depending on the crop and the evapotranspiration and soil water supply. The majority of food grains consumed are produced on soil. Soil is now and will always be a crucial part of food production. Nowadays, the agricultural land availability per capita is decreasing at an alarming rate due to the blast in human population, non-agricultural activities such as urbanization, deforestation, recreation and industrial uses as well as various other soil degrading activities are harmful for ecological restoration.

Soil degradation still remains a significant constraint around the world that threatens the sustainability as well as security of food and water (Bai et al. 2008a; Bai et al. 2008b, FAO 2011, and Temesgen et al. 2012a). Soil health management is the key factor to achieve excellence and security for food and nutrition to mitigate the global hunger issues (Lal 2016). The soil with optimum physical, chemical and biological properties is providing critical overview in case of food and energy availability with optimum utilization efficiencies. Soil and water management policies are adopting a nexus-based approach to address the burning issues like soil degradation, poor soil aggregation (Dexter 1988) as well as runoff issues to counteract the problems like water logging and flooding (Dexter 1988; Temesgen et al. 2012b). Therefore, increasing soil infiltration through various management approaches, decreasing runoff and erosion, and enhancing root zone water content while emphasizing on elevating groundwater concentration will provide stability and mitigate the global issues with a promise of better sustainability.

To understand soil-water interaction, it's important to understand soil hydrology, the retention and transmission of water in the vadose zone. Between 1960 and 2010, global energy consumption tripled to 550 EJ and is expected to triple again by 2040. Water, nutrients, weed and pest management as well as other farm operations such as tillage, seeding and harvesting are considered energy-based inputs into agriculture production systems. As scientific insights about these resources, it is also crucial to integrate other parameters (soil and critical interactions between anthrospheres, climate) under sustainable goals. It is, therefore, imperative that we develop a nexus approach that puts soil at the center as humanity's foundation of the future (Cairns and Krzywoszynska 2016). As a result, it is necessary to develop integrated nexus insights and tools so we can create a better understanding of the food, energy and water nexus that can meet the needs of long-term sustainability. Besides food, energy and water, the food-energy-water-soil nexus tool also has application in soil reclamation and waste management (crucial to reducing environmental risks, recycling waste at the municipality level and maximizing resource efficiency). In the waste management process, soil plays a vital role, especially when it comes to managing municipal solid waste, as well as gray water and black water for nutrient cycling. Therefore, a nexus toolkit includes soil quality and soil management. In the context of global issues during the 21st century, this chapter focuses on the importance of soils as a nexus tool and the role of soil organic carbon in improving soil health.

## 14.2  SOIL AS A MEDIUM FOR PLANT GROWTH AND SUSTAINABILITY

Soil provides the substrate or medium for the establishment of crop for its growth and completes its lifecycle on it. The per capita land resources are declining day by day as it was 0.37 ha in 1961 to 0.197 ha in 2013 and so the production of crops decreasing on it. Crops require all sorts of micro nutrients (i.e., Fe, Mn, Zn, Cu, Mo, B, Cl) and macro nutrients (i.e., N, P, K, Ca, Mg, S) in adequate quantity for their growth, as these nutrients influence plant metabolism, growth, photosynthesis, energy transformations and disease resistance against biotic and abiotic stresses. Inside plants, the transportation of water, nutrients and various metabolites are carried out with the help of elements provided by the soil system plant growth and maintenance. The soil system has micro and macro pore components associated with it, which help in root proliferation and supply water and store nutrients in the pores which then become available to plants in later stages. Adequate nutrient supply to crops can boost the crop yield by enhancing grain content in crops, increasing the nutrient content as well as the quality of grain. Finally, the yield is higher. Crop types and classifications modify the nutrient demand and uptake of nutrients from soil systems. For instance, annual crops grow on soil for a period of one year and nutrient depletion from soil by crops is for that certain period of time, but in case of perennial crop, the time period of crop growth extends. The food crisis is directly linked with the fertility status of the soil.

### 14.2.1 SOIL AS A SOURCE OF NUTRIENT POOL AND MITIGATE NUTRIENT STRESS

The soil system has a large component of organic matter and upon decomposition of organic matter by microbes, the are adsorbed along with associated nutrients and get released into the soil solution from which plants can take nutrients. Chelating nutrients often get exchanged with the soil colloidal phase and enhance nutrient exchange capacity of soil. Some nutrients or ions are not available to crop plants and remain in organic form. So these organic forms of nutrients must be converted to inorganic forms which are available to plants through the process of mineralization, whereas the reverse process is called immobilization. Earth crust in the soil system also acts as a source of nutrients in it, and upon transformation processes, these nutrients are added in the soil solution pool and it is absorbed by plants later on.

### 14.2.2 SOIL AS RESOURCE EFFICIENCY

Several approaches are being taken to develop functional soil maps, and any positive improvements will naturally be relevant to the issues and cause threat to the human civilization as well include (i) food and energy security, (ii) water management, renewability and availability, (iii) mitigation of climate change, (iv) ecosystems restoration and security, (v) enhancement of biodiversity both above and below ground, (vi) human health and (vii) resource conservation and sustainable management (Diamond 2005). The soil as a nexus tool can address these issues given in Figure 14.1 The soil plays a major role in the cycling of elements (N, P, S, $H_2O$), microbial biodiversity, energy security and nutritional availability which overall affect the productivity of the ecosystem with sustainable causes for burning regional/global issues (Lal 2015a; Lal 2015b).

It becomes increasingly difficult to maintain sustainable food and water security. As a result of the environmental changes, crop yields have decreased, livestock

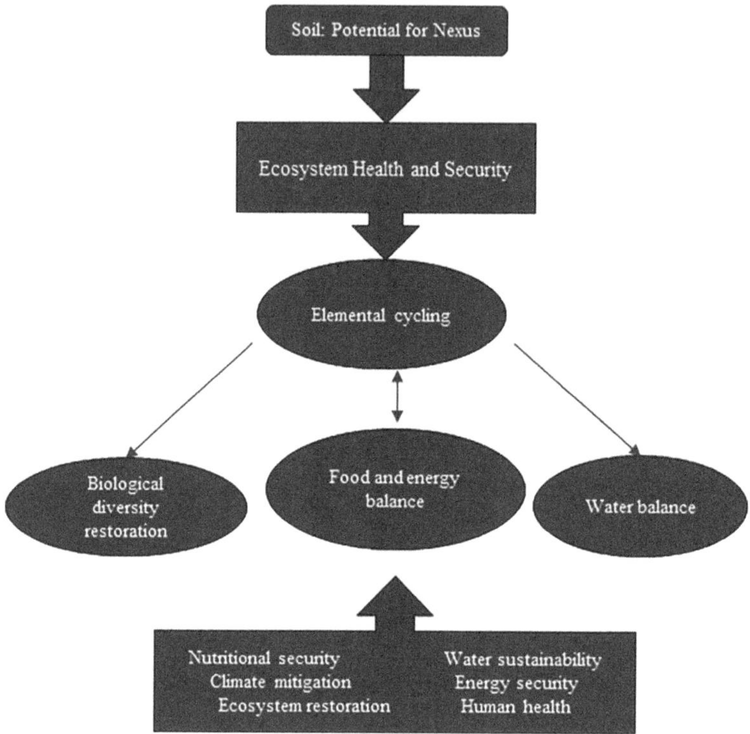


**FIGURE 14.1** Soil as a nexus tool and ecosystem sustainability.

production has declined, water supply has decreased and pesticide uses have increased (McCarl and Reilly 2007). It is becoming difficult to increase productivity and efficiency of agricultural production systems as a result of shrinking agricultural production areas and diminished access to water resources. This is dependent on a number of interrelated factors, including soil micro climate, soil health, water management, abiotic stresses and agronomic practices. In order to improve cropland productivity, we must quantify the impact that agronomic practices have on soil health and ecosystem. An understanding of how agricultural production systems work must be holistic and quantitative. In addition to incorporating the soil-water-plant nexus, the system should be quantified based on its functionality.

## 14.3 CLASSIFICATION AND FUNCTIONS OF NUTRIENTS

Nutrients or ions are basically inorganic in nature and supplement the plants for their growth and proliferation. Different types of plants alter the process of nutrient depletion and uptake in plants as well as demand of nutrients by plant changes, so nutrients are classified as follows:

### 14.3.1 Basic Nutrient

These nutrients make the basic structural component of plants and cells. They are present in large quantities, e.g., C, H, O.

### 14.3.2 Macro Nutrient

Macro nutrients are required by plants in larger quantities, i.e., more than 10 ppm in plants. For example, N, P, K, Ca, Mg and S. Among these nutrients, Ca, Mg and S nutrients are called secondary macro nutrients.

### 14.3.3 Micro Nutrient

Micro nutrients are required by plants in smaller quantities, i.e., less than 10 ppm. Micro nutrients found in the soil system are again divided into certain types:

a. **Cationic:** Nutrients are present in cationic form, e.g., Fe, Mn, Cu, Zn.
b. **Anionic:** Nutrients present in anionic form, e.g., B, Cl, Mo.

### 14.3.4 Criteria of Essentiality

Ions or nutrients are the major components that lead to the increase in food production and maintain the fertility status of soil. Among 118 elements in nature, only 90

**TABLE 14.1**
**Rating Chart of Nutrients in Soil**

| S. No. | Nutrients | Low | Medium | High |
|---|---|---|---|---|
| 1. | Organic Carbon (%) | < 0.50 | 0.50–0.75 | > 0.75 |
| 2. | Available N (kg/ha) | < 250 | 250–500 | > 500 |
| 3. | Available $P_2O_5$ (kg/ha) | < 25 | 25–50 | > 50 |
| 4. | Available $K_2O$ (kg/ha) | < 125 | 125–300 | > 300 |
| 5. | Available Zn (ppm) | < 0.78 | 0.78–1.20 | > 1.20 |
| 6. | Available Cu (ppm) | < 0.60 | 0.60–1.20 | > 1.20 |
| 7. | Available Fe (ppm) | < 7.0 | 7.0–12.0 | > 12.0 |
| 8. | Available Mn (ppm) | < 3.00 | 3.00–5.00 | > 5.00 |
| 9. | Available B (ppm) | < 0.50 | 0.5–1.0 | > 1.0 |
| 10. | Available S (ppm), noncalcareous soil | < 10.0 | 10.0–20.0 | > 20.0 |
| 11. | Available S (ppm), calcareous soil | < 13.0 | 13.0–20.0 | > 20.0 |

*Source:* Sinha and Chiranjeeb (2022)

elements are taken up by the plants. Then again, the major essentiality of elements is given by Arnon and Stout (1939) as follows:

1. The plant must be unable to grow normally or complete its life cycle in the absence of the element.
2. The element is specific and cannot be replaced by another.
3. The element plays a direct role in plant metabolism.

As the soil is more fertile, plants will have ample amount of growth-enhancing factors to overcome adverse conditions. Nowadays the decline in per capita land availability is due to the increasing human population along with the adoption of destructive non-agricultural processes such as rapid urbanization, recreational activities, deforestation, industrial growth and other soil-degrading activities associated with the process.

The heavily denser countries—i.e., Asia, China, and Africa—will have to face the consequences of food crisis and nutritional security due to depletion and improper

**TABLE 14.2**

**Indicator Plants**

| S. No. | Nutrients | Indicator Plant |
| --- | --- | --- |
| 1. | Sodium | Sugarbeet |
| 2. | Phosphorus | Rapeseed |
| 3. | Manganese | Oat, Potato, Sugarbeet |
| 4. | Copper | Wheat |
| 5. | Boron | Sunflower |
| 6. | Potassium and Magnesium | Potato |
| 7. | Nitrogen, Calcium | Cauliflower and Cabbage |

*Source:* Sinha and Chiranjeeb (2022)

**TABLE 14.3**

**Critical Limits of Nutrients**

| S. No. | Nutrients | Critical limit value (ppm) |
| --- | --- | --- |
| 1. | Zn | 0.6 |
| 2. | B | 0.5 |
| 3. | Fe | 4.5 |
| 4. | Mn | 2.0 |
| 5. | Cu | 0.2 |
| 6. | Mo | 0.1 |
| 7. | S | 10 |

*Source:* Sinha and Chiranjeeb (2022)

utilization of the existing natural resources as well as soil degradation with soil health deterioration. The nutrient content and calorie shortage in food materials are directly associated with soil degradation. The significance of enhancement of agroecosystem sustainability cannot be overemphasized anytime. Soil health plays a key role in making progress in food security. The uneven distribution of soil components and land resources also interact with the water component, nutrient balance, etc., in soil and controls the food security and overall production to mitigate global food crisis. Thus, better crop production with enriched food quality can fulfill the aim of sustainability in the field of agriculture food production and can advance a step towards completing UN goals of sustainable development with mitigating global hunger.

## 14.4 SOIL SOURCE FOR MITIGATING WATER STRESS AND WATER IMBALANCE

Water sources on earth occupy about 72%, and it is highest among other existing resources on earth. Water plays key role in the plant growth system. Water exists in nature in various forms like solid (ice), liquid (water) and gas (vapor). The plant system can uptake moisture in these discussed forms, but the stability is more for moisture near the root zone. The soil system contains micro and macro pores. These pores help in the retention of moisture in the structure for the particular time period during plant growth stages. Macro pores are larger in size and may contain moisture for a certain time frame, as due to gravity, the loss of moisture is also more in this case, but the micro pores can contain water for a longer period of time. The natural soil system has a definite and potential filter system which in turn helps in the filtering of water or moisture present in soil and supplying good quality of water to the crop plants for uptake. Harmful and toxic substances are being filtered through this membranous filter system in soil which in turn assures the quality irrigation of water to the crops.

Soil also acts as a natural source of mulch in the agriculture sector and controls the water loss through evaporation and other potential losses. Soil particles themselves form a hydrophilic bond with water molecules and stabilize the soil-water environment. Soil organic matter content also influence the water status in soil as higher surface area in case of finer-sized organic matter particles enhance the adsorption of water molecules on it. Dark-colored organic matter in soil increases the temperature of soil so the contained moisture faces the fate of loss from the soil ecosystem as compared to the light-colored organic matter. Various agricultural inputs play a significant role in determining the yield as well as overall production in the long run. Irrigation is also one of the most important inputs for agriculture, as it influences the crop production during various critical stages, thus controlling the yield and productivity of the crop. More than 85% of freshwater resources are being used for irrigation purposes in India (Bhatia 2007; Mukherjee 2008). Due to the expansion of agricultural land to mitigate global hunger, the irrigated area percentage is also increasing day by day in the last few decades. In India, the total potential of irrigation has enhanced from 81.1 million ha in 1991–92 to 108.2 million ha in March 2010 (Economic Survey 2011–12). The potential sources are rainfall, canals, tanks as well as underground water sources. Most of the cropping system in India depends on rainfall as rainfed system, i.e., around 60% (MoA 2011) of net sown area (140 million hectare) for which it is highly unpredictable in its dimension and declines the peak of production in agriculture (Briscoe and Malik 2007). The government canal system for irrigation is enhancing its boundary and share as compared to private canal. In case of a canal irrigated area, the private canal participation is continuously decreasing from 13.3% in 1950–51 to around 1.5% in 2006–07. Groundwater depletion and its pollution have led our nation to a phase of slower growth in the whole world. The yields obtained from groundwater irrigation are substantially higher than from surface irrigation system as groundwater acts as a source of flexibility, ease of use with control, reliability and as a backup source during stressful conditions (Bhatia 2007; Iyer 2009).

At present, about 231 BMC of groundwater extraction is carried out annually in India (GOI 2009–10), which is the highest volume of annual groundwater extraction in the world (Scott and Shah 2004; Shah 2005). After the intensive tube well installation in 1970, the Indian agricultural system has become the major source of irrigation for crop production (Economic Survey 2011–12). During 2006–07, the net irrigation was like canals, wells, tanks and other sources that accounted about 25%, 59%, 3%, and 12 % respectively. About 36% of the blocks in our country will be deprived of use of groundwater overuse by 2017–18 due to the increase of irrigated areas in agriculture with improper management for crop production (Moench 2002; Dubash 2007). The safe and sustainable use of groundwater in agriculture has enhanced the overall productivity, but the negative usage of groundwater has caused more and more ecological disaster.

### 14.4.1 Quality of Irrigation Water

Good quality irrigation water is necessary for enhancing crop production as well as maintaining good soil health conditions. Irrigation water contains chloride and sulphate salts of calcium, magnesium and sodium along with other cations like carbonate, bicarbonate and other heavy metals which determine the quality of irrigation water as it may affect the crop physiology, growth and sustainability. The irrigation water ensures better photosynthetic activities, nutrients mobilization and grain development, enhancing crop genetic potential.

### 14.4.2 Harmful Impacts of Poor-Quality Water

Due to the heavy irrigation of saline water in the field, the soil turns to saline in nature and restricts crop growth to a large extent. Salt accumulation in soil damages the root system and restricts the uptake of other nutrients by crop plant due to the enhanced osmotic pressure of the soil solution. Excessive sodium accumulation causes the destruction of the soil structure, deflocculation of soil particles and poor circulation of air and water in soil.

## 14.5 BIOENERGY CRISIS AND ITS MANAGEMENT

Soil has provided necessary conditions to grow crops on it, and these crops act as a source of bioenergy. Soil also provides habitat for billions of microbial organisms that play a key role in energy transformations and regulations in soil and atmosphere. The energy cycle is running between soil and atmosphere, and soil plays the role in maintaining stability in this energy cycle. For crops grown in the soil system, when incorporated inside the soil system after harvesting, they get decomposed by microbes, releasing the nutrients associated with the crop upon complete degradation. The soil system aids in moisture conservation, and its management has been depicted in Figure 14.2.

These microbes act as decomposers to degrade the plant and other bioenergy-based materials in soil in response to suitable conditions such as pH, moisture, aeration and the organic matter status of the soil. These decomposing microorganisms get necessary growth conditions in the soil and proliferate inside the soil system. In acidic pH, the fungal microorganism gets dominant, while in case of alkaline and neutral pH, actinomycetes and bacteria species get dominant.

Soil acts as a potential system to mitigate the bioenergy crisis developed in the soil-plant-atmosphere continuum. In this process, soil microorganisms take part in the decomposition and transformation processes. These microbes majorly take part in the oxidation and mineralization processes along with the decomposition reaction

**TABLE 14.4**

**Grouping of Different Types of Irrigation Water**

| S. No. | Main class | Subclass | $EC_{iw}$ | $SAR_{iw}$ | RSC (me L$^{-1}$) |
|---|---|---|---|---|---|
| 1. | Good | ** | < 2 | < 10 | < 2.5 |
| 2. | Saline | Marginally Saline | 2–4 | < 10 | < 2.5 |
| | | Saline | > 4 | < 10 | < 2.5 |
| | | High SAR Saline | > 4 | > 10 | < 2.5 |
| 3. | Alkali | Marginally Alkali | < 4 | < 10 | 2.5–4.0 |
| | | Alkali | < 4 | < 10 | > 4.0 |
| | | Highly Alkali | variable | > 10 | > 4.0 |

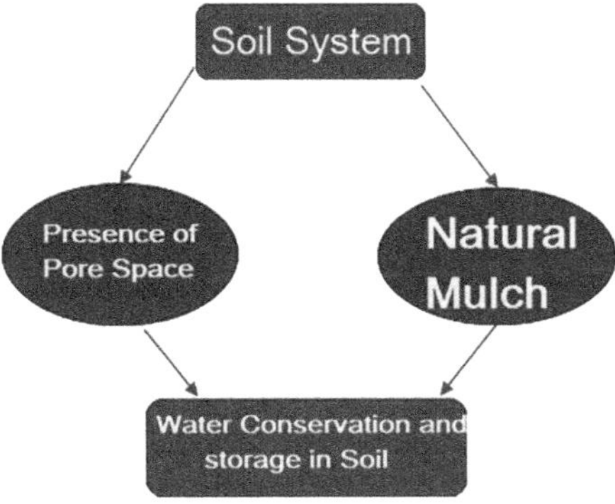


**FIGURE 14.2** Soil system aid in moisture conservation and its management.

by releasing certain acidic substances, which in turn make the nutrients available in the soil solution pool from which plant can easily uptake the nutrients. Microbes may convert these materials to produce biofuel and may convert to produce electrical energy and store it as microbial fuel cell. These fuel cells can store energy for a longer period of time, and the microbial organisms associated with it are *E. Coli* and *yeast*. This fuel cell energy is purely nontoxic and eco-friendly in nature and is produced without hampering the existing natural resources. These biofuel cells act as bioenergy, and the soil system provides the natural habitat for engineered microorganisms like *E. Coli* and other *yeast* species. Apart from this, microbes also take part in the production of bioethanol, methane derivatives as well as other alternatives of existing polluting fuels towards a clean bioenergy system. Soil, by providing a habitat for microbes, clears a pathway for bioenergy source, management and crisis mitigating system.

Soil provides stability and a medium for the growth of a large and diversified biomass of various plants and other crops grown on it. Annual as well as perennial plants have different growth habitats, and the duration of growth with difference in nutrient uptake capacity and all these systems are being regulated by soil. Plants perform photosynthesis and in turn accumulate carbohydrate compounds in it. Plants in these said processes produce or transfer or exchange a huge amount of energy within the associated atmosphere. This energy is also a major part of bioenergy. Other important processes are respiration, evapotranspiration and transpiration, and all these processes regulate the energy exchange within the atmosphere. Plant roots also act as potential habitats for growth and proliferation of microbes as in the form of rhizosphere. Microbes also do the respiration process as well as other organic matter and nutrient transformation processes during which the bioenergy associated with plants and microbes gets exchanged, and indirectly, the soil takes part in energy utilization processes. During natural activities when plant materials get incorporated inside the soil system, it produces coal and other fuel materials which provide energy to the ecosystem and, on burning, get exchanged within the associated atmosphere.

### 14.5.1 Biomass as a Potential Source of Energy

The plants possess photosynthetic pigments in the thylakoid membrane that act as a key component of plant photosynthetic activities. Plants utilize water, dissolved nutrients and carbon dioxide to produce carbohydrate compounds for plant nutrition management and growth. Annual as well as perennial crops have different growth duration and mode of nutrient management during their growth. The residues of these plants upon harvesting are incorporated in the soil for decomposition, and after that, the plants release the accumulated nutrients and energy contained within. This energy released is regulated within the ecosystem and contributes as a large component of biomass energy. Plant biomass on average takes a certain time period to decompose, and microorganisms actually play a key role in the process of decomposition. Harvested plant parts can be utilized for the preparation of manures, compost and other organic substitutes, vermicompost, etc. During the preparation of these substances, certain organisms as well as microorganisms get involved, and microbes take part in the energy exchange process, thereby contributing to the energy pool existing within the ecosystem. There are billions of microorganisms present in the soil, and microbial respiration holds a large component of exchangeable energy between soil and atmosphere. In this way, soil acts as a medium for energy exchange and determines the overall energy pool transformation as well as the energy exchange component in the ecosystem. Soil nutrient kinetics and transformations by microorganisms also contribute to the energy pool formed within the soil, and thus, energy exchange is possible within the ecosystem.

### 14.5.2 Biofuel Production and as the Source of Energy

#### 14.5.2.1 Biogas

Various organic waste materials such as agricultural waste, plant materials, manures, municipal water and food waste materials are converted into useful fuel materials with the help of certain soil microorganisms through the anaerobic process, consisting of carbon dioxide, methane, hydrogen sulfide, etc. This is called biogas. Biogas acts as a source of clean, sustainable fuel energy for multipurpose use. Soil provides the habitat for these microorganisms to grow and carry out the fuel production process, so soil is indirectly related to the nexus of the energy management system. Different microorganisms of the group—i.e., sulphate reducing and methanogens (*Methano bacterium*)—take part in the biogas production process.

#### 14.5.2.2 Biodiesel

Plants after harvesting can be processed to produce various useful substances for agriculture and allied sectors. Microalgae, bacteria and fungi species in soil play a key role in various conversion processes as well as transformation processes to produce energy sources for future use. The microbe residing in soil has the mechanism to produce biodiesel through the process of transesterification and esterification, acting as a sustainable energy pool for use. Most fast-growing microbes have the potential to produce higher quantities of biodiesel in nature through conversion processes. Most of the time, the sugarcane crop is preferred more for the production of sustainable fuels in agricultural sectors. Some microbial genera such as *Botryococcus*,

**TABLE 14.5**

**Composition of Biogas**

| S. No. | Compound | Formula | Percentage (by volume) |
|---|---|---|---|
| 1. | Carbon dioxide | $CO_2$ | 20–45 |
| 2. | Methane | $CH_4$ | 50–70 |
| 3. | Nitrogen | $N_2$ | 0–10 |
| 4. | Hydrogen sulfide | $H_2S$ | 0–0.5 |
| 5. | Oxygen | $O_2$ | 0–2.5 |
| 6. | Hydrogen | H | 0–1 |

*Source:* Makishah (2017)

*Nannochloropsis* and *Schizochytrium* are dominant in nature and thus help in bio-diesel production. Oil produced or extracted from microalgae has a primary constit-uent of unsaturated fatty acids, i.e., linolenic acids, linoleic, oleic and palmitoleic. Saturated fatty acids are extracted in a low concentration, i.e., stearic and palmitic (Meng et al. 2009).

### 14.5.3.3 Bio Ethanol

Global warming and climate change has developed a worldwide concern for the reduction in carbon emission as well as the production of clear energy for sustainabil-ity and development. The latest scientific research is focused on converting biomass into ethanol production through various processes, and in such way, there might be a change to mitigate the problem of biomass decomposition.

### 14.5.4.4 Microbial Lipid

More prevalent microorganisms such as bacteria, yeast, fungi and microalgae have been used widely for the source of lipids and enzymes. From 1960 onwards, these organisms are also used in conjugation with various detergent or cleaning agents. These microbial lipid materials have been used as an alternative resource for biofuel and other value-added products to replace harmful petrochemical uses. The oleagi-nous type of microorganism contains more than 25% of lipid in their own cell mass (Ezeji et al. 2007). There is a group of microorganisms containing oil (% dry basis) and that can be used for oil production such as the following (Ratledge 2004).

a. Microalgae—*Cylindrotheca sp.* (16%–37%), *Schizochytrium sp.* (50%–77%)
b. Bacteria—*Arthrobacter sp.* (> 40%), *Acinetobactercalcoaceticus* (27%–38%)
c. Yeast—*Rhodotorulaglutinis* (72%), *Chlorella sp.* (28%–32%)
d. Fungi—*Mortierellaisabellina* (86%), *Humicola lanuginose* (75%)

## 14.6 CONCLUSION

The security of food, energy and water is one of the top global risks facing the future of the planet. It is more challenging to cope with these global risks due to the nonsta-tionary nature of dynamic stresses affecting natural resources (soil, water and envi-ronment). An integrated approach to resource allocation and management benefited through the interconnection among the food, energy, water, and soil which provides a holistic and multilevel solution of the emerging problems, resulting in the long-term sustainability of the available natural resources. Although integrated water resources management are potential platforms for integrating human efforts to address global risks, it is imperative to pay attention to soil processes (formation, genesis) and prop-erties (associated with soil organic carbon pool), as they play a spectacular role in providing the three primary resources of humanity (food, water and energy), thereby supporting human existence and ecosystem balance.

Despite the integrated approaches to natural resource management, soil has not received enough recognition for its imperative role in supporting food, energy and water security. In this chapter, authors mainly focused on the imperative role of soil for plant growth, nutrient pool source, mitigating nutrient stress and sustainability of the

**TABLE 14.6**

**Details of Microorganisms for Lipid Production for Biofuel**

| S. No. | Microorganism | Substrate used | Lipid production | References |
|---|---|---|---|---|
| 1. | *Mortierellaisabellina* | Switch grass parts or hydrolysate, miscanthus, giant reed and corn stove. | 4.4 g/L | (Wu et al. 2010) |
| 2. | *Cryptococcus curvatus* | Corncob parts | 61.3% | (Ruan et al. 2013) |
| 3. | *Trichosporonfermentas* | Waste sweet potato wine | 9.68 g/L | (Chang et al. 2013) |
| 4. | *Mortirellaisabellianna* | Wheat straw | 39.4% | (Zhan et al. 2013) |
| 5. | *Chlorella protothecoides* | Molasses materials | 40.8 g/L | (Zeng et al. 2013) |
| 6. | *Lipomycesstarkeyi* | Sewage and sludge | 68% | (Yan et al. 2011) |
| 7. | *Rhodotorulaglutinis* | Starch wastewater products | 35% | (Gao et al. 2010) |

*Source:* Makishah (2017)

available natural resources. Soil acts as a resource-efficient source which maintains the water balance and mitigates the water stress along with the bioenergy management for the future of humans and animals on the earth's surface and ecosystem balance. However, researchers among the globe have highlighted the importance of the soil in providing food, water and energy securities to the human creatures. However, to achieve sustainability among the natural resources, there is a need to apply the learned knowledge about soil and soil organic pool into integrated food-water-energy nexus frameworks in the future which provides numerous ecosystem services essential to human well-being and nature conservation. For soil medium and its functioning to be characterized and modeled conceptually, the nexus paradigm must provide a conceptual understanding. In a changing and uncertain climate, this framework enables us to use quantitative characteristics parameters to measure soil functions and management and restoration of soil resources in order to ensure the provision of food, energy and water resources as well as sustain the functions of nature.

## REFERENCES

Al-Saidi M, Elagib NA. Towards understanding the integrative approach of the water, energy and food nexus. *Sci Total Environ.* 2017;574:1131–9.

Arnon DI, Stout PR. The essentiality of certain elements in minute quantity for plants with special reference to copper. *Plant Physiol.* 1939;14:371–5. http://dx.doi.org/10.1104/pp.14.2.371

Bai ZG, Dent DL, Olsson L, Schaepman ME. Proxy global assessment of land degradation. *Soil Use Manag.* 2008a;24(3):223–34.

Bai ZG, Dent DL, Olsson L, Schaepman ME. Global assessment of land degradation and improvement 1: Identification by remote sensing. 2008b; Report 2008/01, FAO/ISRIC-Rome/Wageningen.

Bazilian M, Rogner H, Howells M, Hermann S, Arent D, Gielen D, et al. Considering the energy, water and food nexus: Towards an integrated modeling approach. *Energy Policy.* 2011;39(12):7896–906.

Bhatia R. Water and economic growth. In: Briscoe J, Malik RPS (eds) *India's water economy: Bracing for a turbulent future*. Oxford University Press, New Delhi, 2007; pp. 99–136, 206–246

Briscoe J, Malik RPS. India's water economy: An overview. In: Briscoe J, Malik RPS (eds) *India's water economy: Bracing for a turbulent future*. Oxford University Press, New Delhi, 2007; pp. 1–9.

Cairns R, Krzywoszynska A. Anatomy of a buzzword: The emergence of the water-energy-food nexus' in UK natural resource debates. *Environ Sci Pol*. 2016;64:164–70.

Chang Y, Li GJ, Yao Y, Zhang LX, Yu C. Quantifying the water-energy-food nexus: Current status and trends. *Energies*. 2016;9(2):1–17.

Chang YH, Chang KS, Hsu CL, Chuang LT, Chen CY, Huang FY, Jang HD. A comparative study on batch and fed-batch cultures of oleaginous yeast *Cryptococcus sp.* In glucose-based media and corncob hydrolysate for microbial oil production. *Fuel*. 2013; 105:711–717.

Dexter AR. Advances in characterization of soil structure. *Soil Tillage Res*. 1988; 11:199–238.

Diamond J. *Collapse: How societies choose to fail or succeed*: Penguin Books, Penguin Group Inc., New York, 2005; p. 592.

Dubash NK. The electricity-ground water conundrum: Case for a political solution to a political problem. *Econ Polit Wkly*. 2007;45–55.

Economic Survey 2011–12. Statistical Appendix. https://www.indiabudget.gov.in/budget2012-2013/es2011-12/estat1.pdf

Ezeji T, Qureshi N, Blaschek HP. Butanol production from agricultural residues: Impact of degradation products on *Clostridium beijerinckii*growth and butanol fermentation. *Biotechnol Bioeng*. 2007a;97:6.

FAO. *The state of the world's land and water resources for food and agriculture (SOLAW)—managing systems at risk*. 2011. The Food and Agriculture Organization of the United Nations and Earthscan. Rome, Italy.

FAO. *The state of food and agriculture*. Rome, 2016. Food and Agriculture Organization of the United Nations, Rome.

Gao C, Zhai Y, Ding Y, Wu Q. Application of sweet sorghum for biodiesel production by heterotrophic microalga Chlorella protothecoides. *Appl Energy*. 2010; 87:756–61.

Gerbens-Leenes W, Hoekstra AY, van der Meer TH. The water footprint of bioenergy. *PNAS*. 2009;106(25):10219–23.

Hamiche AM, Stambouli AB, Flazi S. A review of the water-energy nexus. *Renew Sustain Energy Rev*. 2016;65:319–31.

Howard, A. *The soil and health: A study of organic agriculture (Introduction by Wendell Berry)*. The University Press of Kentucky, 1947. https://doi.org/10.2307/j.ctt2jcp63

Iyer RR. *Towards water wisdom: Limits, justice, harmony*. Sage Publications Ltd, New Delhi, 2009.

Lal, R. *The nexus of soil, water, and waste*. Lecture Series #1 UNU-FLORES, Dresden, Germany, 2013.

Lal R. The nexus approach in managing water, soil and waste under changing climate and growing demands on natural resources. In: Kurian M, Ardakanian R, editors. *Governing the nexus: Water, soil, waste change*. Springer, Dordrecht, Holland, 2015a; pp. 39–61.

Lal R. The soil-peace nexus: Our common future. *Soil Sci Plant Nutr*. 2015b;61:566–78.

Lal R. Global food security and nexus thinking. *J Soil Water Conserv*. 2016;71(4):85A–90A.

Lal R, Rabi H, Mohtar, Amjad T, Assi, Ray R, Baybil H, Molly J. Soil as a basic nexus tool: Soils at the center of the food–energy–water nexus. *Curr Sustainable Renewable Energy Rep*. 2017;4:117–29; DOI 10.1007/s40518-017-0082-4

Makishah NH Al. Bioenergy: Microbial biofuel production advancement. *Int J Pharm Res Allied Sci*. 2017;6(3):93–106.

McCarl BA, and Reilly JM. US Agriculture in the climate change squeeze: Part 1: Sectoral Sensitivity and Vulnerability. Report to National Environmental Trust, 2007. http://

agecon2.tamu.edu/people/faculty/mccarl-bruce/papers/1303Agriculture in the climate change squeez1.doc.

Meng et al, Meng X, Yang J, Xu X, Zhang L, et al. Biodiesel production from oleaginous microorganisms. *Renew Energ.* 2009;34:1–5.

Ministry of Agriculture (MoA). Annual Report 2010–2011. Government of India. Department of Agriculture and Cooperation Ministry of Agriculture Government of India.

Moench M. 2002. India's groundwater challenge. Available at www.india-seminar.com/2000/486/486percent20moench.htm. Accessed on 25 Sept 2011

Mohtar RH, Assi AT, Daher BT. Current water for food situational analysis in the Arab region and expected changes due to dynamic externalities. In: Water Security in a New World, Amer K, et al., editors. *The water, energy, and food security nexus in the Arab region.* Springer, New York, 2016; pp. 193–208.

Mukherjee S. *Impact of power subsidy on groundwater extraction for agriculture: A study in Madhya Pradesh.* ISEC, Bangalore, 2008.

Pate R, Hightower M, Cemron C, Einfeld W. *Overview of energy-water interdependencies and the emerging energy demands on water resources.* Report SAND 2007–1349C. Los Alamos National Lab, NM.

Ratledge C. Fatty acid biosynthesis in microorganisms being used for single cell oil production. *Biochimie.* 2004; 86:807–815.

Ruan Z, Zanotti M, Zhong Y, Liao W, Ducey C, Liu Y. Cohydrolysis of lignocellulosic biomass for microbial lipid accumulation. *Biotechnol Bioeng.* 2013; 10:1039–49.

Scott CA, Shah T. Groundwater overdraft reduction through agricultural energy policy: Insights from India and Mexico. *Int J Water Resour Dev.* 2004;20(2):149–64.

Shah T. Community management of ground-water resources: An appropriate response to ground water overdraft in India. IWMI-TATA water policy program: 2005; Comment 4.

Sinha B, Chiranjeeb K. *Conceptual objective agriculture.* 2022; pp. 335–389. ISBN: 978-93-92725-18-0 e-ISBN: 978-93-92725-19-7

Temesgen M, Savenije H, Rockström J, Hoogmoed W. Assessment of strip tillage systems for maize production in semi-arid Ethiopia: Effects on grain yield, water balance and water productivity. *Phys Chem Earth.* 2012a;47:156–65.

Temesgen M, Uhlenbrook S, Simane B, Zaag P, Mohamed Y, Wenninger J, et al. Impacts of conservation tillage on the hydrological and agronomic performance of Fanya juus in the upper Blue Nile (Abbay) river basin. *Hydrol Earth Syst Sci.* 2012b; 16:4725–35.

Wada Y, Florke M, Hanasaki N, Eisner S, Fischer G, Tramberend S, et al. Modeling global water use for the 21st century: The water futures and solutions (WFAS) initiative and its approaches. *Geosci Model Dev.* 2016;9(1):175–222.

Wichelns D. The water-energy-food nexus: Is the increasing attention warranted from either a research or policy perspective? *Environ Sci Pol.* 2017;69:113–23.

Wu S, Hu C, Jin G, Zhao X, Zhao ZK. Phosphatelimitation mediated lipid production by *Rhodosporidiumtoruloides. Bioresour Technol.* 2010;101:6124–9.

Yan D, Lu Y, Chen YF, Wu Q. Waste molasses alone displaces glucose-based medium for microalgal fermentation towards cost-saving biodiesel production. *Bioresour Technol.* 2011;102:6487–93.

Zeng J, Zheng Y, Yu X, Yu L, Gao D, Chen S. Ligno-cellulosic biomass as a carbohydrate source for lipid production by *Mortierellaisabellina. Bioresour Technol.* 2013; 128:385–91.

Zhan J, Lin H, Shen Q, Zhou Q, Zhao Y. Potential utilization of waste sweetpotato vines hydrolysate as a new source for single cell oils production by *Trichosporonfermentans. Bioresour Technol.* 2013; 135:622–29.

# 15 Role of Soils for Satisfying Global Demands for Food, Water, and Bioenergy

*Bhabani Prasad Mondal, G. Bhupal Raj,
Mahima Dixit, Tithli Sadhu, Vaibhav Pandit,
Laxmi Prasanna, Arghya Chattopadhyay,
Suman Dutta, and Suchith Kumar*

## 15.1 INTRODUCTION

Ancient Indian literatures mentioned soil and water as components of "*Pancha-maha-bhuta*" (soil, water, air, fire, and ether), playing a crucial role in fulfilling our demands of food, water, and bioenergy for our sustenance on earth. The agricultural system depends on the management of soil and water in order to produce food grains and other essentials. However, the increasing population, urbanization, industrial expansion, and rising living standards all place enormous demand on our limited natural resources and create a challenge on agricultural production (Wani et al. 2021). Although the food production increases with green revolution, the world is currently confronting food and nutritional insecurity, which ultimately contributes to poverty. This is a problem that is also being exacerbated by the world's seemingly endless population growth (Shukla and Behra 2019). Climate change, competition in global trade, etc., further create more problems in this regard. Till now, the developing countries, like India, Sri Lanka, and some African countries, show the evidences of hidden hunger, malnutrition, etc. According to the National Family Health Survey-5 report (2019–21), India still has malnourished kids, with 35.5% of them stunted and roughly 19.3% of them fading away. Under such challenging scenario, achievement of the Sustainable Development Goals (SDGs) of no hunger, zero poverty and food security is very much an intimidating task. The goals of the SDGs are directly tied to nutrition, demonstrating its importance for improving education, employment, health, and women's empowerment (United Nations 2013). The Sustainable Development Solutions Network (UN-SDSN 2013) gave a brief outline of some of the important SDGs pertaining to the current topic of discussion:

Goal 1: Obliteration of hunger and extreme poverty
Goal 2: Achievement of enhanced nutrition, food security, and sustainable agriculture



DOI: 10.1201/9781003358169-15"

Goal 3: Ensuring livelihood to youth and child through proper learning
Goal 4: Achieving gender equality and equal human rights
Goal 5: Improving health and well-being to all
Goal 6: Improving agricultural production system
Goal 7: Empowering productive cities
Goal 8: Confirming the accessibility to sustainable and affordable energy and
stop climate change
Goal 9: Ensuring sustainable management of water resources
Goal 10: Change governance to promote sustainability

The chapter will focus more on Goals 2, 8, and 9 which are related to the sustainable management of soil, energy, and water resources. Assumed recent trends of increasing human population, urbanization, and earnings, all of which have led to an increase in food demand, as well as of climate change and the exhaustion of natural resource reserves. Concerns about existing and upcoming food security issues are at the top of the agenda of sustainable development (FAO 2011; van Ittersum et al. 2016). The four interrelated pillars of availability, accessibility, utility, including nutrition and safety, and stability all contribute to human food security (FAO 2006). The availability and stability of the food supply are among the most serious threats posed by climate change and soil degradation brought on by variations in land use and poor managing techniques (Gourdji et al. 2015). Large-scale soil erosion and compaction, diminishing soil organic matter levels, nutrient removal, acidification, salinization, and contamination all have an impact on agricultural areas. All of these factors are responsible for declining the productivity of the agriculture system, quality of nutrition, food security, and also the decreased ability to reduce the bad impacts of climate change (FAO 2011). In substantial portions of the developing world, decreased crop harvests, crop failure, and deficiencies of important nutrient elements are caused by poor agricultural management techniques and insufficient soil fertility. All of these potential threats produce the negative impact on food safety, security, and sustainable production and management of water resources and bioenergy, owing to their serious impacts on soil productivity as well as biodiversity. Sustainable management of soil biodiversity is crucial to ensure food availability and security by enhancing environmental amenities which enable more effective utilization of natural wealth, bio-control of soilborne pathogens, bioremediation of degraded lands and ecosystems, mitigation, as well as adaptation to the effect of change of climate (Kaul et al. 2018).

The integrated farming system (IFS) method involving the divergence in cropping patterns, crop rotation, cropping systems, and addition of agricultural and horticultural crops along with agroforestry, livestock, and fisheries is another crucial aspect to address the aforementioned challenges. Moreover, research and application of modern and smart technologies like proximal sensing technologies, airborne and satellite-based sensing, remote sensing technologies, Internet of Things (IoTs), digital soil mapping (DSM), and agricultural robotics should also be initiated in massive scale, especially for developing countries to bring to the use of digital agriculture intensively and to solve the real-world problems. This chapter will definitely be highlighted on the sustainable management of soil or land resources, current issues related to soil, and the application of viable and smart technologies to address those issues to ensure food security, safety, water quality, as well as sustainable production of bioenergy.

## 15.2 SUSTAINABLE LAND MANAGEMENT FOR SDGS

Sustainable land management, including agriculturally important soil management, is necessary to attain the SDGs. The Commission of the European Communities (CEC 2006) mentioned seven functions of soils:

1. Production of biomass, encompassing forestry and agriculture
2. Storing, filtering, buffering, and transforming nutrient elements and water resources
3. Conservation of biodiversity
4. Acting as physical and cultural environment
5. Supplying fresh resources
6. Serving as a source of carbon pool
7. Acting as a storehouse of archaeological as well as geological inheritance

These seven functions are generally used to define the condition of land for a particular area. These functions are also help to define the concept of soil quality, i.e., "ability to perform" any function within a certain boundary (Bouma 2010). The term "soil quality" is still an elusive terminology in the domain of soil science which is not properly defined. Bouma (2002) tried to define it as soil quality indicator in terms of some numbers like 0–100, calculated as ratio of water-limited yield and potential biomass yield on the basis of function 1. However, agricultural systems not only provide food, fiber, and bioenergy on the aboveground land, simultaneously, they also help to recharge groundwater underneath the land surface. Rainwater passes through the "vadose zone" of soil through which the recharging of groundwater takes place. Soil management should ensure the minimization of loss of water as surface runoff and channelize the excess runoff into open aboveground water structure like pond, lake, river, etc. Agricultural interventions should regulate and decrease the generation of potentially harmful greenhouse gases like carbon dioxide ($CO_2$), methane ($CH_4$), and nitrous oxide ($N_2O$) since soils are the medium of direct gaseous exchange with the atmosphere. Soil biodiversity is crucial for all of the activities in the soil that have been linked to the production of food materials and the conservation of groundwater resources.

## 15.3 ROLE OF SOIL BIODIVERSITY IN NUTRITION, FOOD SAFETY, AND FOOD SECURITY

Soil is a habitat of so many organisms, and it may hold more than 10,000 biological species per square meter area influenced by the effect of variable factors like soil temperature and air moisture (Orgiazzi et al. 2016). Soil biodiversity includes taxonomic, phylogenetic, and functional variations among soil species from gene level to community level and also the habitat variation from micro-aggregate to landscape

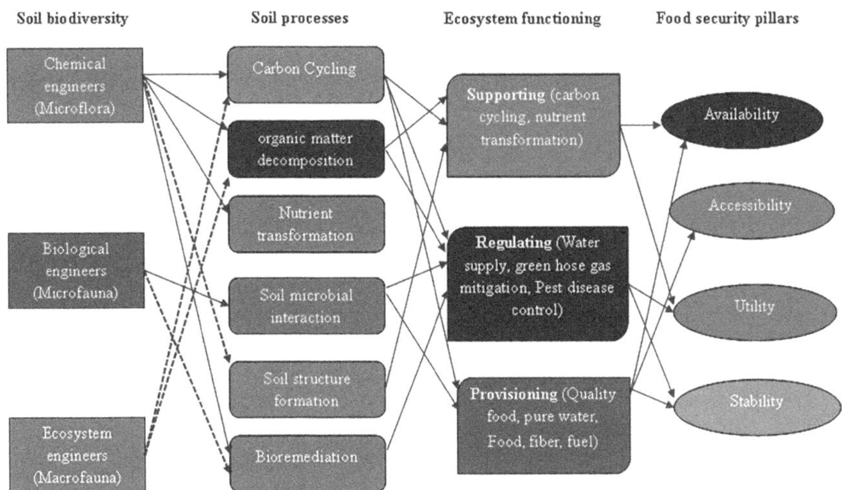


**FIGURE 15.1** Correlation among pillars of food security and soil biological diversity, mediated by soil-based processes and ecological functions (solid line = direct role; dotted line = indirect role).

**Source: El Mujtar et al. (2019), modified by the authors**

level (Naeem et al. 2016). The relations between the organisms of soil and roots of plants control the movement of materials and energy. According to their dominating influence to key soil-related processes and ecological services, soil organisms are divided into three primary functional groupings in Figure 15.1, which illustrate soil biodiversity. Soil biodiversity, or soil microbial community, performs a significant role in the nutrition of plant by affecting the uptake of nutrient elements by the crop plant. Rhizospheric microbes play an important role in the nutrient uptake by roots. Various groups of bacteria and fungi promote nutrient uptake, improve the crop and yield performance, and produce a good effect on nutrition and the quality and safety of the food materials (Table 15.1.).

Mycorrhizal fungi forms a symbiotic relation with the crop plant and enriches the nutritive value in the edible parts of the plants by directly enhancing the nutrient uptake of the crop or modifying the soil or plant system to exploit higher nutrient elements by the plant (Antunes et al. 2012). Human nourishment relies on plant nutrition because the accessibility of soil nutrient elements and their extraction by crop plants controls the human nutrition. However, the deficiency of certain nutrients in soil causes nutritional imbalance in the food product. It has been observed that with the concomitant increase in crop yield or productivity, there is a decrease in nutritional value in the food commodity (Davis 2009). One potential solution of this problem is to fortify the food product (biofortification) by adding the deficient nutrients like flours added with micro nutrients like zinc, iron, vitamin B, or another nutrient-deficient element (Engle-Stone et al. 2017). Phytonutrients, sometimes referred to as nutraceuticals or phytonutrients, are secondary metabolites produced by plants that have biological function in addition to providing macro and micro nutrients and are

**TABLE 15.1**

**Effect of Rhizospheric Microbial Population on Nutrition, Food Quality, and Food Safety**

| Crops | Microbes in rhizosphere | Vegetal parts | Effect on nutrition, quality, and safety of the food materials | References |
|---|---|---|---|---|
| Tomato | Arbuscular mycorrhizal (AM) fungi | Tomato fruit | Increased the mineral nutrient elements like nitrogen (N), copper (Cu), quantity of carotenoids and antioxidants. | El Mujtar et al. (2019) |
| Rice-Wheat | Cyanobacteria (BGA) and Plant Growth-Promoting Bacteria (PGPB) | Grains | Increased micro nutrient concentrations in grains: for rice (Zn, Fe enrichment); for wheat (Cu and Mn enrichment). | El Mujtar et al. (2019) |
| Snap bean | *Rhizobium* sp. and *Glomus* sp. | Fruit | Increased protein content, improved yield, etc. | |
| Strawberry | AM fungi and PGPB | Fruit | Increased the fruit size as well as fruit production. Enrichment with folic acid, ascorbic acid, and sugar content | El Mujtar et al. (2019) |
| Pakchoi | *Rhizophagus-intraradices* | Shoots | Reduction in heavy mental concentration like lead and cadmium in the shoot portions of pakchoi. | El Mujtar et al. (2019) |
| Onion | *Rhizophagus-irregularis* (AM fungi) | Bulb | Enrichment of Vit-B1, increased biomass production. | El Mujtar et al. (2019) |
| Potato | *Glomus irregulare* (AM fungi) | Tuber | Checked the development of *Fusarium sambucinum* and inhibited the secretion of mycotoxin. | El Mujtar et al. (2019) |

essential for promoting human health because these are helpful for stimulating the immune system, detoxification of toxic substances, and reducing oxidative damages within humans (Giovannetti et al. 2013).

Regarding food safety issue, research is ongoing to find out the potential of different soil-inhabiting microbes to reduce the concentration of heavy metallic elements in the plant tissue, especially in edible parts. A scientific study reported that mycorrhizal fungi have the capability to reduce lead (Pb) and cadmium (Cd) content by 20.6%–67.5% and 14.3%–54.1%, respectively, in the edible part of inoculated pakchoi shoots in comparison with non-inoculated shoots (Wu et al. 2016). The most prevalent types of food-borne sickness are caused by *Salmonella typhi*, *Escherichia coli*, etc., which are also regarded to be human diseases spread by soil, notably through infecting fresh tissues of vegetables (Jeffery and Putten 2011). The influence of soil biodiversity, especially the microbial community, on the management of human-infecting pathogens transmitted by soil is not well understood. Theoretically, though, increased struggle for nutrient elements and a diversity of predators might shorten their life span of existence in the soil (Ongeng et al. 2015). Wall et al. (2015)

mentioned that soil biodiversity may limit the growth pathogenic soil microbes and help to produce fresh air, water, and food, so it directly promotes human health.

Food security is not only determined by observing crop yields on an average basis, the other aspects like accessibility or availability of food and quality of food are also taken into consideration for judging this (Foley et al. 2011). Food security can also be achieved through augmenting the governance system which will allow to reduce the stunting and malnutrition of a child and human beings by addressing the root causes like safe access to clean and pure water, good sanitation, women empowerment, and adequate availability of quality foods (Smith and Haddad 2015). By using scarce resources more effectively, lowering production costs, minimizing yield variability, and lowering economic risks for farmers, the integrated management of soil biodiversity has a significant potential to boost agricultural productivity. All of these factors significantly improve food security (El Mujtar et al. 2019).

## 15.4 SUSTAINABLE RESOURCE MANAGEMENT FOR AGRICULTURAL PRODUCTION

### 15.4.1 LAND RESOURCE MANAGEMENT

The land resources mainly comprise of soil, underground minerals, hydrologic components (soil water, groundwater, snow cover, etc.), crops, livestock animals, etc. The majority of these land resources are used to enhance land productivity and to increase the production of food, fodder, fiber, fuel, wood, and bioenergy. The RS-based technology has wide application in land resource management like the assessment of land suitability for crop cultivation and other engineering purposes, assessment of types and severity of land degradation, sustainable land management, agro-technology transfer, proper recommendation of various agricultural inputs, analysis of farming system concept and its development, characterization of agro-ecological region and creation of agro-economic zone for research and land resource development, conservation of natural resources, and so many other applications (Choudhary et al. 2018). Such technology also helps in land use land cover classification (LULC), land capability classification (LCC), etc. In India, the department of space has adopted one mission—i.e., Integrated Mission for Sustainable Development (IMSD)—to generate spatial databases for natural resources and analyze the databases for providing practical and sustainable agro-based land use for 174 districts of the country which covers 45% of the country's geographical area (Choudhary et al. 2018).

### 15.4.2 SOIL AND VEGETATION MANAGEMENT

RS technology can be applied for the assessment of the soil properties related to the fertility of soil and also to identify the soil-related problems like salinity, acidity, etc. Salinity assessment by RS is a very daunting task, as it mainly occurs in the subsurface of soil and is dynamic in nature, and the RS sensors are unable to sense subsurface salinity (Allbed and Kumar 2013). The upper vegetation cover is another challenge which inhibits the RS signal to reach up to the soil surface. Availability of bare land is very difficult to get for RS application in soil analyses. From 2000 to

2013, a time series analysis of thermal RS data and vegetation indices revealed significant dynamics of surface lands in Turkmenistan and Kazakhstan (Conrad et al. 2020). Cropland deterioration in the basin of the Aral Sea mostly pertains to the degradation of soil, and this issue was marked by identifying pessimistic trends in time series data of vegetation indicators that were made available using moderate to coarse resolution RS data, such as Moderate Resolution Imaging Spectroradiometer (MODIS) data (Conard et al. 2020).

### 15.4.3 Sustainable Soil Water Management

Water and soil are necessary for all terrestrial life to exist and survive because these are considered as fundamental natural resources. The world's limited natural resources are under extreme stress as a consequence of the planet's growing population and changing climate. Under such climate change scenario and tremendous anthropogenic activities, it is a significant concern to manage water and soil resources sustainably, particularly in arid and semi-arid climatic conditions. Climate change can produce a higher effect on water yield, while land uses produce effect on transport of nutrient elements (Bai et al. 2019). Varying climatic condition and variable land management practices generate differential impact on water yield and hydrological responses. Modeling approach is one of the best approaches to quantify water yield. However, most of the models like integrated valuation of ecosystem services and tradeoffs models are unable to consider the influence of landscape patterns or the geological makeup of landscapes on hydrological cycles, thus making the equation simple to quantify runoff and water yield which is not comprehensive at all (Sharp et al. 2020). Ecohydrological models were generally superior to the prior model in terms of spatiotemporal resolution and the accuracy of water yield evaluation. The other two models like the soil and water assessment tool model and variable infiltration capacity model have often been utilized to compute water yield (Schmalz et al. 2016). The distributed hydrological soil vegetation model is one of the comprehensive models to simulate hydrological processes in watershed areas or catchment basins and help to manage water resources in an efficient way through calculating water yield of that catchment area (Wigmosta et al. 2002; Huang and Yu 2021).

## 15.5 ROLE OF SOILS IN BIOENERGY PRODUCTION

Soil is not only necessary for food and fiber production; it also plays a significant role in energy production (Gonzalez-Salazar et al. 2016). Due to the population's rapid growth and expansion, demand and consumption of energy are increasing day by day. Simultaneously, the largest energy source—i.e., fossil fuel—is going to be exhausted in the future, and its excessive use threatens our environment, causing climate change by emission of huge amounts of potentially harmful greenhouse gases (Raheem et al. 2018). Another issue is the consistent rise of price of the fossil fuel causing economic burden to the people. The report of the International Energy Agency mentioned that by 2040, energy production must increase by 30% to keep up with the growing demand (Chen et al. 2019). Production of renewable energy source is a viable option and an urgent need in an effort to lessen the impact of climate change. Production of bioenergy, which is a good example of a renewable energy source, is certainly deemed eminent to curtail the harmful effects of fossil fuel energy towards environment and to strengthen the rural economy (Gonzalez-Salazar et al. 2017). Bioenergy, which excludes resources buried in geological foundations and converted into fossil fuels, is energy created by using a green biomass of plant as a carrier material in the natural world (Holmatov et al. 2019). Agricultural wastes like crop straw, animal excrement, and forest debris are the main raw materials used in the production of bioenergy (Kivimaa and Mickwitz 2011). Irrational usages of agricultural wastes like burning of crop residues and discarding of farm produce and livestock excreta cause waste of good organic materials and pollution of soil, water, and air (Li et al. 2020). Efficient utilization of agricultural wastes promotes generations of sustainable bioenergy. Land resource, water resource, and energy resource along with food production are some of the numerous factors that must be taken into account for the generation of sustainable bioenergy (Mirzabaev et al. 2015). Bioenergy can be produced in an agricultural system using special energy crops like sugarcane, oil-bearing crops, etc., as well as crop husks used in food production, which uses water and energy and occupies land. Farmers can benefit from new markets for their conventional waste output and lessen agricultural non-point causes of pollution by using animal dung as a bioenergy substrate. The nexus of land-water-food-energy (LWFE) is a useful tool for addressing the many intricate problems that connect resources of land, water, energy, and food (Ringler et al. 2013). Several studies reported the importance of the components of LWFE for production of bioenergy. Wise et al. (2014) looked at how energy, carbon emissions, land usage, and agriculture might affect worldwide biofuel mandates by the middle of the century. The gross and net output of bioenergy's environmental, land, and carbon footprints were assessed by Holmatov et al. (2019), who also took into account two circumstances with various output stocks of bioenergy and varied components of carrier of energy. Gonzalez-Salazar et al. (2016) developed an optimized model framework to analyze the relationship between the production of bioenergy and the economy, land usage, and greenhouse gas emissions. Additionally, decision-makers frequently encounter difficulties and uncertainties in parameter determination and optimization while maximizing bioenergy output in an agricultural system due to incomplete information, variations in the natural resources, shifts in the socioeconomic environment, and subjective judgement (Purkus et al. 2015). In general, widely used techniques that integrate uncertainty in bioenergy production include programming using fuzzy logic, stochastic mathematical programming, and programming of interval mathematics (Yu et al. 2018). An optimization-assessment strategy for producing sustainable bioenergy is giving decision-makers the flexibility to choose the best policies for water, land, energy, and livestock while taking into account the compromise between the effects on the environment and the economy for the development of bioenergy (Li et al. 2020). Li et al. (2020) developed a conceptual framework of bioenergy production utilizing natural resources and that framework has been represented here (Figure 15.2.).

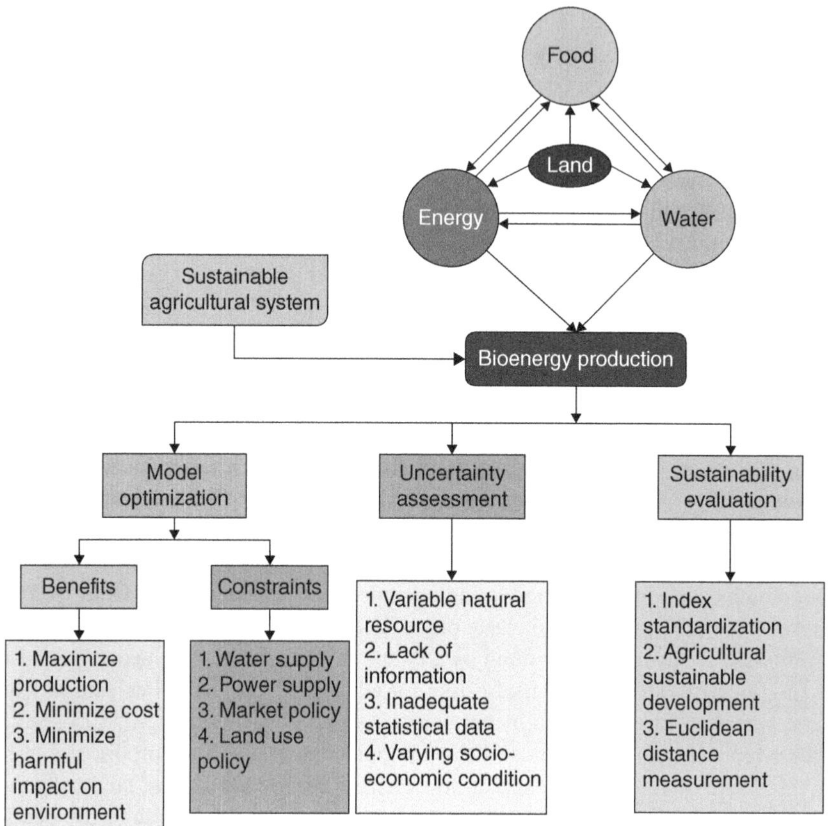



**FIGURE 15.2**    A proposed framework for bioenergy production.

**Source: adapted from Li et al. (2020)**

## 15.6   SMART TECHNOLOGIES FOR DIGITAL AGRICULTURE

Growing agricultural start-up ecosystems, smart sensing techniques, and emerging geospatial and remote sensing technologies are thought to produce workable solutions to the persistent problems of yield stagnation and environmental degradation (Das et al. 2022). One of the good applications of smart sensors in soil-based research is the diffuse reflectance spectroscopy (DRS) technique using visible-near infrared (VIS-NIR) proximal sensor. This technology is rapid, non-invasive, and non-expensive, can supplement the traditional laboratory-based analysis, and used for analyzing large number of samples required for capturing spatial variability (Van der Meer 2018). The capacity to implement these sensors on aerial and satellite platforms for capturing the hyperspectral data for mapping the soil surface property quantitatively is another benefit of VNIR technology (Sahoo 2022). This modern way of analysis for soil texture mapping produced more accurate results than the multi-spectral data (Das et al. 2022). Over the last few years, few Indian laboratories are extensively utilizing the DRS technology for the prediction of several soil properties like soil water potential (Panigrahi and Das 2018), soil hydraulic properties (Santra et al. 2018), macronutrients like potassium (Mondal et al. 2020), secondary nutrients like sulphur (Mondal et al. 2021), and soil salinity parameters (Mahajan et al. 2021). Hati et al. (2022) recently estimated a number of soil parameters utilizing spectral reflectivity information over the region of mid-infrared using the DRS technique.

One of the main global concerns of providing critical zone resources for human benefit is the identification of research to address the physical and structural mechanisms of soil functions and its quantified representation in computational and mathematical frameworks. One of the finest methods for analyzing soils in the critical region is through an integrated study of local watersheds, which is increasingly being used with the help of the critical zone observatories (CZO) established across the world (Gaillardet et al. 2018). Research has been conducted at the Kabini Critical Zone Observatory (CZO) in the southern portion of India to better comprehend the significance of critical factors (such as relief, climate, soil, lithology, vegetation, anthropogenic activities, and so on) in biogeochemical and hydrological phases associated to the tropics, particularly in semi-humid environments (Sekhar et al. 2016). Data on soil, crops, and hydrology at scales greater than those of an agricultural farm are needed for the CZO framework. For the Kabini CZO, efforts to use geospatial and remote sensing technology have produced unique information. For building a crop-hydrology model, it requires accurate estimations of the hydraulic properties of soil and their allocation in a watershed area. For this purpose, extensive soil data are required to build up models through capturing spatial variability. Digital soil mapping (DSM) is a viable technique for supplying larger amounts of soil data at various scales for the predictive mapping of several soil properties essential for hydrological modeling (Lagacherie 2008). The HYDRUS 1-D, a mechanistic soil water model, was used by Montzka et al. (2011) to demonstrate the feasibility of calculating the hydraulic properties of soil utilizing data obtained from remote sensing of surface soil moisture estimation. On the other hand, crop biophysical metrics like the leaf area index (LAI) and aboveground biomass are sensitive to the soil's qualities in the root zone; therefore, these data can be helpful for predicting the soil's deeper layers.

The difficulties of addressing sustainable land management, preventing as well as acclimating the effects of climate change, and ensuring the availability of food (food security) and water will encourage more investigation on studies that incorporate the functions of the soil and how they relate to more broad critical zone processes and enhance field measurements enabled by advances in data products obtained from various sensors (ground-based to airborne to satellite-based sensors). It will encourage the improvement of modern soil-based models that sequentially include all the required physical and biochemical processes. These also help to explore the opportunities to use more new generations and advanced data science and computational tools (such as artificial intelligence, machine learning [ML], deep learning [DL] big data analysis, Internet of Things [IoT], cloud computing, agricultural robotics, and so on). Such smart technologies are also helpful in mitigating soil-related problems like erosion, desertification, salinization, acidification, land degradation as well as maintaining soil health, enhancing resource use efficiency, increasing soil or land biodiversity, and achieving food security.

## 15.7  GOVERNMENT INITIATIVES AND PARTICIPATORY APPROACHES

Government initiatives are very much essential to ensure food security in a country through promoting sustainable development in agricultural production. In the United States, the United States Department of Agriculture (USDA) ensures food security and food safety throughout the country by accessing food to low-income people, increasing the quality of food and healthy diet, providing nutrition education, and building up the confidence among the people. In developing countries like India, agricultural research is conducted by the Ministry of Agriculture and Farmers Welfare, which consists of three important departments such as the Department of Agricultural Research and Education (DARE), the Department of Agriculture, Cooperation & Farmers Welfare (DAC&FW), and the Department of Animal Husbandry, Dairying & Fisheries (DAHD&F). The Indian Council of Agricultural Research (ICAR) under DARE is the apex agricultural organization for India, promoting all kinds of agricultural research for ensuring, food safety, quality, and security in India through adopting or promoting precision agriculture and sustainable agriculture. After Independence, ICAR initiated the All India Coordinated Research Project (AICRP) to popularize the soil test crop response (STCR) correlation approach and target yield approach to increase agricultural production in a sustainable way (sustaining soil quality and health in a long-term basis). The ICAR launched Soil Health Card (SHC) mission in 2015 to provide nutrient status of the farmers' fields and soil test-based fertilizer recommendations to the individual farmer, and this mission covers 30 million farmers each and every year.

Not only national but there are also several international organizations like the Food and Agricultural Organization (FAO) and International Crops Research Institute for the Semi-Arid Tropics (ICRISAT) also that are independently and collaboratively doing research to ensure food safety and security in the various countries across the globe. In India, ICRISAT started Bhoochetana, Bhoosamruddhi, and Rythukosam in the states of Karnataka, Andhra Pradesh, and Telangana to increase the farmers' profit by promoting STCR-based nutrient application and fertilizer recommendation while reducing the load on soil resource. Millions of farmers in India have benefited from stratified soil-analysis-based balanced nutrient management advice with higher earnings and productivity (Wani et al. 2021).

Crop yields and farmer profitability increased when farmers applied balanced amounts of nutrients and raised their rainfall use efficiency (RUE). The maximum returns and RUE were seen in the Integrated Nutrient Management (INM) involving 50% N through vermicompost, demonstrating the need to implement the INM approach in order to increase productivity, profitability, and sustainability. The Government of India's (GoI) "more crop per drop" program showed that balanced nutrient management enhanced crop yields by 14% to 33% while also improving benefit-to-cost ratios to 1.6 to 10 from 1.2 to 9 in farmers' practices (Das et al. 2022).

## 15.8   CONCLUSION

It is a very challenging task to secure food, nutrition, pure water, and bioenergy for the world's growing population. It is assessed that agricultural production should be continued to increase from 60% to 110% by 2050 to feed the 9–10 billion of human population of the world. Thus, the sustainability of the agricultural production is crucial for maintaining the food quality and safety and eradicating hunger of the increasing population. Sustainable management of water resources are also required, as agricultural production faces water scarcity due to the weather and climate change effect. Urgently, a strategic shift is required from the current agricultural production system to the sustainable production system along with the soil-human linkage for modern soil research for meeting the goals of SDGs. It is necessary to understand, assess, and manage soil biodiversity holistically to ensure four key components of food security, availability, stability, and utility, including nutrition and safety. Moreover, to anticipate such problems and to help farmers, application of digital technologies like AI, ML, IOT, and DSM in the agricultural sector is required. The development of modern IT technologies, like mobile apps and decision support systems built by local languages, is necessary to communicate with a large number of farmers to provide them with soil-crop-weather-based agro-advisories for the effective management of resources. Developing an AI-based smart irrigation system is also needed to optimize crop water use. The chapter highlighted all these issues along with sustainable bioenergy production through the optimum utilization of agricultural wastes. Bioenergy production is optimized using modeling approaches, and uncertainty was predicted to address sustainability issues. Precision agriculture needs to be adopted to enable farmers optimal and precise application of agri-inputs in agricultural crop fields, thereby boosting crop production and reducing crop water use and greenhouse gas emission. Partnerships between researchers, decision-makers, and development professionals—such as extension agents, small farm owners, input suppliers, and government and private organizations—are also required in this regard.

## REFERENCES

Allbed, A., and Kumar, L. 2013. Soil salinity mapping and monitoring in arid and semi-aridregions using remote sensing technology: A review. *Adv. Remote Sens* 2, 373–385.

Antunes, P.M., Franken, P., Schwarz, D., Rillig, M.C., Cosme, M., Scott, M., and Hart, M.M., 2012. Linking soil biodiversity and human health: Do arbuscular mycorrhizal fungi contribute to food nutrition? In: *Soil Ecology and Ecosystem Services*. Oxford University Press, pp. 153–172.

Bai, Y., Ochuodho, T., and Yang, J. 2019. Impact of land use and climate change on water related cosystem services in Kentucky, USA. *Ecol. Indic* 102, 51–64.

Bouma, J. 2002. Land quality indicators of sustainable land management across scales. *Agric. Ecosyst. Environ* 88, 129–136.

Bouma, J. 2010. Implications of the knowledge paradox for soil science. *Adv. Agron* 106, 143–171.

CEC, 2006. Communication from the Commission to the Council, the European Parliament, the European Economic and Social Committee and the Committee of Regions—Thematic strategy for soil protection (COM 2006.231), Brussels.

Chen, S., Tan, Y., and Liu, Z. 2019. Direct and embodied energy-water-carbon nexus at an inter-regional scale. *Appl. Energy* 251, 113401.

Choudhary, B., Pandey, P., Kohli, R., Garg, V.K., and Dhawan, A. 2018. *Applications of Remote Sensing and GIS in Land Resource Management.* Pathshala, MHRD GOI.

Conrad, C., Usman, M., Morper-Busch, L., and Schönbrodt-Stitt, S. 2020. Remote sensing-based assessments of land use, soil and vegetation status, crop production and water use in irrigation systems of the Aral Sea Basin. A review. *Water Secur* 11, 100078.

Das, B.S., Wani, S.P., Benbi, D.K., Muddu, S., Bhattacharyya, T., Mandal, B., Santra, P., Chakraborty, D., Bhattacharyya, R., Basak, N., and Reddy, N.N. 2022. Soil health and its relationship with food security and human health to meet the sustainable development goals in India. *Soil Secur* 100071.

Davis, D.R. 2009. Declining fruit and vegetable nutrient composition: What is the evidence? *Hortscience* 44, 15–19.

Department of Economic and Social Affairs, World Economic and Social Survey. 2013, Sustainable Development Challenges, United Nations publication https://www.un.org/en/development/desa/policy/wess/wess_current/wess2013/WESS2013.pdf

El Mujtar, V., Muñoz, N., Mc Cormick, B.P., Pulleman, M., and Tittonell, P. 2019. Role and management of soil biodiversity for food security and nutrition; where do we stand? *Glob. Food Sec* 20, 132–144.

Engle-Stone, R., Nankap, M., Ndjebayi, A.O., Allen, L.H., Shahab-Ferdows, S., Hampel, D., Killilea, D.W., Gimou, M.-M., Houghton, L.A., Friedman, A., Tarini, A., Stamm, R.A., and Brown, K.H. 2017. Iron, zinc, folate, and vitamin B-12 status increased among women and children in Yaoundé and Douala, Cameroon, 1 year after introducing fortified wheat flour. *J. Nutr* 147, 1426–1436.

FAO, 2006. Food security. *Policy Br* 1–4.

FAO, 2011. *The state of the world's land and water resources for food and agriculture -Managing systems at risk.* Food and Agriculture Organization of the United Nations.

Foley, J.A., Ramankutty, N., Brauman, K.A., Cassidy, E.S., Gerber, J.S., Johnston, M., Mueller, N.D., O'Connell, C., Ray, D.K., West, P.C., Balzer, C., Bennett, E.M., Carpenter, S.R., Hill, J., Monfreda, C., Polasky, S., Rockström, J., Sheehan, J., Siebert, S., Tilman, D., and Zaks, D.P.M., 2011. Solutions for a cultivated planet. *Nature* 478, 337–342.

Gaillardet, J., Braud, I., Gandois, L., Probst, A., Probst, J.L., Sanchez-P´erez, J.M., and Simeoni-Sauvage, S. 2018. OZCAR: The French network of critical zone observatories. *Vadose Zone J.* 17 (1), 1–24.

Giovannetti, M., Avio, L., and Sbrana, C. 2013. Improvement of Nutraceutical Value of Food by Plant Symbionts. In: Ramawat, K., and Mérillon, J. (Eds.), *Natural Products Phytochemistry, Botany and Metabolism of Alkaloids, Phenolics and Terpenes.* Springer, pp. 2641–2662.

Gonzalez-Salazar, M.A., Venturini, M., Poganietz, W.R., Finkenrath, M., Kirsten, T., Acevedo, H., and Spina, P.R. 2016. A general modelling framework to evaluate energy, economy, land-use and GHGs emissions nexus for bioenergy exploitation. *Appl. Energy* 179, 223–249.

Gonzalez-Salazar, M.A., Venturini, M., Poganietz, W.R., Finkenrath, M., and Leal, M.R.L.V. 2017. Combining an accelerated deployment of bioenergy and land use strategies: Review and insights for a post-conflict scenario in Colombia. *Renew. Sust. Energ. Rev* 73, 159–177.

Gourdji, S., Läderach, P., Valle, A.M., Martinez, C.Z., and Lobell, D.B., 2015. Historical climate trends, deforestation, and maize and bean yields in Nicaragua. *Agric. For. Meteorol* 200, 270–281.

Hati, K.M., Sinha, N.K., Mohanty, M., Jha, P., Londhe, S., Sila, A., Towett, E., Chaudhary, R.S., Jayaraman, S., Vassanda Coumar, M., and Thakur, J.K. 2022. Midinfraredreflectance spectroscopy for estimation of soil properties of Alfisols from Eastern India. *Sustainability* 14 (9), 4883.

Holmatov, B., Hoekstra, A.Y., and Krol, M.S., 2019. Land, water and carbon footprints of circular bioenergy production systems. *Renew. Sust. Energ. Rev* 111, 224–235.

Huang, T., and Yu, D. 2021. Water-soil conservation services dynamic and its implication for landscape management in a fragile semiarid landscape. *Ecol. Indic* 130, 108150.

Jeffery, S., Putten, W.H., and van der. 2011. Soil borne diseases of humans. JRC Scientific reports, Luxembourg Publications Office of the European Union. p. 55. https://op.europa.eu/en/publication-detail/-/publication/ee3d1682-0ea4-483d-9a3d-92e83812b6c9/language-en

Kaul, S., Gupta, S., Sharma, T., and Dhar, M.K., 2018. Unfolding the role of rhizomicrobiome toward sustainable agriculture. In: Giri, B., Prasad, R., and Varma, A. (Eds.), *Root Biology, Soil Biology*. Springer, pp. 341–365.

Kivimaa, P., and Mickwitz, P. 2011. Public policy as a part of transforming energy systems: Framing bioenergy in Finnish energy policy. *J. Clean. Prod* 19, 1812–1821.

Lagacherie, P. 2008. Digital soil mapping: A state of the art. In: *Digital Soil Mapping with Limited Data*. Springer, pp. 3–14.

Li, M., Fu, Q., Singh, V.P., Liu, D., and Gong, X. 2020. Risk-based agricultural water allocation under multiple uncertainties. *Agric. Water Manag* 223, 106105.

Mahajan, G.R., Das, B., Gaikwad, B., Murgaonkar, D., Desai, A., Morajkar, S., Patel, K.P., and Kulkarni, R.M. 2021. Monitoring properties of the salt-affected soils by multivariate analysis of the visible and near-infrared hyperspectral data. *Catena* 198, 105041.

Mirzabaev, A., Guta, D., Goedecke, J., Gaur, V., Börner, J., Virchow, D., Denich, M., and von Braun, J. 2015. Bioenergy, food security and poverty reduction: Trade-offs and synergies along the water-energy-food security nexus. *Water Int* 40 (5–6), 772–790.

Mondal, B.P., Sahoo, R.N., Ahmed, N., Singh, R.K., Das, B., Mridha, N., and Gakhar, S. 2021. Rapid prediction of soil available sulphur using visible near-infrared reflectance spectroscopy. *Indian J. Agric. Sci* 91(9), 1328–1332.

Mondal, B.P., Sekhon, B.S., Paul, P., Barman, A., Chattopadhyay, A., and Mridha, N. 2020. Vis-nir reflectance spectroscopy as an alternative method for rapid estimation of soil available potassium. *J. Indian Soc. Soil Sci* 68(3), 322–329.

Montzka, C., Moradkhani, H., Weihermüller, L., Franssen, H.J.H., Canty, M., and Vereecken, H. 2011. Hydraulic parameter estimation by remotely-sensed top soil moisture observations with the particle filter. *J. Hydrol* 399 (3–4), 410–421.

Naeem, S., Prager, C., Weeks, B., Varga, A., Flynn, D.F.B., Griffin, K., Muscarella, R., Palmer, M., Wood, S., and Schuster, W. 2016. Biodiversity as a multidimensional construct: A review, framework and case study of herbivory's impact on plant biodiversity. *Proc. Roy. Soc. B Biol. Sci* 283, 20153005.

Ongeng, D., Geeraerd, A.H., Springael, D., Ryckeboer, J., Muyanja, C., and Mauriello, G. 2015. Fate of Escherichia coli O157:H7 and Salmonella enterica in the manure amended soil-plant ecosystem of fresh vegetable crops: A review. *Crit. Rev. Microbiol* 41, 273–294.

Orgiazzi, A., Bardgett, R., Barrios, E., Behan-Pelletier, V., Briones, M., Chotte, J., De Deyn, G., Eggleton, P., Fierer, N., Fraser, T., Hedlund, K., Jeffery, S., Johnson, N., Jones, A., Kandeler, E., Kaneko, N., Lavelle, P., Lemanceau, P., Miko, L., Montanarella, L., Moreira, F., Ramirez, K., Scheu, S., Singh, B., Six, J., van der Putten, W., Wall, D., 2016. *Global Soil Biodiversity Atlas*. European Commission, Publications Office of the European Union.

Panigrahi, N., and Das, B.S. 2018. Canopy spectral reflectance as a predictor of soil water potential in rice. *Water Resour. Res* 54 (4), 2544–2560.

Purkus, A., Röder, M., Gawel, E., Thrän, D., and Thornley, P. 2015. Handling uncertainty in bioenergy policy desigh—A case study analysis of UK and German. *Biomass Bioenergy* 79, 64–79.

Raheem, A., Prinsen, P., Vuppaladadiyam, A.K., Zhao, M., and Luque, R. 2018. A review on sustainable microalgae based biofuel and bioenergy production: Resent developments. *J. Clean. Prod* 181, 42–59.

Ringler, C., Bhaduri, A., and Lawford, R. 2013. The nexus across water, energy, land and food (WELF): Potential for improved resource use efficiency? *Curr. Opin. Environ. Sustain* 5, 617–624.

Sahoo, R.N. 2022. *Sensor-Based Monitoring of Soil and Crop Health for Enhancing Input Use Efficiency. Food, Energy, and Water Nexus.* Springer, pp. 129–147.

Santra, P., Kumar, M., Kumawat, R.N., Painuli, D.K., Hati, K.M., Heuvelink, G.B.M., and Batjes, N.H., 2018. Pedotransfer functions to estimate soil water content at field capacity and permanent wilting point in hot Arid Western India. *J. Earth Syst. Sci* 127 (3), 1–16.

Sekhar, G.C., Sahu, R.K., Baliarsingh, A.K., and Panda, S., 2016. Load frequency control of power system under deregulated environment using optimal firefly algorithm. *Int J. Elec Power Energy Syst* 74, 195–211.

Schmalz, B., Kruse, M., Kiesel, J., Müller, F., and Fohrer, N. 2016. Water-related ecosystem services in western Siberian lowland basins—Analyzing and mapping spatial and seasonal effects on regulating services based on ecohydrological modelling results. *Ecol. Indic* 71, 55–65.

Sharp, R., Douglass, J., Wolny, S., Arkema, K., Bernhardt, J., Bierbower, W., Chaumont, N., Denu, D., Fisher, D., Glowinski, K., Griffin, R., Guannel, G., Guerry, A., Johnson, J., Hamel, P., Kennedy, C., Kim, C.K., Lacayo, M., Lonsdorf, E., Mandle, L., Rogers, L., Silver, J., Toft, J., Verutes, G., Vogl, A.L., Wood, S, and Wyatt, K. 2020. *InVEST3.9.0.post155+ug.ga885d58 User's Guide.* The Natural Capital Project, Stanford University, University of Minnesota, The Nature Conservancy, and World Wildlife Fund.

Shukla, A.K., and Behra, S.K. 2019. All India coordinated research project on micro-and secondary nutrients and pollutant elements in soils and plants: Research achievements and future thrusts. *Indian J. Fert* 15 (5), 522–543.

Smith, L., and Haddad, L. 2015. Reducing child undernutrition: Past drivers and priorities for the post-MDG era. *World Dev* 68, 180–204.

UN-SDSN 2013. An action agenda for sustainable development: Report for the Secretary General of the UN. Available at: http://unsdsn.org/2013/06/06/action-agenda-sustainable-developmentreport (accessed on February 2014).

Van der Meer, F. 2018. Near-infrared laboratory spectroscopy of mineral chemistry: Are view. *Int. J. Appl. Earth Observ. Geoinf* 65, 71–78.

Van Ittersum, M.K., van Bussel, L.G.J., Wolf, J., Grassini, P., van Wart, J., Guilpart, N., Claessens, L., de Groot, H., Wiebe, K., Mason-D'Croz, D., Yang, H., Boogaard, H., van Oort, P.A.J., van Loon, M.P., Saito, K., Adimo, O., Adjei-Nsiah, S., Agali, A., Bala, A., Chikowo, R., Kaizzi, K., Kouressy, M., Makoi, J.H.J.R., Ouattara, K., Tesfaye, K., Cassman, K.G. 2016. Can sub-Saharan Africa feed itself? *Proc. Natl. Acad. Sci.* 113, 14964–14969.

Wall, D.H., Nielsen, U.N., and Six, J. 2015. Soil biodiversity and human health. *Nature* 528, 69–76.

Wani, S.P., Sudi, R., and Pardhasardhi, G. 2021. Scaling-up land and crop management solutions for farmers through participatory integrated demonstrations "seeing is believing" approach. In: *Scaling-up Solutions for Farmers.* Springer, pp. 149–203.

Wigmosta, M.S., Nijssen, B., Storck, P., Singh, V.P., and Frevert, D.K. 2002. The distributed hydrology soil vegetation model. Math. Models Small Watershed. *Hydrol. Appl.* 22(21), 4205–4213.

Wise, M., Dooley, J., Luckow, P., Calvin, K., Kyle, P., 2014. Agriculture, land use, energy and carbon emission impacts of global biofuel mandates to mid-century. *Appl. Energy* 114, 763–773.

Wu, Z., Wu, W., Zhou, S., and Wu, S. 2016. Mycorrhizal Inoculation Affects Pb and Cd Accumulation and Translocation in Pakchoi (*Brassica chinensis L.*). *Pedosphere* 26, 13–26.

Yu, L., Li, Y.P., Huang, G.H., Fan, Y.R., and Nie, S. 2018. A copula-based flexible-stochastic programming method for planning regional energy system under multiple uncertainties: A case study of the urban agglomeration of Beijing and Tianjin. *Appl. Energy* 210, 60–74.

# 16 Agriculture, Rural Poverty, and Natural Resource Management in Less Favored Environments

## Revisiting Challenges and Conceptual Issues

*L. Devarishi Sharma, Rahul Sadhukhan, Hiren Das, L. Banarjee Singh, and Animesh Sarkar*

## 16.1 INTRODUCTION

Natural resources play a crucial role in poor man's life. Fisheries, forests, and agriculture offer approximately 1.3 billion job opportunities. Based on World Bank (2002) data, out of the 1.1 billion poor people, about 90% have an income of < $1/ day, earning their fraction of income from forestry. According to the international development, agencies reckoned that > 90% of people out of 15 million were marginal-scale fishers. It also encompasses millions of poor people fishing in lakes and rivers.

Poor people in rural areas mostly depend on healthy ecosystems because they provide all the necessities of life. Now humankind is associated with ecological processes and so on; that is why poor people rely more on natural resources than the rich people. Out of 10, 7 people of poor background live in rural areas of Africa. The main occupations of these poor people in these areas are marginal-scale farming, hunting, and rearing cattle, hens, and ducks. Their income solely depends on the harvest of the aforementioned activities. The products produced from different activities are these poor people's main income.

Different policies, regulations, and policies were developed to move forward towards development to alleviate poverty in this world. Some of these policies and approaches are the World Bank Poverty Reduction Strategy Papers (PRSPs) and the United Nations (UN) Millennium Development Goals (MDGs). The earlier policies

DOI: 10.1201/9781003358169-16

or approaches help in eradicating poverty by giving ways to generate income for the poor. They are the key driving force in improving the livelihood of poor people. They bridge the gap between the management of natural resources and the reduction of poverty to some extent. They also ascertain the poor man's dependency on natural resources.

Poor people from rural areas earn their income from natural ecosystems. Natural resources are renewable, and they are extensively distributed. They can use these resources from the land they have owned above. The services and goods provided by the ecosystem are valuable or useful for the community and can benefit the rural poor beyond the reach of other income sources. The work nature of rural people is almost similar, i.e., exploration or harvesting of natural resources. Similar work nature brings them together. The community of rural areas was improved and strengthened by the resources obtained.

## 16.2   ENVIRONMENT AND POVERTY NEXUS

Understanding the connections between poverty and environmental resources quality demands data on the spatial allotment of the poor in every region and country. Therefore, a comprehensive poverty-environment assessment must be performed to create policy-relevant data worldwide. Considering the minimal consumption of standard per capita of $1 per day from 1990 to 1999, the overall population of poor people in developing nations fell from 1.3 billion to 1.2 billion, and the pace of poverty dropped as of 29% to 23%. Throughout this time, the overall number of sub-Saharan Africans livelihood in extreme poverty climbed from 242 million during 1990 to 300 million just at the end of the century. Whereas the proportion of the poor fell marginally throughout this time, it was insufficient to halt the actual growth in the number of poor individuals. Sub-Saharan Africa is the sole area in which the actual poor population is predicted to climb, with 345 million people living in poverty in 2015 because of the economy's instability or sluggish growth.

In comparison, South Asia is expected to fall from 490 million in 1990 to 279 million by 2015. However, many emerging economies are seeing more globalization and structural reforms (e.g., Brazil, China, India, and Mexico). Rapid financial development in several regions beyond sub-Saharan Africa has resulted in an astonishing drop in poverty rates and relative poverty numbers. Even though there is significant variance amongst countries, it suggests that poverty seems to be likely to intensify as economic marginalization increases in some underdeveloped areas at which deep-seated systemic issues such as poor facilities, higher costs, unfavorable weather patterns, disease prevalence, and a lack of human capital hinder higher capital massive influx and minimize trade competitiveness (World Bank 2002a, 2002b). According to the latest data, the number of poor people is higher in rural areas than in metropolitan ones (World Bank 1990). As a result, many rural poor people reside in locations with limited agricultural potential (Leonard et al. 1989, World Bank 1990). In developing countries, more than 75% of the poor lived in rural regions in 1996, with 47% located in marginal locations with little agricultural production. The geographical distribution throughout the continents bears some resemblance, with Asia having a slightly higher percentage (49%). An estimated 995 million (three-quarters) of the impoverished in emerging nations live in countryside (Ryan and Spencer, 2001).

Furthermore, around 38% (379 million) of the poor were concentrated in arid as well as in semi-arid areas, and 50% (500 million) live in humid as well as sub-humid regions, with the remainder in areas of temperate regions. The statistics also show that the actual figure of poor in each agro-ecological zone was marginally greater in rainfed areas than those in irrigated ones. Furthermore, in emerging countries where agro-based rural livelihoods are prevalent, agriculture occupies the bulk of land utilization, and most poor livelihoods are directly dependent on natural resource utilization. As a result, the deterioration of these resources directly impacts the daily activities of rural populations, resulting in either a decrease in the production of the resources on which they depend or adverse health consequences. For example, soil erosion, loss of vegetation cover, and overexploitation limit agricultural land production, whereas increasing soil sediments deposition, pollutants, and additional toxins in fresh water may augment the prevalence of waterborne ailments. Natural resource deterioration might even raise the amount of time required for home production, such as collecting fuelwood or obtaining water from remote sites, as well as competing with the human resources required for commercial agriculture and conservation expenditures. In case the poor are clustered in places with limited land, agricultural output is little, and ecological deterioration is prevalent (Leonard et al. 1989, World Bank 1990) and relies directly or indirectly on farming for a living.

Moreover, it may imply a significant degree of linkage between poverty processes and the incapacity to make investments that develop or preserve the natural resource base. The data generally serves as the foundation for evolving ideas on poverty-environment connections and a rigorous micro-level inquiry to comprehend the methods through which poor people connect with their surroundings and the accompanying conditions that could guide to resource base enhancement or degradation. As a result, considerable research suggests a two-way relationship between environmental deterioration and poverty (World Bank 1992, WCED 1987, Reardon and Vosti 1995, Cleaver and Schreiber 1994). A poverty index based on well-being criteria may exclude certain households that can meet basic consumption demands but cannot make essential resource-enhancing investments in imperfect marketplaces. As a result, being able to spend in resource development necessitates that households be over the "investment poverty" threshold, which deposits that resource users are accessible to critical assets that are desired to undertake substantial expenditures above and over what is required to meet essential requirements. Others contend that even when institutionalized solutions for smoothing utilization (livelihood security) are lacking, the poor's primary goals may be instant survivability and food sustainability. As a result, programs that are designed are short and subjective rates of subjective value are high. When current survival is endangered, families with limited assets may be unable to forego present consumption to make resource-saving or boost investments required to preserve future consumption. Because of the constructive link between significant discount rates and poverty or low earners (Holden et al. 1998), asset or conservation expenditures with extended gestation and durations may be mitigated. If certain assumptions are correct, many sections of the semi-arid tropics will have adverse biophysical constraints. Poor socioeconomic facilities may

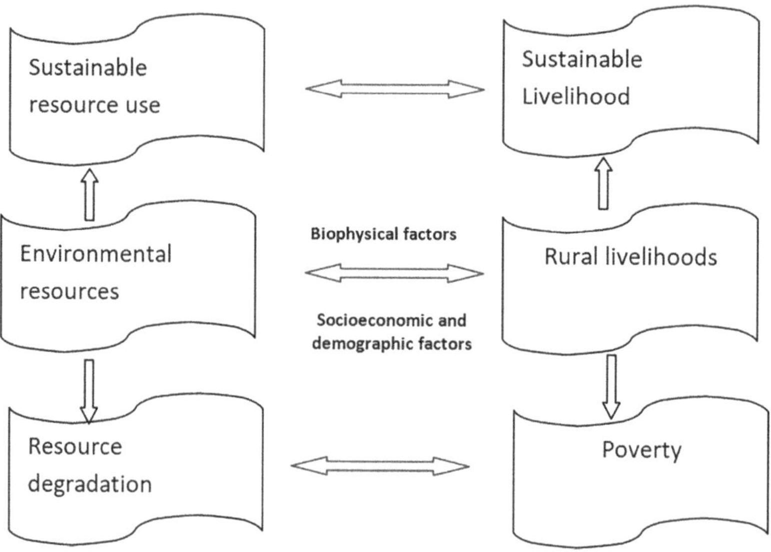



**FIGURE 16.1**　The linkage between poverty and the environment in rural areas.

be an example of the deep interconnections between poverty with resource deterioration. In exceptional cases, the poverty-environment cycle may result in a "path of growth" that precludes future choices for sustainable food production and livelihood protection. Small-scale farmers and peasant farmers people in poor and degraded places may thus become locked in a self-perpetuating treadmill of poverty and biodiversity loss (Figure 16.1). Establishing such a link requires ongoing individual and ecological capital expenditures, improved agricultural research technologies for stress tolerance, conservation of water, management of pest and disease, enhanced market access, and improved off-farm employment opportunities. Nevertheless, it should be noted that the connection between the environment and the poverty is complicated and is not always a negative cycle (Scherr 2000); eradication of the poor is not always beneficial to the atmosphere.

## 16.3　FARMER INVESTMENT PLANS

Growing population, technical possibilities, infrastructural development, government policies, and market access and institutions are all factors to consider to determine the rural community's livelihood plans and resource utilization patterns (Reardon and Vosti 1995). The degree and nature of poverty-environment linkages in a given context and farmers' investment plans relies on the degree and extent of poverty, the baseline level of resources, and the output effects of degradation. Depending on socioeconomic, regulatory, and meteorological variables, the farm-level efficiency of output and restoration methods and viable investment alternatives fluctuate among areas and countries (Binswanger and Rosenzweig 1986, Pender et al. 1999a, Shiferaw and Holden 1999). This indicates that technological advancement and management techniques for intensive farming should account for changes in both socioeconomic and biophysical characteristics among ecoregions. The next section discusses farmers' technological choices and investment plans influenced by markets and legislation, poverty, property rights, and biophysical factors.

## 16.4  GOVERNMENT POLICIES RELATED TO MARKET

The implementation of innovative conservation techniques will be influenced by the comparative returns and income stability that new options bring when compared to old alternatives. As a result, small-scale farmers are often averse to risk (Binswanger 1980). However, crop plant failures and risk-averse households are more likely due to land degradation. Therefore, the selection of technologies and investment tactics will be based on concerns about profitability and risk (income stability). With the resource rights and assets of livelihood established by governmental policies, accessibility to regional institutions, options for off-farm jobs, participation in social organizations, and biophysical factors, the ability to regulate and disperse risk rises. Food safety and security concerns come first because emerging innovations are seen as perilous, which can prevent the adoption of profitable solutions. Beyond risk, the capacity to access financing and remove capital restrictions also influences the adoption of technology and farmer investment decision. Inputs that are projected to produce benefits soon, like enhanced seeds and fertilizer, are given credit across many developing nations. Finally, credit markets provide weak support for investments in conservation and resource improvement, which frequently have benefits over the medium to long term. For example, farmers may be discouraged from implementing such alternatives because of the high price of capital credit, if it is even accessible at all. This is because the return rate on conservation expenditures may also be greater. However, since future advantages are not well known and long-term rewards are frequently uncertain, risk aversion may deter such long-term profitable investments in a world without perfect foresight and imperfect circumstances. This demonstrates that product improvement and the profitability of different choices cannot be the primary priority for farmers when making technology selections. Income stability in the presence of pest, disease, and drought stress, as well as access and availability to inputs required in the process of production, are critical factors for farmers. Concerning profitability, the operation of local markets impacts the amount of fertilizer, human resources, and certain other inputs used in the production line. The growing season in semi-arid environments is quite short, and farming operations must be accomplished in a short time frame. Imperfections in financing and labour markets also make it difficult to alleviate such limitations substantially. This demonstrates the continued growth of labor-saving choices in farming and the importance of giving labour demand implications careful thought during the design and development of technologies. Leveling and terracing are two soil and water conservation techniques that frequently demand significant labour expenditures per unit of managed land.

The best solutions are those for managing water and soil that demand the fewest resources and require the least labour. In the context of growing land availability, vegetative approaches such as grasses, legumes, and agroforestry systems that do not interact heavily with existing cropland are becoming more popular. It provides additional advantages in terms of enhanced food, forage, and firewood production, as well as lessened wind and water erosion, which are viable choices that necessitate additional consideration in NRM studies and development. In some circumstances, governmental policies support specific inputs (for example, subsidies of fertilizer in India) or the public service accounts for a large portion of national and local supply. A few of these subsidies may send mixed signals towards resource consumers and distort individual attempts to make resource-saving or improvement initiatives. Farmers may be discouraged from implementing innovations that decrease soil erosion and conserve water resources if they get fertilizer and irrigation water incentives. There is a lack of substitute methods for supplementing and managing nutrients for plant growth.

## 16.5 EXTERNALITIES AND PROPERTY RIGHTS

Another aspect that has recently garnered more attention in the scientific literature is the security of resource access rights (Place and Hazell 1993, Feder and David 1991, Besley 1995). In contrast, privatization claims to the resource are not necessary to protect rights. Instead, what appears to be most important for investors is the level of security and durability of use provided by a specific property rights system to the resource consumer. For example, the projected profits through resource-enhancing investment may be exceptionally low when the usage of rights is minimal or the chances of maintaining rights are limited. This results in the effect of reducing the resource user's planning horizon.

Inadequate property ownership and non-exclusion issues also inhibit investment. A quintessential example is the failure of the market mainly for resources that are open access. This also encompasses for public goods that are inferior. For instance, a farmer constructing some structures associated with flood control at the upstream will surely benefit the farmers residing at the downstream areas. The occurrence of market failure is very common, as all the costs are borne by a private individual, but they all share the benefits. The resources managed by a social group commonly suffer from non-exclusion constraints because the improvement of resources is made by collective work. Sometimes the investment made for improvement are not suffice to tackle the degradation associated with resources. Several farmers are also linked through the flow of externality in various directions. Since it is a collectively taken action, that is why there will be no complete control by a single user for the happening taken place. Henceforth, it requires mutual understanding and cooperations among one another in order to achieve optimal private investments. A problem may arise if the portion of the private investment benefits the society by the build-up of finance over time. If the benefits of private benefits are lesser in comparison to social benefits, then the investment level will be less taken up by the individual as compared to the community. In order to boost acceptability at the social level, the private investment needs the influence of the public by sharing the cost and other subsidies. The cost incurred and benefits gained by investing are not distributed uniformly and vary for different groups of people residing in different geographical areas. The earlier said problems were common in farmers residing in upper reaches during the investments made for soil and water conservation in watershed areas.

However, this investment gives greater benefits to those farmers living in the watershed areas of lower reaches. So the farmers both from upper reaches as well as lower reaches should design innovative arrangements to compensate the farmers residing at the upper reaches. Consequently, these arrangements can minimize the losses met by the farmers living in the upper reaches.

## 16.5.1 SOCIAL DIMENSIONS AND GOVERNANCE

The division, exercise, and boundaries of authority are all part of governance. In order to sustain social harmony, establish law and order, encourage or create the circumstances for economic growth, ensure equality and social justice, defend rights (civil, political, economic, social, and cultural), and provide a basic degree of social security, good governance is implied. By deciding who owns, has access to, decides how to use, and has power over resources like forests, land, and fisheries, environmental governance systems establish a relationship between natural resources and poverty. Laws, institutions, political structures, cultural norms, social networks, and policies that specify how important resources should be used, owned, and controlled make up these "rules of the game." The term "corruption" refers to the abuse of governmental authority for personal benefit and denotes poor governance.

It is generally acknowledged that less corrupt governments have more effective bureaucracies and deliver better policies. Assuring everyone's rights to information access, decision-making participation, and swift access to justice, good governance increases efficiency, effectiveness, and equity. In spite of their heavy dependence on natural resources, the poor frequently have had petite influence over environmental choices. The creation of sustainable livelihood plans and the reduction of poverty depend on effective institutions. Institutions have a range of effects on asset access. Households risk losing access to land and resources if land ownership laws are unclear and/or contradictory as a result of inadequate institutions protecting property rights. In this situation, they might choose to focus on short-term (typically low-return asset uses) strategies rather than longer-term (perhaps more lucrative but riskier) ones. Institutions can influence additional returns on assets and incentives to amass them in this way. Poor people are less able to fully utilize this asset (such as using it as collateral for a bank loan) and may be less motivated to improve their holdings of natural resources as a result of complicated government procedures that discourage them from getting property titles (e.g., adopting soil conservation measures). In terms of the law, tenure describes a collection of ownership, possession, and use requirements.

The "capacity to rely upon a collective to back one's claim to a benefit stream" has been described as its definition. Traditional customs and social networks that control resource ownership and access can also be considered forms of tenure. This is particularly right for pooled resources owned by tribes or communities, which frequently give informal access to fishing areas, woods, or grazing pastures to chosen members of the community. In many nations, impoverished and indigenous people may not have access to native forests, and their access rights may not be recognized by government legislation, which frequently leads to egregious inequalities in the distribution of land. Two factors are pertinent to a pro-poor tenure policy in this situation: tenure must be secure, and equitable land allocation is required. Persistent poverty is characterized by unequal access to land and other productive resources. The effects of unstable and unfair tenure laws are especially severe for the poor, whose livelihoods depend significantly on natural resources. Growing global integration has an impact on two broad trends that affect resource tenure. First, governments tend to play a smaller role in relation to the private sector and civil society, favoring private property and individual responsibility. Second, decentralization empowers institutions at the municipal, tribal, and community levels to exercise greater assertiveness in administering regional resources. The ability of the poor to obtain environmental income from natural resources is expected to change as a result of these two changes. Decentralization is the process through which a central government delegates some of its authority or responsibilities to a local authority, institution, or leader. Theoretically, decentralization has advantages like the following: Decentralization encourages more participation in public decision-making in democracies. Decentralization improves managerial and economic effectiveness (efficiency). Benefits are distributed more democratically and with greater retention thanks to equity decentralization. Decentralization puts governmental decision-making in the people's direct line of sight (accountability).

A distinct chain of accountability from decision-makers to the local populace is necessary for effective decentralized governance. At the Rio de Janeiro Earth Summit in 1992, 178 countries approved Principle 10, pledging to give individuals better access to environmental information, chances to participate in environmental decision-making processes, and access to redress and remedy (i.e., justice) to uphold their rights. The three procedural rights of access to information, participation, and justice—also known as the access principles—help define the qualities of effective governance. Encourage active citizenship. The poor's capacity to use natural resources sustainably and build wealth is frequently constrained by a lack of control over the resources at hand and an incapacity to engage in decision-making processes. Policymakers can strike a balance between traditional systems and contemporary legal definitions by acknowledging the multiple nature of tenure systems—formal and informal agreements—and the dynamic aspect of ownership. In order to achieve this, policymakers must establish a clear system for registering, transferring, and enforcing resource rights, strengthen weak government institutions that result in onerous land regulations, confusing registration processes, and a lack of access to land titles for the poor, translate laws and policies into local languages, offer legal assistance and training on property rights, and recognize and respect pre-existing local rights. Encourage democratic processes and accountability.

Local governments should be held responsible for implementing efficient governance through transparent election processes or other legal means, such as civil service regulations. These representatives should offer services that are driven by demand and based on democratic, inclusive procedures. In order to select priorities and carry out development projects, qualifying institutions must have access to sufficient funding and discretion. However, they also need to be able to show that they are financially capable. Pro-poor policies should also be incorporated into resource management plans by local institutions.

## 16.6 RESOURCE TENURE AND PROPERTY RIGHTS

Tenure can be defined as the holding of something like land, position, property or office, manner, acts and rights, etc. At this point, tenure of natural resources can be interpreted as the arrangement to make natural resources accessible to the people, to formulate the dos and don'ts regarding the use of resources, to exhibit the benefits derived from the natural resources by participation, and to establish institutions and other related processes from resource management. Legally, tenure describes the rights related to property holding. It also indicates a group or a person for what they can do with their property, and property entails a piece of land that is utilized for farm, forestry, grazing, recreational water bodies, fishpond, wildlife sanctuaries, and other resources, i.e., minerals.

Tenure gives an idea about the property of the person and their utilization of property in a manner with proper legal. From the viewpoint of environmental safety, the framework of the tenure arrangement should be focused on natural resource sustainability rather than on overexploitation of the natural resources. Ambiguous tenure arrangement may aggravate the extraction level of resources. In another way, tenure is the cornerstone in making environmental policies that emphasizes on pollution control which includes in-depth discussions on data related to past pollution. In this case, the framed tenure is equivocal, and the liability regimes made on environmental sustainability would be unworkable. Besides, inclusion of climate change in the liability aids in making proper tenure of natural resources. The crucial factors like forest management, land management, and water are some of the essentials required for both mitigation and adaptation measures in many countries.

Any policy developing tenure rights plays a significant role in fostering and suppressing economic growth, uniform distribution of resources, empowerment of users in relation to resources, and the basis of resource sustainability of environment and climate. A brief discussion about the linkages between areas—i.e., urban areas and rural areas and resources like land and water—requires a clear policy and tenure arrangement. Undoubtedly, maximum achievements will be there, in case water sources are considered during the allocation of land rights. Contrastingly, awareness on water sources identification may wreak havoc on land right distribution. The absence of information of water sources results in the confrontation and degradation of resources in the past. Convergence coupled with coordination of different systems of tenure are the prime reasons to inscribe the interdependence.

Generally, natural resource tenure can be classified into four types:

1. Property related to state or public: the state has all the rights in which all the rights over the resource is exercised by the public sector.
2. Property related to private: the individual has all the right over the resource, e.g., corporations.
3. Property commonly shared by two or more people: a group of people jointly holds all the rights to use the resource, e.g., a community or group.
4. Open access property: in such type, rights are not specifically mentioned. No one has the right to impose any laws or rules. Moreover, resource depletion will occur in open access if the resource is in the brink of exhaustion because there are no specific rights mentioned for anyone and everyone has unlimited rights for resource utilization.

**TABLE 16.1**

**Classification of the Bundles of Right**

| Sl. No. | Type of rights | Description |
|---|---|---|
| 1 | Right of access | The right to enter a defined physical property: to use the property in a non-consumptive way. |
| 2 | Right of withdrawal | The right to obtain benefits of a resource: to use the property productively for profit. |
| 3 | Right of management | The right to regulate internal-use patterns and transform the resource by making improvements: to set up and modify rules for the use of a property. |
| 4 | Right of exclusion | The right to determine who will have an access right: to exclude some users and set rules for access to a property. |
| 5 | Right of alienation | The right to sell, lease, or inherit property with the rights mentioned previously. |

*Source:* Schlager and Ostrom (1992)

## 16.7  NATURAL RESOURCES TENURE POLICY

1. Policies of tenure must ensure the utilization of natural resources in a sustainable manner and promote the growth of the economy.

    Provision of incentives should be included so as to make investment in the long term which will thereby boost the productivity, e.g., fishing. Enforcing accountability for any bad implications on the environment would not be possible in case the tenure is not well defined. Lastly, utilization of natural resources in a sustainable manner is only possible when the policy tenure is well defined.

2. Privatization of utilization of natural resource brings gain.

    The tenure optimum form is primarily dependent on the traditions of the country, resource characteristics, model, and rules of the institutions, whether it can be formal or informal. The privatization suits well when there are genuine competitions in the market and when the government is unable to convey any calamities information related to the environment which may cause failures of market.

3. Amendments of tenure based on transparency, participation, and specific context should be done. Political processes play a key role in the amendments of tenure. It includes different sections of stakeholders and people with diversified interest. The only way to success is primarily dependent on equitable cost distribution, yet it requires involvement of participation, various transparent organizations such as ministries, civil societies, business firms, etc. Consequently, this will fend off the capture of the resource by elite persons.

4. Proper provision of incentives and appropriately executed policies are pivotal to sustain the natural resource utilization.

Tenure rights for natural resources should be augmented by well-defined rules, regulations, and circumstances. These set of rules and regulations should convey the market failure constraints and certain countermeasures so as to improve the efficacy of the economy—for instance, reforestation commitments and transparency in revenue works to set up an evaluation committee on its effect on environmental standards and permissible levels of emissions.

5. Statutory regulations. Strict rules and regulations will help to alleviate the problems of bribing, corruption, and other illegal activities. Resources like fish, mineral, and timber are regarded as resources of high value, and these resources are the common focuses of corrupt behavior along with the over-exploitation of these resources in a very unsustainable way. The effectiveness of the policy incentives by large have direct impact on the sustainability of the utilization of natural resources. Corruptive behavior or activities can be easily quashed in case the transparency is increased, enhancing communication or information systems coupled with practically achievable or simple rules and regulations. Enforcing regulatory and establishment needs should be separated institutionally. Besides, direct government support and regulatory functions need to be separated. The best way to curtail the problems of conflict of interest is to avoid the overlapping of responsibilities and functions of public institutions.

## 16.8   THE ASSET-BASED APPROACH

Overall, poverty affects the well-being, capabilities, and potentialities of an individual, owing to an intricate and multiple deprivation of essential needs. Plutocrats do not face the earlier problems, as they have sufficient basic needs. Important components required for the development of sustainable livelihood are land, community, educational institutions, health facilities, judiciary systems, banks, etc. Poverty is often explained as the lack of freedom, such as the lack of social and political freedom and freedom of choice, which in turn has gruesome effects on their livelihood. For instance, political liberty aids in securing a fair resource right. Freedom, or liberty, is pivotal in this environment. It will pave the way to development.

It involves more than just the plutocrat and income. It's a complicated and multi-faceted privation that affects individualities, different capabilities, and their overall well-being. Access to land, education, health, justice, family and community support, credit, other productive coffers, and a voice in institutions are all important in developing sustainable livelihoods. Poverty has been described as the dispossession of different types of freedoms—profitable, political, social, and choices which affect livelihoods. For an illustration, political freedom can help secure better resource rights administrations, leading to lesser wealth and equity. In this environment, freedom is both the ends and means of development.

The poor themselves frequently take a broad view of poverty that not only includes income, consumption, and physical means but also nontangible social and political means similar to kinship systems, a sense of community, the capability to share in decision timber, and the capability to impact factors that affect livelihoods. Poor people, like others, may seek to gain fresh security from crime and conflict, representing yet another dimension of poverty reduction. The asset-grounded approach to poverty reduction focuses on developing the stock of wealth available to the poor and on their capability to manage threat and vulnerability and to achieve sustainable long-term advancements and well-being. The significance of natural capital, within the total stock of capital available to homes, tends to vary equally with situations of income. The poorer the country or the population, the more significant the part natural capital plays in discouraging mining poverty issues. Natural coffers contribute to livelihoods by furnishing a buffer against temporary salutary and profitable faults, serving as sources of cash income and employment in times of extremity, and serving as a readily convertible capital asset.

Possible solutions are as follows:

- Governments or other nongovernmental organizations should inform particular communities or segments of people about the impact of overdependence on natural resources.
- Governments or nongovernmental organizations should provide alternate income opportunities to the poor segments of people.
- Governments need to make pro-poor policies to strengthen weaker sections of people.
- They should be supported by providing various inputs.
- Different agencies or institutions must conduct various skills development and training programs.
- Governments should motivate them to form self-help groups and other cooperative groups and develop better income options. For that, governments need to support them by providing financial assistance.

## 16.9　CONCLUSION

Poor people and emerging countries have the propensity to be dependent to a large extent on natural resources for growth, poverty reduction, and commission. The relationship between natural resources management and poverty is complex and dynamic. It is dodgy to synthesize these connections, which are not possible to explain in a few words. There are some general considerations to integrate natural resources and poverty reduction in development programs and strategies. Poor and weaker people depend greatly on natural resources because they are easily accessible and freely available in nature. If the financial social status of that segment of people does not improve, then the burden on natural resources and environment will never slow down.

## REFERENCES

Besley, T. 1995. Property rights and investment incentives: Theory and evidence from Ghana. *Journal of Political Economy* 103:903–937.

Binswanger, H.P. 1980. Attitudes towards risk: Experimental measurements in rural India. *American Journal of Agricultural Economics* 62:395–407.

Binswanger, H.P., and Rosenzweig, M.R. 1986. Behavioral and material determinants of production relations in agriculture. *The Journal of Development Studies* 22(3):503–539.

Cleaver, K.M., and Schreiber, G.A. 1994. *Reversing the spiral: The population, agriculture and environmental nexus in sub-Saharan Africa.* Washington, DC, USA: The World Bank.

Feder, G., and David, F. 1991. Land tenure and property rights: Theory and implications for development policy. *World Bank Economic Review* 5:135–153.

Holden, S.T., Shiferaw, B., and Wik, M. 1998. Poverty, credit constraints, and time preferences: Of relevance for environmental policy? *Environment and Development Economics* 3:105–130.

Leonard, H.J., Yudelman, M., Stryker, J.D., Browder, J.O., De Boer, A.J., Campbell, T., and Jolly, A. 1989. *Environment and the poor: Development strategies for a common agenda.* USThird World Policy Perspectives, No. 11. Washington, DC, USA: Overseas Development Council.

Pender, J., Place, F., and Ehui, S. 1999a. *Strategies for sustainable agricultural development in the East African highlands.* EPTD Discussion Paper No. 41. Washington, DC, USA: International Food Policy Research Institute.

Place, F., and Hazell, P. 1993. Productivity effects of indigenous land tenure systems in sub-Saharan Africa. *American Journal of Agricultural Economics* 75:10–19.

Reardon, T., and Vosti, S.A. 1995. Links between rural poverty and the environment in developing countries: Asset categories and investment poverty. *World Development* 23(9):1495–1506.

Ryan, J.G. and Spencer, D.C., 2001. *Future challenges and opportunities for agricultural R&D in the semi-arid tropics.* International Crops Research Institute for the Semi-Arid Tropics. Patancheru, Hyderabad, Telangana, India.

Scherr, S. 2000. A downward spiral? Research evidence on the relationship between poverty and natural resource degradation. *Food Policy* 25:479–498.

Schlager, E., and Ostrom, E. 1992. Property-rights regimes and natural resources: A conceptual analysis. *Land Economics* 68(3):249–262.

Shiferaw, B., and Holden, S.T. 1999. Soil erosion and smallholder conservation decisions in the highlands of Ethiopia. *World Development* 27:739–752.

WCED (World Commission on Environment and Development). 1987. *Our common future.* Oxford, UK: Oxford University Press.

World Bank. 1990. *World Development Report 1990. Poverty.* Oxford, UK: Oxford University Press.

World Bank. 1992. *World Development Report 1992. Development and the environment.* Oxford, UK: Oxford University Press. 21.

World Bank. 2002. *The environment and the millennium development goals.* Washington, DC: World Bank.

World Bank. 2002a. *Global economic prospects and the developing countries.* Washington, DC, USA: The World Bank.

World Bank. 2002b. *Globalization growth and poverty: Building an inclusive world economy.* A World Bank Policy Research Report. Washington, DC, USA: The World Bank.

# 17 Environmental Resource Management with Reference to Global Climate Change

*Suwa Lal Yadav, Dileep Kumar, Manish Yadav,*
*K.C. Patel, Devilal Birla, Lakshman,*
*and Praveen Singh*

## 17.1 INTRODUCTION

The increasing pollution of surface and groundwater resources, particularly in developing countries, has made water quality a major concern worldwide. There are several causes of water pollution, including soil contamination, agricultural waste runoff, improper solid waste management, reusing wastewater for irrigation after inadequate or incomplete treatment, and open defecation as a result of poor sanitation. The urbanization of the world, as well as the changes in lifestyle, will have an inevitable impact on the production of solid and liquid waste, resulting in soil and water pollution caused by the leaching of organic compounds and other pollutants, such as heavy metals. In many developing nations, it is common practice to discharge garbage directly into bodies of water and land. A significant threat to the quality of groundwater is posed by surface runoff and seepage of fertilizers and biocides into groundwater.

Another rising worry is soil contamination, which can be caused by unsustainable land use, the use of excessive fertilizers or biocides, and improper waste disposal, which can release harmful organic compounds and heavy metals into the soil. In order to safeguard water quality, soil, wastewater, and waste need to be managed comprehensively based on the nexus approach. A highly dynamic landscape environment is characterized by a variety of types of soil, and precipitation supplying both surface and groundwater, including significant amounts of water in the unsaturated zone of the soil, fuels the complicated water cycle. Every nexus must, therefore, be able to articulate broad principles and generalize them in order to develop a useful nexus method. Research must be inter- and transdisciplinary, defining "options" rather than nonexistent "solutions" and considering trade-offs between frequently differing demands for the different competing aspects.

Natural resources are being degraded because of human meddling and irresponsible management. These will influence the composition and functions of ecosystems, as well as the sustainability of the environment generally and globally. The most



DOI: 10.1201/9781003358169-17

prominent and urgent problem today for decision-makers, stakeholders, scientists, and academics on various national and international forums is climate change. As a result, we must manage these natural resources sustainably, considering the future generation.

Using the nexus approach to manage natural resources and adapt to planetary change is one of the objectives of this book. It encourages a discussion of various points of view on the application of the nexus approach. This introduction offers a concise but comprehensive overview of many thought leaders' opinions. They discuss how the nexus method can be used to manage trash, soil, and water. The nexus approach to environmental resource management will be clearly and unbiasedly discussed in this book. Furthermore, we believe that this will help shape the much-needed nexus thinking in the future.

## 17.2 POPULATION TRENDS AROUND THE WORLD ARE CHANGING

By 2025, the world's population is expected to grow by nearly 1 billion from 7.20 billion in mid 2013. It is projected to reach 9.60 billion by 2050 and 10.90 billion by 2100. In this recurrent upsurge, the main contributor is the rapid growth of the population of developing countries, particularly in Africa, where the population is expected to increase from 1.10 billion to 2.40 billion by 2050. It is still a matter of concern that several nations in sub-Saharan Africa are experiencing rapid population growth and high fertility rates. India, Indonesia, Pakistan, the Philippines, and United States are other countries with similar trends. According to projections, India's population will surpass China's by 2030, becoming the world's most populous nation. The Nigerian population is expected to surpass that of the United States in 2045, making it the third most populous country in the world. In the next century, Nigeria's population is likely to rival that of China (Montanarella *et al.*, 2015).

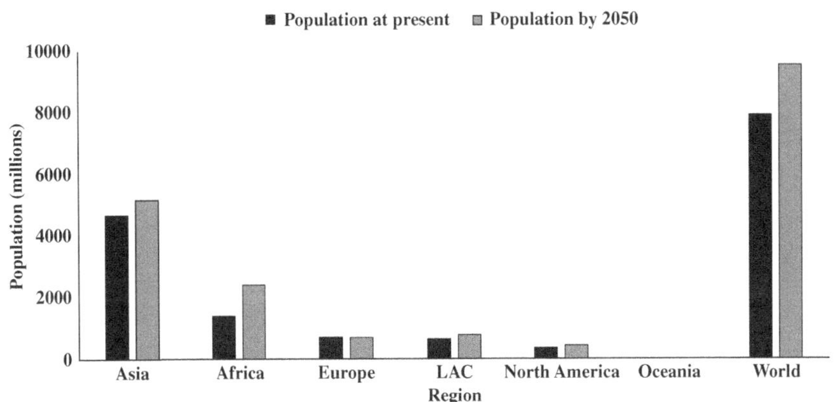


**FIGURE 17.1** World population growth projection by 2050.

**Source: Gerland et al. (2014)**

Population aging poses a challenge to Europe, as fertility declines and life expectancy rises. It is also likely that other developing countries with young populations but lower fertility (e.g., China, Brazil, and India) will face the challenges of an aging society by the end of this century (Gerland *et al.*, 2014).

## 17.3 PROJECTIONS OF WATER, WASTE, AND SOIL DEGRADATION BY 2050

### 17.3.1 WATER DEMAND BY 2050

Population increase, economic expansion, and consuming habits are all contributing factors to rising water demand. Globally, the demand for water has increased by 600% over the past century. This translates into a growth rate of 1.80% per year. There is a claim that the current yearly growth rate is less than 1%, but this estimate may be optimistic. In the next two decades, water demand will increase in all three sectors: industrial, residential, and agricultural. Even though home and industrial demand will increase faster than agricultural demand, agricultural demand will remain the highest. The need for growth in industries other than agriculture will outpace that of the agricultural sector (Boretti and Rosa, 2019).

By in 2050, there will be an increase of 20.0% to 30.0% in the global water demand for all applications, reaching 5,500 km$^3$ to 6,000 km$^3$ annually. The current global water demand for all applications is around 4,600 km$^3$. By 2025, there will be a 60% rise in the global water demand for agriculture.

The 70% of all water used worldwide is currently used for agriculture. The irrigation sector is the foremost user of water. Estimates and projections at the global level are hazy. The demand for food will rise by 60% by 2050, and to meet this growth, more arable land and more production will be needed. This will result in more water being used. Industrial water consumption accounts for 20% of the world's water supply. The remaining 25% of the industry is made up of manufacturing, while the remaining 75% is made up of energy generation. Except for North America and Western Europe, water demand for the industry will increase everywhere by 2050. In Africa, where there has been very low water use in the industrial sector, the demand for water will increase in the industry by 800%. In Asia, the water needed by the sector (industry) will rise by 250%. The need for water in the industry will rise by 400% globally (Boretti and Rosa, 2019).

Globally, the use of water for the energy sector will rise by 20% between 2010 and 2035 and by 85% by 2050. Currently, 10% of all water use worldwide is domestic. Almost everywhere in the world, except for Western Europe, the demand for domestic water is predicted to rise dramatically between 2010 and 2050. In Africa and Asia, there will be a 300% increase. The rise in Central and South America will be by 200%. The expansion of water delivery services to urban populations is blamed for this rise.

### 17.3.2 WATER RESOURCES BY 2050

Water availability cannot be greater than water demand. Water resources are depleting, and pollution is causing a shortage despite the rising demand. While surface water availability will remain relatively constant at the continental level, its quality will diminish and its spatial and temporal distribution will change. In coastal locations, aquifers will be more likely to dry up, resulting in salt intrusion that will be noticeable. Contrarily, there will be the uneven worldwide expansion in the population, the GDP, and the need for water. The magnitude of the changes will be substantially greater at the subregional level than at the national or global levels.

Water scarcity circumstances currently exist in many nations. By 2050, many more nations will experience decreased access to surface water resources. Around 1.90 billion people live in places that would be severely water scarce by the mid-2010s. A total of 2.70 billion to 3.20 billion people will live in the world by 2050, a rise of 42% to 95%. If monthly variability is considered rather than yearly variability, 3.6 billion people globally, or just under 50%, live in potentially water-scarce areas at least half the year. From 33% to 58%, this figure will rise to 4.8 billion to 5.7 billion by 2050. Asia is currently home to more than 73% of those suffering from water scarcity (Boretti and Rosa, 2019).

Global groundwater consumption in the 2010s totaled 800 $km^3$ $year^{-1}$. The 67% of the world's extractions were made in China, Iran, Pakistan, India, and the United States. Global groundwater depletion is mostly caused by water withdrawals for irrigation and other uses. By 2050, groundwater extraction will have increased by 1,100 $km^3$ $year^{-1}$. Water depletion at the basin level may get worse overall if irrigation water use is made more efficient. Current worldwide withdrawals are already close to their maximum sustainable limits at roughly 4,600 $km^3$ annually. Currently, there are problems with more than 30% of the greatest groundwater systems on earth.

### 17.3.3  WASTE GENERATION BY 2050

Approximately 2.01 billion tonnes of municipal solid waste are generated each year worldwide, with at least 33% produced conventionally and not environmentally friendly. Across the globe, people generate 0.74 kg of waste per day on average, but the range is wide, ranging from 0.11 to 4.54 kg. Although they only make up 16.0% of the world's population, high-income countries generate 34% of the world's waste, or 683 million tonnes (Kaza *et al.*, 2018).

By 2050, it is predicted that worldwide waste generation will increase to 3.40 billion tonnes, more than double the growth in population during the same period. In general, waste generation and income level are positively correlated. In high-income countries, daily per capita waste generation is projected to rise by 19% by 2050, compared to roughly 40% in low- and middle-income countries. Initially, waste generation declines at the lowest income levels and then rises more rapidly there than at higher income levels when income levels fluctuate incrementally. Low-income nations are expected to produce three times as much waste by 2050 as high-income nations. East Asia and the Pacific produce 23% of global waste, while the Middle East and North Africa produce 6%. By 2050, it is expected that the amount of waste generated in sub-Saharan Africa, South Asia, and the Middle East and North Africa will quadruple, double, and double, respectively. This region currently disposes of more than half of its waste openly, and the trajectory of waste increase will negatively impact the environment, human health, and economic growth (Kaza *et al.*, 2018).

### 17.3.4 LAND DEGRADATION BY 2050

According to a comprehensive, evidence-based analysis, over 75% of earth's land is noticeably degraded, harming 3.2 billion people. Destruction of forests, pollution, and deforestation are all contributing factors to the extinction of species on these lands. Unless this trend continues, by 2050, 95% of earth's land area will be degraded. As food production collapses in many places, hundreds of millions of people could be forced to migrate, the report warns (Leahy, 2018).

## 17.4 MANAGEMENT OF SOIL IN A CLIMATE CHANGE CONTEXT

### 17.4.1 LAND USE CHANGES COUPLED WITH CLIMATE CHANGE

In recent years, land has been used more extensively than ever just now. Demand for land has increased due to the growth of markets, an increase in population, economic development, and higher earnings, which has resulted in an unprecedented change in land usage. Forest cover declines, farming intensifies, and the urbanization of areas have seen the most significant changes (UNEP, 2007).

Many nations rely on the production of animals and crops for their economies. Several nations have implemented renewable energy promotion programs as well as higher and more unpredictable oil prices, which have led to increased incentives for biofuel production (Rulli *et al.*, 2013). North America dominates biofuel production with 48% of the global market. According to the OECD/FAO, Brazil produced 24% of global biofuel production in 2011. The expansion of biofuel production is causing deforestation and other changes in land use.

There are three major causes of inefficiency: lack of clearly defined property rights (Besley, 1995), greater bargaining power exercised by various buyers (Ghebru and Holden, 2012), and a lack of adequate functioning of insurance markets to absorb risk and uncertainties in nature (like climate change) (Dayton-Johnson, 2006) and environmental factors.

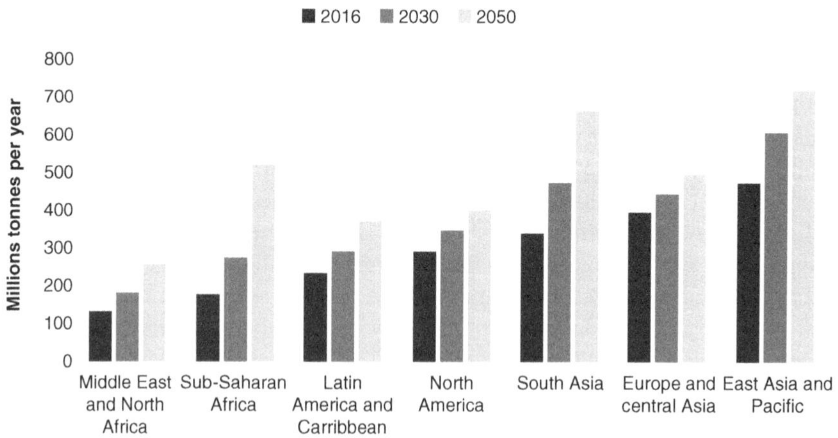


**FIGURE 17.2** Projected waste generation by region (millions of tonnes/year).

Large-scale land acquisitions, or "land grabbing," first emerged as a result of food price increases in 2007–08. The phenomenon has gotten worse since then (IMF, 2008). National governments often support foreign states, businesses, and domestic investors in order to meet the demands of the food supply and the energy sector. Unregulated "land grabs" have caused the eviction of farmers in Africa, Eastern Europe, South America, and South and Southeast Asia (Rulli *et al.*, 2013). Having weak regulatory systems and power imbalances may increase poverty, violence, and social conflict in nations with fertile land as a finite resource (Nolte and Ostermeier, 2015).

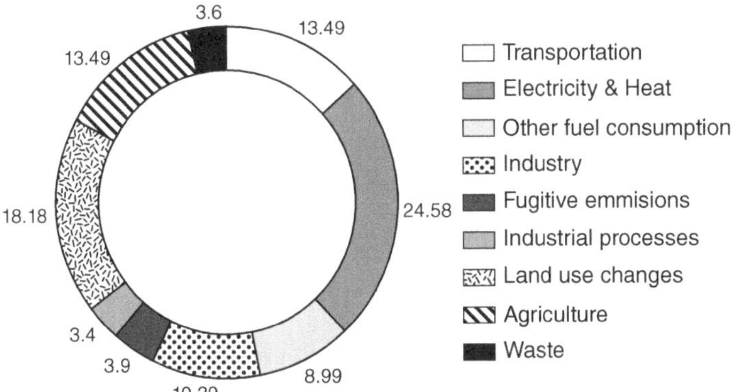



**FIGURE 17.3**  Greenhouse gas emission from different sectors.

**Source: Leahy (2018)**

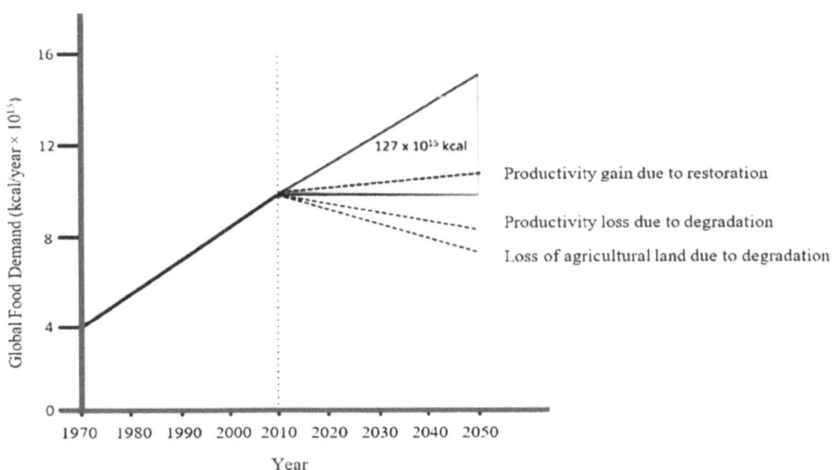



**FIGURE 17.4**  The area of the food wedge as a function of soil changes.

**Source: Keating et al. (2014)**

**TABLE 17.1**

**Sustainable Development and Soil Governance Milestones.**

| | |
|---|---|
| 1982 | "Charter of the FAO World Soil Organization" |
| 1988 | "Intergovernmental Panel on Climate Change (IPCC)" |
| 1992 | "Environment and Development Conference of the United Nations" |
| | "Declaration of Rio" |
| | "The 21st century agenda" |
| | "Environment Facility of the Global Environment" |
| | "Desertification Convention of the United Nations (UNCCD)" |
| | "UNFCCC (United Nations Framework to Combat Climate Change)" |
| | "Convention on Biological Diversity (CBD)" |
| 1997 | "Kyoto Protocol" |
| 2000 | "Millennium Development Goals (MDGs)" |
| 2005 | "Millennium Ecosystem Assessment" |
| 2008 | "UNCCD's Zero Net Land Degradation" |
| 2011 | "Global Soil Partnership initiated (FAO/EU)" |
| 2012 | "Rio+20" |
| | "Sustainable Development Goals (SDGs) and Post-2015 Development Agenda" |
| | "Intergovernmental Technical Panel on Soils (ITPS) of the GSP" |
| | "Updated FAO World Soil Charter" |
| | "Land and Soils integrated in the Open Working Group of the Sustainable Development Goals" |
| | "Regional Soil Partnerships of the GSP" |
| 2015 | "International Year of Soils declared by the UN General Assembly" |

*Source:* Montanarella et al. (2015)

Approximately 39 million hectares of land have been sold to foreign investors since 2000. The Land Matrix Global Observatory database also shows 200 deals covering 16 million ha. Agricultural production remains the primary application of large-scale land acquisitions, with 40% of deals related to food crops and livestock farming. Forestry projects have increased by 50%, and agrofuels rank second with 190 deals (Land Matrix Newsletter, 2014). There were also purchases made for infrastructure, mining, urban development, and tourism (Nolte and Ostermeier, 2015).

### 17.4.2 Soil Global Partnership Actions Based on Five Pillars

"World Commission on Environment and Development, 1987" is most closely associated with the concept of sustainable development in its 1987 report, often referred to as the Brundtland Commission after its chairperson, Gro Harlem Brundtland of Norway. Sustainable growth is growth that meets present requirements without jeopardizing the capacity of future generations to satisfy their own needs. As defined by the World Soil Charter, sustainable soil management means taking steps to maintain or improve soil functions that support, provision, regulate, and provide cultural benefits without adversely affecting biodiversity or soil functions that enable these functions. The "Global Soil Partnership's" first pillar is centered on the idea of soil sustainability: "Promote sustainable management of soil resources for soil protection, conservation, and sustainable production."

1. Protect, conserve, and enhance soil productivity through sustainable soil resource management.
2. Investments, collaborations, policy, and public awareness should be promoted.
3. Identify and develop soil research and development programs that combine environmental, productive, and social development initiatives in order to identify gaps and priorities.
4. Collect, analyze, validate, report, monitor, and integrate soil data and information for better quality and quantity.
5. Conservation and management of soil resources require alignment of techniques, metrics, and indicators.

### 17.4.3 LAND-COVER CHANGE ASSESSMENT CLIMATE CHANGE

Land use trends reveal how land is transferred from one land use to another, thereby influencing soil features as a function of management. In 2010, Earth Observation

---

**TABLE 17.2**

**An Overview of the Effects of Worldwide Tendencies of Soil Management on Water Ecosystem Services**

| Management (global trend) | Provisioning | Regulating | Cultural |
| --- | --- | --- | --- |
| Land use change (from agriculture to urbanization) | Decreased biomass, decreased availability of water for agricultural use | Increased impervious surface, decreased water infiltration, storage, and regulation by soils | Decreased natural environment |
| Increase in intensive grassland (increase in arable land) | Increase in intensive grassland (increase in arable land) | A greater need for water, increased C sequestration, and a stress on downstream waterways' ecosystem health | – |
| Intensification of irrigation | Increased biomass production over dryland agriculture, decrease of urban water supplies | Enhanced C sequestration but decreased filtration potential | Infrastructure alters landscape |
| Increased drainage (increased marginal land) | Decreased soil saturation, increased biomass, and reduced wetland area | Attenuation of floods, denitrification, and carbon sequestration | The potential for recreation (e.g., ecotourism) has decreased |

*Source:* Schlager and Ostrom (1992)

data was used to study land-cover change based on land-cover data from 1994/1995, 2000, and 2005. A study was conducted on changes in five different land-cover classes—urban, mining, forestry, agriculture, and other.

As a result of the land-cover change, there was an increase of 1.2% in changed land, specifically related to urban, cultivation, forestry, and mining. There has been an increase in transformed land in South Africa from 14.5% in 1994 to 15.7% in 2005. Within the past ten years, urban, forestry and plantation, and mining regions have all grown, while cultivation areas have shrunk. There has been a decline in cultivation from 12.4% to 11.9% but an increase in urban areas from 0.8% to 2%, in forestry and plantations from 1.2% to 1.6%, and in mining from 0.1% to 0.2%.

### 17.4.4 CRITERIA AND PRACTICES FOR SUSTAINABLE SOIL MANAGEMENT

However, without complementary strategies and coordinated efforts at regional, national, district, and municipal levels, international consensuses on soil and land resources will not matter. Natural resources, cultural acceptance, and economic viability must be incorporated into effective strategies based on the local context. There is a high level of biological productivity when compared to the potential restrictions imposed by climate and water availability. Earth and its atmosphere are biodiverse. It is the root zone that effectively receives and stores rainwater. In the absence of this kind of information, policymakers and land managers do not have any way of knowing whether they are moving forward or regressing in terms of sustainability. As agricultural practices become more automated and technologically sophisticated, it is generally recommended to avoid mechanical tillage as much as possible to minimize soil disturbance while preserving soil organic matter, soil structure, and overall soil function (FAO, 2013). Improve and maintain the soil's surface with cover crops and crop leftovers, which will shield it, preserve nutrients, and encourage soil biological activity. Including trees, shrubs, pastures, and crops in associations, sequences, and rotations will increase crop nutrition and system resilience. Use well-adapted types with increased nutritional content, resilience to biotic and abiotic stress, and planting parameters such as timing, seedling age, and spacing. By using fertilizer wisely, both organic and inorganic, crop nutrition and soil function can be improved.

Most civilizations have strong ties to the land, and several means of honouring the soil exist (Churchman and Land, 2014). In addition to sustainable soil management, the following may also be considered:

- A soil and land use education program
- Developing and extending soil research
- Payments for ecosystem services, public goods, and private benefits

### 17.4.5 SOIL C POOLS

As a result of the large mixture of organic compounds in the soil, soil C is often divided into several pools (usually between two and five) (Smith *et al.*, 2010). According to Parton *et al.* (1987), several soil C models, including the century model, use a beneficial three-pool structure for soil C (excluding litter) that divides soil C

**TABLE 17.3**

**Global Losses of Soil Carbon (PgC) from 1860 to 2010**

| Model | Tropical | Temperate | Boreal | Global |
|---|---|---|---|---|
| LPJ-GUESS | 12.63 | 15.01 | 0.37 | 29.85 |
| LPJmL | 34.86 | 25.99 | 0.05 | 61.86 |
| ISAM | 17.24 | 37.83 | 5.28 | 60.35 |
| **Mean** | **21.58** | **26.28** | **1.90** | **50.69** |

*Source:* Montanarella et al. (2015)

(including litter) into a labile pool, an intermediate pool, and a recalcitrant (stable) pool. This labile pool of metabolites includes rapidly degradable plant matter, microbial biomass, and labile metabolites, and it may change over the course of a few months or years as the plants degrade. Essentially, the intermediate pool is an accumulation of organic material that has been subjected to microbial processing and has been partly stabilized on mineral surfaces or protected within aggregates, with turnover intervals ranging from decades to several centuries. A recalcitrant pool is composed of mineral complexes that are highly stable and pyrogenic carbon that have the potential to remain in soils for a long period of time. There is often a misconception that individual model pools do not constitute measurable pools on their own (as opposed to the entire stock of C). For estimating the kinetics of the model conceptual pools, laboratory incubations, C dating studies, and tracer studies are used, as well as long-term field experiments. Despite this, solutions to harmonize "measurable" and "modellable" datasets have been discussed for some time (Dungait *et al.*, 2012). To better understand SOC dynamics, this reconciliation is still a desirable objective (Schmidt *et al.*, 2011).

### 17.4.6 Factors Influencing Soil C Storage

In fundamental terms, the amount of soil carbon that accumulates—primarily through exudates and residues—and the amount of carbon that leaves the soil as a result of mineralization (as $CO_2$), fueled by microbial processes, as well as leaching out as dissolved organic carbon (DOC), determine the amount of soil carbon stored. At the local, landscape, and regional scales, soil C can be lost or gained through soil erosion or deposition. Hence, plant residues serve as an important factor in determining how much SOC can be stored in an ecosystem. Through the breakdown of plant biomass (necro mass) during senescence and death, organic C is added to the soil system. Plant residues will, therefore, generally support higher SOC stores when there are higher amounts of their input and vice versa. This is also influenced by the pedoclimatic state. Fertilizers, which are essential to maintaining plant production in agriculture, have an impact on the C levels of many soils as well.

Climate variables including soil temperature and water content have a significant impact on C storage by influencing microbial activity, in addition to productivity and plant C inputs. Water affects soil C accumulation through several mechanisms. Those soils are rich in moisture but well-aerated to facilitate microbial growth. As soils dry out, decomposition rates consequently decline. However, flooded soils often produce soils with extreme quantities of soil C due to decreased aeration (for example, depletion of oxygen in flooded soils) and slower rates of organic matter breakdown (for example, peats and mucks). C may also move down the soil profile as particulate and dissolved organic matter as a result of high precipitation. Under extreme conditions, such as drought, SOM decomposition may initially decrease, but after rewetting, it may increase. In the short term, fire may decrease soil carbon storage, but in the long run, fire may enhance it by promoting plant growth and adding pyrogenic carbon to the soil (Knicker, 2007).

The amount and chemical makeup of soil C may be further impacted by animal or plant bioturbation. Bioturbation is typically used in biologically active areas to boost the transformation and incorporation of organic chemicals into soil, thus enhancing organo-mineral interactions and carbon storage (Wilkinson *et al.*, 2009).

Through the "priming effect," the addition of labile organic materials can either increase or decrease the rate of microbial breakdown of SOM. Positive priming is the mineralization of normally stable C caused by changes in the makeup of the microbial population.

### 17.4.7 THE EFFECT OF PESTICIDES ON SOIL BIODIVERSITY UNDER CLIMATE CHANGE

Pesticides used wisely can affect soil biodiversity directly or indirectly. According to De *et al.* (2014), herbicides account for 47.50% of pesticide use globally, followed by insecticides at 29.50%, fungicides at 17.50%, and others at 5.50%. This increase is largely due to the expansion of agriculture. Studies on the impact of pesticides on soil biodiversity have produced conflicting findings. Variables such as chemical composition, rate of application, soil buffering ability, and target soil organisms affect the effects.

## 17.5 MANAGEMENT OF WASTE IN THE CONTEXT OF CLIMATE CHANGE

As a result of the treatment and disposal of solid waste, 1.6 billion tonnes of carbon dioxide ($CO_2$) equivalent greenhouse gas emissions were produced in 2016, or about 5% of the world's emissions. It is calculated based on the volume of waste produced, its composition, and how it is managed. In open dumps and landfills without landfill gas collection equipment, waste is disposed of in open dumps and landfills. Food waste is responsible for nearly half of all emissions. By 2050, solid waste-related emissions are predicted to increase by 2.38 billion tonnes of carbon dioxide equivalent. Most nations have local governments responsible for managing solid wastes, and almost 70% have institutions for developing policies and monitoring regulations. Solid waste management legislation and rules vary wildly across countries, but almost two-thirds have been adopted. The central government usually only participates in regulatory monitoring or monetary transfers, while 70% of waste services are directly supervised by local authorities. In the waste management industry, at least half of the services are provided by public organizations, and around a third are provided by public–private partnerships.

In order to finance solid waste management systems, operational costs must be considered upfront, more so than capital investments. In high-income countries, integrated waste management expenses typically exceed $100 for each tonne of waste transported, treated, and disposed. In lower-income countries, waste operations cost about $35 per tonne and can sometimes be higher, but they have a much harder time recovering their costs. Transport alone costs $20 to $50 per tonne, and waste management requires a lot of labour. Waste service cost recovery varies greatly across income levels. In low-income countries, user fees range between $35 and $170 per year, with full or near-full cost recovery mostly restricted to high-income countries.

Although waste is a modest source of world emissions, it can become a significant source of emissions savings if it transforms into a recycling. Even when minor emissions are created during waste treatment and disposal, waste prevention and recovery (i.e., recycling wastes) prevent emissions in all other economic sectors. Through a holistic approach to waste management, energy, forestry, agricultural, mining, transportation, and manufacturing sectors can reduce GHG emissions. The "United Nations Environment Programme" (UNEP) has instructed the "International Environmental Technology Centre" (IETC) to take action in waste management.

As a result of waste management methods, greenhouse gases (GHG) are produced, both directly (through process emissions) and indirectly (through energy consumption). As a result, the waste management system's total climate impact or benefit

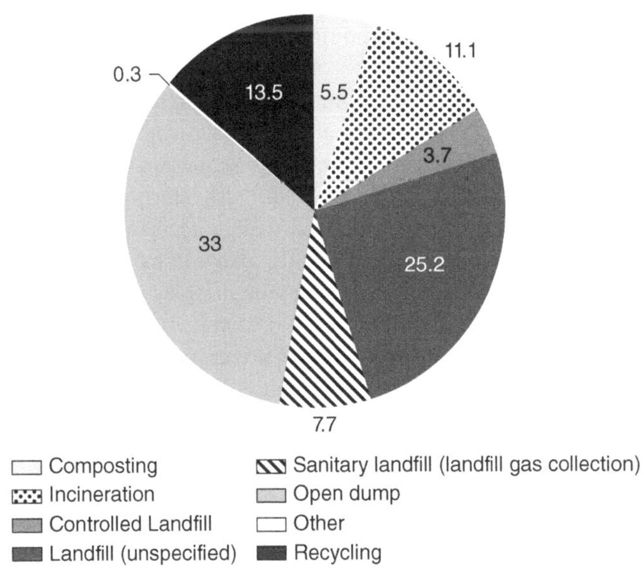


**FIGURE 17.5**   Global percent waste composition.

**Source: Kaza et at. (2018)**

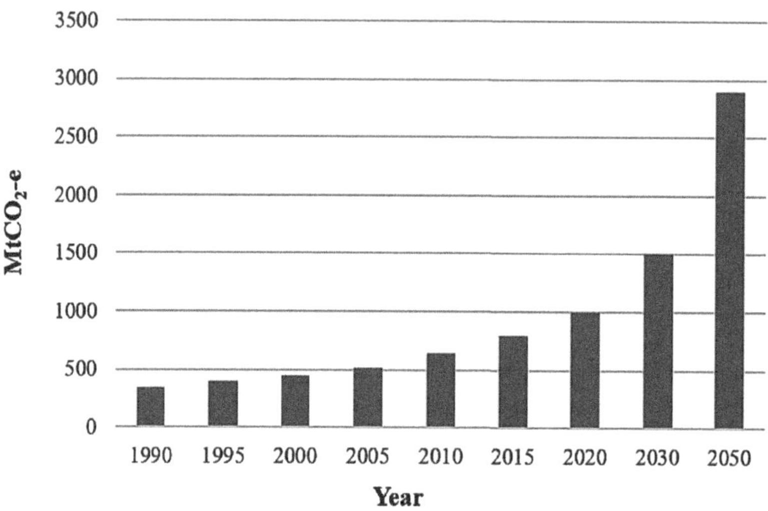

**FIGURE 17.6**   Projected greenhouse gas (GHG) emission from waste.

**Source: UNEP (2010)**

will be determined by net GHGs, meaning both emissions as well as indirect, downstream GHG savings. Due to incomplete information on global waste creation, composition, and management, as well as errors in emissions models, it is challenging to estimate the exact quantity of these emissions. The "Organisation for Economic Co-operation and Development" (OECD) projected that developing countries will produce much more of it as landfill conditions become more anaerobic and conducive to the production of methane.

Using recycled materials and energy instead of virgin materials and fossil fuels to generate energy, reducing the extraction and manufacturing of raw materials, reducing the use of raw materials and manufacturing, sequestering carbon in the soil by applying compost, and storing carbon in landfills by storing recalcitrant waste are all contributing to a sustainable future. It is widely accepted that waste avoidance and recycling are much more environmentally viable than waste treatment, even if energy is recovered during the process. Even though waste prevention is at the top of the "ladder of waste management," it receives relatively little attention and resources.

International organizations, notably UNEP, are currently in charge of several initiatives centered on waste and climate change. UNEP is involved in several partnerships and programs, including integrated waste management, cleaner production, and sustainable consumption and production. Projects under the Clean Development Mechanism (CDM) are also receiving a lot of attention in the waste industry.

By partnering with existing organizations to ensure more successful implementation of projects around the world, UNEP is ideally situated to catalyze increased action to combat climate change in the waste sector. Due to its role as a recognized authority within the United Nations system on environmental issues, UNEP plays a crucial role in fostering partnerships in waste management and climate change.

A variety of stakeholders must contribute to the development of the framework plan to put the suggested method into action.

## 17.5.1  Conventional Hierarchy of Integrated Solid Waste Management

The earth's greenhouse gases (GHGs) have increased dramatically in recent years, which is causing global warming. The reason for this is that global warming has the potential to cause several changes, particularly in the world's climate, which is now understood to pose a serious threat to human civilization. Another global concern is solid waste, which is a significant source of GHGs. Conventional solid waste management needs to be changed to resource management in order to reduce GHG emissions from solid waste (UNEP, 2010).

### 17.5.1.1  Recycling

In the production process, recycled materials are employed in place of virgin ones. Calculating the avoided GHG emissions from remanufacturing using recycled inputs (including the process of collecting and transporting recyclables) by comparing the emissions generated by manufacturing an equivalent amount of the material from 100% virgin inputs and the emissions generated by manufacturing an equivalent amount from 100% recycled inputs (accounting for loss rates) from 100% virgin inputs is based on the difference between the two emissions. During the waste management process, no GHG emissions occur since recycled material is diverted from waste treatment facilities. As a result of the composting, burning, or dumping of the residues made from recycled material, the GHG emissions would be attributed to the item.

### 17.5.1.2  Composting

Composting is a practical solution to treat the organic portion of municipal solid waste (MSW) since it stabilizes biodegradable organic matter and reduces the need to burn it. Composting plants and adding compost to soil produces biogenic $CO_2$ emissions. Carbon molecules that do not break down, however, store carbon for long periods of time. Trees and other plants produce all the waste products that can be composted, including leaves, grass, brush, food scraps, and newspapers. As stated earlier in "$CO_2$ Emissions from Biogenic Sources," biogenic $CO_2$ released during the composting process is not included in the calculation of greenhouse gas emissions. Composting does, however, enhance soil carbon storage because it produces more humic material (natural organic polymers that disintegrate slowly) and several other causes.

The gases like $CO_2$ and $N_2O$ are the two GHGs released during the combustion of waste. The GHG emissions linked to combustion include non-biogenic $CO_2$, but non-biogenic $CO_2$ is emitted during combustion. By subtracting utility GHG emissions from gross GHG emissions, net GHG emissions are calculated. This is because most waste combustors produce electricity that may be used in place of electricity generated by utilities. Combustion or incineration has a lower relative global warming potential (GWP) than composting and landfilling, but it is expensive.

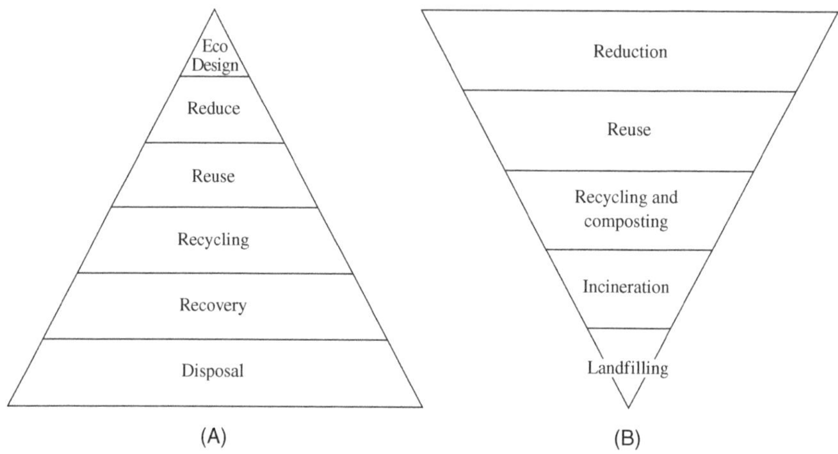



**FIGURE 17.7**   (a) New hierarchy of integrated solid waste management and (b) conventional hierarchy of integrated solid waste management combustion.

### 17.5.1.3   Landfilling

Landfill gas, which is made up of $CH_4$ and $CO_2$ in roughly equal amounts, is a by-product of bacterial breakdown when organic waste is dumped in landfills. Some organic material never even begin to break down; instead, it turns into carbon that is stored. In order to reduce the emissions of these hazardous gases from landfills, gas extraction systems must be installed. However, despite technical efforts to reduce landfill gas emissions, substantial amounts of non-controlled emissions are released into the atmosphere from landfill surfaces. Credit is awarded for the avoided GHG emissions by the electric utility when $CH_4$ is burnt for energy recovery. Some organic waste can receive credit for landfill carbon storage regardless of what happens with $CH_4$. By first collecting $CH_4$ and then burning it off, GHG emissions can be cut by 65%. $CH_4$ can reduce carbon dioxide emissions by 69% when utilized to generate power.

### 17.5.2   New Hierarchy of Integrated Solid Waste Management

### 17.5.2.1   Eco-Design

At all stages of the product development process, eco-design considers environmental aspects, striving to produce products with the lowest possible environmental impact. In eco-design, products are designed with special consideration given to their environmental impact throughout their entire lifecycle. This reduces pressure on solid waste treatment and disposal downstream.

### 17.5.2.2   Reduce

The reduction of GHG emissions through source reduction is generally a very effective way of reducing them. Using power-efficient raw materials reduces GHG emissions more than alternative solutions because they require less waste management and energy-related $CO_2$ emissions.

### 17.5.2.3 Recycling

Most materials can be recycled with the lowest GHG emissions. The recycling of these materials reduces both waste management emissions and energy-related $CO_2$ emissions throughout the manufacturing process (albeit not as significantly as source reduction). Carbon from forests is more effectively sequestered when the paper is recycled.

### 17.5.2.4 Recovery

Solid waste is recovered by obtaining metals and energy produced by the chemical reaction of the waste. It is possible to compost food waste and yard debris as a management strategy. Composting consumes less water and thereby emits fewer greenhouse gases (GHG) than landfilling, but it retains more carbon than landfilling when dealing with yard trimmings (landfilling assumes the carbon storage that occurs when yard trimmings do not decompose completely). Because of the uncertainty associated with the analytical process, the emission factors for composting and burning these items are comparable. Although it could not be the primary method for producing electricity as solid waste incineration, it may be seen as a supplement to traditional power generation in addition to being a waste management solution.

### 17.5.2.5 Disposal

For example, a landfill is used to dispose of waste that cannot be recovered and waste that remains after recycling and recovery. In a perfect world, these wastes would store carbon, such as carbonate, instead of releasing GHGs.

## 17.6 WATER MANAGEMENT IN THE CONTEXT OF GLOBAL CLIMATE CHANGE

Most semi-arid regions with limited water supplies have experienced dramatic increases in water resources over the last few decades, encouraging population expansion, irrigation, and tourism. Adapting to variable water resources in a time of rapid economic and demographic change is a history of Southern Europe, California, and Australia. Freshwater resources play an important role in maintaining the quality of life in these regions, so water management encompasses a wide range of activities (water supply for urban and irrigation demands, hydropower, water quality, flood control, and ecosystem preservation). Pressure on these resources has continuously increased due to withdrawals from various sectors, raising legitimate concerns about traditional water management's sustainability and requiring further work to deal with climate change's upcoming challenges.

But future difficulties will demand more work. The evidence that climate change is having an impact on water resources has only gotten stronger. Future scenarios for these regions may threaten the sustainability of existing water uses by predicting a decline in mean annual runoff and changes in seasonal and interannual variability. Conflicts between users who want a say in water management decisions are frequently caused by pressure on water resources. The need for the proper conservation of aquatic ecosystems is rising as environmental consciousness raises, thus limiting the amount of water available for consumptive activities. For water-scarce places, achieving sustainability under climatic uncertainty presents new problems that call for further broadening the paradigm to consider not just socioeconomic factors but also ecosystem functions and services.

Water scarcity has been understood and practiced more effectively in the past 30 years due to significant advances in both understanding the root causes and developing practical solutions to minimize its negative effects. During the past 30 years, significant progress has been made in understanding the root causes of water scarcity and developing practical solutions to mitigating.

### 17.6.1   Water Availability and Climate Change

As a result, water supply is of paramount importance to arid and semi-arid regions. Many approaches have been taken to studying water availability in the literature. The average runoff is a direct result of variations in water availability as it is the difference between precipitation and evapotranspiration. Geophysical approaches are particularly useful for global comparative research because they can produce indicators of water stress (Alcamo *et al.*, 2007). Water availability may be affected more by geographical and temporal variability than average values in areas with limited water resources.

Streamflow variability contributes to water scarcity in numerous cases. Urban water supply and irrigation are the primary uses of water in semi-arid environments. Regardless of the situation, a consistent supply is necessary to maintain water use, and it is assessed by using appropriate indicators (Hashimoto *et al.*, 1982). The water availability would be zero or very low without infrastructure since it would be based on long-term minimum flows. Since the turn of the century, these areas have been

---

**TABLE 17.4**

**Water Resource Impacts Projected by Climate Change**

| Projected impact | Potential negative effects and consequences for water resources |
| --- | --- |
| All changes in the hydrologic cycle | An assessment of all impacts resulting from the development. Decreased availability of water Risks of water quality loss. Depletion of groundwater. Conflicts between users of water |
| Changes in extremes (floods and water scarcity) | Extreme events are occurring more frequently and in greater magnitude, seasonality is changing, and water shortages have increased |
| Increased water requirements | Urbanization has increased the demand for irrigation, particularly in areas that are already water scarce |
| Effect on water ecosystems | Desertification increases as a result of a decrease in water quality, which has a significant impact on water-scarce countries |

*Source:* Garrote (2017)

---

subjected to substantial infrastructure expenditures. The role of hydraulic infrastructure in supplying water to users is crucial since it overcomes the spatial and temporal abnormalities of natural regimes through the basic tasks of regulation and transportation.

- A close relationship exists between climate and water resources because of the hydrologic cycle, which describes the movement of water from the ocean to groundwater and surface water storage. Hydrological cycles are responsible for the following:
- Evaporation of ocean water
- Condensation of water vapor occurs
- Rain and snow as precipitation
- Lake and surface water runoff
- Upon infiltration into the ground
- Aquifer percolation
- Condensation of water vapor occurs by evapotranspiration at the surface

Here is a compilation of some of the most significant IPCC climate change projections (2007). Water resources in the Caribbean are expected to be affected in varying degrees by these predictions.

The research specifically states the following:

- There is a possibility of a temperature rise between 1.1°C and 6.4°C (33.98°F–43.52°F) by the end of the century, with a probable temperature rise between 1.8°C and 4.0°C (35.24°F–39.2°F)
- Sea levels are expected to upsurge by 28 cm–43 cm

---

**TABLE 17.5**

**Projections of Climate Change in the Future**

| Trends and phenomena | Future trends based on SRES scenarios projections for the 21st century |
|---|---|
| Warmer and fewer cold days and nights over most land areas | Virtually certain |
| Warmer and more frequent hot days and nights over most land areas | Virtually certain |
| Warm spells/heat waves. Frequency increases over most land areas | Very likely |
| Heavy precipitation events. Frequency (or proportion of total rainfall from heavy falls) increases over most areas | Likely |
| Area affected by drought increases | Likely |
| Intense tropical cyclone activity increases | Likely |
| Increased incidence of extremely high sea levels (excludes tsunamis) | Likely |

*Source:* Farrell et al. (2007)

### 17.6.2  CLIMATE CHANGE STRATEGIES FOR MANAGING WATER RESOURCES

There are two categories of water management strategies in the context of global climate change:

1. By adoption
2. By area of interest

## 17.7  CONCLUSION

Human intervention and unsustainable management of natural resources are causing land, water, and other natural resources to degrade. Global environmental sustainability and climate change will be impacted by these changes. There has been a significant increase in groundwater and surface water pollution worldwide, especially in developing countries. There are several major factors causing water pollution, including soil pollution, agricultural runoff, improper solid waste management, reuse of untreated or inadequately treated wastewater, and open defecation due to poor sanitation. A growing concern is soil pollution, which may result from the use of non-sustainable land, the application of excessive fertilizers, and the improper disposal of waste. Hence, a holistic approach based on the nexus model is needed for better wastewater, waste, and soil management to protect water quality.

Human intervention and unsustainable management of natural resources are causing land, water, and other natural resources to degrade. Global environmental sustainability and global climate will be affected by these factors. In waste-soil-water nexus studies, waste is commonly added to soil, and its effects on soil and water are measured with static indicators. Nexus requires a holistic approach to waste generation, soil processes, and water regimes in soils and landscapes. Physical, chemical, and biological soil processes are largely responsible for waste transformation, especially in unsaturated soil. There should be more research on these processes. In numerous studies examining the effects of waste disposal on soil, the waste-soil-water nexus is highlighted as a source of soil erosion and water contamination. In addition

---

**TABLE 17.6**

**Strategy for Adapting to Climate Change by Water Utilities**

| By adaptation action | By area of interest |
|---|---|
| • Build new infrastructure | • Drought |
| • Improve the efficiency of the system | • Infiltration of salt water |
| • Risk modeling for climate change | • Increasing sea levels |
| • Land use should be modified | • A flood occurs |
| • Change the water demand | • Preparedness for general utility services |
| • Operational capabilities monitoring | • Runoff from stormwater |
| • Climate change plan (CCP) | • Sedimentation and erosion |
| • Facility repairs and retrofits | • Blooms of algae |

to enhanced soil and water quality, increased aquifer feeding, and reduced erosion, future efforts should focus on enhancing biomass production.

Future studies should place a stronger emphasis on socioeconomic situations. Applying existing knowledge may accomplish a lot, but the execution is too frequently weak due to stakeholders' lack of interest. The nexus concept advocates for new communication as well as transdisciplinary and multidisciplinary approaches. Overall, the nexus method is very helpful in creating a correct balance between the three components by elucidating the linkages between waste, water, and soil. The position paper's conclusion is that, as a result of global climate change, greater attention must be paid to managing water, soil, and waste.

## REFERENCES


Alcamo, J., Flörke, M., and Märker, M. 2007. Future long-term changes in global water resources driven by socio-economic and climatic changes. *Hydrological Sciences Journal*, 52(2), 247–275.

Besley, T. 1995. Property rights and investment incentives: Theory and evidence from Ghana. *Journal of Political Economy*, 103(5), 903–937.

Boretti, A., and Rosa, L. 2019. Reassessing the projections of the world water development report. *NPJ Clean Water*, 2(1), 1–6.

Churchman, G. J., and Landa, E. R. (Eds.). 2014. *The soil underfoot: Infinite possibilities for a finite resource*. CRC Press.

Dayton-Johnson, J. 2006. *Natural disaster and vulnerability*. OECD Development Centre Policy Briefs 29, OECD Publishing. doi: 10.1787/202670544086

De, A., Bose, R., Kumar, A., and Mozumdar, S. 2014. *Targeted delivery of pesticides using biodegradable polymeric nanoparticles* (pp. 5–6). Springer India.

Dungait, J. A., Hopkins, D. W., Gregory, A. S., and Whitmore, A. P. 2012. Soil organic matter turnover is governed by accessibility not recalcitrance. *Global Change Biology*, 18(6), 1781–1796.

FAO. 2013. *Statistical analysis of select food and agricultural indicators for the near East and North Africa*. FAO.

Farrrell, D., Nurse, L., and Moseley, L. 2007. *Managing water resources in the face of climate change: A Caribbean perspective* (pp. 26–28). UWI.

Garrotc, L. 2017. Managing watcr rcsourccs to adapt to climatc change: Facing uncertainty and scarcity in a changing context. *Water Resources Management*, 31(10), 2951–2963.

Gerland, P., Raftery, A. E., Ševčíková, H., Li, N., Gu, D., Spoorenberg, T., and Wilmoth, J. 2014. World population stabilization unlikely this century. *Science*, 346(6206), 234–237.

Ghebru, H.H., and Holden, S.T. 2012. Reverse share-tenancy and urbanizatio inefficiency: Bargaining power of landowners and the sharecropper's productivity. *International Association of Agricultural Economists* (IAAE) Triennial Conference, Foz do Iguaçu, Brazil.

Hashimoto, T., Stedinger, J. R., and Loucks, D. P. 1982. Reliability, resiliency, and vulnerability criteria for water resource system performance evaluation. *Water Resources Research*, 18(1), 14–20.

IMF, F. 2008. *Macroeconomic impact, and policy responses*. Fiscal Affairs, Policy Development and Review, and Research Departments.

Kaza, S., Yao, L., Bhada-Tata, P., and Van Woerden, F. 2018. *What a waste 2.0: A global snapshot of solid waste management to 2050*. World Bank Publications.

Keating, B. A., Herrero, M., Carberry, P. S., Gardner, J., and Cole, M. B. 2014. Food wedges: Framing the global food demand and supply towards 2050. *Global Food Security*, 3, 125–132.

Knicker, H. 2007. How does fire affect the nature and stability of soil organic nitrogen and carbon? A review. *Biogeochemistry*, 85(1), 91–118.

Land Matrix Newsletter. Land Matrix Newsletter. October 2014. (Also available at www.land matrix. org)

Leahy, S. 2018. 75% of earth's land areas are degraded. *National Geographic*, 26.

Montanarella, L., Badraoui, M., Chude, V., Costa, I. D. S. B., Mamo, T., Yemefack, M., and McKenzie, N. 2015. *Status of the world's soil resources: Main report*. Embrapa Solos-Livrocientífico (ALICE).

Nolte, K., and Ostermeier, M. 2015. Land Investments: A new type of territorial expansion. In Heinrich Böll Foundation, ed., *Soil atlas 2015- Facts and figures about earth, land and fields*. Pp. 38–39. Institute for Advanced Sustainability Studies (IASS).

OECD/FAO. 2011. *OECD-FAO agricultural outlook 2011–2020*. OECD Publishing & FAO.

Parton, W. J., Schimel, D. S., Cole, C. V., and Ojima, D. S. 1987. Analysis of factors controlling soil organic matter levels in Great Plains grasslands. *Soil Science Society of America Journal*, 51(5), 1173–1179.

Rulli, M. C., Saviori, A., and D'Odorico, P. 2013. Global land and water grabbing. *Proceedings of the National Academy of Sciences*, 110(3), 892–897.

Schmidt, M. W., Torn, M. S., Abiven, S., Dittmar, T., Guggenberger, G., Janssens, I. A., and Trumbore, S. E. 2011. Persistence of soil organic matter as an ecosystem property. *Nature*, 478(7367), 49–56.

Smith, P., and Olesen, J. E. 2010. Synergies between the mitigation of, and adaptation to, climate change in agriculture. *The Journal of Agricultural Science*, 148(5), 543–552.

UNEP. 2007. *Global environmental outlook, GEO 4*. UNEP.

UNEP. 2010. *Waste and climate change: Global trends and strategy framework*. UNEP.

Wilkinson, M. T., Richards, P. J., and Humphreys, G. S. 2009. Breaking ground: Pedological, geological, and ecological implications of soil bioturbation. *Earth-Science Reviews*, 97(1–4), 257–272.

World Commission on Environment and Development. 1987. *Report of the World Commission on environment and development: Our common future*. Oxford University Press. 383 pp.

# 18 Management of Urban Polluted Soil to Enhance Resource Use Efficiency in Developing Cities

*Yamini S., V.K. Paswan, Neha, and S. Rohith*

## 18.1 INTRODUCTION

According to United Nations and Social Affairs (2019), the global population is projected to cross 9 billion by 2050. Most of the increase will be seen in urban areas. The average annual rate of change of the percentage urban according to Nations (2018) globally will be 0.58% by 2045–2050. Urban land areas are expected to grow up to 80%. Principal growth will be visible in developing countries (Mahendra and Seto 2019). This rapid growth rate of urbanisation and industrial sector in and around cities has affected one of the most important integrant of earth—i.e., soil—and converted it to a human-dominated ecosystem from being a natural ecosystem. Urbanisation no doubt has several advantages like health care, sanitation, transportation, and convenience, but it also negatively affects natural systems which cannot be overlooked (Yu et al. 2012).

Urban soil, also called as anthropic soil, is strongly influenced by human activities and differs from natural soil in terms of texture, buffering capacity, microbial diversity, heterogeneity, etc. The application of urban soil includes activities related to the urban environment and industry, forestry and agriculture, and urban green spaces such as parks, gardens, etc.

Urban soil gets contaminated through a variety of anthropogenic activities such as emission from transportation, coal combustion, and industrial activities such as mining, metallurgy and waste disposal, incineration, etc. Soil contaminants can be divided into organic or inorganic.

Organic pollutants include PAHs (polycyclic aromatic hydrocarbons), PCBs, (polychlorinated biphenyls), OCPs (organochlorine pesticides), PAEs (phthalic acid esters), and BFRs (brominated flame retardants), and inorganic pollutants include mostly heavy metals such as cadmium (Cd), arsenic (As), chromium (Cr), copper (Cu), zinc (Zn), nickel (Ni), and lead (Pb).

People living in urban cities get exposed to contaminated urban soil either directly or indirectly; two general pathways seen are soil-human pathway and soil-plant-human pathway. Children and elderly are at more risk from contaminants in soil than adults (Li et al. 2018).

DOI: 10.1201/9781003358169-18

Assessment of contaminants is important to know their bioavailability and risks associated with them (Li et al. 2018). Contaminants can enter in human body through ingestion or oral intake and inhalation (Adimalla 2020). Traditional methods are not reliable methods of assessment. Nowadays, *in vivo* bioassay and *in vitro* chemical method are mostly adopted. *In vivo* bioassays are reliable and accurate but very difficult to perform due to the complexity of metabolism, species, and extraction of contaminants.

Till date, *in vivo* has been performed on animal models for the assessment of pollutants, whereas *in vitro* methods are fast, cheap, and ethical. Examples of *in vitro* methods include SHIME (simulation of the human intestinal microbial ecosystem), PBET (physiologically based simulation test), and FOREShT (fed organic estimation human simulation test) (Juhasz et al. 2016, Li et al. 2018).

Urbanisation has led to decrease interaction between human and natural environment including natural soil microbiota. "Biodiversity hypothesis" states that plenty of natural microbiota in soil protects human from many types of allergies and auto-immune disorder, for instance, people living in rural areas have more diversity of bacteria on their skin which might offer beneficial effects on their immune system than people living in urban areas or apartments. A decrease in natural microbial biodiversity poses harmful effects on public health like asthma and other atopic diseases.

Furthermore, soil microbial biodiversity is an indicator of soil health. Another factor that contributes to decreased interaction is the sedentary lifestyle of major urban populations (von Hertzen and Haahtela 2006, Li et al. 2018). To maintain the quality of the soil and to prevent its degradation, a soil policy must be put in place. One such policy is the Global Soil Partnership (GSP) launched by the United Nations and Agriculture Organization in 2012.

To prevent damage of soil quality, it is important to plan activities carefully that can have detrimental effects on soil quality such as during urban management practices, land use planning, proper disposal system, etc. As contaminated soil not only harms humans but also hampers plant and animal productivity, lowers water and air quality, and adversely affects the natural ecosystem (Li et al. 2018).

## 18.2　URBAN SOIL

Urban soil or anthropic soil is formed due to human activities such as mixing, importing, and exporting material. Urban soils are characterized by heterogeneity due to the presence of various exogenous materials in the original soil. Urban soils are distributed in parks, roads, sport fields, urban and peri-urban areas, near buildings, etc. (Li et al. 2018).

Urban soils are comprised of three parts:

1. Soils that are formed due to mixing, importing, and exporting or by contamination and are very different from agricultural or forest areas;
2. Soils closer to agricultural soil but vary to some extent in properties like composition and use. They are mainly found in parks and gardens; and
3. Soils formed due to construction activities in cities.

In cities, the center region has more soil affected by anthropic activities than the periphery region (Prabhakaran et al. 2022).

The soil formation process generally involves three main processes: weathering, transporting, and accumulating. Three factors that majorly affect the formation process include climatic condition, parent rock, and topography. The same process and pathway are followed by urban soil, too, except that urban soil formation is greatly influenced by human activities (Elgarahy et al. 2021). Urban soil composition depends on the parent material from which it is formed. Urban soil in cities near parks and gardens support the growth of plants and trees for which air and water supply is an important requirement. Bulk density, amount of mottles (medium, coarse, or very coarse), organic matter content, and texture affect the growth rate (Joimel et al. 2021).

Parks, gardens, green corridors, wetlands, plant nurseries, and so on all come under urban green space. They are supported by urban soil; also, urban waste management plays an important role for the betterment of green space. For instance, the production of biochar improves the physiochemical properties of soil that will eventually show positive effects on plant and tree growth (Li et al. 2018). As already mentioned regarding the significance of human and microbial interaction, microbes also

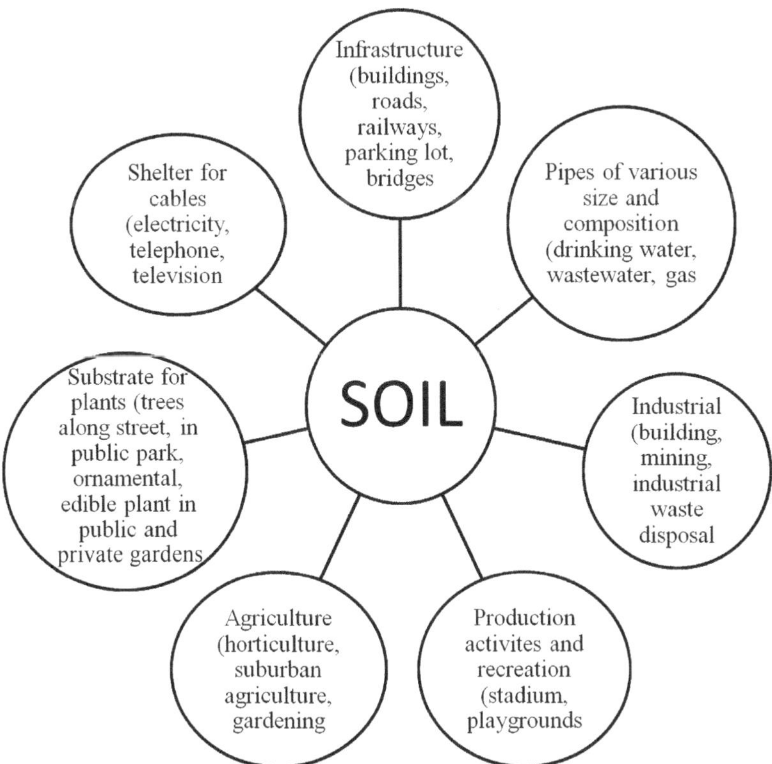


**FIGURE 18.1**  Shows the potential use of soil in urban areas.

play an important role in plant growth. The urban heat island effect is the increase in ambient temperature of urban environments which also increases the soil temperature near the vicinity, considerably affecting the microbial activity in soil. Soil has a good buffering capacity against perturbation, but due to anthropic activities, the urban environment gets overloaded and disturbs the balance (Li et al. 2018).

## 18.3 URBAN AGRICULTURE

Refers to food production in urban areas or densely populated areas and plays a notable role in social, economic, and ecological quality. It can be a continuous and secure source of clean food for all the people of the world, especially in view of global pandemics like COVID-19 which hampered the transportation, distribution, and availability of food commodities across the world. In developed countries, urban agriculture is seen more as a leisure activity or ecological activism. Even though it promotes food security along with social, cultural, and ecological factors. Apart from traditional methods like soil or soil-based medium, other advanced systems are also used, such as hydroponics combined with vertical gardening and LED light systems (Salomon et al. 2020, Sarkar et al. 2020).

## 18.4 CLASSIFICATION OF URBAN SOIL CONTAMINANTS/POLLUTANTS

Urban soil contamination can occur through a number of activities, and most of them are due to anthropic activities, such as transportation, coal combustion, industrial activities like mining or metallurgy, waste disposal either from industry or household, etc. Also, air and water can be carriers for the accumulation of contaminants in urban soil through wind or water runoff. Generally, urban soil contaminants are grouped into two: organic pollutants and inorganic pollutants.

### 18.4.1 Organic Pollutants

Organic pollutants found in soil include polychlorinated biphenyls (PCBs), polycyclic aromatic hydrocarbons (PAHs), organochlorine pesticides (OCPs), phthalic acid esters (PAEs), brominated flame retardants (BFRs), organophosphates, organic fuels, insecticide, pesticide, herbicides, and others (McKone and Maddalena 2007).

1. **PAHs**
   PAHs are introduced into the soil mostly as a result of environmental and anthropogenic factors that have existed in soil for a long period (Belykh et al. 1998). PAHs are classified as follows, as illustrated in Figure 18.2.
2. PCBs (polychlorinated biphenyls), like PAHs, remain in soil for an extended period of time; they are not naturally present in soil or the environment but are entirely introduced into the environment as a result of anthropogenic activities such as waste combustion, transformer and capacitor improper disposal, and accidental spills (Ross 2004). PCB manufacture was prohibited in 1977.

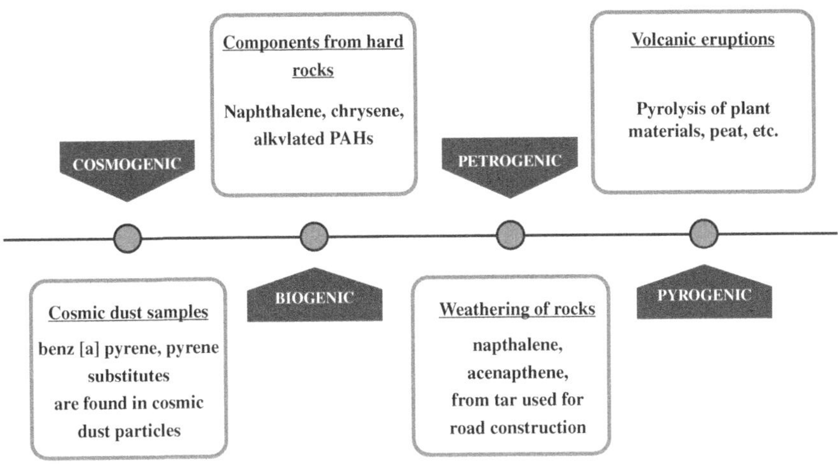



**FIGURE 18.2**  Sources of polycyclic aromatic hydrocarbons (PAHs).

3. Next is OCPs (organo-chlorine pesticides). It includes hexachlorocyclohexane isomers and dichlorodiphenyltrichloroethane and metabolites (Erickson and Kaley 2011). They enter into the environment either by evaporation or as dust and can be transported to far distances (Adu-Kumi et al. 2010).

## 18.4.2  HEAVY METALS IN URBAN SOIL

Soil heavy metals (toxic metals) get incorporated in urban soil due to human activities and threaten not only the environment but also human beings. Heavy metals enter the human body through ingestion, inhalation, and dermal contact. To assess the heavy metal contamination of the soil index of geo-accumulation, $I_{geo}$ is used, given by Muller in 1969; $I_{geo}$ is grouped into seven classes (summarized in Table 18.1).

Heavy metal concentration occurs in a heterogeneous way and differs in concentration along vertical and horizontal distribution.

The horizontal distribution is divided into three categories:

1. From industry and mining processes, i.e., point pollution;
2. Road transport, i.e., belt pollution; and
3. Dust or particle discharge from industry or due to fuel combustion termed as non-point pollution.

For vertical distribution, the highest concentration is found in the upper layer of urban soil (Yu and Li 2012, Yu et al. 2012). Natural soil has better buffering capacity than contaminated soil, so heavy metal in natural soil is more stabilized due to the fact that they get strongly bound to either organic or inorganic ligands and form metal complexes, becoming less bioavailable (Li et al. 2018).

Also, another process that inhibits mobilization of heavy metals is a shift in pH to alkaline, creating a geochemical barrier in the topsoil (Morel et al. 2005).

**TABLE 18.1**

**Index of Geo-Accumulation ($I_{geo}$)**

| Class | Nature | Range |
|---|---|---|
| Class 0 | Unpolluted | $I_{geo} \leq 0$ |
| Class 1 | Unpolluted to moderately polluted | $0 \leq I_{geo} \leq 1$ |
| Class 2 | Moderately polluted | $1 \leq I_{geo} \leq 2$ |
| Class 3 | Moderately to heavily polluted | $2 \leq I_{geo} \leq 3$ |
| Class 4 | Heavily polluted | $3 \leq I_{geo} \leq 4$ |
| Class 5 | Heavily polluted to extremely polluted | $4 \leq I_{geo} \leq 5$ |
| Class 6 | Extremely polluted | $I_{geo} \geq 5$ |

*Source:* Muller (1969)

The decreasing order of heavy metal concentration found is industry and traffic > residential, commercial, and administrative areas > recreational and scenic areas (Yu et al. 2012, Cai et al. 2013). Potential sources of metal in soil include industries, urban cities, emissions from traffic, power plant, brake abrasion, and wear of tyres. Also, fly ash produced during the combustion of coal carry heavy metals that can accumulate on topsoil surfaces either through wet or dry deposition (Li et al. 2018).

Antibiotic-resistant genes (ARGs) are found in urban soil. Due to the shortage of water supply in urban cities and the ever-increasing population, reclaimed water are used in irrigation. Reclaimed water contains both antibiotics and heavy metals which increase the antibiotic resistance in the environment and also develop antibiotic resistance in bacteria through co-selection (Li et al. 2018). It is also evident from research that reclaimed water may contain human pathogens such as *Enterococcus faecium* that might potentially harbour some ARGs (Arias and Murray 2009).

## 18.5 URBANISATION

The change in the size, density, and heterogeneity of cities is referred to as urbanisation. Segregation, population mobility, and industrialisation are all typically associated with urbanisation (Leviton et al. 2000). Urbanisation, for instance, may involve the creation (or destruction) of new structures or neighbourhoods, the creation or destruction of transit links, as well as the inflow and outflow of people, which alters the racial and ethnic makeup. The urbanisation process results in distinctive characteristics of urban environments that demand further research, such as jobs, housing, food, water, sewage, transportation, and health care.

Urbanisation has its own set of advantages and disadvantages. Urban cities offer a better standard of living, better facilities and accessibility to education, health safety, and employment opportunities, but excessive urbanisation often leads to deforestation, disease outbreak, environmental degradation, and global warming.

### 18.5.1  Advantages of Urbanisation—the possibilities

- Increased employment opportunities.
- Increased efficiency in land use and service delivery processes, resulting in lower per-person costs and labour to provide essential services like power and water.
- Encourages the expansion of the economy, trade, and tourism.
- Provides a setting for social integration, allowing people from different backgrounds, organisations, religions, and socioeconomic levels to coexist and work together.
- Serves as information hubs by offering the tools necessary to educate and develop human resources, enabling people to exchange ideas and follow the professions of their choosing and enhancing their economic circumstances.

### 18.5.2  Disadvantages

- The rapid rise in India's urban population has resulted in a number of unanticipated or unplanned consequences, including an increase in slums, a lower standard of living, an increase in air pollution, urban sprawl and traffic demand, and environmental degradation, as well as a demand for rising income and comfortable living.
- There are also the traditional issues that arise from an unmanaged and unexpected population increase, such as unemployment, changes in family and social structures, and an increase in crime rates. In such scenarios, cities that are unprepared for such an onslaught are bound to face a crisis, all of which has a negative impact on citizens' quality of life, perpetuating a vicious cycle.

## 18.6  URBANISATION IS HAZARDOUS TO HUMAN HEALTH

The earlier notion of urbanisation providing quality of life and happiness is now being challenged due to problems associated with it, such as the following:

1. Climate change
2. Changes in land use pattern
3. Accentuates loss of biodiversity
4. Change in the food habits of humans which is proven to be hazardous for the social, physical, and mental health of humans

Every megacity has its own set of issues, such as Delhi, which scored poorly on the Ease of Living Index due to air pollution, overcrowding, water and sanitation issues, and a flawed solid waste management system.

Mumbai and Chennai are not immune to floods and draughts at different times of the year. This will only result in 50% of India's population living in urban areas by 2030.

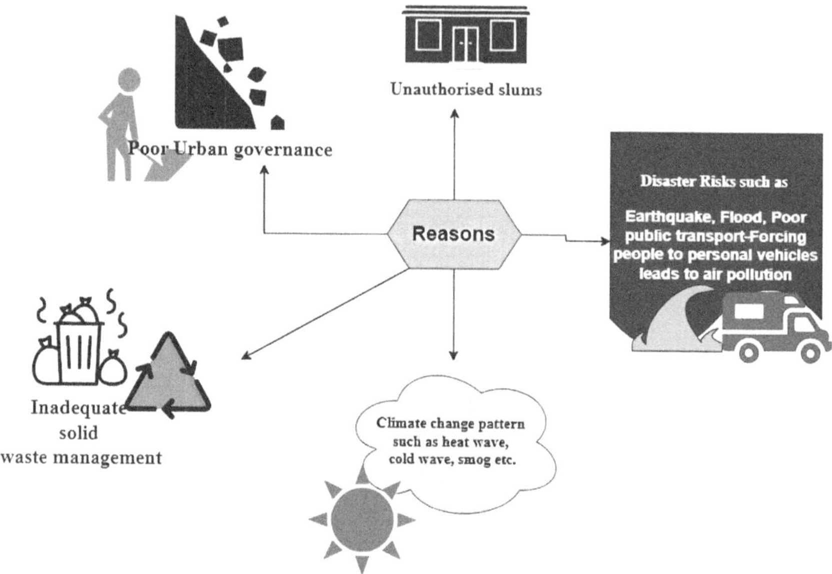


**FIGURE 18.3**    Major reasons involved in an unsustainable development model.

There are several significant factors that have contributed to blind urbanisation, an unsustainable development paradigm, and a risk to human health as shown in Figure 19.3.

All of these concerns have resulted in major health conditions that have reduced life expectancy by 8.5 years, and India, which accounts for 18% of the world population, is responsible for 26% of worldwide premature death, according to the Global Burden of Disease Report 2017.

People living in urban areas cannot avoid exposure to harmful materials in the urban soil. The hazard of exposure to urban soils can be assessed in two different stages: the assessment of exposure and the hazard from urban soil pollutants. These are exposed in two different pathways: the soil-human pathway and the soil-plant-human pathway.

- Soil can affect human health in many ways; these ways can be direct or indirect. These soils may be contaminated naturally or through anthropogenic activities. Excess of any element can be toxic to humans. There is an optimal level of all essential elements, and crossing this optimal level can even make essential elements toxic. These soils may be contaminated naturally or through anthropogenic activities.
- Pollutants in urban soil are a major risk to residents, as they are easily exposed to them, especially children who are at a high risk of consuming soil. Emphasis must be laid on the risks to children, as they are the most sensitive to pollutants and are at a greater risk than adults. Children, given their less body weight, are at higher exposure to soil pollutants per unit of body.

There is also more hand-to-mouth behavior than adults (Li et al. 2018). The typical issue for most children and a widely investigated issue is lead poisoning from exposure through inhalation of dust and soil ingestion. Lead exposure happens mostly from ingestion and is absorbed in the intestine.

- More than 50% of ingested lead enters into children's tissues as compared to 5% in adults. This is due to their poorly developed gastrointestinal system. Lead exposure can lead to complex problems in children like mental retardation, learning disorders, attention deficit hyperactivity disorder, and so on. A strong relation has been identified by epidemiologists in lead-polluted indoor dust and blood lead concentration in children. Considering the sensitivity of children towards lead poisoning, a national-level action should be taken to reduce prenatal and childhood exposure to lead.

- In a paper titled "Heavy metals pollution assessment and its associated human health risk evaluation of urban soils from Indian cities: A review," Narsimha Asimall states that the non-carcinogenic health risk from copper (Cu), zinc (Zn), nickel (Ni), and lead (Pb) in most urban regions was lower than threshold value (HI < 1) which indicates no non-carcinogenic health risk for adults and children. Arsenic (As) and chromium (Cr) non-carcinogenic risk was higher in children than in adults. The risk value of arsenic and chromium was also higher than threshold level. This shows that As and Cr in urban soil posed a considerable non-carcinogenic health risk on urban residents. The total carcinogenic risk due to Pb in most urban regions was lower than the recommended limit of 1.00E-04. Cr and As did show potential cancer risk for both children and adults, making As and Cr sole heavy metals the cause for potential health risk in most urban regions in India. This was also proved in a similar work conducted by Adimalla et al. (2020).

## 18.7  POLICY FOR URBAN SOIL MANAGEMENT

- With rapid urbanisation, more and more agricultural land is being converted into commercial land (Deng et al. 2009). Urbanisation is increasing the risk of soil pollution through acid and waste deposition. Untreated sewage of industrial and municipal refuses, excessive use of agrochemicals and their leaching into soil, and irrigation of sewage and polluted water are some of the major causes of soil pollution.

- Pre- and post-contamination management strategies can be followed to reduce soil pollution.

- Some of these strategies are generation of eco-friendly approaches to farming where focus must be laid on crop rotations, crop residues, usage of organic manures, cultivation of legumes, green gram, etc. Formulants made from natural ingredients derived either from animals or plants or microorganisms should be encouraged to be used by farmers close to urban areas.

- Recycling and recovery of useful material can be done to reduce the amount of waste ending up in landfills. Hung et al. used gamma radiations to sterilize medical equipment in Hanoi City and reused the waste as a carrier material of inoculants. Composting is one of the economic and environmental-friendly methods that can be adopted in poor and developing nations with limited resources.

- Nutrient-rich manures produced with the help of macro fauna and micro-flora will help in restoring the soil condition. Sanitary landfilling is another good option for disposing municipal solid waste (MSW).
- Thin layers of compacted solid waste are disposed of and then they are covered with liners made of plastic or a suitable earth substance using foam as defenses against surface or groundwater contamination, dust, wind-borne trash, odour, exposed fire, bird threats, rodents or pests, greenhouse gases emissions, unstable slopes, and erosion. This leachate is sent for treatment, and the produced methane gas can be further used for electricity generation. As they bind with metal ions and alter their speciation forms in soils, humic substances (HS), which are pervasive in the natural environment and have a high level of stability, can be used to remediate soils contaminated with heavy metals. The effectiveness of soluble HS and ethylenediamine-tetraacetic acid was compared by Borggaard et al. (2011).
- The removal of cadmium (Cd), copper (Cu), nickel (Ni), and lead (Pb) from a heavily polluted calcareous urban soil was studied by Sarkar et al. Their main goal was to replace expensive manufactured chemicals with less expensive naturally occurring HS compounds, as cleaning agents had been discovered to extract up to 45%, 54%, 17%, and 4% Cu, Ni, and Pb combined, respectively.
- Government should also focus on plantation, social, agroforestry, and watershed programs. Awareness in public must be created on following conservation agriculture, crop rotation, conservation tillage, livestock production, etc., to protect the dwindling indigenous technologies. Participatory rural appraisal programs can be conducted to encourage local to people to suggest proper solutions. Bioengineering techniques like micro-remediation, vermi-remediation, and phytoremediation can be followed for cleaning polluted lands and are also useful in stabilizing the eroded lands.

## 18.8 CONCLUSION

Urban soils are an essential part of the urban ecology and have a direct impact on health. The usual organic and inorganic contaminants as well as antibiotic-resistant genes (ARGs) have been recognised as issues with urban soils. Although the links between contaminants, ARGs, and human health are widely recognised, there is still considerable work to be done. Inhalation, skin contact, and ingestion of soil and food cultivated in urban soil were found to be the main routes of exposure to pollutants, according to earlier research on pollutants in urban soils that concentrated on their distribution, bioavailability assessment, and health hazards.

However, in urban soils, people are typically exposed to many chemicals at once rather than just one pollutant. Our comprehension of their comprehensive dangers can be improved by including multi-chemical exposure and bioavailability contaminants in urban soils in the context of health risk assessment. It is unknown how long-term, low-dose exposure to contaminants through urban soil will affect one's health.

It is necessary to do epidemiological research in order to clarify the long-term dose exposure of contaminants in urban soils.

Future research should consider not only traditional chemical pollutants but also newly emerging ones, such as ARGs, as well as the links between urban green spaces and human health, even though some indirect effects have already been observed. To sustain human health in the context of the rapid urbanisation of the world, proper urban soil management and policy should be developed.

## REFERENCES


Adimalla, N. (2020). Heavy metals pollution assessment and its associated human health risk evaluation of urban soils from Indian cities: A review. *Environmental Geochemistry and Health*, *42*(1), 173–190.

Adimalla, N., Chen, J., & Qian, H. (2020). Spatial characteristics of heavy metal contamination and potential human health risk assessment of urban soils: A case study from an urban region of South India. *Ecotoxicology and Environmental Safety*, *194*, 110406.

Adu-Kumi, S., Kawano, M., Shiki, Y., Yeboah, P. O., Carboo, D., Pwamang, J., . . . Suzuki, N. (2010). Organochlorine pesticides (OCPs), dioxin-like polychlorinated biphenyls (dl-PCBs), polychlorinated dibenzo-p-dioxins and polychlorinated dibenzo furans (PCDD/Fs) in edible fish from Lake Volta, Lake Bosumtwi and Weija Lake in Ghana. *Chemosphere*, *81*(6), 675–684.

Arias, C. A., & Murray, B. E. (2009). Antibiotic-resistant bugs in the 21st century—a clinical super-challenge. *New England Journal of Medicine*, *360*(5), 439–443.

Belykh, L. I., Seryshev, V. A., Penzina, E. E., Belogolovova, G. A., & Khutoryanskii, V. A. (1998). Benzo[a] pyrene content in some soils of Irkutsk Oblast. *Eurasian Soil Science*, *31*(3), 334–341.

Borggaard, O.K., Holm, P.E., Jensen, J.K., Soleimani, M. & Strobel, B.W., 2011. Cleaning heavy metal contaminated soil with soluble humic substances instead of synthetic polycarboxylic acids. *Acta Agriculturae Scandinavica, Section B-Soil & Plant Science*, *61*(6), 577–581.

Cai, Q. Y., Mo, C. H., Li, H. Q., Lü, H., Zeng, Q. Y., Li, Y. W., & Wu, X. L. (2013). Heavy metal contamination of urban soils and dusts in Guangzhou, South China. *Environmental Monitoring and Assessment*, *185*(2), 1095–1106.

Deng, J. S., Wang, K., Hong, Y., & Qi, J. G. (2009). Spatio-temporal dynamics and evolution of land use change and landscape pattern in response to rapid urbanisation. *Landscape and Urban Planning*, 92(3–4), 187–198.

Elgarahy, A. M., Akhdhar, A., & Elwakeel, K. Z. (2021). Microplastics prevalence, interactions, and remediation in the aquatic environment: A critical review. *Journal of Environmental Chemical Engineering*, *9*(5), 106224.

Erickson, M. D., & Kaley, R. G. (2011). Applications of polychlorinated biphenyls. *Environmental Science and Pollution Research*, *18*(2), 135–151.

Hung, N. M., Nhan, D. D., Quynh, T. M., & Thuan, V. V. (1998). Gamma radiation sterilization of municipal waste for reuse as a carrier for inoculant. *Radiation Technology for Conservation of the Environment*, 331–337.

Joimel, S., Cortet, J., Consalès, J. N., Branchu, P., Haudin, C. S., Morel, J. L., & Schwartz, C. (2021). Contribution of chemical inputs on the trace elements concentrations of surface soils in urban allotment gardens. *Journal of Soils and Sediments*, *21*(1), 328–337.

Juhasz, A. L., Scheckel, K. G., Betts, A. R., & Smith, E. (2016). Predictive capabilities of in vitro assays for estimating Pb relative bioavailability in phosphate amended soils. *Environmental Science & Technology*, *50*(23), 13086–13094.

Leviton, L. C., Snell, E., & McGinnis, M. (2000). Urban issues in health promotion strategies. *American Journal of Public Health, 90*(6), 863.

Li, G., Sun, G. X., Ren, Y., Luo, X. S., & Zhu, Y. G. (2018). Urban soil and human health: A review. *European Journal of Soil Science, 69*(1), 196–215.

Mahendra, A., & Seto, K. C. (2019). *Upward and outward growth: Managing urban expansion for more equitable cities in the global south.* Working Paper, World Resources Institute, Washington, DC. Available online at www.citiesforall.org.

McKone, T. E., & Maddalena, R. L. (2007). Plant uptake of organic pollutants from soil: Bioconcentration estimates based on models and experiments. *Environmental Toxicology and Chemistry: An International Journal, 26*(12), 2494–2504.

Morel, J. L., Schwartz, C., Florentin, L., & De Kimpe, C. (2005). Urban soils. In *Encyclopedia of soils in the environment*, Academic Press, Cambridge, Massachusetts, USA, (pp. 202–208).

Muller, G. M. M. G. M. G. M. G. P. (1969). Index of geoaccumulation in sediments of the Rhine River. *Geojournal, 2*, 108–118.

Prabhakaran, A., Meenatchi, R., Pal, S., Hassan, S., Bramhachari, P. V., Kiran, G. S., & Selvin, J. (2022). Soil microbiome: Characteristics, impact of climate change and resilience. In *Understanding the microbiome interactions in agriculture and the environment* (pp. 285–313). Springer.

Ross, G. (2004). The public health implications of polychlorinated biphenyls (PCBs) in the environment. *Ecotoxicology and Environmental Safety, 59*(3), 275–291.

Salomon, M. J., Watts-Williams, S. J., McLaughlin, M. J., & Cavagnaro, T. R. (2020). Urban soil health: A city-wide survey of chemical and biological properties of urban agriculture soils. *Journal of Cleaner Production, 275*, 122900.

Sarkar, S. L. G., Pineda-Martos, R., Timpe, A., Pölling, B., Bohn, K., Külvik, M., . . . Junge, R. (2020). Urban agriculture as a keystone contribution towards securing sustainable and healthy development for cities in the future. *Blue-Green Systems, 2*(1), 1–27.

United Nations, D. o. E., & P. D. Social Affairs (2019). *World population prospects: The 2019 revision, key findings and advance tables.* Working Paper No. ESA/P/WP/248."

vonHertzen, L., & Haahtela, T. (2006). Disconnection of man and the soil: Reason for the asthma and atopy epidemic? *Journal of Allergy and Clinical Immunology, 117*(2), 334–344.

Yu, S., & Li, X. D. (2012). The mobility, bioavailability, and human bioaccessibilityof trace metals in urban soils of Hong Kong. *Applied Geochemistry, 27*(5), 995–1004.

Yu, S., Zhu, Y. G., & Li, X. D. (2012). Trace metal contamination in urban soils of China. *Science of the Total Environment, 421*, 17–30.

## 19 Engineered Nanoparticles for Plant-Originated Environmental Consequence

*Sushmita Thokchom, Guntamukkala Sekhar,
Prem Kumar Bharteey, and Sumit Rai*

### 19.1 INTRODUCTION

Nanotechnology is an emerging tool to improve crop productivity and quality in agriculture. It brings the next industrial revolution, and its implementation is important for pest management, plant growth, nutrient shortage and climate change. Through nanotechnology, the nanomaterial developed nanosensors which can detect pathogens at levels as low as parts per billion and has the potential and revolutionary action in several ways. Due to their incredible properties of engineered nanomaterial with its potential of slow and targeted release with a large surface area, it has the power to replace the current methods of pest management like chemical pesticides using nanosensors, nanopesticides and nanoherbicides (Joshi and Somdutt 2019). Any material when convert to nano size changes into a new property which lacks a macro-scale form which includes great surface to volume ratio and increases biochemical and reactivity activity. It also has the property of quantum mechanical effects. The physical properties of nanomaterial make the melting point lower for smaller particles. Engineered nanomaterial has several changes like settling, dissolution and agglomeration when exposed which are difficult to estimate accurately. It increases surface area and can easily absorb nutrient medium molecules of organic and inorganic ions which cause phytotoxicity like chlorosis and wilting (Begum et al. 2012; Slomberg et al. 2012). The biologically engineered nanoparticles targeted the drug delivery, also in the analysis of DNA and gene therapy, enhanced reaction rates, magnetic resonance imaging (MRI) and in cancer treatment (Das et al. 2017). The uses of inorganic and organic salts are less, but in chemical synthesis of engineered nanomaterials, it is high. Hence, the process of biological synthesis of engineered nanomaterials is called "green synthesis". Plants' responses to engineered nanomaterials involve chemical interactions that involve the disruption of ion cell membranes for

DOI: 10.1201/9781003358169-19

**TABLE 19.1**

**Types of Engineered Nanomaterial used in Plants**

| Engineered nanomaterial | Uses | Reference |
| --- | --- | --- |
| Carbon nanomaterials | Multiple uses | Elmer and White 2018 |
| Metalloids Si, B | Fungicides/Bactericides/Nanofertilizers genetic material delivery system | Elmer and White 2018 |
| Metallic oxides Au, Ag, Cu, Al, Fe, Ni, Mg, Ti, Zn and Mn | Fungicides/Bactericides/Nanofertilizers genetic material delivery system | Elmer and White 2018 |
| Nonmetal S | Fungicides/Bactericides/Nanofertilizers genetic material delivery system | Elmer and White 2018 |
| Multi-walled and single-walled nanotubes | Genetic material delivery system and antimicrobial | Elmer and White 2018 |
| Fullerenes | Genetic material delivery system and antimicrobial | Elmer and White 2018 |
| Nanobiosensor | Purpose for research and diagnostic | Elmer and White 2018 |
| Quantum dots | Purpose for research and diagnostic | Elmer and White 2018 |
| Graphene oxide sheet | Genetic material delivery system and antimicrobial | Elmer and White 2018 |
| Nanoshell | Purpose for research and diagnostic | Elmer and White 2018 |

transport activity, lipid peroxidation and oxidation damage. The responses emitted by the plants to engineered nanomaterials depend on the growth stages, plant species type and nature of engineered nanomaterials. The response of engineered nonmaterials has different variations because of the proteins of different species which makes it different in their physicochemical characteristics. Thus, the nanomaterial interaction could be valuable or destructive and are highly used in pathogen management, such as bactericides and fungicides along with fertilizers to enhance plant health.

## 19.2 ENTRY OF ENGINEERED NANOMATERIAL TO THE PLANTS

Engineered nanomaterial passes in the plants via the root system which moves via the pericycle and cortex and finally reaches the xylem. The entry nature of nanomaterial occurs on the size similar with cell wall pores and crosses the outermost boundaries of the cell and reaches the plasma membrane (Navarro et al. 2008). Later, the plasma membrane produces a structure like cavity and surrounds the engineered nanomaterial which will bind into diverse groups of organelles like endoplasmic reticulum, Golgi bodies, etc. The large-size nanomaterial enters through hydathodes, flower stigmas and stomata, and the smaller penetrates through the seed coat through the intercellular space. After that, a new pore is created in the intake seed coat by regulating the appearance of aquaporin which is a membrane protein and helps in the transport of water across the cellular membrane. Those fungal cell walls (negative charge) react with the positive-charged nanomaterial and prevent the mycelial growth due to electrostatic interaction. The uptake and transference of engineered nanomaterial towards root cells comprise both the apoplastic pathway and the symplastic pathway in which the apoplastic pathway frequently happens outside the plasma membrane across the extracellular spaces besides also adjacent cells and its cell wall and xylem vessels, and it is very vital for the radial movement inside the plant tissue which permits the engineered nanomaterial to enter the root of the central cylinder and also the tissues of the vascular moving towards the aerial part (Larue et al. 2012; Zhao et al. 2012; Sun et al. 2014).

During the symplastic pathway, the engineered nanomaterial is transported towards the organs as well as the non-photosynthetic tissues (Wang et al. 2016; Raliya et al. 2016). Sometimes, engineered nanomaterial can be stopped and built up near the Casparian strip (Lv et al. 2015). The engineered nanomaterial at a lower dose of concentration has no adverse effect on the plants, but at a higher dose of concentration, it carries out to stress or sometimes toxicity and facilitates the production of ROS and, finally, cell death. Shapes are another significant influencer, as it transports professional and consequential patterns. Nanomaterial whose size is greater than 100 nm and rods in shape show higher translocation than cubes, spheres and cylinders, but for nanomaterial less than size 100 nm, spheres have improved translocation than rods (short axis rods when compared to the long axis rods). The accumulation of a detailed shape of engineered nanomaterial inside the cells depends on the net translocation and exocytosis. Also, the duration and energy of the cell required to wrap the particles either endocytosis or exocytosis fairly depends upon the length ratio and width ratio of the engineered nanomaterial. When the soil is rich in humic acid and other organic matter, it makes a good bioavailability of engineered nanomaterial, and also, salt ions help stimulate precipitation and trigger an effect (Navarro et al. 2008). Engineered nanomaterial into the soil influences the bioavailability and is applied from time to time to the soil by injection or spraying method. The existence of microorganisms and others can control the uptake of engineered nanomaterial with the symbiosis with plants mainly by fungal mycorrhizal (Feng et al. 2013 and Wang et al. 2016). There are several factors for influencing the engineered nanomaterial absorption, translocation, transportation and penetration to the plant and uptake. When the engineered nanomaterial is translocating into the plants either via irrigation and foliar spray, it interacts with the microorganisms like rhizobacteria, mycorrhiza, salt ions and organic matter which facilitate or harm the absorption. When the engineered nanomaterial is sprayed by irrigation, then it translocates to the root hairs, epidermis, cortex, endodermis and xylems. On the contrary, if the application is via foliar spray, then it reaches through the epidermis, palisade mesophyll, cuticle, spongy mesophyll, stoma and vascular bundle. After that, it will move through the symplastic and apoplastic pathway to the plants.

During spraying, the nanomaterial can breach through the tissue of plants and reach the areas which are extremely far from the point of application, even though the movements are small. Also, when the engineered nanomaterial is entered by foliar and shoot entry, it spreads to the cuticle, epidermis, stomata, hydathodes, lenticels and wounds. When it enters through the root, it goes to the root tips, rhizodermis or cortex, lateral root junctions and wounds. Sometimes due to the rise of the engineered nanoparticles' properties and surface area, they captivate molecules of organic ions and inorganic ions from the nutrient medium ensuing indirect other

**TABLE 19.2**

**Use of Plant Species for the Green Synthesis of Engineered Nanoparticles**

| Plant | Plant material | Engineered nanoparticles | Mechanism | References |
|---|---|---|---|---|
| *Azadirachta indica* | Kernel | Au, Ag | Azadirachtin | Das et al. 2017 |
| *Geranium* | Leaves | Ag | Terpenoids | Das et al. 2017 |
| *Aloe vera* | Leaves | Ag | Not mention | Das et al. 2017 |
| *Camellia sinensis* | Leaves | Au | Catechins, theaflavins and thearubigins | Das et al. 2017 |
| *Avena sativa* | Stems | Au | Not mention | Das et al. 2017 |
| *Jatropha curcas* | Latex | Pb | Curcacycline A and Curcacycline B | Das et al. 2017 |
| *Eucalyptus* | Leaves | $Fe_2O_3$ | Epicatechin and quercetin-glucuronide | Das et al. 2017 |
| *Cinnamomum camphora* | Leaves | Au, Ag | Polyol components and heterocyclic components | Das et al. 2017 |
| *Asparagus racemosus* | Tuber cortex | Pb | Bioactive compounds | Das et al. 2017 |
| *Syzygium aromaticum* | Flower buds | Cu | Eugenol | Das et al. 2017 |
| *Nephelium lappaceum* | Peels | NiO | Nickel-ellagate complex formation | Das et al. 2017 |
| *Pinus eldarica* | Bark | Ag | Phenolic compounds | Das et al. 2017 |
| *Cycas* | Leaf | Ag | Ascorbic, dehydroascorbic acid | Das et al. 2017 |
| *Ginkgo biloba* | Leaves | Ag, Au | Proteins and metabolites | Das et al. 2017 |
| *Pinus densiflora* | Leaves | Ag | Not mentioned | Das et al. 2017 |
| *Adiantum philippense* | Leaves | Au, Ag | Not mentioned | Das et al. 2017 |
| *Adiantum capillus-veneris* | Whole plant | Ag | Not mentioned | Das et al. 2017 |
| *Nephrolepis exaltata* | Leaflet extract | Ag | Not mentioned | Das et al. 2017 |
| *Adiantum caudatum* | Leaves | Ag | Not mentioned | Das et al. 2017 |
| *Riccia liverworts* | Mature thalli | Ag | Not mentioned | Das et al. 2017 |
| *Fissidens minutus* | Thallus | Ag | Not mentioned | Das et al. 2017 |
| *Chlorella vulgaris* | Cell extract | Au | Not mentioned | Das et al. 2017 |
| *Phormidium tenue NTDM05* | Cell extract | CdS | The C-phycoerythrin pigment/ | Das et al. 2017 |
| *Sargassum muticum* | Cell extract | ZnO | Not mentioned | Das et al. 2017 |
| *Bifurcaria bifurcata* | Cell extract | CuO | Not mentioned | Das et al. 2017 |
| *Anabaena sp. 66–2* | Cell extract | Ag | C-phycocyanin, polysaccharide | Das et al. 2017 |
| *Arthrospira platensis IPPAS B-256* | Cell extract | Au | Not mentioned | Das et al. 2017 |
| *Plectonema boryanum UTEX 485* | Cyanobacteria culture | Au | Not mentioned | Das et al. 2017 |
| *Sargassum bovinum* | Cell extract | Pb | Not mentioned | Das et al. 2017 |
| *Aphanizomenon sp. 127–1* | Cell extract | Ag | Not mentioned | Das et al. 2017 |
| *Oscillatoria spp.* | Cell extract | Ag | Not mentioned | Das et al. 2017 |

**TABLE 19.3**

**Use of Microorganism Species for Green Synthesis of Engineered Nanoparticles**

| Microorganism | Location | Engineered nanoparticles | References |
| --- | --- | --- | --- |
| *Aquaspirillum magnetotacticum* | Intracellular | $Fe_3O_4$ | Mann et al. 1984 |
| *Desulfovibrio desulfuricans* | Cell surface | Pb | Yong et al. 2002 |
| *Lactobacillus sp.* | Intracellular | Ti | Nair and Pradeep 2002 |
| *Bacillus subtilis 168* | Inside cell wall | Au | Beveridge and Murray 1980 |
| *Klebsiella pneumoniae* | Cell surface | CdS | Smith et al. 1998 |
| *Corynebacterium sp. SH09* | Cell wall | Ag | Zhang et al. 2005 |
| *F. oxysporum* | Extracellular | Si, Ti, Au, Ag | Mukherjee et al. 2002 |
| *Aspergillus flavus* | Extracellular | Ag | Vigneshwaran et al. 2007 |
| *V. luteoalbum* and isolate 6–3 | Extracellular | Au | Gericke and Pinches 2006 |
| Tobacco mosaic virus (TMV) | Nanotubes on surface | $SiO_2$, $Fe_2O_3$, PbS and CdS | Shenton et al. 1999 |

toxicity symptoms like chlorosis and wilting. The engineered nanomaterial existence makes the organic acid present in plant root exudates decline the pH of the media which altered the supply nutrients and properties of engineered nanomaterials.

## 19.3 MECHANISM OF ENGINEERED NANOMATERIAL TO PROTECT PLANTS

Several interactions have occurred between engineered nanomaterials and plants. The mechanism of engineered nanomaterials and plants can be either chemical or physical interactions. In chemical interactions, it occurs by the generation of ROS, interference of activity of cell membrane ion transport, oxidation exchange and lipid peroxidation. The engineered nanomaterial protects the plants by the dissolution potential of the particle, i.e., the nanomaterials should have the ability to form any solution. Net accumulation: endocytosis/exocytosis, formation of aquaporin and new holes and following permeability increase, improved carbon accumulation in shoot which increases the photosynthesis due to the protection of the cellular membrane from the ultraviolet radiation and, lastly, reaction environment, i.e., pH, zeta potential. The cytotoxic effects caused by active oxygen radicals can be protected by the antioxidant enzymes like CAT peroxidase, superoxide dismutase, POD and ascorbate peroxidase besides lesser molecular weight antioxidants like glutathione, proline, ascorbate, phenolics and carotenoids as well as by the non-enzymatic components like tocopherol, carotenoids and ascorbate (Getnet et al. 2015 and Ozyigit et al. 2016).

Engineered nanomaterial acts as carriers for delivery of pesticides and can be sprayed or drenched, foliar tissue or roots, seeds soaking and provide crop protection.

❖ENGINEERED NANOMATERIALS
❖ENTRY BASIS
❖ENTRY MECHANISM
❖RESPONSE
- Negative Effect: Reduced germination, seedling vigor, plant growth
- Positive Effect: Increase biomass, chlorophyll content, photosynthetic reactions

**FIGURE 19.1** Mechanism of plant responses.

It also performs as a carrier, producing shelf life boosted, better quality for the solubility of lowly water-soluble pesticides, reducing the toxicity to the plants and boosting the site-specific translocation into the target pest. Engineered nanomaterial also proliferates the legume-rhizobium symbiosis and upgrades the plant growth by increasing nitrogen metabolism and enzymes. The biotransformation of engineered nanomaterial occurs inside the plants. When plants are exposed to ZnO nanomaterials, it reduces the content of non-enzymatic glutathione concentration inside the fungal cells (Suman et al. 2015). Cifuentes et al. (2010) reported that a greater quantity of carbon-coated ion engineered nanomaterial is created in the roots of peas and has better uptake to the aerial parts but is found less in wheat and sunflower. The application of engineered nanomaterial is being used for new agrochemical delivery, sensor to monitor soil condition, identity preservation, to target genetic engineering and for seed germination. The mechanism of plant responses has been presented in Figure 19.1.

## 19.4 PLANT RESPONSE TO ENGINEERED NANOMATERIAL

Most of the plants response to engineered nanomaterial by the types of engineered nanomaterial, entry basis, entry mechanism and the response the plant produce, i.e., positive effect and negative effect. The types of engineered materials like nano silver, titanium dioxide NPs, nano alumina silicate, Zn NPs, copper NPs, mesoporous silica NPs, molybdenum NPs and chitosan NPs. Physical as well as chemical properties of the nanomaterial are also necessary in which the melting point is lower for smaller particles. The harmfulness of nanopesticides and nanofertilizers when airborne may deposit on the flowering parts and leaves and destroy the stomata by creating a physical layer and toxic barrier layer on the stigma which prevent the germination of pollen grains into the stigma and tube penetration. The toxicity of the plants reaches the vascular tissues and impairs the uptake of water, translocation of minerals and photosynthesis. Most of the time, $CeO_2$-engineered nanomaterial affects the nitrogen-fixing bacteria in soybean plants which dismisses the nitrogen-fixing potential and plant growth (López-Moreno et al. 2010). $FeO_2$-engineered nanomaterial reduces macro nutrients which are Ca, K, Mg and S in sunflower shoots because of the water-blocking effect of engineered nanomaterial which reduces the water nutrients (Martínez-Fernández et al. 2016). Small-seed species like rapeseed and lettuce have an increased effect because their large surface area or volume ratio comparing to wheat. The seed coat has the chemical nature that transports selective permeability and facilitates different entries to countless engineered nanomaterial which influences a response.

Endocytosis helps the engineered nanomaterial inside the cell by invaginating the plasma membrane and producing a structure of vesicle that can pass through diverse regions of the cell. The carrier proteins occasionally help the engineered nanomaterial enchain to surround the cell membrane protein and act as a carrier for internalization and translocation inside the cell (Nel et al. 2009). Aquaporins have their own tiny size ranging between 2.8 A°–3.4 A° (Wu and Beitz 2007) which conveys the engineered nanomaterial to transport inside the cell (Rico et al. 2011). Plasmodesmata are also one way for entering the engineered nanomaterial, but it should be already present in the simplest form (Zhai et al. 2014). Another is the ion channel, and they usually have a size of 1 nm which is difficult for the engineered nanomaterial to cross them without some modification (Schwab et al. 2015).

The response of engineered nanomaterial is different for plant species and observed variations because the proteins of plant species have different physiochemical characteristics which facilitate the expression of different plant responses. Majorly, plants response to engineered nanomaterial positively and negatively. Positive responses include photo sterilization, secondary metabolites, specific enzyme conjugate formation, scavenging free radicals, improves N fixation and assimilation and hormonal balance. Negative responses include lipid peroxidation, oxidative stress and disturbance of bio membrane transport, ROS production and change in the metabolic level of cell. When ZnO is applied on *Allium cepa* in hydroponic condition, it can cause a greater release of ROS (Kumari et al. 2011). Yoon et al. (2014) found that ZnO-engineered nanomaterial showed a negative response in soybean plants by affecting the developmental stages and reproduction. Yang et al. (2015) reported phytotoxic responses of Al- and CuO-engineered nanomaterials in rice and maize, and a higher dose of ZnO-engineered nanomaterial causes net photosynthesis reduction, reduction of chlorophyll a, b, reduction of carotenoids and transpiration rate. A positive response from the spinach plant is helping in the regulation at numerous points of diverse pathways through the application of nanoTiO$_2$. It may be through ROS generation which facilitates antimicrobial activity and accelerates growth and development to the spinach plant. Another negative and positive response were observed when CeO$_2$-engineered nanomaterial when applied to *Zea mays* does not occur lipid peroxidation and any other physiological changes, but the activity of ascorbate and catalase were hampered, and finally, the up regulation of protein occurred (Zhao et al. 2012). The engineered nanomaterial at small concentration does not have any adverse effect on the plants, but at larger doses of concentration, it brings out stress or plant toxicity and augments the formation of ROS and, finally, cell death. Accumulation of the Ag nanoparticle in fungal cell on *R. solani* causes swelling and breakdown of the fungal cell wall (Figure 19.2).

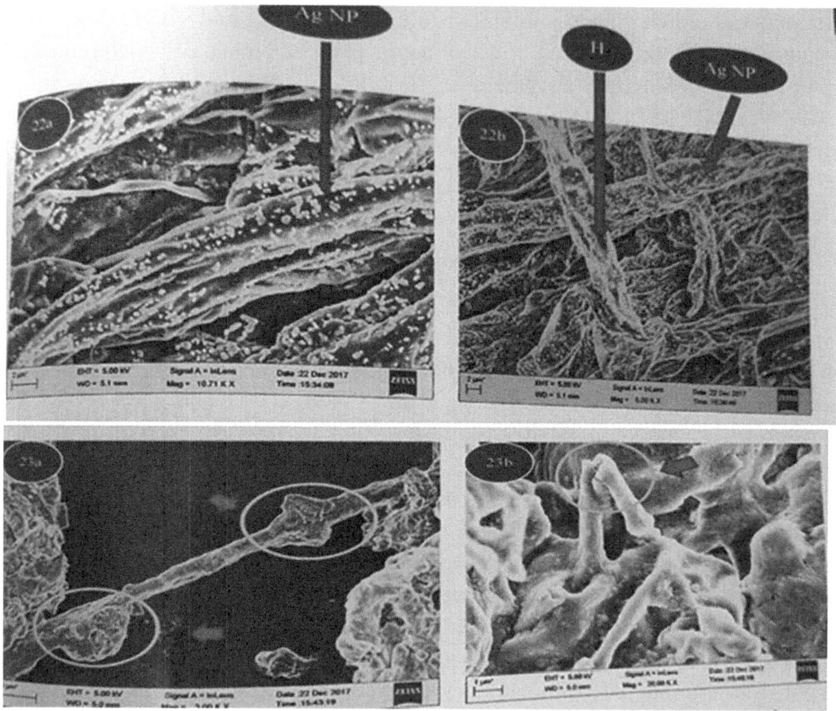

**FIGURE 19.2**　Accumulation of the Ag nanoparticle in fungal cell on *R. solani* causes swelling and breakdown of the fungal cell wall.

Source: Ankita (2018)

## 19.5　ROLE OF ENGINEERED NANOMATERIAL IN INCREASING THE IMMUNITY OF PLANTS

Engineered nanomaterial boosts the plant immunity in different ways. Introduction of engineered nanomaterial to plants helps in the expression of different defense enzymes PO (peroxidase), PPO (polyphenol oxidase), β 1–3 glucan and PAL (phenylalanine ammonia lyase), antibiotic enzymes like SOD (superoxide dismutase), CAT, glutathione, carotenoids, tocopherols and phenolics. Phenolics metabolites participate in the activity of the antioxidant that entirely depends on the number of hydroxyl group and availability in the molecules. The peroxidase and polyphenol oxidase participate in lignin biosynthesis which increases protection towards different pathogens by supporting the barrier of plant cell walls. Increasing the activity of enzymes like α, β-amylase, peroxidase, ascorbate and catalase enhances malondialdehyde content which gives a wonderful protective system by the application of nanomaterial. This is due to the biotic and abiotic stresses disruption of ETC and giving rise to ROS, which acts as strong oxidizing and detrimental agents for the cells. Reactive oxygen species destroy the plant cell membrane through peroxidation of lipid leading to the leakage of ion and disrupting the cell metabolism, causing cell death.

During the flooding stress, the enzyme glyoxalase II 3, for glyoxalase detoxification pathway was high which is reduced by the application of Ag-engineered nanomaterial. An Ag-engineered nanomaterial (15 nm at 20 ppm) being applied to a soybean plant causes oxygen deprivation which is important for the growth of soybean during flooding stress, but Ag-engineered nanomaterial (15 nm at 20 ppm) causes phytotoxicity in soybean seedlings (Mustafa et al. 2015). When the level of ROS increases, then oxidative stress occurs, and if the level of ROS decreases due to environmental stresses, this causes the peroxidation of lipids and the death of the cells. The biotransformation of engineered nanomaterial depends on redox reaction, sulfidation, phosphorylation and modification of molecules which may provide toxicity or detoxification of living systems (Lowry et al. 2012). DHAR (dehydroascorbate reductase), GR and APX in the AA-GSH sway the ROS in plants. Oxidative stress by-products and reactive oxygen species can give rise to covalent modification of a protein which is known as protein oxidation. The damaged tissues attacked by oxidative stress contain large amounts of carbonylated proteins and can be utilize further as a marker for protein oxidation. Low concentration of reactive oxygen species acts as a secondary messenger and helps to mediate several responses in plant cells. Oxidized GSH, APX and CAT contain the important antioxidant pathway of enzyme which helps to detoxify $H_2O_2$. GPX is usually a catalyzer which uses GSH to decrease lipid hydroperoxide and organic hydroperoxides which hold up plants in the environment stress. The plants' defense proteins like L-ascorbate peroxidase, superoxide dismutase and glutathione S-transferase can facilitate the construction of reactive oxygen species (Mirzajani et al. 2014). GST (glutathione S-transferase) and GSH (glutathione) are metabolites which spring antioxidant defenses against ROS-induced oxidative stress. They also perform as donors of proton in the free radicals organics in the presence of ROS and help in lowering oxidized glutathione and disulfide form.

Moreover, high-antioxidant enzymes should be maintained to facilitate tolerance to plants in stress, combating oxidative stress and inducing the defense mechanism in plants. More exposure of engineered nanomaterial leads to reactive oxygen generation, redox homeostasis disruption, mitochondrial impaired function, peroxidation of lipid and damage of membrane. More generation of reactive oxygen species can bring on the modification of amino acid site specific, peptide chain fragmentation, cross-linked reaction products aggregation and electric charge modification and also expand proteins to proteolysis susceptibility (Sharma et al. 2012). Ag-engineered nanomaterial phytotoxicity on *Oryza sativa* found that in Ag-engineered nanomaterial, there are responsive proteins which can connect pathways of oxidative stress response, signaling and $Ca^{2+}$ regulation, degradation of protein, transcription synthesis of cell wall, cell division and apoptosis (Mirzajani et al. 2014). Engineered nanomaterials like $TiO_2$, ZnO, Cerium oxide and Ag NP when placed on the cell's and organelle's surface can develop oxidative stress to the cell by the induction of signaling oxidative stress (Buzea et al. 2007). The production of reactive nitrogen species, reactive oxygen species and $H_2O_2$ when exposed to ZnO and Ag-engineered nanomaterial on *Spirodela punctate* causes phytotoxicity of ZnO and Ag by its ionic forms (Thwala et al. 2013). Engineered nanomaterials introduced to plants stimulates an antioxidant system which improves the plants' resistance and produces

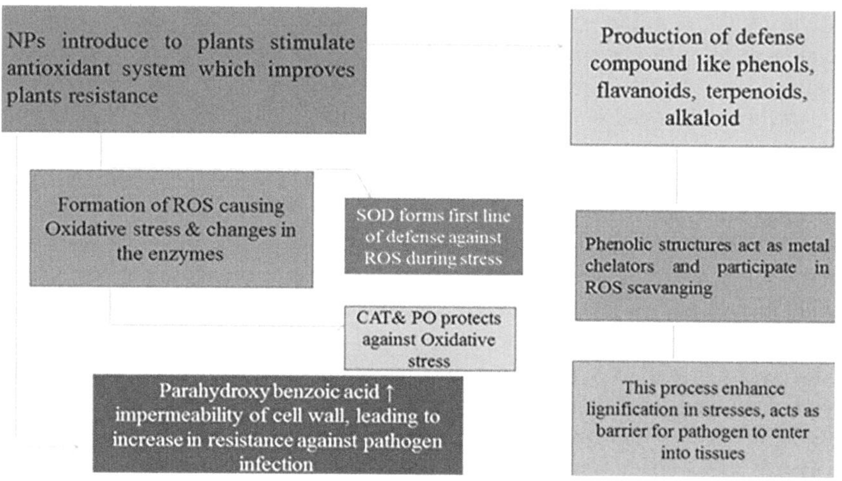



**FIGURE 19.3** Engineered nanoparticle boosting the innate immunity of the plant.

defense compounds like phenols, flavonoids, terpenoids and alkaloid. POD acts as a scavenger of superoxide dismutase. ROS and CAT combined convert $O_2^-$ and $H_2O_2$ to water and oxygen and largely decrease OH free radical. The enzymes of SOD form the defense line against reactive oxygen species during stress and CAT and PO protect against oxidative stress. Para hydroxybenzoic acid accelerates impermeability of cell wall which gives resistance against pathogens (Yasur and Rani 2013). Also, Ag-engineered nanomaterial when associated with membrane proteins triggers signaling pathways which show cell proliferation inhibition (Roh et al. 2012). Engineered nanoparticle boosting the innate immunity of the plant has been presented in Figure 19.3.

## 19.6 ROLE OF ENGINEERED NANOPARTICLE IN SUPPRESSING THE PLANT DISEASE

Studies found that applying Ag-engineered nanoparticle helps in controlling soil-borne diseases like *Meloidogyne spp.*, *Phytophthora parasitica* and *Fusarium spp.* Single application of Ag-engineered nanoparticle at 10 μg/mL or 100 μg/mL at 24 hours prior to inoculating *Artemisia absinthium* helps in suppressing *Phytophthora parasitica* by 78%–98% (Ali et al. 2015). Ag-engineered nanoparticle at 800 μg/mL helps in controlling *Fusarium* wilt in *Crossandra* spp. flower by 75% (Mallaiah 2015). Application of Ag-engineered nanoparticle on *Meloidogynes* pp. on stage 2 juvenile can completely eradicate all the juvenile stages of the nematode (Cromwell et al. 2014; Abdellatif et al. 2016; Nassar 2016). DNA-directed, Ag-engineered nanoparticle equipped on graphene oxide helps to control *Xanthomonas perforans* at 100 μg/mL on tomatoes (Ocsoy et al. 2013). Cu-engineered nanoparticle helps to suppress *Verticillium* wilt of brinjal and *Fusarium* wilt of tomato when applied at 500 μg/mL during the young seedling stage (Elmer and White 2016; Elmer and

White 2018). Applying Cu-engineered nanoparticle was also reported to have antifungal agents towards *Botrytis cinerea, F. oxysporum, Phoma destructive, A. alternate* and *Curvularia lunata* (Ouda 2014; Bramhanwade et al. 2016; Van Viet et al. 2016). Recent studies found that super paramagnetic iron oxide–engineered nanoparticles help detect the pathogen and also in medicine (Li et al. 2011). Encapsulate liposomes with herbal extracts and chitosan help inhibit gram-negative bacteria because of the phenolic contents (Matouskova et al. 2016). Zn-engineered nanoparticle has the antimicrobial properties to suppress bacterial diseases like *F. oxysporum, Rhizopus stolonifer, A. alternata, Penicillium expansum, B. cinerea, Sclerotinia sclerotiorum, Mucor plumbeus* and *Rhizoctonia solani* (Indhumathy and Mala 2013; Graham et al. 2016; Kaushik and Dutta 2017). Single-walled carbon nanotubes inhibit *Phlebia tremellosa* and *Trametes versicolor* by disrupting the oxidative enzymes of the fungus (Berry et al. 2014).

## 19.7 CONCLUSION AND FUTURE PROSPECTS

Agriculture is the backbone of any place in the globe. Adopting the advance nanotechnology will give revolution in agricultural systems by increasing productivity and reducing agricultural waste that will load out the environment pollution. Recently, nanotechnology has been acquired in physics, chemistry, environment science and medicine, but their mechanisms with plant and interactions are still needed. Arthropods are related for destroying crops of nearly 18%–20% throughout the globe with an amount of more than $470 billion (Sharma et al. 2017). To minimize the agricultural products' losses against various pests, 2.26 MT of pesticides are used every year. Among all, 40% contributed to herbicides, 17% to insecticides and 10% to fungicides in the globe.

Nanotechnology is an emerging tool to improve crop productivity and quality in agriculture. The application of engineered nanomaterial improves the growth of plants and also boosts the immunity of plants. Majorly, plant response has been done in the germination stage when the plant system is not fully matured. More findings are essential to know the mechanism of engineered nanomaterial and the factors which influence them. Most of the green-synthesis-engineered nanomaterials are studied in the laboratory, so it is important to study them as commercialization processes which can contribute to sustainable agriculture. There are scant studies about the proteins responsible and mechanisms involved in the enzymes of engineered nanomaterials. Limited studies are done on plants' innate immune responses, and more findings are essential to see if the metallic nanomaterial produces their toxicity as a result of their exceptional properties or by the metal ions released. More research can be followed on the proteomic study of Ag-engineered nanomaterial which can induced phytotoxicity and important for the type and cellular response. Studies on "omics", which is based on high throughput technique like transcriptomics and metabolomics, can possibly determine the toxicity of nanoscale material (García-Sánchez et al. 2015 and Schnackenberg et al. 2012). More studies on matrix-assisted laser desorption/ionization imaging MALDI mass spectrometry to investigate proteins, peptides and nanotoxicology which give snapshots of engineered nanomaterial distribution in tissue for characterizing and learning nanomaterial-based toxicity (Thwala et al. 2013).

A more detailed investigation of engineered nanomaterials is required for the environment and human health.

### 19.7.1 Disclosure Statement

No potential conflict of interest was reported by the author.

## REFERENCES


Abdellatif KF, Abdelfattah RH and El-Ansary MSM (2016). Green nanoparticles engineering on root-knot nematode infecting eggplants and their effect on plant DNA modification. *Iranian Journal of Biotechnology* 14(4):250–259.

Ali M, Kim B, Belfield KD, Norman D, Brennan M and Ali GS (2015). Inhibition of Phytophthoraparasitica and P. capsici by silver nanoparticles synthesized using aqueous extract of Artemisia absinthium. *Phytopathology* 105(9):1183–1190.

Ankita D (2018). Study the antimicrobial activity of silver and gold nanoparticles against *Xanthomonasoryzae*pv. oryzae (Ishiyama) Swings *et al.* and *Rhizoctoniasolani* Kuhn of rice. Msc. Thesis. Assam Agricultural University, Jorhat, Assam, India.

Begum P and Fugetsu B (2012). Phytotoxicity of multi-walled carbon nanotubes on red spinach (*Amaranthus tricolor* L) and the role of ascorbic acid as an antioxidant. *Journal of Hazardous Material* 243:212–222.

Berry TD, Filley TR and Blanchette RA (2014). Oxidative enzymatic response of white-rot fungi to single walled carbon nanotubes. *Environmental Pollution* 193:197–204.

Beveridge T, Murray R (1980). Sites of metal deposition in the cell wall of Bacillus subtilis. *Journal of Bacteriology* 141(2):876–887.

Bramhanwade K, Shende S, Bonde S, Gade A and Rai M (2016). Fungicidal activity of Cu nanoparticles against Fusarium causing crop diseases. *Environmental Chemistry Letters* 14(2):229–235.

Buzea C, Pacheco II and Robbie K (2007). Nanomaterials and nanoparticles: Sources and toxicity. *Biointerphases* 2:17–71.

Cifuentes Z, Custardoy L, de la Fuente JM, Marquina C, Ibarra MR and Rubiales D (2010). Absorption and translocation to the aerial part of magnetic carbon-coated nanoparticles through the root of different crop plants. *Journal of Nanobiotechnology* 8:26.

Cromwell WA, Yang J, Starr JL and Jo YK (2014). Nematicidal effects of silver nanoparticles on root-knot nematode in bermudagrass. *Journal of Nematology* 46(3):261–266.

Das RK, Pachapur VL, Lonappan L, Naghdi M, Pulicharla R, Maiti S, Cledon M, Dalila LMA, Sarma SJ, Brar SK. (2017). Biological synthesis of metallic nanoparticles: Plants, animals and microbial aspects. *Nanotechnology for Environmental Engineering* 2:18. https://doi.org/10.1007/s41204-017-0029-4

Elmer W and White JC. (2018). The future of nanotechnology in plant pathology. *Annual Review of Phytopathology* 56:111–133. https://doi.org/10.1146/annurev-phyto-080417-050108

Elmer W, De La Torre-Roche R, Pagano L, Majumdar S, Zuverza-Mena N, Dimkpa C, Gardea-Torresdey J and White JC. (2018). Effect of metalloid and metal oxide nanoparticles on Fusarium wilt of watermelon. *Plant Disease* 102(7):1394–1401.

Feng Y, Cui X, He S, Dong G, Chen M and Wang J (2013). The role of metal nanoparticles in influencing arbuscularmycorrhizal fungi effects on plant growth. *Environmental Science and Technology* 47:9496–9504.

García-Sánchez S, Bernales I and Cristobal S (2015). Early response to nanoparticles in the Arabidopsis transcriptome compromises plant defence and root-hair development through salicylic acid signaling. *BMC Genome* 16:341–356.

Gericke M and Pinches A (2006). Microbial production of gold nanoparticles. *Gold Bulletin* 39(1):22–28.

Getnet Z, Husen A, Fetene M and Yemata G (2015). Growth, water status, physiological, biochemical and yield response of stay green sorghum (*Sorghum bicolor* (L.) Moench) varieties—a field trial under drought-prone. *Journal of Agronomy* 14(4):188–202.

Graham JH, Johnson EG, Myers ME, Young M, Rajasekaran P, et al. (2016). Potential of nano-formulated zinc oxide for control of citrus canker on grapefruit trees. *Plant Disease* 100(12):2442–2447.

Indhumathy Mand Mala R (2013). Photocatalytic activity of zinc sulphatenano material on phytopathogens. *International Journal of Agriculture Environment and Biotechnology* 6(4S):737–743.

Joshi H and Somdutt PCS (2019). Future prospects of nanotechnology in agriculture. *International Journal of Chemical Studies* 7(2):957–963.

Kaushik Hand Dutta P (2017). Chemical synthesis of zinc oxide nanoparticle: Its application for antimicrobial activity and plant health management. Paper presented at the 109th Annual Meeting of the American Phytopathological Society, San Antonio, TX, August.

Kumari M, Khan SS, Pakrashi S, Mukherjee A and Chandrasekaran N (2011). Cytogenetic and genotoxic effects of zinc oxide nanoparticles on root cells of *Allium cepa*. *Journal of Hazardous Material* 190:613–621.

Larue C, Veronesi G, Flank A M, Surble S, Herlin-Boime N and Carrière M (2012). Comparative uptake and impact of TiO$_2$ nanoparticles in wheat and rapeseed. *Journal of Toxicology and Environmental Health* 75:722–734.

Li XM, Xu G, Liu Y and He T (2011). Magnetic Fe$_3$O$_4$ nanoparticles: Synthesis and application in water treatment. *Nanoscience & Nanotechnology-Asia* 1:14–24.

López-Moreno ML, de la Rosa G, Hernández-Viezcas JA, Castillo-Michel H, Botez CE, Peralta-Videa JR and Gardea-Torresdey JL (2010). Evidence of the differential biotransformation and genotoxicity of ZnO and CeO$_2$ nanoparticles on soybean (*Glycine max)* plants. *Environmental Science and Technology* 44:7315–7320.

Lowry GV, Gregory KB, Apte SC and Lead JR (2012). Transformations of nanomaterials in the environment. *Environmental and Science Technology* 46:6893–6899.

Lv J, Zhang S, Luo L, Zhang J, Yangc K and Christie P (2015). Accumulation, speciation and uptake pathway of ZnO nanoparticles in maize. *Environmental Science and Nanotechnology* 2:68–77.

Mallaiah B (2015). Integrate approaches for the management of Crossandra (CrossandrainfundibuliformisL.Nees) wilt caused by Fusariumincarnatum (Desm.) Sacc. PhD Thesis. Tamil Nadu Agricultural University, Madurai, India.

Mann S, Frankel RB and Blakemore RP (1984). Structure, morphology and crystal growth of bacterial magnetite. *Nature* 310(5976):405.

Martínez-Fernández D, Barroso D and Komárek M (2016). Root water transport of *Helianthus annuus* L. under iron oxide nanoparticle exposure. *Environment Science and Pollution Research* 23:1732–1741

Matouskova P, Marova I, Bokrova J and Benesova P (2016). Effect of encapsulation on antimicrobial activity of herbal extracts with lysozyme. *Food Technology and Biotechnology* 54(3):304–316.

Mirzajani F, Askari H, Hamzelou S, Schober Y, Römpp A, Ghassempour A and Spengler B (2014). Proteomics study of silver nanoparticles toxicity on *Oryza sativa* L. *Ecotoxicology and Environment Safety* 108:335–339.

Mukherjee P, Senapati S, Mandal D, Ahmad A, Khan M I, Kumar R and Sastry M. (2002). Extracellular synthesis of gold nanoparticles by the fungus Fusariumoxysporum. *Chem Bio Chem* 3(5):461–463.

Mustafa G, Sakata K, Hossain Z and Komatsu S (2015). Proteomic study on the effects of silver nanoparticles on soybean under flooding stress. *Journal of Proteomics* 122:100–118.

Nassar AM (2016). Effectiveness of silver nano-particles of extracts of Urticaurens (Urticaceae) against root-knot nematode *Meloidogyne incognita*. *Asian Journal of Nematology* 5:14–19.

Nair Band Pradeep T (2002). Coalescence of nanoclusters and formation of submicron crystallites assisted by Lactobacillus strains. *Crystal Growth & Design* 2(4):293–298.

Navarro E, Baun A, Behra R, Hartmann NB, Filser J and Miao AJ (2008). Environmental behavior and ecotoxicity of engineered nanoparticles to algae, plants, and fungi. *Ecotoxicology* 17:372–386.

Nel AE, Mädler L, Velegol D, Xia T, Hoek EM and Somasundaran P (2009). Understanding biophysicochemical interactions at the nano-bio interface. *Nature Materials* 8:543–557.

Ocsoy I, Paret ML, Ocsoy MA, Kunwar S, Chen T, et al. (2013). Nanotechnology in plant disease management: DNA-directed silver nanoparticles on graphene oxide as an antibacterial against *Xanthomonasperforans*. *ACS Nano* 7(10):8972–8980.

Ouda SM (2014). Antifungal activity of silver and copper nanoparticles on two plant pathogens, *Alternariaalternata* and *Botrytis cinerea*. *Research Journal of Microbiology* 9(1):34–42.

Ozyigit II, Filiz E, Vatansever R, Kurtoglu KY, Koc I, Öztürk MX and Anjum NA (2016). Identification and comparative analysis of $H_2O_2$-scavenging enzymes (ascorbate peroxidase and glutathione peroxidase) in selected plants employing bioinformatics approaches. *Frontiers in Plant Science* 7:301.

Raliya R, Franke C, Chavalmane S, Nair R, Reed N and Biswas P (2016). Quantitative understanding of nanoparticle uptake in watermelon plants. *Frontiers in Plant Science* 7:1288

Rico CM, Majumdar S, Duarte-Gardea M, Peralta-Videa JR and Gardea-Torresdey JL (2011). Interaction of nanoparticles with edible plants and their possible implications in the food chain. *Journal of Agricultural and Food Chemistry* 59:3485–3498.

Roh JY, Eom HJ and Choi J (2012). Involvement of *Caenorhabditiselegans* MAPK signaling pathways in oxidative stress response induced by silver nanoparticles exposure. *Toxicology Research* 28:19–24.

Schnackenberg LK, Sun J and Beger RD (2012). Metabolomics techniques in nanotoxicology studies. *Methodsin Molecular Biology* 926:141–156.

Schwab F, Zhai G, Kern M, Turner A, Schnoor JL and Wiesner MR (2015). Barriers, pathways and processes for uptake, translocation and accumulation of nanomaterials in plants. *Critical Reviewin Nanotoxicology* 10:257–278

Sharma P, Bhatt D, Zaidi MG, Saradhi PP, Khanna PK and Arora S (2012). Silver nanoparticle-mediated enhancement in growth and antioxidant status of *Brassica juncea*. *Applied Biochemistry and Biotechnology* 167:2225–2233.

Sharma S, Kooner R and Arora R (2017). Insect pests and crop losses. In *Breeding Insect Resistant Crops for Sustainable Agriculture*. Springer, Singapore, 45–66.

Shenton W, Douglas T, Young M, Stubbs G, Mann S. (1999). Inorganic–organic nanotube composites from template mineralization of tobacco mosaic virus. *Advanced Materials* 11(3):253–256

Slomberg DL and Schoenfisch MH (2012). Silica nanoparticle phytotoxicity to *Arabidopsis thaliana*. *Environmental Science and Technology* 46:10247–10254.

Smith PR, Holmes JD, Richardson DJ, Russell DA and Sodeau JR. (1998). Photophysical and photochemical characterisation of bacterial semiconductor cadmium sulfide particles. *Journal of the Chemical Society, Faraday Transactions* 94(9):1235–1241

Suman TY, Radhika RSR and Kirubagaran R (2015). Evaluation of zinc oxide nanoparticles toxicity on marine algae *Chlorella vulgaris* through flow cytometric, cytotoxicity and oxidative stress analysis. *Ecotoxicology and Environmental Safety* 113:23–30.

Sun D, Hussain HI, Yi Z, Siegele R, Cresswell T and Kong L (2014). Uptake and cellular distribution, in four plant species, of fluorescently labeled mesoporous silica nanoparticles. *Plant Cell Report* 33:1389–1402.

Thwala M, Musee N, Sikhwivhilu L and Wepener V (2013). The oxidative toxicity of Ag and ZnO nanoparticles towards the aquatic plant *Spirodelapunctuta* and the role of testing media parameters. *Environmental Science: Process and Impacts* 15:1830–1843.

Van Viet P, Nguyen HN, Cao TM and van Hieu L (2016). Fusarium antifungal activities of copper nanoparticles synthesized by a chemical reduction method. *Nanomaterials* 2016: e1957612.

Vigneshwaran N, Ashtaputre NM, Varadarajan PV, Nachane RP, Paralikar KM and Balasubra-manya RH. (2007). Biological synthesis of silver nanoparticles using the fungus Asper-gillusflavus. *Materials Letters* 61(6):1413–1418.

Wang F, Liu X, Shi Z, Tong R, Adams CA and Shi X (2016). *Arbuscularmycorrhizae* alleviate negative effects of zinc oxide nanoparticle and zinc accumulation in maize plants-A soil microcosm experiment. *Chemosphere* 147:88–97.

Wu B and Beitz E (2007). Aquaporins with selectivity for unconventional permeants. *Cellular and Molecular Life Sciences* 64:2413–2421.

Yang Z, Chen J, Dou R, Gao X, Mao C and Wang L (2015). Assessment of the phytotox-icity of metal oxide nanoparticles on two crop plants, maize (*Zea mays* L.) and rice (*Oryza sativa* L.). *International Journal of Environmental Research and Public Health* 12:15100–15109.

Yasur J and Rani PU (2013). Environment effects of nanosilver: Impact on castor seed germi-nation, seeding growth and plant physiochemical. *Environmental Science and Pollution Research* 20:8636–8648.

Yong P, Rowson NA, Farr JPG, Harris IR and Macaskie LE (2002). Bioreduction and biocrys-tallization of palladium by Desulfovibriodesulfuricans NCIMB 8307. *Biotechnology and Bioengineering* 80(4):369–379.

Yoon SJ, Kwak JI, Lee WM, Holden PA and An YJ (2014). Zinc oxide nanoparticles delay soy-bean development: A standard soil microcosm study. *Ecotoxicology and Environmental Safety* 100:131–137.

Zhai G, Walters KS, Peate DW, Alvarez PJ and Schnoor JL (2014). Transport of gold nanopar-ticles through plasmodesmata and precipitation of gold ions in woody poplar. *Environ-mental Science and Technology Letters* 1:146–151.

Zhang H, Li Q, Lu Y, Sun D, Lin X, Deng X, He N and Zheng S. (2005). Biosorption and bioreduction of diamine silver complex by Corynebacterium. *Journal of Chemical Tech-nology & Biotechnology* 80(3):285–290

Zhao L, Peng B, Hernandez-Viezcas JA, Rico C, Sun Y, Peralta-Videa JR, Tang X, Niu G, Jin L, Varela-Ramirez A, Zhang JY and Gardea-Torresdey JL (2012). Stress response and tolerance of *Zea mays* to $CeO_2$ nanoparticles: Cross talk among $H_2O_2$, heat shock pro-tein, and lipid peroxidation. *ACS Nano* 6:9615–9622.

# Index

## S



## T



## U



## W